TO George Sokera
with my best
wishes
Myron
8-30-76

Gluconeogenesis: Its Regulation in Mammalian Species

Henry A. Lardy

Gluconeogenesis: Its Regulation in Mammalian Species

Edited by

RICHARD W. HANSON, Ph.D.
Fels Research Institute and
Department of Biochemistry
Temple University Medical School
Philadelphia, Pennsylvania

MYRON A. MEHLMAN, Ph.D.
National Institutes of Health
Bethesda, Maryland

A WILEY-INTERSCIENCE PUBLICATION

JOHN WILEY & SONS, New York • London • Sydney • Toronto

Library of Congress Cataloging in Publication Data:

Main entry under title:

Gluconeogenesis, its regulation in mammalian species.

"A Wiley-Interscience publication."
Dedicated to Henry A. Lardy.
Includes bibliographies and index.
1. Glucose synthesis. 2. Mammals—Physiology.
I. Hanson, Richard W. II. Mehlman, Myron A.
III. Lardy, Henry A. [DNLM: 1. Gluconeogenesis.
2. Mammals—Physiology. QU75 G567]

QP702.G56G56 599′.01′9294 75-41489
ISBN 0-471-59045-2

Printed in the United States of America

10 9 8 7 6 5 4 3 2 1

Contributors

George A. O. Alleyne
Department of Medicine
University of the West Indies
Mona, Kingston, Jamaica

T. T. Aoki
Joslin Research Laboratories
Harvard University Medical School
Boston, Massachusetts

F. John Ballard
CSIRO
Division of Human Nutrition
Adelaide, S. A., Australia

Gregory J. Barritt
Department of Biochemistry, Faculty of Science
The Australian National University
Canberra, ACT, Australia

Barry S. B. Block
Department of Medicine
Temple University School of Medicine
Philadelphia, Pennsylvania

George F. Cahill, Jr.
Peter Bent Brigham Hospital
Harvard University School of Medicine
Boston, Masschusetts

Neil Cornell
Laboratory of Alcohol Research—NIAAA
St. Elizabeth Hospital
Washington, D.C.

Charles Elson
Departments of Medicine and Nutritional Science
Veterans Administration Hospital
Madison, Wisconsin

R. W. Guynn
Laboratory of Alcohol Research—NIAAA
St. Elizabeth Hospital
Washington, D.C.

Richard W. Hanson
Fels Research Institute and the Department of Biochemistry
Temple University School of Medicine
Philadelphia, Pennsylvania

Jochen Kleineke
Abteilung für Klinische Biochemie
Department für Innere Medizin der Universität Göttingen
Göttingen, Germany

Hans A. Krebs
Metabolic Research Laboratory
Nuffield Department of Clinical Medicine
Radcliffe Infirmary
Oxford, United Kingdom

Thomas H. Kreulen
Department of Medicine
Temple University School of Medicine
Philadelphia, Pennsylvania

J. W. R. Lawson
Laboratory of Alcohol Research—NIAAA
St. Elizabeth Hospital
Washington, D.C.

I. D. Longshaw
Department of Biochemistry
University of Bristol
Bristol, United Kingdom

Patricia Lund
Metabolic Research Laboratory
Nuffield Department of Clinical Medicine
Radcliffe Infirmary
Oxford, United Kingdom

Maria A. Mozzoli
Department of Medicine
Temple University School of Medicine
Philadelphia, Pennsylvania

Robert C. Nordlie
Department of Biochemistry
University of North Dakota Medical School
Grand Forks, North Dakota

Oliver E. Owen
Fels Research Institute and the Department of Medicine
Temple University School of Medicine
Philadelphia, Pennsylvania

Mulchand S. Patel
Department of Biochemistry and the Department of Medicine
Temple University School of Medicine
Philadelphia, Pennsylvania

Christopher I. Pogson
Biological Laboratory
University of Kent
Canterbury, United Kingdom

Paul D. Ray
Department of Biochemistry
University of North Dakota School of Medicine
Grand Forks, North Dakota

Fredrick A. Reichle
Department of Surgery
Temple University School of Medicine
Philadelphia, Pennsylvania

Ann Roobol
Biological Laboratory
University of Kent
Canterbury, United Kingdom

Neil Ruderman
Joslin Research Laboratories
Department of Medicine
Harvard University School of Medicine
Boston, Massachusetts

Earl Shrago
Departments of Medicine and Nutritional Science
University of Wisconsin
and the Veterans Administration Hospital
Madison, Wisconsin

Austin Shug
Departments of Medicine and Nutritional Science
University of Wisconsin
and the Veterans Administration Hospital
Madison, Wisconsin

S. A. Smith
Biological Laboratories
University of Kent
Canterbury, United Kingdom

Hans-Dieter Söling
Abteilung für Klinische Biochemie
Department für Innere Medizin der Universität Göttingen
Göttingen, Germany

Paul A. Srere
Pre-Clinical Science Unit
Veterans Administration Hospital and Department of Biochemistry
The University of Texas Health Science Center
Dallas, Texas

Marion Stubbs
Metabolic Research Laboratory
Nuffield Department of Clinical Medicine
Radcliffe Infirmary
Oxford, United Kingdom

Shirley M. Tilghman
Fels Research Institute and Department of Biochemistry
Temple University School of Medicine
Philadelphia, Pennsylvania

Merton F. Utter
Department of Biochemistry
School of Medicine
Case Western Reserve University
Cleveland, Ohio

Richard L. Veech
Laboratory of Alcohol Research—NIAAA
St. Elizabeth Hospital
Washington, D.C.

Carlo M. Veneziale
Department of Molecular Medicine
Mayo Medical School
Rochester, Minnesota

Paul Walter
Medizinisch-chemisches Institut
Universität Bern
Bern, Switzerland

John R. Williamson
Johnson Research Foundation
University of Pennsylvania
Philadelphia, Pennsylvania

Gary L. Zander
Department of Biochemistry
School of Medicine
Case Western Reserve University
Cleveland, Ohio

Dedication

This volume is dedicated to Professor Henry A. Lardy for his outstanding contributions to the science of biochemistry, especially in the area of gluconeogenesis. We also pay tribute to the man himself and his personal, very rare qualities of simplicity, honesty, and integrity that have left everlasting impressions on all who have had the privilege to be associated with him.

We would like to express personal thanks and sincere appreciation to each of the individual contributors to this volume. We would also like to acknowledge all those friends and colleagues of Professor Lardy whose suggestions and comments have been incorporated: to Dr. Rainer M. Zahlten for his pictures and biographical details, and to Professor Robert C. Nordlie for his partial description of Professor Lardy's role in the development of the Enzyme Institute.

A BRIEF BIOGRAPHY

Henry A. Lardy was born on August 19, 1917, in the small town of Roslyn, South Dakota. He received his bachelor's degree in 1939 from the State University of South Dakota and his master's degree and doctorate in 1941 and 1943 from the University of Wisconsin. From 1944 to 1945 he was a National Research Council Fellow in chemistry at the Banting Institute, University of Toronto. In 1945 he returned to the University of Wisconsin, where he is now a Professor of Biochemistry and Chairman of Section II of the Institute for Enzyme Research.

AFFILIATIONS AND HONORS

Professor Lardy is a member of several professional societies, has served on numerous scientific advisory councils and editorial boards and has been the recipient of many honors:

Memberships:	American Chemical Society American Society of Biological Chemists (President, 1964) National Academy of Science American Academy of Arts and Sciences Harvey Society Biochemical Society (Great Britain) American Diabetes Association Society for the Study of Reproduction
Scientific advisory councils:	American Vitamin Association Life Insurance Medical Research Fund National Science Foundation Panel on Molecular Biology Oregon Regional Primate Laboratory National Heart Institute of the National Institutes of Health American Cancer Society
Editorial boards:	*Archives of Biochemistry and Biophysics* *Journal of Biological Chemistry* *Biochemical Preparations* *Methods of Biochemical Analysis* *Biochemistry* *FEBS Letters*
Honors:	Paul Lewis Laboratories Award in Enzyme Chemistry, American Chemical Society Carl Neuberg Award, American Society of European Chemists Stieglitz Lecturer, University of Chicago Reilly Lecturer, University of Notre Dame Vilas Research Professor of Biological Sciences

MAJOR CONTRIBUTIONS TO BIOCHEMISTRY

The Institute for Enzyme Research was founded at the University of Wisconsin in 1948 with two defined purposes—to make possible long-range programs of research on fundamental problems of enzyme chemistry, and to provide for quality training at the postdoctoral level within this country. The Institute was and is unique in the United States.

Also unique is its senior staff and co-director. Professor Lardy has excelled in both missions of the Institute. The science of biochemistry has advanced significantly and meaningfully over the past 26 years as a direct consequence of research carried out in an imposingly diverse group of areas by Professor Lardy and his associates. The following is a partial list of the major contributions from the laboratory of Professor Lardy at the Institute:

1943	Interpretation of the metabolic effects of 2,4-dinitrophenol as resulting from "an interference of the energy-coupling mechanism with the result that oxidation and glycolysis run rampant, while the energy is lost as heat rather than as work."
1947	Discovery of the role of biotin in bicarbonate utilization.
1950	Demonstration of the role of folate in formate utilization.
1950	Elucidation of the mechanism by which L-glyceraldehyde inhibits glycolysis.
1952	Establishment of the role of inorganic phosphate and phosphoryl acceptor concentrations in the control of respiratory rate.
1952	Discovery of the carboxylation of propionate to permit its metabolism by way of succinate.
1953	First solubilized the fatty acid oxidizing system from mitochondria.
1956	Discovery of the inhibition of phosphofructokinase by ATP.
1956	Discovery of the uncoupling of oxidative phosphorylation and stimulation of ATPase by free fatty acids.
1957–1961	Discovery of the pathway of itaconate and mesaconate metabolism in mammalian mitochondria.
1950 et seq.	The systematic exploitation of toxic antibiotics as tools for studying metabolic processes. Some of the useful agents found in this manner include usnic acid, antimycin A, oligomycin, nigericin, dianemycin, valacidin, aurovertin, nonactin and monoactin, stendomycin, venturicidin, peliomycin, ossamycin, monensin, and monazomycin.
1963	Demonstration of the metabolic and hormonal control of phosphoenolpyruvate carboxykinase in liver.
1965	Elucidation of the role of aspartate and malate in the pathway of carbon in gluconeogenesis.
1965	Suggestion that malic enzyme generator TPNH for fat synthesis and that this enzyme is not involved in pyruvate carboxylation in mammals.

1966–1967	Discovery of the unique effect of administered tryptophan in activating liver phosphoenolpyruvate carboxykinase and in blocking gluconeogenesis by way of quinolinic acid formation.
1968–1973	Work on the mode of action of hypoglycemic agents: quinaldic acid, 5-methoxyindole-2-carboxylic acid, hydrazine, and diphenylene iodonium.
1971	Suggestion that ferrous iron is the natural activator of phosphoenolpyruvate carboxykinase in liver.
1971	Discovery of the dramatic enhancement of sperm respiration and motility by phosphodiesterase inhibitors and cyclic nucleotides.
1972	Discovery of the glucagon-stimulated phosphorylation of mitochondrial and lysosomal membrane proteins.
1972	Demonstration that glutamine is an effective therapeutic agent in one form of Leigh's encephalomyelopathy in children.
1973	Development of a system for measuring the fructose diphosphatase-phosphofructokinase substrate cycle and demonstration that it is a source of heat in bumblebee flight muscle and in malignant hyperpyrexia.
1974	Demonstration that glucagon and epinephrine can enhance gluconeogenesis by inhibiting phosphofructokinase and stimulating of fructose-1,6-diphosphatase.

More than 290 papers have been published describing such discoveries as those given here. Perusal of the References of various chapters in this volume, for example, illustrates the impact Professor Lardy's work has had on one important area—gluconeogenesis.

Perhaps the most significant measure of a teacher's greatness is the contributions and accomplishments of his students. In thirty states, four provinces, and eighteen countries, the more than 170 of us who have had the privilege and pleasure of association with Henry Lardy over the years are now attempting in some small measure to instill in our own students the interest, enthusiasm, vigor, insight, and zeal for research that have been our legacy from him. Who of our number will cease to remember fondly the "Camelot days" at the Enzyme Institute Group II at Madison? Who will never ponder his impressions of Professor Lardy the scientist and always the teacher? And who among us will ever forget Hank Lardy, the man?

MYRON A. MEHLMAN

Bethesda, Maryland
December 1975

Preface

The maintenance of blood glucose in mammalian organisms is a unique process. Despite large fluctuations in metabolism, food intake, and physical activity, the concentration of glucose in the blood is maintained at a remarkably constant level. The concentration of other metabolites—for example, ketone bodies—may vary by several orders of magnitude, whereas the concentration of glucose, except in specific diseases, does not.

One part of this delicate glucose homeostasis is the *de novo* synthesis of glucose from nonhexose precursors. This process is confined to the liver and, under some physiological conditions, the kidney cortex. The available carbohydrate reserves, which can be used to support body function, are stored primarily in the liver; they can account for no more than 8 to 10 hours of glucose requirement by peripheral tissues such as the brain. Thus during starvation gluconeogenesis plays a key role in replenishing the blood glucose concentration.

From the elegant studies of Cahill and his co-workers, we now know that mechanisms exist to spare glucose during periods of prolonged starvation. This sparing is essential to preserve muscle protein, which provides amino acids, the major precursors for glucose synthesis. Rather than minimize the importance of gluconeogenesis, these findings underline the highly sophisticated regulatory systems that have evolved in mammalian organisms.

Detailed studies of the pathway of glucose synthesis and its regulation are relatively recent. The great burst of interest in this area began in the early 1960s and included the pioneering work of Krebs and his associates, the efforts of the man to whom this volume is dedicated, Henry Lardy, and the work of Lardy's students. Two important questions have been posed by this research: (1) what is the pathway of gluconeogenesis from various substrates, and (2) how is it regulated? We know a good deal about the former and far less about the latter question.

One more recent observation that has a significant bearing on our understanding of gluconeogenesis is the finding that regulation of this process varies among species. The choice of animal models for the study of hepatic gluconeogenesis has been restricted to relatively few species. In fact, most of the work has been with a single species, that rat. This choice is probably related to the large body of information collected over the years using rat tissues. Kinetic constants for specific enzymes, substrate concentrations, effects of hormones on the liver, and the functioning of isolated mitochondria are all available in more detail for rat liver than for any other single organ. Also, rat liver is a convenient size for perfusion studies. These studies, until recently, have been an important experimental approach for studying gluconeogenesis.

The species differences in the regulation of hepatic gluconeogenesis are partly due to the intracellular distribution of the enzyme P-enolpyruvate carboxykinase (PEP-CK). It was Nordlie and Lardy who first showed that this enzyme is present in the mitochondria and cytosol in varying proportions depending on the species. Subsequent work has clearly demonstrated that these two forms of the enzyme are separate proteins, regulated in different manners. It is now possible to group animals into classes related to the intracellular distribution of hepatic P-enolpyruvate carboxykinase: first, animals with predominantly cytosolic enzyme (rat, mouse, hamster); second, animals with the enzyme predominantly mitochondrial (chicken, pigeon, rabbit); and third, animals with the enzyme almost equally distributed between cytosol and mitochondria. The latter group includes most of the species thus far investigated, such as the guinea pig, sheep, ox, dog, cat, and certain primates including man.

Since the pathways of hepatic gluconeogenesis were based largely on studies with rat liver, considerable variations in the regulation of the pathway have been noted. Söling and co-workers were the first to report that fatty acids inhibit gluconeogenesis from lactate in guinea pig liver, in contrast to rat liver in which fatty acids stimulate glucose synthesis. A detailed discussion of this area by Söling and Kleineke and by Ray, included in this book, marks the first attempt to bring the available information together.

The pathway of glucose synthesis in liver involves reactions in the cytosol and mitochondria, and a complex set of metabolite shuttles may account for the formation and movement of specific anions between intracellular components. This work is based primarily on the direct measurement of the flux of anions such as aspartate, malate, and P-enolpyruvate from isolated mitochondria, or on the use of specific inhibitors in whole tissue studies.

The evidence supporting the specific routes of carbon flow during gluconeogenesis from various substrates is reviewed in several chapters. The involvement of both the mitochondria and the cytosol in gluconeogenesis has been a focal point of studies of this process. The compartmentation of the major anaplerotic enzyme in the sequence, pyruvate carboxylase, within the mitochondria (whereas P-enolpyruvate carboxykinase has significant cytosolic activity in most species) necessitates the transport of a variety of anions from the mitochondria to the cytosol.

The original studies of Chappell and his associates first demonstrated the presence of specific transport processes that facilitate the exchange of intermediates across the inner mitochondrial membrane. Much of the information on these transporters has been reviewed by Williamson in his chapter on the role of anion transport in the regulation of metabolism. It is clear from this review that although there is considerable information concerning the kinetics and substrate specificities of the various transporters, the integration of these exchange processes with other metabolic reactions remains elusive. It is possible, for example, that the acyl-CoA inhibition of adenine nucleotide transport into the mitochondria may also be related to the negative effect of fatty acids on P-enolpyruvate formation from mitochondria of species such as the guinea pig, the rabbit, and man. In their chapter on the regulation of cell metabolism by mitochondrial transport systems, Shrago and co-workers present evidence that supports control mechanisms operative at the level of metabolite transport.

A major problem involving the regulation of gluconeogenesis is the mechanism by which hormones exert their effects. Glucagon, for example, is known to cause a rapid increase in hepatic gluconeogenesis, an effect mediated by cyclic AMP. The exact mechanism of this effect is not clear, although two major sites in the gluconeogenic sequence have been suggested. These studies, based on the so-called crossover analysis, have indicated regulation by glucagon at a point between pyruvate and P-enolpyruvate (presumably at P-enolpyruvate carboxykinase). Veneziale reviews this area, with special emphasis on the second site of glucagon regulation of hepatic gluconeogenesis, the fructose-1,6-diphosphatase-phosphofructokinase steps.

Since the effects of hormones such as glucagon and cyclic AMP are rapid and can be demonstrated within minutes using isolated, perfused livers, it is unlikely that they increase hepatic gluconeogenesis over the short term by stimulating *de novo* enzyme synthesis. It has been suggested that cyclic AMP increases mitochondrial uptake of pyruvate or that it causes lipolysis, thereby stimulating gluconeogenesis indirectly by way of increased fatty acid oxidation.

Further complicating this area are the species differences in the response of hepatic gluconeogenesis to glucagon. Söling and co-workers were the first to demonstrate that gluconeogenesis in guinea pig liver does not respond to glucagon in the same manner as rat liver. The reason for this is still not clearly understood, nor have we identified the specific mechanism for the immediate increase in glucose synthesis caused by glucagon.

It is necessary to differentiate between short- and long-term effects of hormones on hepatic processes. Hormones such as glucagon and insulin have the capacity to regulate metabolic flux over a given pathway by altering the actual concentration of specific enzymes in the sequence. The key adaptive enzyme in the pathway of glucose synthesis is cytosolic P-enolpyruvate carboxykinase. This enzyme is induced in the liver by glucagon, epinephrine, and norepinephrine (all acting via cyclic AMP), and by glucocorticoids. Insulin decreases the activity of the enzyme.

All hormonal effects on the concentration of the enzyme in rat liver described to date are due to changes in the rate of enzyme synthesis without changing enzyme degradation. One interesting aspect of investigations with this enzyme is its response to peptide hormones, a response that may make hepatic P-enolpyruvate carboxykinase an important model enzyme for studies of hormone action. The synthesis of this enzyme drops rapidly (half-time of 40 min.) when insulin is present or when cyclic AMP is removed from cells in culture. Polyribosomes isolated from the livers of animals with low rates of enzyme synthesis also have a lower rate of elongation, termination, and release of newly formed P-enolpyruvate carboxykinase. This suggests that messenger RNA (mRNA), specifically coding for the synthesis of the enzyme, is decreased after insulin treatment or after reduction in the concentration of cyclic AMP. A number of studies of this type are reviewed in the chapter by Tilghman, Hanson, and Ballard, including the development of an assay for mRNA coding for P-enolpyruvate carboxykinase using a heterologous (rabbit reticulocyte lysate) protein synthesizing system.

Liver is not the sole gluconeogenic tissue. The kidney cortex can also play a significant role in maintaining the level of blood glucose, especially during prolonged starvation. Glutamine is an important precursor for renal glucose synthesis, and during periods of acidosis, the ammonia produced by the deamination of glutamine (and glutamate) is used to titrate acidity in the urine. After the deamination of glutamine in the kidney cortex, the resulting α-ketoglutarate is converted to glucose by way of the pathway involving P-enolpyruvate carboxykinase. This enzyme is important, therefore, for ammoniagenesis as well as for gluconeogenesis in the kidney cortex.

It is of particular interest that the modes of regulating the levels of P-enolpyruvate carboxykinase differ in the kidney and in the liver. These differences are discussed in the chapter by Pogson and his collaborators. Renal P-enolpyruvate carboxykinase is induced by acidosis and by glucocorticoids. Cyclic AMP, which increases the activity of hepatic P-enolpyruvate carboxykinase, does not affect the renal enzyme. Insulin, on the other hand, markedly decreases hepatic but not renal P-enolpyruvate carboxykinase. The mechanism by which acidosis increases the activity of the enzyme in the renal cortex has not been fully resolved. Pogson *et al.* present evidence that the increased content of renal P-enolpyruvate carboxykinase is due to an acid-associated decrease in the rate of enzyme degradation.

Work by Iynedjian, Ballard, and Hanson, reviewed by Tilghman *et al.*, indicates that P-enolpyruvate carboxykinase levels in the kidney cortex are regulated by changes in the rate of enzyme synthesis alone, with no observable alterations in the degradation rate, especially after induction of acidosis in the rat. This important and fundamental difference regarding the mechanism of enzyme induction in the kidney cortex requires further work for verification of either contention.

The terminal step in gluconeogenesis is the splitting of glucose-6-phosphate by glucose-6-phosphatase. This enzyme was first described in 1945, and as discussed by Nordlie in his chapter on glucose-6-phosphatase, it has been the subject of a substantial research effort. Glucose-6-phosphatase is not only a gluconeogenic enzyme, it also plays a key role in glycogenolysis and, unlike other enzymes in the pathway, it is bound to the membranes of the endoplasmic reticulum and the nucleus. The enzyme is adaptable, increasing in activity during starvation, diabetes, and high-protein, low-carbohydrate diets. Of considerable potential interest is the possible role of glucose-6-phosphatase as a "third glucose kinase." Weinhouse and co-workers first suggested that the enzyme might operate in the reverse direction and phosphorylate glucose by way of a transferase reaction involving inorganic pyrophosphate. Veech, Lawson, Guynn, and Cornell also give support to this role of glucose-6-phosphatase in their chapter on the potential control by pyrophosphate of hepatic glucose metabolism.

Perhaps the most intensively studied enzyme of the gluconeogenic pathway is pyruvate carboxylase. Utter and his co-workers have amassed considerable evidence concerning the regulation of the enzyme by activators such as acetyl CoA. These studies have provided a critical analysis of the subunit structure of the enzyme and have shown that acetyl CoA activates pyruvate carboxylase by displacing the equilibrium between active and inactive tetrameric forms of the enzyme.

Despite the careful work done with purified pyruvate carboxylase, far less is known about the regulation of this enzyme *in vivo*. In fact, its intracellular location has been the subject of some controversy. It is now generally considered to be totally mitochondrial, and the chapter by Walter reviews the evidence for the intracellular localization of pyruvate carboxylase. In their treatment of the regulation of pyruvate carboxylase, Barrett, Zander, and Utter bring together a large body of information on the regulation of this enzyme in mitochondria. Kinetic data for the regulation of pyruvate carboxylase are presented, with emphasis on the actual alterations in key intermediates measured experimentally in liver mitochondria. Estimates of the concentration of free and bound acetyl CoA in the mitochondria is integrated with the known physiological properties of pyruvate carboxylase from kinetic studies with the isolated purified enzyme.

Analysis of interactions at this level of complexity are enormously complicated; however, the results presented in the chapter by Utter and coworkers represent a beginning attempt to correlate data from diverse experiments. It is particularly noteworthy that this effort is being made by an enzymologist who has insight into the regulation of isolated pyruvate carboxylase. As pointed out by Srere in his chapter on apparent conundrums in metabolic regulation, "It is the goal of enzymologic research to be able to explain and predict the metabolic behavior of cells and tissues using data gathered from studies on pure enzymes." Srere argues persuasively that apparent discrepancies exist between calculated rates of substrate flux over certain reactions and the actual rates as measured experimentally *in situ*. This may mean abandoning many of our assumptions about "near-equilibrium" data gathered so laboriously by so many investigators using "freeze-stop" techniques. Alternatively, our assumptions about metabolite compartmentation may be in error. There may be microenvironments within cellular organelles such as the mitochondria which create special concentration gradients for intermediates like oxalacetate and acetyl CoA. Srere's chapter using citrate synthase as a model raises a number of important issues and suggests some answers.

The determination of the concentration of various intermediates in specific organelles within the cell has always been difficult. A major technical advance was the detailed paper of Williamson, Lund, and Krebs on the reduction oxidation state of free nicotinamide-adenine denucleotide in the cytoplasm and mitochondria of rat liver.[1] The work in this paper was based on earlier suggestions of Bucher and his colleagues that the concentration of an intermediate located in a specific compartment of a cell could be determined if it took part in a reaction that was sufficiently active to reach near-equilibrium. Thus if the equilibrium constant for a reaction such as lactate dehydrogenase is known, and the concentration of the reactants lactate and pyruvate are determined experimentally, it should be possible to calculate the ratio of NAD to NADH. Furthermore, this ratio should reflect the *free* nucleotide in the cytosol, assuming that the majority of both pyruvate and lactate are present in this cellular compartment. This general technique has been of major significance in determining changes in metabolite concentrations in both mitochondria and cytosol and in assessing the oxidation-reduction state of both compartments after various experimental manipulations.

One example of the usefulness of the procedure has been its application to the partitioning of intermediates such as malate and oxalacetate. A direct determination of malate and oxalacetate in "freeze-stopped" liver will give the whole cell concentration of both metabolites, but not the concentration

[1] Williamson *et al.*, *Biochem. J.* **103,** 514 (1967).

of either compound in the mitochondria or cytosol. In his chapter, Williamson uses the near-equilibrium technique for the purpose of partitioning intermediates. This novel and potentially important method is finding increasing use. Veech *et al.* also present an extension of this technique for the calculation of the intracellular concentration of free cytosol pyrophosphate. They note a near-equilibrium between glucose/glucose-6-phosphate and fructose-6-phosphate/fructose-1,6-diphosphate couples and the PP_i/P_i couple. They extend these basic observations to suggest that substrate recycling between fructose-6-phosphate and fructose-1,6-diphosphate may occur without energy loss.

The entire field of metabolic regulation is limited by the availability of techniques that are sufficiently penetrating to answer specific questions. In an important conference held in 1970 on the subject of the regulation of gluconeogenesis (Ninth Conference of the Gesellschaft für Biologische Chemie, Göttingen, Germany), the question of new techniques in this area was discussed. Interestingly, the participants stressed the need for a more integrative approach in which information from several systems is correlated simultaneously. Since that time, the only major new approach has been the substitution of isolated cells for perfused organs or tissue slices.

Several new inhibitors, such as aminoxyacetic acid (which blocks aspartate aminotransferase) and 3-mercaptopicolinate (an inhibitor of P-enolpyruvate carboxykinase), have been applied to studies of the regulation of gluconeogenesis. However, the integration of information from a number of individual approaches is still the major methodological device for analyzing control processes. This analysis is, of course, no better than the individual experimental approaches used, and significant advances in our understanding of the regulation of gluconeogenesis in mammalian tissues will depend largely on the development of new ways of getting basic information from the cell. The chapters in this book reflect the "state of the art" in this area, taken together, they indicate both the major advances in our knowledge over the past decade and the major questions yet unanswered. Those who are actively working in this field know only too well that a full understanding of the regulation of a pathway as complex as gluconeogenesis remains elusive.

One major goal of work in this area is its application to the alleviation of human disease. The work of Cahill and his collaborators typifies the interface between basic information and its direct verification in clinical studies with humans. To a large extent, clinical studies have changed our view of the dynamics of metabolic flux in all animals, not just man. A good example is the study of Owen and Cahill with obese human subjects; their work demonstrated unequivocally that the brain can use ketone bodies and at the same time spare glucose. Since the brain was previously thought to have an obligatory requirement for glucose, using approximately 600 kcal/day (for a 75 kg man), this finding had a great impact on our understanding of overall body fuel economy. During prolonged starvation when the hepatic glycogen reverse has been depleted, gluconeogenesis, especially from amino acids, must provide the carbohydrate energy source for brain metabolism. This switch by the brain from glucose to ketone body utilization is now viewed as an essential adaptive mechanism for sparing body protien during starvation.

The fact that a finding this basic was first made in studies of the human and not in work with animal models is of interest. Biochemists knew for some time that the brain slices *could* use ketone bodies, but prior to the experiments of Owen and Cahill no one had clearly demonstrated that the brain did *in fact* use ketone bodies. Information on metabolic regulation thus flows two ways, and clinical studies are an important source of what actually happens to various fuels. Another example is the "glucose-alanine" cycle first described from work with humans by Felig, Cahill, and associates.

The chapter by Ruderman, Aoki, and Cahill on gluconeogenesis and its human disorders brings together information of the kinetics of gluconeogenesis during starvation. Figure 1 in their chapter is a compilation of the five phases of glucose homeostasis during starvation, showing the origin of blood glucose and the tissues that use glucose during each of the phases. They also include substrate levels in the blood of humans who fasted for up to 42 days. A number of hypoglycemic disorders associated with impaired human gluconeogenesis are presented and discussed in detail. Owen and associates have studied gluconeogenesis in cirrhotic human liver using catheterization techniques that measure the arteriohepatic venous differences in the concentration of various intermediates. Their data indicate surprisingly high rates of gluconeogenesis even by cirrhotic liver and show that during ketoacidosis, glucose output by the liver is not much greater than after an overnight fast. These findings stress the importance of glucose utilization by peripheral tissues in glucose homeostasis.

Despite the size and scope of this volume, we have not included chapters by many investigators who have contributed significantly to the field. The omissions do not negate the importance of the contributions but reflect a limitation in space. There is some overlap in several of the chapters, an unavoidable result of allowing contributors the freedom to write whatever seemed appropriate to the central theme. The reader will also note a certain variation in presentation in the chapters; some are lengthy reviews of a subject area; others resemble short technical papers. Again, the editors left the choice of format to the contributors, insisting only that whatever was presented be clearly written. We are indebted to each of the contributors for the excellent cooperation they afforded us during the editing of this volume.

We wish to thank the many individuals who helped us with this book, especially Dr. Vern Schramm, Temple University Medical School, for providing detailed reviews of several chapters, and Ms. Kathy Guyer for her invaluable assistance with the editorial details.

RICHARD W. HANSON
MYRON A. MEHLMAN

Philadelphia, Pennsylvania
Bethesda, Maryland
October 1975

Contents

Gluconeogenesis: Its Regulation in Mammalian Species

PART I

Enzymes and Gluconeogenesis

CHAPTER ONE

The Regulation of Pyruvate Carboxylase Activity in Gluconeogenic Tissues

GREGORY J. BARRITT, GARY L. ZANDER, AND MERTON F. UTTER

I. INTRODUCTION

The most important pathway for phosphoenolpyruvate synthesis for gluconeogenesis in mammalian and avian species is considered to be that which involves pyruvate carboxylase and phosphoenolpyruvate carboxykinase. However, the situation in mammalian species is complicated by the different intracellular locations of these enzymes. In all species so far examined, pyruvate carboxylase is located in the mitochondria (Bottger *et al.*, 1969; Brech *et al.*, 1970; Landriscinda *et al.*, 1970; Walter and Anabitarte, 1971; Yamazaki *et al.*, 1975), whereas phosphoenolpyruvate carboxykinase is found mainly in the mitochondria in avian liver (Utter, 1959; Gevers, 1967), mainly in the cytosol in the rat liver (Nordlie and

This work was supported in part by AEC contract AT-(11-1)-1242 and by NIH grant AM 12245.

Lardy, 1963; Ballard and Hanson, 1967a), and in both compartments in the guinea pig (Holten and Nordlie, 1965), rabbit (Garber and Hanson, 1971; Johnson *et al.*, 1973), and human liver (Brech *et al.*, 1970; Wieland *et al.*, 1968). Lardy *et al.* (1965) have shown that when phosphoenolpyruvate carboxykinase is located in the cytosol, the synthesis of phosphoenolpyruvate in this compartment from oxalacetate formed in the mitochondria involves the transport of four carbon units from the mitochondria to the cytosol.

Initial investigations of the tissue distribution and properties of pyruvate carboxylase from mammalian and avian sources pointed to a major role for this enzyme in supplying oxalacetate for gluconeogenesis in the liver and kidney (Utter and Keech, 1963; Ling and Keech, 1966). However, significant levels of the enzyme have been found in adipose tissue (Wise and Ball, 1964; Ballard *et al.*, 1970), brain (Salganicoff and Koeppe, 1968), and a number of other tissues (Ballard *et al.*, 1970). The role of pyruvate carboxylase in the nongluconeogenic tissues appears to be twofold. First, the oxalacetate formed in the reaction appears to play a catalytic role in lipid and cholesterol synthesis (Rognstad and Katz, 1966; Ballard and Hanson, 1967b), and second, it has been proposed that pyruvate carboxylase participates in an anaplerotic sequence in which the oxalacetate formed in the reaction replaces 4-carbon intermediates of the citric acid cycle which are used in other biosynthetic reactions (Ballard *et al.*, 1970). Although the major role of pyruvate carboxylase in the liver and kidney appears to be in the gluconeogenic pathway, it is likely that the enzyme in these tissues also participates in lipogenesis and in the anaplerotic synthesis of oxalacetate. An anaplerotic role for pyruvate carboxylase also appears likely for this enzyme in insect flight muscle (Newsholme *et al.*, 1972) and in a wide variety of microorganisms (Kornberg, 1966).

The reaction catalyzed by pyruvate carboxylase can be described by the following equation:

$$\text{pyruvate} + HCO_3^- + \text{MgATP}^- \xrightleftharpoons{\text{Acyl-CoA}, Mg^{2+}, K^+} \text{oxalacetate} + \text{MgADP}^- + \text{phosphate}^{2-}$$

The enzyme from mammalian and avian species exhibits a requirement for the presence of an acyl-CoA activator for maximal activity (Utter and Scrutton, 1969). Species of both monovalent and divalent cations are essential for enzyme activity, with K^+ and Mg^{2+} being the most effective in this capacity (Utter and Scrutton, 1969). The structure and mechanism of action of the pyruvate carboxylase enzymes have recently been discussed in some detail (Utter *et al.*, 1975).

With one or two exceptions, the present discussion is confined to pyruvate carboxylase enzymes of the gluconeogenic tissues of mammalian and avian species, centering on the regulation of the activity of the enzyme. Since the discovery of pyruvate carboxylase, there has been much interest in the factors that may control the activity of the enzyme, hence the rate of gluconeogenesis. This has led to the discovery of many parameters that influence the activity of the enzyme. Furthermore, it has been proposed that most of these parameters may contribute to the regulation of oxalacetate formation *in vivo*. However, many of the experiments that have led to these proposed regulatory mechanisms have been conducted *in vitro*, often with the use of isolated or purified enzymes. This raises the question of whether the mechanisms of control, which have been proposed chiefly on the basis of studies conducted *in vitro*, are applicable to the situation *in vivo*. In this chapter we describe the proposed control mechanisms in some detail, with the aim of evaluating the evidence that either supports or argues against them as possible control mechanisms. Although there are gaps in the experimental data, it is hoped that this analysis will point to directions in which further investigations may be made.

II. THE REGULATION OF PYRUVATE CARBOXYLASE ACTIVITY: GENERAL CONSIDERATIONS

A. Evidence Indicating that the Reaction Catalyzed by Pyruvate Carboxylase Is Subject to Regulation

One of the most important observations made in connection with the discovery of pyruvate carboxylase was that the enzyme from chicken liver exhibits an essentially absolute requirement for activation by acetyl-CoA (Utter and Keech, 1963). This led to the proposal that pyruvate carboxylase may catalyze a regulatory step in the gluconeogenic pathway (Utter *et al.*, 1964). Activation of pyruvate carboxylase by acetyl-CoA provides a link between gluconeogenesis and the pathways of fatty acid oxidation and synthesis as well as the citric acid cycle (Scrutton and Utter, 1968). The idea that pyruvate carboxylase is a control point in the gluconeogenic pathway was strengthened by the observation that this enzyme exhibits a number of the properties that are characteristic of the allosteric enzymes, including nonlinear double reciprocal initial velocity plots for acetyl-CoA and pyruvate, as well as a multisubunit structure (Scrutton and Utter, 1967; Utter and Fung, 1971; Valentine *et al.*, 1966). Furthermore, it has been shown that activation of the enzyme by acetyl-CoA is accompanied by conformational changes in the enzyme (Scrutton and Utter, 1967; Frey and Utter, 1974).

A regulatory role for pyruvate carboxylase is also indicated by teleological arguments. In gluconeogenic tissues, the reaction catalyzed by the enzyme can be considered to be the first reaction in the synthesis of glucose from pyruvate, and by analogy with other systems, it could be regarded as a likely regulatory step. Moreover, the reaction is not readily reversible under physiological conditions (Scrutton and Utter, 1965) and is part of one of the "bypass" steps in the gluconeogenic pathway. Such reactions are considered to be probable sites of regulation (Scrutton and Utter, 1968; Stadtman, 1966; Newsholme and Gevers, 1967). Finally, when they are all present in the same cell, pyruvate carboxylase, phosphoenolpyruvate carboxykinase, and pyruvate kinase constitute a cycle that if completely unrestrained, could result in the hydrolysis of excessive amounts of ATP (Scrutton and Utter, 1968; Newsholme and Gevers, 1967). The control of pyruvate carboxylase activity may contribute to the regulation of this cycle.

Direct evidence that pyruvate carboxylase catalyzes a rate-limiting step in the gluconeogenic pathway, hence is probably involved in the control of this pathway, has been obtained by application of the "crossover" theorem. The crossover theorem, which describes the relationship between a change in flux through a given step in a multistep sequence and the observed changes in the concentrations of intermediates of the sequence when the flux through the overall pathway is altered, has been used to determine the rate-limiting step(s) in multistep pathways. The theorem was initially applied to the sequence of reactions constituting the electron transport chain (Chance *et al.*, 1958); later it was extended to other systems, including the gluconeogenic pathway (J. R. Williamson *et al.*, 1969a; Exton and Park, 1969).[1] Experiments with perfused liver systems have shown that activation by oleate (J. R. Williamson *et al.*, 1969a) or glucagon (Exton and Park, 1969) of glucose synthesis from 3-carbon precursors is accompanied by a well-defined crossover between pyruvate and phosphoenolpyruvate. Similarly, a crossover at the same point is observed when glucose synthesis is inhibited by 4-pentenoic acid (J. R. Williamson *et al.*, 1970). A crossover between pyruvate and phosphoenolpyruvate can be attributed to a regulatory effect exerted on pyruvate carboxylase (J. R. Williamson *et al.*, 1969a). However, the same data can be interpreted in terms of an effect on phosphoenolpyruvate carboxykinase (Exton *et al.*, 1970) or even as a reflection of a reaction involved only secondarily in the formation of phosphoenolpyruvate (e.g., the contribution of GTP to this process through the action of succinyl thiokinase). It is evident that caution must be exercised in the interpretation of data obtained with such a complex system (Scrutton and Utter, 1968; Rolleston, 1972).

[1] A description of the nature, development, and application of the crossover theorem is given by Rolleston (1972).

B. Regulation of the Intracellular Concentration of Pyruvate Carboxylase

Studies of the regulation of pyruvate carboxylase activity have been chiefly directed toward the elucidation of mechanisms by which the activity of the enzyme is controlled, rather than on the regulation of the synthesis and degradation of the enzyme. However, the intracellular concentration of pyruvate carboxylase has been reported to change under certain conditions. These include biotin deficiency, in which the proportion of holoenzyme is found to decrease (Madappally and Mistry, 1970), and certain stages of development (Ballard and Hanson, 1967a). Thus pyruvate carboxylase activity cannot be detected in the livers of fetal rats, but the activity increases dramatically just before and immediately following birth (Ballard and Hanson, 1967a). Furthermore, it has been reported that the activity of the enzyme in the livers of rat (Prinz and Seubert, 1968; Henning *et al.*, 1966) and sheep (Filsell *et al.*, 1969; Taylor *et al.*, 1971) increases during fasting or alloxan diabetes. However, other investigators (Krebs, 1966; Shrago and Lardy, 1966; Brech *et al.*, 1970; Yamazaki *et al.*, 1975) have not observed increased activity of this enzyme, particularly in diabetes. The reason for these varying results is not clear, but it can be noted that widely differing methods have been used to elicit and estimate the activity of this easily inactivated enzyme. An answer to the question of possible regulation of pyruvate carboxylase through altered rates of synthesis and degradation probably must await more definitive studies of the turnover of this enzyme.

One other possible mechanism for the control of the concentration of pyruvate carboxylase is regulation of the concentration of an active form of the enzyme by a cycle of phosphorylation and dephosphorylation. Such cycles appear to be important in the control of a number of regulatory enzymes in carbohydrate metabolism (Cori and Green, 1943; Fischer and Krebs, 1955; Villar-Palasi and Larner, 1961; Carlson and Kim, 1974) including the pyruvate dehydrogenase complex (Linn *et al.*, 1969; Reed, 1969). However, there is no evidence at present to indicate that such a mechanism applies to pyruvate carboxylase.

C. Regulation of the Activity of Pyruvate Carboxylase

The present discussion concentrates on factors that control the activity of pyruvate carboxylase at fixed intracellular concentrations of the enzyme. Although activation of pyruvate carboxylase by acetyl-CoA is considered to be one of the most important modes of control (Utter *et al.*, 1964; Scrutton and Utter, 1967), a number of other factors have been implicated in the regulation of this enzyme. These include the availability of substrates (Utter and Fung, 1971; Adam and Haynes, 1969; McClure and Lardy, 1971) and the influence of adenine nucleotides (Stucki *et al.*, 1972), other acyl-CoA derivatives (Utter and Fung, 1971), and metal ions (Gevers and Krebs, 1966). We now consider these ideas in more detail in the hope that an assessment of the validity of each proposed control mechanism can be made. In view of its potential importance in providing a link between the control of gluconeogenesis and the activity of related pathways, the activation of pyruvate carboxylase by acetyl-CoA is treated first.

III. CONTROL OF PYRUVATE CARBOXYLASE ACTIVITY THROUGH CHANGES IN ACETYL–CoA CONCENTRATION

A. Activation of Purified Pyruvate Carboxylase by Acetyl-CoA

Pyruvate carboxylase enzymes from all mammalian and avian species studied are activated by acetyl-CoA (Utter and Scrutton, 1969). The enzyme from chicken liver exhibits an essentially complete dependence on acetyl-CoA for activity under all conditions investigated (Utter and Scrutton, 1969; Scrutton and Utter, 1967). However, this rigid requirement is not shared by the enzymes from sheep kidney (Ashman *et al.*, 1972) and rat liver (Scrutton and White, 1972), which display significant activity in the absence of acetyl-CoA when assayed under appropriate conditions (e.g., very high concentrations of potassium and bicarbonate ions). Pyruvate carboxylase from yeast is stimulated two- or threefold by the addition of acyl-CoA compounds under normal assay conditions but may become almost fully active in the absence of acyl-CoA at very high concentrations of potassium and bicarbonate ions (B. Tolbert and M. F. Utter, unpublished results). Ashman *et al.* (1973) have provided additional evidence for the independence of the catalytic and control functions of the pyruvate carboxylase protein by showing that the enzyme from sheep kidney has residual activity after treatment with trinitrobenzene sulfonate but no longer responds to activation by acetyl-CoA. The same reagent has basically similar effects on the enzymes from rat liver and yeast (Scrutton and White, 1973a). In contrast, the latter workers found that the trinitrobenzene sulfonate simply inactivates avian pyruvate carboxylase (Scrutton and White, 1973a), again emphasizing the essentially absolute dependence for activity of this variety of the enzyme on the presence of acetyl-CoA or a similar compound.

The activation of pyruvate carboxylase from chicken liver by acetyl-CoA has been shown to be a rapid and reversible process (Scrutton and Utter, 1967) in which the activator has a catalytic role and does not participate in the chemical reaction (Utter and Keech, 1963). Studies with pyruvate carboxylase from chicken liver have revealed a number of other acyl-CoA derivatives that can activate this enzyme, although these are less effective than acetyl-CoA (Utter and Scrutton, 1969; Scrutton and Utter, 1967); C.-H. Fung and M. F. Utter, unpublished results). The possibility that some of these alternative activators may be important in the control of pyruvate carboxylase activity *in vivo* is discussed later (Section IV.4B).

Steady-state kinetic studies of the activation of purified pyruvate carboxylase from avian and mammalian sources by acetyl-CoA yield plots of the initial reaction velocity as a function of acetyl-CoA concentration that are sigmoidal or, in double reciprocal form, nonlinear (Scrutton and Utter, 1967; Barritt *et al.*, 1966; McClure *et al.*, 1971). When these data are plotted according to the Hill equation (Hill, 1913), values close to 3 are obtained for n_H, the Hill coefficient, for the enzyme from chicken liver (Scrutton and Utter, 1967) and about 2 for the enzyme from rat liver (McClure *et al.*, 1971) and sheep kidney (Barritt *et al.*, 1966). It appears likely that the sigmoidal initial velocity plot for acetyl-CoA is the result of interactions between multiple binding sites for acetyl-CoA. There appear to be four binding sites for acetyl-CoA on pyruvate carboxylase from chicken liver, and the binding isotherm indicates that the combination of acetyl-CoA with the enzyme is a cooperative process (Frey and Utter, 1974). The dissociation constant for the combination of acetyl-CoA with the enzyme at pH 7.2 is 13 μM, which is essentially identical to the K_a value (12 μM) at this pH (Frey and Utter, 1974). In general, the K_a values for activation of the avian liver enzyme by acetyl-CoA vary from a low value of 2 μM to perhaps 15 μM depending on the pH, ionic strength, and other conditions (Scrutton and Utter, 1967; Utter and Fung, 1971; McClure *et al.*, 1971; Utter, 1969). There is also a considerable species difference in that K_a values for the enzyme from rat liver vary from about 15 μM under optimal conditions to about 70 μM under other circumstances (M. C. Scrutton, private communication; C.-H. Fung, unpublished results).

These properties of purified pyruvate carboxylase are consistent with its proposed regulatory role in gluconeogenesis. Moreover, the sigmoidal relationship between initial velocity and the concentration of acetyl-CoA that is observed with purified pyruvate carboxylase would permit the activity of the enzyme to respond sharply to changes in the concentration of acetyl-CoA over a narrow range.

B. Evidence for the Activation of Pyruvate Carboxylase by Acetyl-CoA *in Vivo*

Although there is no doubt that acetyl-CoA activates purified pyruvate carboxylase, it has been more difficult to show conclusively that this activation is important in the control of gluconeogenesis *in vivo*. Some experimental evidence, however, is consistent with this proposal. Metabolites that are precursors of acetyl-CoA, including acetate, acetoacetate, and short-chain fatty acids, stimulate the formation of glucose from lactate in rat kidney cortex slices, and this effect is accompanied by an increase in tissue levels of acetyl-CoA (Krebs *et al.*, 1965). The addition of oleate to the perfusion medium in the perfused rat liver leads to both an increase in the rate of gluconeogenesis and in tissue levels of acetyl-CoA (J. R. Williamson *et al.*, 1968a). Use has also been made of 4-pentenoic acid, an inhibitor of fatty acid oxidation, to obtain variations in the tissue levels of acetyl-CoA in perfused liver. When pyruvate carboxylase appears to be the rate-limiting step in the gluconeogenic pathway, increases in the concentration of 4-pentenoic acid in the perfusion medium lead to decreased tissue levels of acetyl-CoA corresponding to a decrease in the rate of gluconeogenesis (J. R. Williamson *et al.*, 1970; Toews *et al.*, 1970). Under other conditions, such as the stimulation of gluconeogenesis in perfused rat liver by glucagon, the increase in the total tissue levels of acetyl-CoA is not statistically significant (J. R. Williamson *et al.*, 1969b).

Although it can be demonstrated that an increase in the rate of gluconeogenesis is often accompanied by a rise in the tissue level of acetyl-CoA when precursors of this metabolite stimulate gluconeogenesis, calculation of the intramitochondrial concentration of acetyl-CoA from values obtained for the acetyl-CoA content of whole tissue (J. R. Williamson *et al.*, 1968a) or from a direct estimation of the amount of acetyl-CoA present in isolated rat liver mitochondrial (J. R. Williamson *et al.*, 1968a; Garland *et al.*, 1965), assuming that the volume of the mitochondrial matrix is of the order of one microliter per milligram of mitochondrial protein (Harris and Van Dam, 1968), yields values between about 50 and 1000 μM for the intramitochondrial concentration of acetyl-CoA. This concentration range is much higher than the apparent K_a values for acetyl-CoA of 20 μM and 2 μM for pyruvate carboxylase from rat (Utter and Fung, 1971; McClure *et al.*, 1971) and chicken liver (Utter and Fung, 1971), respectively, when measured under optimal conditions. Although some variation might be expected in both the calculated intramitochondrial acetyl-CoA concentration and in the apparent K_a values for acetyl-CoA, it is unlikely that variations of this type are sufficient to explain marked differences between the two sets of values. If the calculated intramitochondrial concentrations of acetyl-CoA are the true concentrations present *in vivo*, pyruvate carboxylase would always be saturated with acetyl-CoA, hence would not respond to changes in the level of this metabolite (Utter and Scrutton, 1969; Utter and Fung, 1971; J. R. Williamson *et al.*, 1968a; Sols and Marco, 1970).

Similar problems, in which there are large differences between the estimated concentrations of a metabolite *in vivo* and the apparent K_a or K_M values of the metabolite for a purified enzyme have been encountered with other enzyme systems. These include the activation of the L-form of pyruvate kinase by fructose-1,6-diphosphate and the reactions of oxalacetate with enzymes of the mitochondrial matrix (Sols and Marco, 1970).

The difficulties associated with the comparison of the properties of purified pyruvate carboxylase with those of pyruvate carboxylase *in vivo* have been described previously (Utter and Scrutton, 1969; J. R. Williamson *et al.*, 1969a), and a number of ideas have been advanced to explain how the differences in properties of the two systems might affect the range of acetyl-CoA concentrations that activate pyruvate carboxylase. These explanations fall into three areas: (1) the hypothesis that a significant proportion of acetyl-CoA in the mitochondrial matrix may be bound to proteins (Utter and Scrutton, 1969; Sols and Marco, 1970), (2) considerations of the differences in properties of pyruvate carboxylase and acetyl-CoA in dilute solution on the one hand and in the mitochondrial matrix on the other (Utter and Scrutton, 1969; Utter and Fung, 1971; J. R. Williamson *et al.*, 1969a), and (3) the possibility that the presence of inhibitors of pyruvate carboxylase may increase the apparent K_a value for acetyl-CoA sufficiently to bring it into the observed intramitochondrial concentration ranges (Utter, 1969). Since it appears that the binding of acetyl-CoA to mitochondrial proteins may be of major importance, we turn next to these ideas.

C. Estimation of the Amount of Free and Bound Acetyl-CoA Present in the Mitochondrial Matrix

1. The Intramitochondrial Concentration of Pyruvate Carboxylase. The concentration of enzymes in cells is often high in relation to the apparent K_M (K_a or K_i) value for substrates or effectors, thus significant amounts of a given metabolite may be bound to intracellular proteins (Sols and Marco, 1970; Srere, 1968). Such may be the case for acetyl-CoA in the mitochondrial matrix (Utter and Scrutton, 1969; Sols and Marco, 1970). In view of the relatively large apparent difference between intramitochondrial acetyl-CoA levels and the apparent K_a value of purified avian liver pyruvate carboxylase for acetyl-CoA, this possibility has been considered more closely.

The concentration of total acetyl-CoA in any system is the sum of the concentrations of free and bound acetyl-CoA, namely,

$$[A_T] = [A] + [A_B]$$

where [A] represents the concentration of free acetyl-CoA and the subscripts

T and B refer to total and bound acetyl-CoA, respectively. Under the conditions used for kinetic studies of isolated pyruvate carboxylase (Scrutton and Utter, 1967; McClure *et al.*, 1971), the concentration of enzyme used has generally been in the range of 0.01–0.1 μM. In this situation, where the concentration of pyruvate carboxylase is low in relation to the concentrations of total acetyl-CoA employed, the amount of acetyl-CoA bound would be negligible (Reiner, 1959). However, if the concentration of pyruvate carboxylase were higher (of the order of the apparent K_a value), a significant difference between the total and free concentration of acetyl-CoA would be observed (Reiner, 1959; Cha, 1970). At high concentrations of the enzyme, the situation is analogous to that found in most binding studies, where the concentration of the "free" ligand is the important parameter in the calculations of the dissociation constants and the number of binding sites. Under these conditions, the discrepancy between free and total ligand may be considerable for the reasons mentioned previously.

The activity (expressed in relation to the wet weight of liver) of pyruvate carboxylase in the liver is quite high (Scrutton and Utter, 1968), which suggests that high concentrations of this enzyme may be present in the mitochondrion. This observation, coupled with the low apparent K_a values for acetyl-CoA, indicates that in the mitochondrial matrix, the enzyme may bind significant amounts of acetyl-CoA. In this connection, it is important to note that the direct measurement of acetyl-CoA in tissues or isolated mitochondria yields a value for the concentration of total acetyl-CoA, whereas the activity of the enzyme will respond to the concentration of free acetyl-CoA.

In mammalian and avian liver, pyruvate carboxylase is found in the mitochondrial fraction, and evidence has been presented that the enzyme is located in the matrix space of the mitochondrion (Bottger *et al.*, 1969; Landriscinda *et al.*, 1970; Taylor *et al.*, 1971). The concentration of pyruvate carboxylase in the matrix can be estimated in the following manner. When isolated chicken liver mitochondria are treated with Triton X-100 in order to elicit the activity of pyruvate carboxylase, the total activity is found to be 0.47 μmoles/min/mg of mitochondrial protein at 30°C. If it is assumed that the volume of the matrix space of chicken liver mitochondria is the same as that of rat liver mitochondria, which is of the order of 1 μl/mg protein (Harris and van Dam, 1968), the activity of pyruvate carboxylase in the matrix space is 470 units/ml. The specific activity of purified pyruvate carboxylase from chicken liver is about 24 units/mg protein at 30°C (Scrutton *et al.*, 1969). If this is also the specific activity of the pyruvate carboxylase protein present in isolated mitochondria, the concentration of the enzyme in the mitochondrial matrix is 19 mg of protein per milliliter. The molar concentration of the enzyme, which can be calculated using the value of 5×10^5 for the molecular weight of pyruvate carboxylase from chicken liver (Barden *et al.*, 1975), is 40 μM, with a binding capacity for acetyl-CoA of 160 μM sites. This is much higher than the concentrations (0.01–0.1 μM) of the enzyme used in kinetic studies, and significantly higher than the apparent K_a of purified pyruvate carboxylase from children liver for acetyl-CoA. The intramitochondrial concentration of pyruvate carboxylase in chicken liver, estimated by this method, is also higher than that of 10 μM (40 μM acetyl-CoA binding sites) calculated in a different way by Sols and Marco (1970), although the difference is probably within the experimental error of the assumptions made in the calculations.

Calculation of the pyruvate carboxylase content of rat liver mitochondria gives a value about three-fourths that estimated previously for chicken liver mitochondria. The estimation is carried out in a somewhat different way. The pyruvate carboxylase content of normal rat liver (Wistar strain) appears to average about 15 units/g wet weight of liver at 30°C (Yamazaki *et al.*, 1975). Let us now assume that the specific activity of pure rat liver pyruvate carboxylase is also 24 units/mg protein—a not unreasonable value (Scrutton and White, 1972), that the enzyme is all located in the mitochondrial matrix, and that there are about 45 mg of mitochondrial protein per gram wet weight of liver (Leighton *et al.*, 1968). Thus it can be calculated that there are 14 mg of pyruvate carboxylase per gram wet weight of liver. Since the molecular weight of the rat liver pyruvate carboxylase also appears to be about 5×10^5 (McClure *et al.*, 1971), the concentration of pyruvate carboxylase is then about 28 μM with a binding site capacity of about 112 μM.

2. *The Combination of Acetyl-CoA with Enzymes of the Mitochondrial Matrix.*

(a) Pyruvate Carboxylase. The calculations just presented indicate that pyruvate carboxylase is present in the mitochondrial matrix at relatively high concentrations and may bind significant amounts of acetyl-CoA. The combination of acetyl-CoA with purified pyruvate carboxylase from chicken liver is a cooperative process in which the activator binds to four sites per molecule of enzyme with an observed Hill coefficient[2] of 2.5 (Frey and Utter, 1974). Thus the equation that completely describes the binding of acetyl-CoA to pyruvate carboxylase is complex (cf. Cornish-Bowden and

[2] Discussion of the definition of the Hill coefficient and the evaluation and meaning of experimental values obtained for this parameter is presented by Rubin and Changeux (1966) and Cleland (1970).

Koshland, 1970). To facilitate calculations of the amount of acetyl-CoA bound to pyruvate carboxylase, it has been assumed that the combination of acetyl-CoA with the four sites on this enzyme is an highly cooperative process which can be described by the equation:

$$E + 4A \overset{K}{\rightleftharpoons} EA_4 \tag{1}$$

where K, the equilibrium constant is

$$K = \frac{[EA_4]}{[E][A]^4}$$

The concentration of free enzyme [E] is given by:

$$[E] = [E_T] - [EA_4]$$

where the subscript T refers to total concentrations. These equations can be used to derive the following expression for the fractional saturation of pyruvate carboxylase by acetyl-CoA:

$$\frac{[EA_4]}{[E_T]} = \frac{K[A]^4}{1 + K[A]^4} \tag{2}$$

If the rate-limiting step of the catalytic reaction is the conversion of the complex EA_4 to products, the relative velocity is

$$\frac{v}{V_{max}} = \frac{[EA_4]}{[E_T]} \tag{3}$$

and the total concentration of acetyl-CoA is

$$[A_T] = 4[EA_4] + [A] \tag{4}$$

Equations 2 through 4 were used to construct theoretical plots of relative initial velocity as a function of the concentrations of free and total acetyl-CoA. Details of the method of calculation and the resulting plots appear in Fig. 1 (curves A and B). Values of 40 and 13 μM were used for the concentration of pyruvate carboxylase and for the value for half-saturation of pyruvate carboxylase by acetyl-CoA, respectively. The latter figure was determined from binding studies conducted at pH 7.2 in the presence of 200 μM KCl (Frey and Utter, 1974). The theoretical plots (curves A and B of Fig. 1) show that under the circumstances used to generate the data of Fig. 1, the acetyl-CoA concentration (expressed in terms of concentration of total acetyl-CoA) required to activate pyruvate carboxylase (curve B) is much higher than the concentration of free acetyl-CoA required to achieve the same degree of activation (curve A).

The concentration of total acetyl-CoA that gives half saturation of pyruvate carboxylase $K_{a(T)}$ can readily be calculated. At half-saturation:

$$[EA_4] = 0.5[E_T] \tag{5}$$

and

$$[A] = K_a \tag{6}$$

Equations 4 through 6 can be used to determine the value of $K_{a(T)}$ for any given values of total enzyme concentration and K_a. Thus if (E_T) represents the total enzyme concentration expressed in terms of moles of enzyme:

$$K_{a(T)} = 2[E_T] + K_a$$

In the case of the theoretical data of Fig. 1 (curve B), where the total enzyme concentration $[E_T]$ is 40 μM (160 μM binding sites for acetyl-CoA), the value of $K_{a(T)}$ is 93 μM, which is about seven times the value of K_a.

It has been assumed that the binding of acetyl-CoA to pyruvate carboxylase is a highly cooperative process. To assess whether the conclusions reached depend on the rate equation used, calculations have been performed assuming that the binding of acetyl-CoA to multiple sites on pyruvate carboxylase occurs in an independent manner. The equations describing this type of combination of acetyl-CoA with the enzyme are given in the legend of Fig. 2. These expressions were used to construct plots of relative initial velocity as a function of the concentrations of free and total acetyl-CoA. It can be seen that in changing the conditions from independent to interacting sites, there is no change in the value of $K_{a(T)}$, although the shape of the plot is altered (Fig. 2). Thus the conclusion that the value of $K_{a(T)}$ is much higher than K_a appears to be valid for conditions of both independent and interacting sites; therefore it is likely to hold in the case of the true equation for the binding of acetyl-CoA to pyruvate carboxylase.

The concentration of pyruvate carboxylase used in these calculations was 160 μM acetyl-CoA binding sites. It is of interest to determine what the effect of varying the concentration of total enzyme at a fixed value of K_a has on the value of $K_{a(T)}$. Table I summarizes the correponding values of $K_{a(T)}$ when K_a is fixed at 13 μM and the total enzyme concentration is varied from 0.1 to 500 μM binding sites. It can be seen that the value of $K_{a(T)}$ is significantly greater than the value of K_a (13 μM) only when the ratio of total enzyme concentration (expressed in terms of binding sites) to K_a begins to approach the value of K_a. In the case of the actual estimated intramitochondrial concentration of pyruvate carboxylase (160 μM sites), the ratio of their two concentrations is 12.

(b) Citrate Synthetase and Hydroxymethylglutaryl-CoA (HMG = CoA) Synthetase. The previous discussion involved the combination of acetyl-CoA with pyruvate carboxylase. Since acetyl-CoA is a participant in reac-

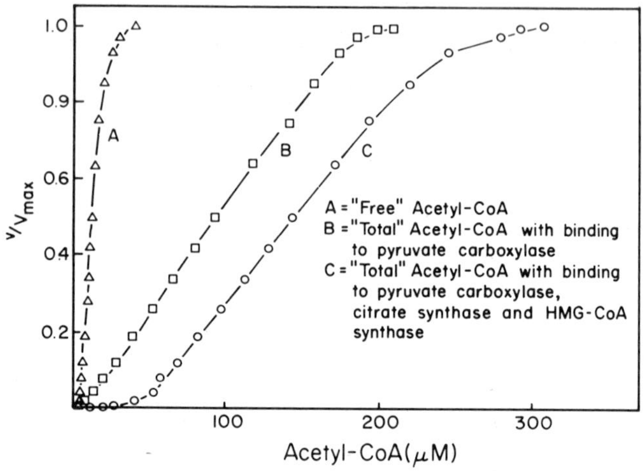

Figure 1. Theoretical plots of the relative initial velocity v/V_{max} of the reaction catalyzed by pyruvate carboxylase as a function of the concentration of acetyl-CoA in the absence and presence of enzymes of the mitochondrial matrix which bind acetyl-CoA. The concentrations of acetyl-CoA are expressed as free acetyl-CoA (curve A); or total acetyl-CoA, assuming that acetyl-CoA binds only to pyruvate carboxylase (curve B) or to pyruvate carboxylase, citrate synthetase and HMG-CoA synthetase (curve C). The data for the relative velocity as a function of the concentration of free acetyl-CoA were calculated using Eqs. 2 and 3. For the data expressed in terms of the concentration of total acetyl-CoA, the concentration of acetyl-CoA bound to pyruvate carboxylase $[A_B]$ was calculated the following equation:

$$[A_B] = \frac{4K[A]^4[E_T]}{1 + K[A]^4}$$

The value of the equilibrium constant K used in this equation and in Eq. 2 was estimated from the relationship $K = K_a^{-4}$, which was derived from Eq. 2 under the condition that the concentration of free acetyl-CoA is equal to the value of K_a for acetyl-CoA. The concentrations of acetyl-CoA bound to citrate synthetase and HMG-CoA synthetase were calculated assuming that the combination of acetyl-CoA with each of these enzymes can be described by the equation:

$$E + A \underset{}{\overset{K}{\rightleftharpoons}} EA$$

where the association constant K is:

$$K = \frac{[EA]}{[E]\,[A]}$$

and the concentration of free enzyme is:

$$[E] = [E_T] - [EA]$$

From these two equations it can be shown that the concentration of bound acetyl-CoA is given by

$$[EA] = \frac{K[A]\,[E_T]}{1 + K[A]}$$

Table I. Theoretical data illustrating the relationship between total pyruvate carboxylase concentration, the concentration of free acetyl-CoA required for half-saturation of this enzyme (K_a), and the concentration of total acetyl-CoA required for half-saturation ($K_{a(T)}$)

Concentration of total enzyme (sites) (μM)	Ratio of the concentration of total enzyme to K_a	$K_{a(T)}$ [a] (μM)
0.1	0.008	13
5	0.4	16
13	1	20
50	4	38
100	8	63
160	12	93
200	15	113
500	39	263

[a] Values for $K_{a(T)}$ were calculated from the concentration of total enzyme $[E_T]$, expressed in terms of binding sites for acetyl-CoA, according to the equation:

$$K_{a(T)} = \frac{[E_T]}{2} + K_a$$

which was derived from Eq. 4 by setting [A] equal to the K_a. A value of 13μM was used for K_a.

tions catalyzed by a number of enzymes of the mitochondrial matrix, however, it is likely that some of these enzymes will also bind significant amounts of acetyl-CoA. Enzymes that may be important in this respect are listed in Table II. The concentration of these enzymes in the matrix space has been estimated by a procedure similar to that employed to determine the concentration of pyruvate carboxylase, using data available in the literature. From a consideration of (1) the concentration of acetyl-CoA binding sites for each enzyme, (2) the K_M (K_i or K_a) value for the combination of acetyl-CoA with a given enzyme (Table II), and (3) the range of total

The last equation was used to calculate the concentration of acetyl-CoA that is bound to each of the enzymes citrate synthetase and HMG-CoA synthetase at given values of free acetyl-CoA, assuming in each case that the association constant K is equal to the reciprocal of the apparent K_M value for acetyl-CoA. The total concentration of acetyl-CoA was then obtained from the sum of free acetyl-CoA plus that bound to each of the enzymes considered. The values used for total enzyme concentrations (expressed in terms of acetyl-CoA binding sites) were 160, 40, and 470 μM for pyruvate carboxylase, citrate synthetase, and HMG-CoA synthetase, respectively, and for the K_a and K_M values for the combination of acetyl-CoA with the given enzyme, 13, 3, and 290 μM, respectively (Table II).

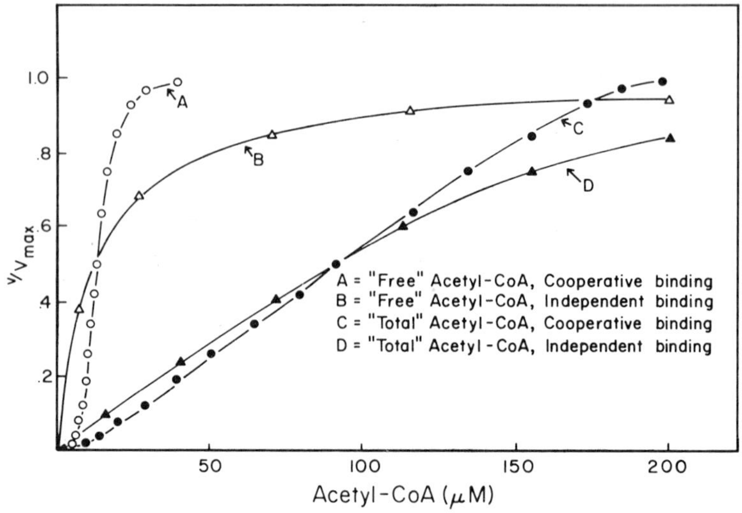

Figure 2. Theoretical plots of the relative initial velocity v/V_{max} of the reaction catalyzed by pyruvate carboxylase as a function of the concentration of acetyl-CoA, assuming that acetyl-CoA binds either to four interacting (curves A and C) or four independent (curves B and D) binding sites on pyruvate carboxylase. The concentrations of acetyl-CoA are expressed as free (curves A and B) or total (curves C and D) acetyl-CoA. In the case of interacting sites, values for relative velocity and the concentration of total acetyl-CoA were calculated as described in the legend of Fig. 1. In the case of independent binding sites, it has been assumed that the combination of acetyl-CoA with pyruvate carboxylase can be described by the equation:

$$E + A \underset{}{\overset{K}{\rightleftharpoons}} EA$$

where the association constant, K is

$$K = \frac{[EA]}{[E]\,[A]}$$

and the concentration of free enzyme is

$$[E] = [E_T] - [EA]$$

These equations were used to derive the following expressions for the relative velocity:

$$\frac{v}{V_{max}} = \frac{K[A]}{1 + K[A]}$$

(assuming that the rate-limiting step of the catalytic reaction is the breakdown of the complex EA to yield the reaction products) and the concentration of bound acetyl-CoA:

$$[EA] = \frac{K[A]\,[E_T]}{1 + K[A]}$$

These two equations were used to calculate theoretical values of the relative velocity as a function of the concentration of either free or total acetyl-CoA, using values of 160 μM acetyl-CoA binding sites for the total concentration of pyruvate carboxylase $[E_T]$ and 7.7×10^4 M^{-1} for the association constant for acetyl-CoA (i.e., determined from the dissociation constant of 13 μM).

Table II. Estimated concentrations of the enzymes of the mitochondrial matrix that bind acetyl-CoA and the K_M (K_a or K_i) values for the combination of acetyl-CoA with each enzyme[a]

Enzyme	Estimated concentration of binding sites for acetyl-CoA (μM)	K_M (K_a or K_i) for acetyl-CoA (μM)	References
Pyruvate carboxylase	157	13	Barden *et al.* (1975), Frey and Utter (1974)
Citrate synthetase	38	3	Moriyama and Srere (1971)
β-Ketothiolase[b]	11	9	Clinkenbeard *et al.* (1971), Gehring *et al.* (1968)
HMG-CoA synthetase[c]	473	290	D. H. Williamson *et al.* (1968), Sugiyama *et al.* (1972)
Pyruvate dehydrogenase complex[d]	4	5	Weiland *et al.* (1972), Linn *et al.* (1972), Bremer (1969)

[a] The estimated mitochondrial concentration of each enzyme was calculated from the observed activity of the enzyme in the mitochondrial fraction of chicken (pyruvate carboxylase and β-ketothiolase) or rat (citrate synthetase, pyruvate dehydrogenase complex, and HMG-CoA synthetase) liver, assuming, where necessary, that 1 g of rat liver (wet weight) yields about 60 mg of mitochondrial protein (Scholz and Bucher, 1965), and the reported molecular weight and specific activity of the purified enzyme, following the procedure used for the estimation of the concentration of pyruvate carboxylase, as described in the text. Data for the specific activity, molecular weight, and kinetic properties of the purified enzyme are those for the enzyme isolated from liver mitochondria, except where other sources of the enzyme are indicated.
[b] The apparent K_M value for acetyl-CoA is that for the enzyme isolated from pig heart.
[c] The specific activity, physical properties, and all kinetic data are for the cytosolic form of the enzyme isolated from chicken liver.
[d] The specific activity and physical properties are those for the enzyme isolated from pig heart.

acetyl-CoA concentrations in mitochondria, which is estimated to be 0.1 to 1.5 mM (Section III.D), it can be concluded that in addition to acetyl-CoA that is bound to pyruvate carboxylase, significant amounts would also be bound to citrate synthetase and HMG-CoA synthetase. The amount of acetyl-CoA bound to these enzymes at given values of free acetyl-CoA was calculated by a procedure similar to that used for the estimation of acetyl-CoA bound to pyruvate carboxylase, although in this case it has been assumed that acetyl-CoA binds to independent sites on these enzymes. These data were then used to generate a theoretical plot of the relative initial velocity of pyruvate carboxylase as a function of total acetyl-CoA concentration, which takes into account the binding of acetyl-CoA to pyruvate carboxylase, citrate synthetase, and HMG-CoA synthetase (Fig. 1, curve C). Under these conditions, the value of $K_{a(T)}$ for acetyl-CoA is 145 μM. This is somewhat higher than the value of 93 μM obtained for this parameter when the combination of acetyl-CoA to only pyruvate carboxylase was considered.

*3. **Evaluation of the Theoretical Calculations.*** The calculations just described involve a number of assumptions that are potential sources of error. Thus it has been assumed that the specific activity of an enzyme present in the mitochondrial matrix is the same as that of the purified enzyme, although this may be a simplification of the true situation. Furthermore, the data used in Table II were not all obtained in the same laboratory, and in some cases the required parameters for the relevant enzyme from liver have not yet been determined. The values given in Table II also represent different species, particularly chicken and rat. In the case of pyruvate carboxylase, the concentration of enzyme appears to be relatively similar in these two species as already discussed, but the K_a value for acetyl-CoA for the enzyme from rat liver may be perhaps five times as large as that for the enzyme from chicken liver. It should be noted that the values for pyruvate carboxylase in Tables I and II are those for the avian liver enzyme. Errors may arise in some of the kinetic parameters used because these would be susceptible to differences in pH, ionic strength, and the presence of other metabolites.

The enzymes that bind acetyl-CoA are assumed to be in contact with the pool of intramitochondrial acetyl-CoA. This assumption seems to be justified in the cases of pyruvate carboxylase (Bottger *et al.*, 1969), citrate synthetase, and the pyruvate dehydrogenase complex (Lardy and Ferguson, 1969), which are located in the matrix space or on the inside of the inner mitochondrial membrane. However, the locations of β-ketothiolase and HMG-CoA synthetase are not known, although it is likely that these enzymes are also in contact with the matrix space. It has also been assumed that the number of binding sites per mole of enzyme, as well as the affinity of these sites for acetyl-CoA, is the same for a given enzyme, regardless of whether this enzyme is in the mitochondrial matrix (where it may be bound to other proteins or membranes) or in a purified form. Moreover, acetyl-CoA that is bound to one enzyme has been considered to be inactive with respect to the action of another enzyme.

Given these assumptions, it would be of interest to determine the percentage of free and bound intramitochondrial acetyl-CoA experimentally as in the case of studies of intracellular sodium (Cope, 1970; Czeisler *et al.*, 1970) and NADH (Bucher *et al.*, 1972). Quite good agreement has been obtained between the experimentally determined concentration of bound NADH present in the cytosol of rat liver cells and the estimated number of potential binding sites of NADH on NAD-linked dehydrogenase enzymes. In this case, the concentration of free NADH is about 1% of the total concentration of NADH (Bucher *et al.*, 1972). On the basis of the calculations presented (Fig. 1, curve C), the concentration of free acetyl-CoA in liver mitochondria would be about 10% of the total acetyl-CoA concentration when pyruvate carboxylase is half-saturated with acetyl-CoA.

D. The Activation of Pyruvate Carboxylase by Acetyl-CoA in Intact Liver Mitochondria

The limitations of theoretical calculations in predicting how the activity of pyruvate carboxylase might be expected to change as a function of intramitochondrial acetyl-CoA concentrations *in vivo* have been discussed. The measurement of pyruvate carboxylase activity when the enzyme is present in its native environment (e.g., in isolated mitochondria, isolated liver cells, or perfused organs) overcomes many of these difficulties. Indirect evidence consistent with the activation of pyruvate carboxylase by acetyl-CoA in perfused liver or kidney slices has been described (Section III.B). However, there are at least two limitations in the interpretation of these results in terms of a specific effect of acetyl-CoA on pyruvate carboxylase. Thus the parameters measured are the rate of the overall pathway of gluconeogenesis, rather than the rate of the reaction catalyzed by pyruvate carboxylase, and the amount of acetyl-CoA in whole tissues, rather than the intramitochondrial concentration of acetyl-CoA. These limitations are substantially reduced in experiments conducted with isolated mitochondria, where more direct measurements of the rate of pyruvate carboxylation and the intramitochondrial concentration of acetyl-CoA can be made. In the isolated mitochondrion, pyruvate carboxylase is situated in an environment that closely resembles the intracellular environment of this enzyme with respect to protein and membrane interactions, ionic strength, and the presence of other metabolites. Although it is unlikely that the properties of isolated mitochondria are the same as those of mitochondria *in vivo*, studies with these organelles are an essential step in the elucidation of the factors that regulate pyruvate carboxylase *in vivo*. Accordingly, it was decided to measure the activity of pyruvate carboxylase and the levels of intramitochondrial acetyl-CoA in isolated chicken liver mitochondria under a variety of metabolic conditions, with the aim of determining whether a correlation between these two parameters can be obtained.

When chicken liver mitochondria are incubated in the presence of pyruvate, ATP and ^{14}C-labeled bicarbonate, the rate of pyruvate carboxylation can be estimated from the amount of ^{14}C isotope that is incorporated into di- and tricarboxylic acids of the citric acid cycle and their related compounds (Walter *et al.*, 1966). In the presence of 7.5 mM pyruvate, intramitochondrial levels of actyl-CoA are high (Fig. 3*b*) and correspond to maximal rates of pyruvate carboxylation (Fig. 3*a*). The inclusion in the incubation medium of arsenite, which is an inhibitor of the oxidation of pyruvate and α-ketoglutarate (Krebs, 1933; Sanadi, 1963), reduces both the rate of pyruvate carboxylation and the intramitochondrial level of acetyl-CoA (Fig. 3). The original rate of pyruvate carboxylation can be restored by the addition of palmitoylcarnitine to the incubation medium, and this is accompanied by an initial increase in the intramitochondrial concentration of acetyl-CoA (data not shown). This method of controlling the acetyl-CoA level is not very satisfactory, however, because in the presence of palmitoylcarnitine, the level of acetyl-CoA begins to fall after about one minute of incubation, with a corresponding decrease in the rate of pyruvate carboxylation.

The addition of acetylcarnitine to the incubation medium has been employed as a more direct method of increasing the acetyl-CoA concentration within the mitochondria. As Fig. 4*b* indicates, in the presence of arsenite the inclusion of 5 mM acetylcarnitine in the reaction mixture increases the acetyl-CoA content of the mitochondria approximately sixfold. The addition of acetylcarnitine produces a five- to sixfold increase in the rate of pyruvate carboxylation (Fig. 4*a*). The level of intramitochondrial acetyl-CoA and the rates of pyruvate carboxylation remained approximately constant over the duration of this experiment, either in the presence or absence of acetylcarnitine.

It has also proved possible to incubate mitochondria with different concentrations of acetylcarnitine (0.1 to 2 mM), yielding intramitochondrial acetyl-CoA levels that increase as the concentration of acetylcarnitine is raised. The rate of pyruvate carboxylation, measured as a function of the acetyl-CoA concentration, also increases as shown in Fig. 5*a*. In our experience, experiments with different batches of chicken liver mitochon-

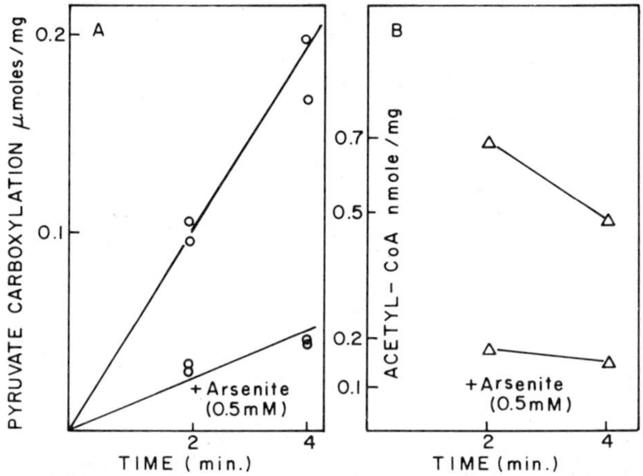

Figure 3. Effects of arsenite on the rate of pyruvate carboxylation and the level of acetyl-CoA in chicken liver mitochondria. Mitochondria (3–6 mg/2 ml) were incubated at 30°C in a medium (final volume of 2 ml) containing KCl, 100 mM, morpholinopropane sulfonic acid, 50 mM; potassium phosphate, 5 mM; EGTA, 1 mM; $MgCl_2$, 5 mM; defatted serum albumin, 10 mg/ml; pyruvate, 7.5 mM; and glutamate, 5 mM, pH 7.4. Pyruvate carboxylation was initiated by the addition of ATP, 1 mM, and ^{14}C-$KHCO_3^-$, 10 mM (0.5 mCi/mmole). Mitochondria were preincubated for 3 minutes with arsenite (0.5 mM) in those samples containing arsenite. The reaction was terminated by the addition of $HClO_4$ (6% (w/v) final concentration) at 0°C. The perchloric acid extracts were centrifuged to remove precipitated protein, adjusted to pH 6 by the addition of K_2CO_3 and 2(*N*-morpholino)-ethane sulfonic acid (75 μmole) and finally centrifuged to remove the $KClO_4$ precipitate. The amount of pyruvate carboxylated was estimated from the amount of ^{14}C label incorporated from ^{14}C-HCO_3^- into di- and tricarboxylic acids present in the neutralized extracts. The products of the carboxylation reaction were mainly malate and aspartate, with some phosphoenolpyruvate (data not shown). Acetyl-CoA levels in separate neutralized extracts were measured by a modification of the method of Prinz *et al.* (1966) in which the amount of ^{14}C-citrate formed after the condensation of excess ^{14}C-oxalacetate and acetyl-CoA in the reaction catalyzed by citrate synthetase was determined. The ^{14}C-oxalacetate was generated from ^{14}C-aspartate and α-ketoglutarate in the reaction catalyzed by glutamate-oxaloacetate transaminase and the ^{14}C-citrate separated from the other ^{14}C-labeled compounds by ion exchange chromatography (see note 2, p. 00).

dria have given basically similar results, although relatively modest errors in the estimation of acetyl-CoA and the average rate of pyruvate carboxylation can lead to sizable differences in the apparent K_a values. Therefore, the results depicted in Fig. 5 represent three different experiments whose V_{max} values have been normalized. The data appear in the form of a double reciprocal plot in Fig. 5*b*. The K_a has been calculated to be 320 μM for acetyl-CoA from these experiments. The double reciprocal plot appears to be essentially linear, although when the data are plotted according to the Hill equation (Fig. 5*c*), the Hill coefficient may be about 1.5.

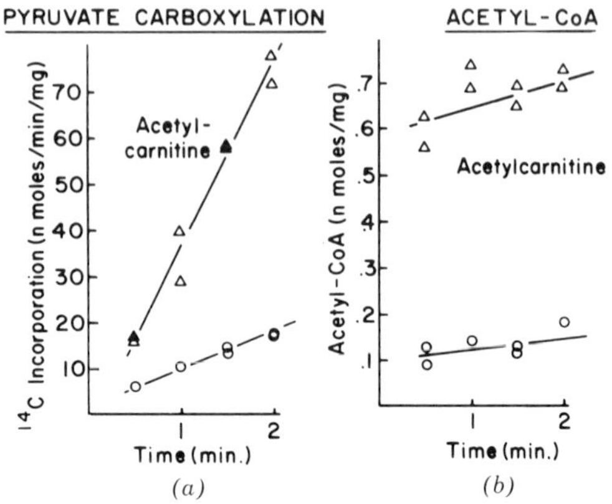

Figure 4. The effects of acetylcarnitine on the rate of pyruvate carboxylation and the levels of acetyl-CoA in intact chicken liver mitochondria. The incubation conditions were those of Fig. 3 with the following changes: arsenite (0.5 mM) was present in all samples and glutamate was absent; acetyl-carnitine (5 mM) was added 0.5 minute before the addition of ATP and HCO_3^-. The preparation of perchloric acid extracts, estimation of pyruvate carboxylation, and the levels of acetyl-CoA were conducted as described in the legend of Fig. 3.

The foregoing data, which are typical of a larger number of experiments, show clearly that in the presence of increasing concentrations of acetylcarnitine, the level of mitochondrial acetyl-CoA and the rate of pyruvate carboxylation, as measured by the incorporation of $^{14}HCO_3$, both increase. The $K_{a(T)}$ value obtained for the mitochondrial system (320 μM) is markedly higher than the K_a for the isolated enzyme (13 μM) and also somewhat higher than the value calculated in Fig. 1 (curve C) on the basis of binding by the various intramitochondrial enzymes (145 μM). The apparent discrepancy between the calculated and observed values of $K_{a(T)}$ may be due to any one or a combination of a number of factors that include imprecision in the calculation of the binding capacity of the intramitochondrial proteins, imprecision in the actual measurement, and the presence within the mitochondria of other substances that can alter the apparent affinity of pyruvate carboxylase for acetyl-CoA. For example, if the amounts of possible inhibitory substances such as succinyl-CoA vary from one mitochondrial preparation to the next, this control factor could be superimposed on that involving acetyl-CoA. The net result may well be to simply move the range of effective acetyl-CoA concentrations.

It is important to establish that the increase in the rate of pyruvate carboxylation that accompanies an increase in the intramitochondrial concentration of acetyl-CoA is not associated with a decrease in the concentration of one or more inhibitors of this reaction. In preliminary ex-

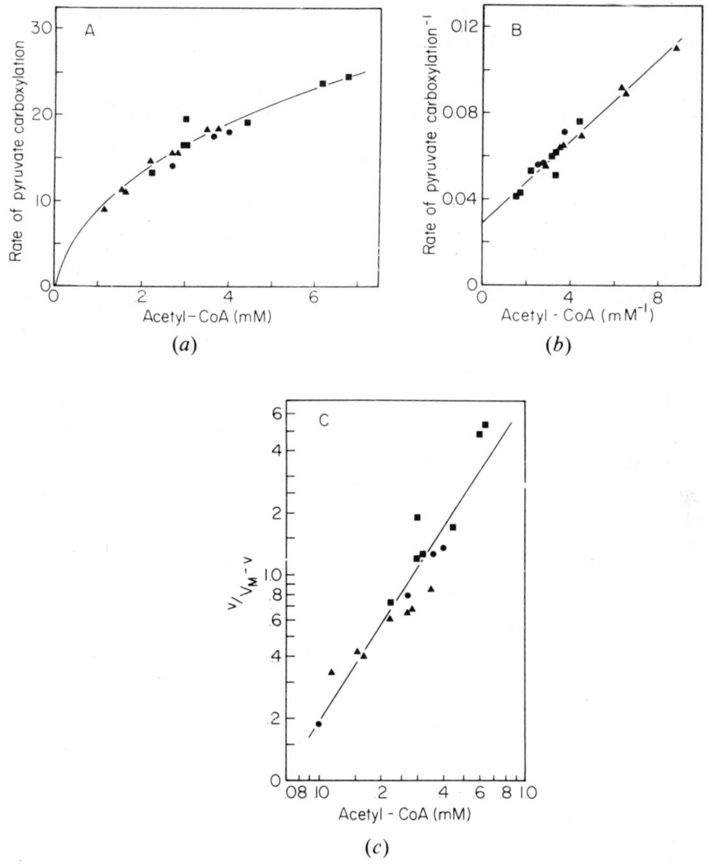

Figure 5. The rate of pyruvate carboxylation in intact chicken liver mitochondria as a function of the intramitochondrial concentration of acetyl-CoA. The levels of acetyl-CoA were varied by the addition of increasing amounts of acetylcarnitine (0.1–2 mM) to the incubation medium. The data are expressed as a plot of the rate of pyruvate carboxylation as a function of the intramitochondrial concentration of acetyl-CoA (*a*), in double reciprocal form (*b*), and in the form of a Hill plot (*c*). The incubation conditions and experimental procedure were the same as those described in the legend of Fig. 3.7. In each case the rate of pyruvate carboxylation was determined from the slope of the plot of the amount of pyruvate carboxylated as a function of time. In the calculation of the intramitochondrial concentration of acetyl-CoA, it has been assumed that 1 mg of mitochondrial protein is equivalent to 1 μl of matrix space. The results represent data from three experiments (indicated by the different symbols) the maximum velocities of which have been normalized. The value of the reciprocal of the normalized maximum velocity is shown in the double reciprocal plot (*b*).

periments, no significant change could be detected in the levels of intramitochondrial ADP with increasing levels of acetylcarnitine under the same conditions characterizing the experiment described in Fig. 5. Further experiments are required to establish this observation more firmly and to determine that there are no significant changes in the concentrations of other possible inhibitors of pyruvate carboxylase.

The data obtained with isolated chicken liver mitochondria (Fig. 5) indicate that pyruvate carboxylase is activated by acetyl-CoA. Furthermore, the apparent $K_{a(T)}$ value for the mitochondrial system is well within the range of observed changes in the intramitochondrial concentrations of acetyl-CoA. Therefore, it seems apparent that either the concentration of free acetyl-CoA at the site of pyruvate carboxylase changes within a range of concentrations that includes the apparent K_a value of purified pyruvate carboxylate for acetyl-CoA, or *in vivo* the apparent K_a value of pyruvate carboxylase for acetyl-CoA is much higher than that of the purified enzyme. In view of the results of calculations presented earlier, which indicate that a significant concentration of intramitochondrial acetyl-CoA may be bound to proteins, the former explanation seems to be the most likely one. It is of interest to note that the apparent K_M value of citrate synthetase for acetyl-CoA in isolated rat liver mitochondria is of the order of 500 μM, compared with the value of 10 μM for the apparent $\geqq_M$ of purified citrate synthetase (Garland *et al.*, 1969).

A noticeable feature of the data presented in Fig. 5 is the linearity, or near- linearity, of the double reciprocal plot, which is reflected in the relatively low value for the Hill coefficient n_H (1.5). This is in marked contrast to the Hill coefficient of about 3, which is observed for the activation by acetyl-CoA of purified pyruvate carboxylase from chicken liver (Scrutton and Utter, 1967). An explanation for this behavior may lie in the relatively narrow range of intramitochondrial acetyl-CoA levels that have been used in the experiments with isolated mitochondria. So far, it has not been possible to obtain data at either very low or very high levels of intramitochondrial acetyl-CoA, although these are the regions in which deviation of the double reciprocal plot from linearity might be expected. Furthermore, enzyme-bound acetyl-CoA, which the theoretical calculations have shown is likely to be present in the mitochondrial matrix (Section III.C), may alter the slope of the Hill plot observed for the activation of pyruvate carboxylase in the isolated mitochondria compared with that found for the purified enzyme. When the theoretical data of Fig. 1, expressed in terms of total acetyl-CoA (curve C) are plotted in the form of the Hill equation, the resulting plot has asymptotes with slopes greater than 4 at both high and low acetyl-CoA concentrations with an intermediate region near the value of $K_{a(T)}$, where the plot is linear and has a slope of 2.6. Although the numerous assumptions made in the calculations do not permit a direct comparison to be made between the experimental (Fig. 5) and theoretical data (Fig. 1, curve C), the results of the calculations indicate that a Hill plot with a slope considerably less than that observed in kinetic experiments with purified pyruvate carboxylase might be expected when a significant proportion of the acetyl-CoA is bound to proteins (e.g., in isolated mitochondria), especially when data for both high and low levels of acetyl-CoA are not available. Another possible explanation for the apparent lower cooperativity observed in the mitochondrial experiments could lie in an undetected variation in the concentration of an inhibitor of the pyruvate carboxylase reaction. For example, if the concentration of succinyl-CoA increases along with acetyl-CoA as the concentration of acetylcarnitine increases, the effect may be to flatten the curve (Fig. 5*a*) thereby decreasing the apparent cooperativity of the response to acetyl-CoA. Finally, when considering the allosteric properties of pyruvate carboxylase *in vitro,* it is also worth noting that at present there is no compelling evidence that a given enzyme necessarily exhibits the same kinetic behavior *in vivo* (e.g., interaction between active sites) and in a purified form.

E. Other Possible Differences Between the Environment of Isolated Pyruvate Carboxylase and that Present *in Vivo*

In comparing the properties of pyruvate carboxylase in isolated mitochondria with those of the purified enzyme, it has been assumed that the apparent K_a value for acetyl-CoA is the same in each situation. However, the high concentration (19 mg/ml) of pyruvate carboxylase, which is estimated to be present in the mitochondrial matrix, and possible interactions of the enzyme with membranes or other proteins, may lead to altered kinetic and regulatory properties *in vivo*. Indeed, Spivey and Hill (1973), using stopped flow methods, have presented evidence suggesting that the apparent K_a value for acetyl-CoA for the enzyme from chicken liver increases by a factor of tenfold when the concentration of the enzyme is increased from 0.004 mg/ml, which is in the same range as that normally used in steady-state kinetic studies, to a concentration of 0.5 mg/ml. This observation would provide an additional or alternative explanation for the difference between the apparent K_a value of purified pyruvate carboxylase and the observed intramitochondrial acetyl-CoA concentration. The hypothesis that the binding properties of avian pyruvate carboxylase for acetyl-CoA are altered at high concentrations of the enzyme is not supported by the studies of Fung and Utter (1974). Binding studies carried out at enzyme concentrations of 2 mg/ml gave values of 13 μM for the dissociation constant for acetyl-CoA; however the activation constant for the catalytic reaction was 12 μM, although the cocentration of protein used in the latter experiments was about three orders of magnitude lower than those in the binding studies. These results suggest that the binding properties for acetyl-CoA are not influenced by protein concentrations over very wide ranges, but it must be noted that the binding studies were carried out in the absence of the substrates of the reaction, whereas substrates were present in the kinetic studies including those of Spivey and Hill (1973) and Fung and Utter (unpublished results). Another possible explanation for the apparently different results obtained with the stopped flow and binding techniques may lie in the observation that only certain preparations of the avian liver enzyme show the apparent increase in K_a for acetyl-CoA at high concentrations of enzyme (H. O. Spivey, private communication). Kinetic studies reveal that the activation constant of acetyl-CoA for avian liver pyruvate carboxylase is not altered appreciably by very wide fluctuations in the concentrations of any of the substrates (Fung and Utter, unpublished results). It is also pertinent to note that the dissociation constant for the binding of substrates to other enzymes, including glutamate dehydrogenase from beef liver (Colen *et al.*, 1972), lactate dehydrogenase from pig heart (Heck, 1969), and creatine kinase (Hammes and Hurst, 1969), whose properties have also been studied at high enzyme concentrations, do not appear to be altered significantly at high protein concentrations.

Two other points should also be mentioned. First, in the estimation of the intramitochondrial concentrations of acetyl-CoA it has been assumed that this metabolite is distributed evenly within the experimentally determined volume of the mitochondrial matrix, which is about 1 μl/mg mitochondrial protein for rat liver mitochondria (Harris and van Dam, 1968). However, it has been reported that there may be at least two separate pools of acetyl-CoA in the matrix (Delisle and Fritz, 1967). If this is indeed the case, it is quite possible that the concentration of acetyl-CoA at the site of action of pyruvate carboxylase differs considerably from values calculated on the assumption that acetyl-CoA is equally distributed within the matrix. Second, it has been assumed that the value of the activity coefficient of acetyl-CoA *in vivo* is the same as that in dilute solution. The chemical reactivity of any metabolite is a function of both its analytical concentration and its activity coefficient, and it has been pointed out that the environment of acetyl-CoA in the mitochondrial matrix may be such that the activity coefficient is very much lower than its value in dilute solution (Garland *et al.*, 1969). If this were true, the chemical reactivity of acetyl-CoA (e.g., toward pyruvate carboxylase) would be lower than the observed intramitochondrial concentrations of acetyl-CoA, which would only reflect the analytical concentrations of this metabolite.

IV. OTHER MECHANISMS FOR THE CONTROL OF PYRUVATE CARBOXYLASE ACTIVITY

A. Availability of Substrates

The preceding discussion has dealt with the activation of pyruvate carboxylase by acetyl-CoA. Yet other modes of control may be of equal importance. The discussion of such modes is concerned with two different regulatory situations. The first is the specific situation in which acetyl-CoA has been generated within the mitochondria by the addition of acetylcarnitine to the reaction mixture as in the experiments described in Fig. 5. In this case, it is important to ask whether these circumstances may have influenced the intramitochondrial concentration of substrates or possible inhibitors, thus complicating the interpretation of the results in which accelerated pyruvate carboxylation was observed. In the second situation, the more general case, the possibility that the substrate or potential inhibitor may play a role under any physiological conditions is considered.

Perhaps the most important of the alternative modes of control of pyruvate carboxylase is that which is mediated through changes in the level of substrate concentration, in particular, of pyruvate. Three lines of evidence indicate that pyruvate carboxylase may respond to fluctuations in the intracellular level of pyruvate. Liver mitochondria isolated from rats that have been treated with epinephrine, glucogen, or cortisol exhibit much higher rates of pyruvate carboxylation and oxidation than mitochondria isolated from control rats (Adam and Haynes, 1969). However, the activity of pyruvate carboxylase assayed in sonicated extracts of mitochondria was found to be the same for mitochondria from both treated and control rats. These observations led to the hypothesis that the activities of pyruvate carboxylase and the pyruvate dehydrogenase complex are limited by the rate of entry of pyruvate to the mitochondrial matrix (Adam and Haynes, 1969; Haynes, 1972). Confirmation of this proposal awaits measurement of the influence of hormones on the rate of pyruvate uptake by isolated mitochondria. Haynes and co-workers have obtained further evidence that although indirect, is consistent with their argument. When the rate of pyruvate carboxylation in isolated rat liver mitochondria is measured as a function of the extramitochondrial pyruvate concentration, values of 0.04 to 0.08 mM are found for the apparent K_M for pyruvate (Haynes, 1972). This is about threefold lower than the value of 0.22 mM obtained for the apparent K_M of the pyruvate carboxylase reaction measured in sonicated extracts of mitochondria (Haynes, 1972) or the value of 0.14 mM for the K_M for purified rat liver pyruvate carboxylase (McClure and Lardy, 1971), and it may represent the half-saturation point of a carrier system for pyruvate. The existence of such a carrier system for pyruvate has been claimed recently on the basis of inhibition of pyruvate transport into liver mitochondria by α-cyano-4-hydroxycinnamate (Halestrap and Denton, 1974). The pertinent question in connection with the present experiments is whether the addition of acetylcarnitine to the medium increases the rate of pyruvate entry into the mitochondria. The concentration of pyruvate present in these experiments (7.5 mM) is far above the K_M noted by Haynes (1972), and the effect of acetylcarnitine could hardly be interpreted as one in which the K_M of the carrier system was made more favorable.

Steady-state kinetic studies of pyruvate carboxylase purified from chicken liver (Utter and Fung, 1971), rat liver (McClure *et al.*, 1971), and sheep kidney (Taylor *et al.*, 1969) reveal that double reciprocal plots of velocity as a function of substrate concentration are nonlinear and exhibit an apparent activation by high concentrations of pyruvate. In the case of pyruvate carboxylase from chicken liver, linear double reciprocal plots are obtained when pyruvate is varied in the range 0.08–2 mM, but a deviation from linearity occurs at pyruvate concentrations above 2 mM (Utter and Fung, 1971). Although these kinetic properties may indicate that a fine control of pyruvate carboxylase activity (activation of the reaction by high pyruvate concentrations) may be exerted through changes in the intramitochondrial concentration of pyruvate, it seems unlikely that the behavior of pyruvate carboxylase at high pyruvate concentrations (greater than about 2 mM) is related to the control of enzyme activity, because the intracellular concentration of pyruvate appears to be well below this value (Utter and Fung, 1971; McClure and Lardy, 1971). Nevertheless, the apparent K_M values for pyruvate determined from the linear part of the double reciprocal plot (i.e., at the low range of pyruvate concentrations) are about 0.3 and 0.14 mM for the enzymes from chicken (Utter and Fung, 1971) and rat liver (McClure and Lardy, 1971) respectively. This is within the range of about 0.05–1 mM for intramitochondrial pyruvate concentrations estimated from data obtained for freeze-clamped liver (J. R. Williamson *et al.*, 1968b; Greenbaum *et al.*, 1971), assuming an even distribution of pyruvate in the cell; thus it is likely that the activity of pyruvate carboxylase will respond to alterations in the intramitochondrial pyruvate concentration.

Control of pyruvate carboxylase activity through the availability of one of the substrates, pyruvate, may well be of primary importance (Utter and Fung, 1971; McClure and Lardy, 1971; Haynes, 1972). However, it should be emphasized that pyruvate carboxylase activity can still be controlled by the concentration of other effectors even though the activity of the enzyme may be limited by the availability of pyruvate (Utter and Fung, 1971). Studies with the purified avian liver enzyme have shown in fact that the ac-

Table III. Some inhibitors of pyruvate carboxylase purified from chicken liver[a]

Inhibitor	Apparent K_i [b] (mM)	Concentration required to Double the apparent K_a for acetyl-CoA (mM)
Succinyl-CoA	.25	.46
Malonyl-CoA	.45	.45
Methylmalonyl-CoA	.46	.46
HMG-CoA	.14	.27
Acetoacetyl-CoA	.20	.66
CoA-S-S-CoA	.03	.27

[a] Scrutton and Utter (1967); Fung and Utter (unpublished results). These data were obtained under conditions that are optimal for the assay of pyruvate carboxylase from chicken liver (i.e., at pH 7.8, in the presence of saturating concentrations of the substrates).
[b] The values for apparent K_i represent the concentration of inhibitor that passed 50% inhibition of pyruvate carboxylase at acetyl-CoA concentrations that equal the apparent K_a value for acetyl-CoA.

tion of pyruvate and acetyl-CoA are essentially independent of each other (Utter and Fung, 1971). Accordingly, these two regulatory mechanisms may well be superimposed.

B. Other Metabolite Effectors

1. Acyl-CoA Derivatives.[3] Extensive investigations of pyruvate carboxylase from chicken liver have revealed a large number of acyl-CoA derivatives that are either activators or inhibitors of enzyme activity (Utter and Scrutton, 1969; Scrutton and Utter, 1967). Some of these compounds are metabolic intermediates and might contribute to the control of pyruvate carboxylase activity *in vivo*. With the exception of acetoacetyl-CoA and possibly CoA-SH, all the CoA derivatives that are inhibitors (Table III) are competitive with respect to acetyl-CoA (Utter and Scrutton, 1969; Scrutton and Utter, 1967), and it is likely, although not certain, that they bind at the same site as acetyl-CoA. Quite a large number of chemical modifications can be made to the acetyl moiety of acetyl-CoA to ensure that the resulting molecule will still activate or inhibit pyruvate carboxylase, although there

[3] In addition to the references cited, this section relies on unpublished results of C.-H. Fung and M. C. Srutton.

appear to be more stringent requirements for activation than for inhibition. Therefore, it is not surprising that intermediates of a number of metabolic pathways are included among these effectors of pyruvate carboxylase, although this does not necessarily imply that such compounds are important in the metabolic control of the enzyme. Our current knowledge relating to whether any given inhibitor of pyruvate carboxylase is important in metabolic control is quite limited. However, we attempt to provide some answers to this question by considering both the likely intramitochondrial concentration of the effector in question and the rationale for metabolic control by this effector.

(a) Propionyl-CoA. The apparent K_a value of 14 μM for the activation of avian liver pyruvate carboxylase by propionyl-CoA (Utter and Scrutton, 1969) under optimal conditions is well within the 10–60 μM range calculated from the data of Söling and Volkman (1973) for the levels of propionyl-CoA measured in liver tissue (assuming that this metabolite is distributed evenly in the cell) and 0.4–1.2 mM calculated from the data of Smith *et al.* (1973) for isolated rat liver mitochondria that are oxidizing propionylcarnitine. Therefore, it is quite likely that the intramitochondrial level of propionyl-CoA is high enough to activate pyruvate carboxylase under conditions of increased flux through propionyl-CoA carboxylase (e.g., during the oxidation of amino acids). J. R. Williamson *et al.* (1969a) have suggested that changes in the ratio of propionyl-CoA to methylmalonyl-CoA may provide an additional control of the activity of pyruvate carboxylase at fixed levels of acetyl-CoA. However, the range of intramitochondrial concentrations of methylmalonyl-CoA is not known. Although there are no experimental data to verify the assumption, it seems somewhat unlikely that activation by propionyl-CoA or inhibition by methylmalonyl-CoA plays any significant role in the experiments such as those described in Fig. 5, since no apparent source of propionyl-CoA was present under these conditions.

(b) Succinyl-CoA and CoA-SH. The levels of CoA-SH and succinyl-CoA in mitochondria isolated from chicken (Table IV) and rat liver (Smith *et al.*, 1973) lie between 0.1 and 2 nmole/mg of protein, or about 0.1 to 2 mM. The concentration of succinyl-CoA is quite high, thus although it is a relatively weak inhibitor of pyruvate carboxylase from chicken liver (Table III), it may inhibit this enzyme *in vivo*. Some caution should be exercised in the interpretation of values obtained for intramitochondrial levels of CoA-SH. This metabolite is readily oxidized, especially in the presence of perchloric acid, although reducing agents such as dithioerythritol are generally included in mitochondrial extracts to reduce the oxidation (Williamson and Corkey, 1969). Under these conditions, however, other labile acyl-CoA de-

rivatives, especially succinyl-CoA, may be hydrolyzed to yield free CoA-SH (G. J. Barritt and M. F. Utter, unpublished results). As a result, some estimations of the intramitochondrial CoA-SH levels may be greater than the true values. If this kind of error is not serious, the intramitochondrial concentrations of CoA-SH are sufficiently high to cause it to be considered to be a potential inhibitor.

The concentrations of free CoA-SH and succinyl-CoA in chicken liver mitochondria have been estimated by making calculations analogous to those used to estimate the concentration of free acetyl-CoA. Enzymes which could bind CoA-SH and succinyl-CoA are listed in Tables V and VI, respectively. These tables also list the estimated concentration of binding sites and the value of K_M or K_i for the combination of CoA-SH or succinyl-CoA to each enzyme. Examination of the estimated concentration of binding sites, the corresponding values of K_M or K_i (Tables V and VI) and the measured intramitochondrial concentration of total CoA-SH and succinyl-CoA (Table IV) indicates that significant amounts of CoA-SH may be bound by palmitoyl-carnitine transferase, citrate synthetase, and pyruvate carboxylase, and comparable amounts of succinyl-CoA may be bound by citrate synthetase and pyruvate carboxylase. Assuming that the inhibitors bind in significant amounts only to these enzymes, it is possible to calculate the concentration of free inhibitor that would be present in the mitochondrial matrix from the observed values for the concentration of total inhibitor under these conditions. The results indicate that the estimated concentration of free inhibitor is somewhat lower than the total concentration (Table IV), although not by a factor of tenfold as in the case of acetyl-CoA. These calculations involve the same assumptions and sources of error as those made in the estimation of the concentration of free acetyl-CoA. Nevertheless, the results can be used as a guide to the relative magnitude of the proportion of succinyl-CoA and CoA-SH bound to intramitochondrial proteins, indicating that the estimated concentration of the free species of each metabolite is high enough to inhibit pyruvate carboxylase *in vivo*.

The factors influencing the concentrations of succinyl-CoA and CoA-SH within liver mitochondria are not clear, although the level of the former appears to be higher during state 4 respiration (Smith *et al.*, 1973). Whether either or both these inhibitors is present in sufficiently high concentrations under the experimental conditions used in Fig. 5 is not known with certainty, although preliminary experiments suggest that this may be the case. If so, it becomes very important to know whether the concentration of these inhibitors is changed by the addition of acetylcarnitine, and such experiments must be carried out. Both these inhibitors appear to exert their effect mainly by changing the apparent K_a value for acetyl-CoA to some extent (C. -H. Fung and M. F. Utter, unpublished results). However, the

Table IV. The measured total concentrations and estimated free concentrations in chicken liver mitochondria of three inhibitors of pyruvate carboxylase

Inhibitor	Measured total concentration of inhibitor[a] (μM)	Estimated concentration of free inhibitor[b] (μM)	K_i for inhibition of pyruvate carboxylase[c] (μM)
CoA-SH	650 ± 60	320	500
Succinyl-CoA	1130 ± 220	980	250
Acetoacetyl-CoA	48 ± 27	0.1	200

[a] Chicken liver mitochondria (10 mg/2 ml) were incubated under aerobic conditions at 30°C for 30 seconds in 2 ml of medium that contained KCl, 100 mM; morpholinopropane sulfonic acid, 50 mM; potassium phosphate, 5 mM; EGTA, 1 mM; $MgCl_2$, 5 mM; defatted serum albumin, 10 mg/ml; pyruvate, 5 mM; malate, 5 mM and ADP, 2.5 mM. The reactions were initiated by addition of mitochondria and stopped by addition of $HClO_4$, which in the case of CoA-SH determinations contained dithioerythritol, to give a final concentration of 6% (w/v) perchloric acid, and when present, 1 mM dithioerythritol. The samples were neutralized as described in the legend of Fig. 3 and lyophylized to a final volume of 0.5 ml. The amount of CoA-SH present in the perchloric acid extracts was determined by measuring the release of ^{14}C-CO_2 from ^{14}C α-ketoglutarate in the reaction catalyzed by α-ketoglutarate dehydrogenase complex[8], of succinyl-CoA by coupling this reaction to that catalyzed by CoA transferase, and of acetoacetyl-CoA by coupling the reactions used for the determination of acetyl-CoA (described in the legend of Fig. 3) to the reaction catalyzed by β-ketothiolase. It has been assumed that 1 mg of mitochondrial protein is equivalent to 1 μl of matrix space. The value given for the intramitochondrial concentration of each inhibitor is the mean and standard deviation of 4 to 8 determinations.

[b] In each case, the concentration of free CoA-derivative (R-S-CoA, a derivative of CoA-SH) was estimated by calculating the amount bound to enzymes and subtracting this from the measured total R-S-CoA concentration. In the calculations, it has been assumed that significant amounts of CoA-SH are bound only to palmitoylcarnitine transferase, citrate synthetase, and pyruvate carboxylase; of succinyl-CoA to citrate synthetase and pyruvate carboxylase; and of acetoacetyl-CoA to HMG-CoA synthetase. The parameters used to calculate the amount of each R-S-CoA bound to these enzymes are listed in Tables V to VII, respectively. In each case, it has been assumed that the combination of R-S-CoA with a given enzyme can be described by the expression:

$$E + \text{R-S-CoA} \xrightarrow{K} \text{E.R-S-CoA}$$

main effect of the presence of these inhibitors would be to shift the range over which acetyl-CoA was effective as an activator (Utter, 1969; Utter and Fung, 1971).

Consideration of the metabolic relationship between succinyl-CoA and the reaction catalyzed by pyruvate carboxylase suggests that inhibition of this reaction by succinyl-CoA may play a role in the control of the anaplerotic function of this enzyme. Pyruvate carboxylase from chicken liver is also inhibited by α-ketoglutarate (Scrutton and White, 1973b), and it is conceivable that these compounds act together to inhibit oxalacetate formation when replenishment of intermediates of the citric acid cycle is not required. The apparent K_i of pyruvate carboxylase for α-ketoglutarate (about 3.5 mM) is quite high, although intramitochondrial levels of α-ketoglutarate have been reported to be within this range (Rolleston, 1972). It seems unlikely that α-ketoglutarate plays an important inhibitory role in experiments of the type illustrated in Fig. 5. No immediate source of the keto acid has been added. Moreover, measurments of oxygen uptake during these experiments show a very low rate of oxygen utilization that is not stimulated by addition of acetylcarnitine. Thus it seems unlikely that α-ketoglutarate is formed in significant quantities from acetyl-CoA and oxalacetate through the reactions of the citric acid cycle.

(c) Malonyl-CoA. Inhibition of pyruvate carboxylase by malonyl-CoA is an attractive idea in terms of the proposed role of pyruvate carboxylase in lipogenesis (Rognstad and Katz, 1966; Ballard and Hanson, 1967b). However, data presently available indicate that such inhibition is unlikely to occur *in vivo*. The level of malonyl-CoA in rat liver is reported to be between 4 and 25 nmole per gram of wet weight (Guynn *et al.*, 1972) or 4 to 25 μM, assuming an even distribution of malonyl-CoA in the cell. This is well below the apparent K_i value for inhibition of pyruvate carboxylase (Table III). Furthermore, the bulk of intracellular malonyl-CoA is likely to be in the cell cytosol with little, if any, contribution from the mitochondria (Guynn *et al.*, 1972).

(d) Acetoacetyl-CoA. Acetoacetyl-CoA is one of the most interesting inhibitors of purified avian liver pyruvate carboxylase because unlike the inhibitors discussed so far (with the possible exception of CoA-SH), it is a

---

The amount of R-S-CoA bound was calculated by using Eq. 10 (legend of Fig. 1) and the reciprocal of the value of K_M (K_i or K_a), for the combination of R-S-CoA with the enzyme in question, for the value of K, the association constant.

[c] The values for apparent K_i represent the concentration of inhibitor that inhibits the activity of pyruvate carboxylase by 50% at acetyl-CoA concentrations that equal the apparent K_a value for acetyl-CoA.

Table V. Estimated concentrations of enzymes of the mitochondrial matrix that bind CoA-SH and the corresponding K_M or K_i values for the combination of CoA-SH with each enzyme[a]

Enzyme	Estimated concentration of binding sites for CoA-SH (μM)	K_M or K_i (μM)	References
Palmitoylcarnitine transferase[b,c]	960	1000	Hoppel and Tomec (1972), West *et al.* (1971)
Citrate synthetase	38	24	Moriyama and Srere (1971), Srere (1972a)
Pyruvate dehydrogenase complex[c]	4	3	Weiland *et al.* (1972), Linn *et al.* (1972), Bremer (1969)
α-Ketoglutarate dehydrogenase complex[d]	1	0.1	Seikevitz (1952), Tanaka *et al.* (1972), Srere (1972b), Sanadi (1969)
β-Ketothiolase	11	21	Clinkenbeard *et al.* (1973), Middleton (1973)
Succinyl-CoA thiokinase[d]	0.2	10	Held *et al.* (1964), Barman (1969)
Pyruvate carboxylase	157	500	Fung and Utter, unpublished results

[a] The estimated mitochondrial concentration of each enzyme was calculated from the observed activity of the enzyme in the mitochondrial fraction of rat liver, or in the case of pyruvate carboxylase and β-ketothiolase, chicken liver, following the procedure described in footnote *a* of Table II.
[b] The portion of the activity of palmitoylcarnitine transferase that is associated with the inner mitochondrial membrane was used in the calculation.
[c] The specific activity and physical properties are those for the enzyme purified from pig heart.
[d] The specific activity and physical and kinetic properties are those for the enzyme purified from pig heart.

Table VI. Estimated concentrations of enzymes of the mitochondrial matrix that bind succinyl-CoA and the corresponding K_M (or K_i) values for the combination of succinyl-CoA with each enzyme[a]

Enzyme	Estimated concentration of binding sites for succinyl-CoA (μM)	K_M or K_i (μM)	References
α-Ketoglutarate dehydrogenase[b]	1	10	Siekevitz (1952), Tanaka *et al.* (1972), Srere (1972b), J. R. Williamson *et al.* (1972)
Succinyl-CoA thiokinase[b]	0.2	1	Held *et al.* (1964), Barman (1969)
Citrate synthetase[c]	38	130	Moriyama and Srere (1971), Smith and Williamson (1971)
Pyruvate carboxylase	157	250	Fung and Utter, unpublished results

[a] The estimated mitochondrial concentration of each enzyme calculated from the observed activity of the enzyme in the mitochondrial fraction of rat liver, or in the case of pyruvate carboxylase, chicken liver, following the procedure described in footnote *a* of Table II.

[b] The specific activity and physical and kinetic properties are those for the enzyme purified from pig heart.

[c] The apparent K_i value for inhibition by succinyl-CoA is for the enzyme from rat kidney.

noncompetitive inhibitor with respect to acetyl-CoA, and it causes a marked decrease in the value of the Hill coefficient for acetyl-CoA (Utter and Fung, 1971).

Previous attempts to measure the intracellular concentration of acetoacetyl-CoA have been largely unsuccessful, chiefly because the levels of this metabolite appear to be very low (Greville and Tubbs, 1968). Using a sensitive radioactive assay, however, the level of acetoacetyl-CoA has been measured in isolated chicken liver mitochondria incubated under a variety of conditions, and is found to be about 0.5 nmole/mg protein, or 50μ M[8]. These values represent total acetoacetyl-CoA concentrations. However, it is likely that a considerable proportion of the acetoacetyl-CoA in the mitochondrial matrix is bound to proteins, which may include β-ketothiolase, HMG-CoA synthetase, and pyruvate carboxylase. Inspection of the estimated number of binding sites for acetoacetyl-CoA as well as the K_M or K_i values for the combination of acetoacetyl-CoA with these enzymes (Table VII) indicate that significant amounts of acetoacetyl-CoA would be bound by HMG-CoA synthetase and pyruvate carboxylase. Based on the hypothesis that acetoacetyl-CoA binds only to these two enzymes, the concentration of free intramitochondrial acetoacetyl-CoA is calculated to be 100 nM (Table IV). If this is the case, only about 0.2% of the total acetoacetyl-CoA would be present in the free form. Both the measured intramitochondrial levels and the estimated free concentrations of acetoacetyl-CoA are well below the apparent K_i value for the inhibition of avian liver pyruvate carboxylase (Table

Table VII. Estimated concentrations of enzymes of the mitochondrial matrix that bind acetoacetyl-CoA and the corresponding K_M (or K_i) values for the combination of aceto-acetyl-CoA with each enzyme[a]

Enzyme	Estimated concentration of binding sites for acetoacetyl-CoA (μM)	K_M or K_i (μM)	References
β-Ketothiolase	11	7	Middleton (1973), Clinkenbeard *et al.* (1973)
HMG-CoA Synthetase[b]	473	3	D. H. Williamson *et al.* (1968), Sugiyama *et al.* (1972)
Pyruvate carboxylase	157	200	Fung and Utter, unpublished observations

[a] The estimated mitochondrial concentration of each enzyme was calculated from the observed activity of the enzyme in the mitochondrial fraction of chicken liver following the procedure described in footnote *a* of Table II.
[b] The specific activity and physical and kinetic properties are those for the cytosolic form of the enzyme from chicken liver.

IV); hence it is unlikely that acetoacetyl-CoA acts as an inhibitor of this enzyme *in vivo*.

Less extensive investigations have been conducted on the effect of acyl-CoA derivatives on the activity of pyruvate carboxylase enzymes isolated from sources other than avian liver. In the case of pyruvate carboxylase purified from rat liver, McClure and Lardy (1971) have reported little or no inhibition of activity by a number of acyl-CoA derivatives that have been found to inhibit the enzyme from avian liver. These authors have concluded that the activity of this species of pyruvate carboxylase is unlikely to be controlled by the intramitochondrial concentrations of acyl-CoA inhibitors. It might be added that some caution should be exercised in testing the effect of acyl-CoA compounds on enzyme activity unless steps have been taken to purify the compounds. Trace contaminants of other acyl-CoA derivatives (e.g., acetyl-CoA) that may be formed during the preparation of a particular acyl-CoA derivative, may interfere with the effect of the modifier being tested.

*2. **Adenine Nucleotides.*** Purified pyruvate carboxylase enzymes are inhibited by $MgADP^-$, which acts as a product inhibitor of the reaction (Utter *et al.*, 1964; McClure and Lardy, 1971; Barden *et al.*, 1972). Stuki *et al.* (1972) have presented evidence that increases in the intramitochondrial concentration of ADP are accompanied by a decrease in the rate of pyruvate carboxylation; and they have noted a correlation between the increase in the rate of pyruvate carboxylation and an increase in the ratio of ATP to ADP. However, this type of experiment raises the question of whether the observed changes in the intramitochondrial adenine nucleotide concentration are governed by the activity of pyruvate carboxylase itself. It has also been shown in studies with purified pyruvate carboxylase that maximum inhibition of the reaction by $MgADP^-$ is observed at high pyruvate concentrations (McClure and Lardy, 1971), and it has been suggested that inhibition by $MgADP^-$ may be of relatively little importance at the lower concentrations of pyruvate likely to be present in the mitochondria (McClure and Lardy, 1971). Nevertheless, present data do not exclude the response of pyruvate carboxylase activity to changes in adenine nucleotide concentration as a control mechanism. However, in the experiments presented in Fig. 5, preliminary data indicate that no significant change in the intramitochondrial concentration of ADP occurs at various levels of acetylcarnitine, indicating that it is unlikely that alterations in the levels of ADP are responsible for the observed changes in pyruvate carboxylase activity.

It has been proposed that control of pyruvate carboxylase activity by changes in the intramitochondrial levels of $MgADP^-$ may be mediated at least in part through control of the rate of adenine nucleotide translocation (Stuki *et al.*, 1972; Shug and Shrago, 1973; Lopes-Cardoza, 1972). Thus it is envisioned that fatty acids may activate pyruvate carboxylase by inhibiting the entry of ADP to the mitochondrial matrix (hence the efflux of ATP), with a resulting increase in the intramitochondrial ATP/ADP ratio.

C. The Influence of Metal Ions

Pyruvate carboxylase requires both monovalent and divalent cations for activity (Utter and Scrutton, 1969; McClure *et al.*, 1971), and it has been proposed that the activity of this enzyme may be controlled in part by changes in the concentrations of cations, especially calcium (Gevers and Krebs, 1966). The purified enzyme is inhibited by low concentrations of calcium (McClure and Lardy, 1971; Wimhurst and Manchester, 1970), and Kimmich and Rasmussen (1969) have reported that calcium inhibits pyruvate carboxylation in isolated rat liver mitochondria. However, isolated mitochondria have the ability to accumulate relatively large amounts of calcium (Lehninger *et al.*, 1967). It is therefore possible that in experiments with isolated mitochondria, the intramitochondrial concentration of calcium may be much higher than that which occurs *in vivo* (McClure and Lardy, 1971; Morikofer-Zwez *et al.*, 1973).

The possibility that changes in the level of free magnesium may provide a fine control of pyruvate carboxylase activity has been considered (McClure and Lardy, 1971). However, recent experiments of Veloso *et al.* (1973) provide evidence that intracellular components may act as a buffer for magnesium ions, and the results suggest that there may be little change in the concentration of free magnesium *in vivo*. It has also been reported that the incubation of isolated rat liver mitochondria in the presence of elevated levels of inorganic phosphate or potassium ions diverts pyruvate metabolism from the formation of citrate as the major end product to the formation of acetoacetate (Schaefer and Veneziale, 1973). This has been interpreted as evidence that an increase in the concentration of these ions inhibits pyruvate carboxylation (Schaefer and Veneziale, 1973). However, a rather indirect assay for pyruvate carboxylase was used in these experiments, and the possibility that the influence of phosphate and potassium ions has a secondary effect on pyruvate carboxylase has not been excluded.

V. CONCLUSIONS

The data obtained with isolated chicken liver mitochondria demonstrate a correlation between the rate of pyruvate carboxylation and the intramitochondrial concentration of acetyl-CoA generated from exogenous acetylcarnitine in the presence of arsenite. This is consistent with the idea that the activation of pyruvate carboxylase by acetyl-CoA contributes to the control of the activity of this enzyme *in vivo*. However, the intramitochondrial concentration of acetyl-CoA which gives half-saturation of the rate of pyruvate carboxylation in isolated mitochondria is much higher than the apparent K_a of purified pyruvate carboxylase for acetyl-CoA. Calculations of the theoretical amount of acetyl-CoA that could be bound to proteins of the mitochondrial matrix indicate that it is likely that a significant proportion of the total acetyl-CoA is present in the bound form. This would result in a concentration of free acetyl-CoA lower than the observed intramitochondrial acetyl-CoA concentration, which is considered to represent total acetyl-CoA, thus providing a reasonable explanation for the seeming discrepancy between the apparent activation constant for acetyl-CoA in the mitochondrial system and that for the purified enzyme. Since a large number of assumptions are necessary in the calculations used to derive the theoretical plots of the rate of the pyruvate carboxylation as a function of total acetyl-CoA concentration, it is concluded that such calculations cannot be used to make quantitative predictions of the behavior of pyruvate carboxylase *in vivo*. The best approach now available appears to be the more direct one in which the activity of the enzyme in isolated mitochondria or cells is measured under various metabolic conditions. The chief difficulty associated with this type of approach is the problem of contolling a large number of variable parameters—in this case, other effectors of pyruvate carboxylase.

Measurement of the intramitochondrial concentrations of other effectors of pyruvate carboxylase indicates that under appropriate conditions, the activity of pyruvate carboxylase *in vivo* may be influenced by the intramitochondrial concentrations of pyruvate, propionyl-CoA, CoA-SH, succinyl-CoA, α-ketoglutarate, MgADP, and possibly calcium. Except in a few cases, however, there is little evidence that these effects have or do not have a regulatory significance. Further evaluation of this problem appears to require the same type of approach described for the investigation of the role of acetyl-CoA in the control of enzyme activity.

REFERENCES

Adam, P. A. J. and Haynes, R. C. (1969). *J. Biol. Chem.* **244,** 6444–6450.

Ashman, L. K., Keech, D. B., Wallace, J. C., and Nielsen, J. (1972). *J. Biol. Chem.* **247,** 5818–5824.

Ashman, L. K., Wallace, J. C., and Keech, D. B. (1973). *Biochem. Biophys. Res. Commun.* **51,** 924–931.

Ballard, F. J. and Hanson, R. W. (1967a). *Biochem. J.* **104,** 866–871.

Ballard, F. J. and Hanson, R. W. (1967b). *J. Lipid Res.* **8,** 73–79.

Ballard, F. J., Hanson, R. W., and Reshef, L. (1970). *Biochem. J.* **119,** 735–742.

Barden, R. E., Fung, C. -H., Utter, M. F., and Scrutton, M. C. (1972). *J. Biol. Chem.* **247,** 1323–1333.

Barden, R. E., Taylor, B. L., Frey, W. H., II, Isohashi, F., Zander, G. L., and Utter, M. F. (1975), in press, *Proc. Nat. Acad. Sci. U.S.A.*

Barman, T. E. (1969). *Enzyme Handbook,* Vol. 2, Springer-Verlag, Berlin, p. 878.

Barritt, G. J., Keech, D. B., and Ling, A. -M. (1966). *Biochem. Biophys. Res. Commun.* **24,** 476–481.

Bottger, I., Wieland, O., Brdiczka, D., and Pette, D. (1969). *Eur. J. Biochem.* **8,** 113–119.

Brech, W., Shrago, E., and Wilken, D. (1970). *Biochim. Biophys. Acta* **201,** 145–154.

Bremer, J. (1969). *Eur. J. Biochem.* **8,** 535–540.

Bucher, T., Brauser, B., Conze, A., Klein, F., Langguth, O., and Sies, H. (1972). *Eur. J. Biochem.* **27,** 301–317.

Carlson, C. A. and Kim, K.-H. (1974). *Arch. Biochem. Biophys.* **164,** 478–489.

Cha, S. (1970). *J. Biol. Chem.* **245,** 4814–4818.

Clinkenbeard, K. D., Sugiyama, T., Moss. J., Reed, W. D., and Lane, M. D. (1973). *J. Biol. Chem.* **248,** 2275–2284.

Chance, B., Holmes, W., Higgins, J. C., and Connelly, C. M. (1958). *Nature* **182,** 1190-1193.

Cleland, W. W. (1970). In *The Enzymes,* Vol. 2, 3rd ed. (P. D. Boyer, ed.). Academic Press, New York, pp. 1–65 (p. 58).

Colen, A. H., Prough, R. A., and Fisher, H. F. (1972). *J. Biol. Chem.* **247,** 7905–7909.

Cope, F. W. (1970). *Biophys. J.* **10,** 843–858.

Cori, G. T. and Green, A. A. (1943). *J. Biol. Chem.* **151,** 31–38.

Cornish-Bowden, A. and Koshland, D. E. (1970). *Biochemistry* **9,** 3325–3336.

Czeisler, J. L., Fritz, O. G., and Swift, T. J. (1970). *Biophys. J.* **10,** 260–268.

Delisle, G. and Fritz, I. B. (1967). *Proc. Nat. Acad. Sci. U.S.* **58,** 790–797.

Exton, J. H. and Park, C. R. (1969). *J. Biol. Chem.* **244,** 1424–1433.

Exton, J. H., Mallette, L. E., Jefferson, L. S., Wong, E. H. A., Friedmann, N., Miller, T. B., Jr., and Park, C. R. (1970). *Rec. Progr. Horm. Res.* **26,** 411–457.

Filsell, O. H., Jarrett, I. G., Taylor, P. H., and Keech, D. B. (1969). *Biochim. Biophys. Acta* **184,** 54–63.

Fischer, E. H. and Krebs, E. G. (1955). *J. Biol. Chem.* **216,** 121–132.

Frey, W. H., II, and Utter, M. F. (1974). *Fed. Proc.* **33,** 1428.

Garber, A. J. and Hanson, R. W. (1971). *J. Biol. Chem.* **246,** 5555–5562.

Garland, P. B., Shepherd, D., Nicholls, D. G., Yates, D. W., and Light, P. A. (1969). In *Citric Acid Cycle: Control and Compartmentation* (J. M. Lowenstein, ed.), Dekker, New York, pp. 163–212.

Garland, P. B., Shepherd, D., and Yates, D. W. (1965). *Biochem. J.* **97,** 587–594.

Gehring, U., Riepertinger, C., and Lynen, F. (1968). *Eur. J. Biochem.* **6,** 264–280.

Gevers, W. and Krebs, H. A. (1966). *Biochem. J.* **98,** 720–739.

Gevers, W. (1967). *Biochem. J.* **103,** 141–152.

Greenbaum, A. L., Gumaa, K. A., and McClean, P. (1971). *Arch. Biochem. Biophys.* **143,** 617–663.

Greville, G. D. and Tubbs, P. K. (1968). In *Essays in Biochemistry,* Vol. 4 (P. N. Campbell and G. D. Greville, eds.), Academic Press, New York, pp. 155–212.

Guynn, R. W., Veloso, D., and Veech, R. L. (1972). *J. Biol. Chem.* **247,** 7325–7331.

Halestrap, A. P. and Denton, R. M. (1974). *Biochem. J.* **138,** 313–316.

Hammes, G. G. and Hurst, J. K. (1969). *Biochemistry* **8,** 1083–1094.

Harris, E. J. and van Dam, K. (1968). *Biochem. J.* **106,** 759–766.

Haynes, R. C. (1972). In *Energy Metabolism and the Regulation of Metabolic Processes in Mitochondria* (M. A. Mehlman and R. W. Hanson, eds.), Academic Press, New York, pp. 239–252.

Heck, H. d'A (1969). *J. Biol. Chem.* **244,** 4375–4381.

Held, H. -W., Jacobs, H., and Klingenberg, M. (1964). *Biochem. Biophys. Res. Commun.* **17,** 130–135.

Henning, H. V., Stumpf, B., Ohly, B., and Seubert, W. (1966). *Biochem. Z.* **344,** 274–288.

Hill, A. V. (1913). *Biochem. J.* **7,** 471–480.

Holten, D. D. and Nordlie, R. C. (1965). *Biochemistry,* **4,** 723–731.

Hoppel, C. L. and Tomec, R. J. (1972). *J. Biol. Chem.* **247,** 832–841.

Johnson, D. C., Brunsvold, R. A., Ebert, K. A., and Ray, P. D. (1973). *J. Biol. Chem.* **248,** 763–770.

Kimmich, G. A. and Rasmussen, H. (1969). *J. Biol. Chem.* **244,** 190–199.

Kornberg, H. L. (1966). In *Essays in Biochemistry,* Vol. 2 (P. N. Campbell and G. D. Greville, eds.), Academic Press, New York, pp. 1–31.

Krebs, H. A. (1933). *Hoppe-Seyler's Z. Physiol. Chem.* **217,** 191–227.

Krebs, H. A., Speake, R. N., and Hems, R. (1965). *Biochem. J.* **94,** 712–720.

Krebs, H. A. (1966). In *Advances in Enzyme Regulation,* Vol. 4 (G. Weber, ed.), Pergamon Press, New York, pp. 339–353.

Landriscinda, C., Papa, S., Coratelli, P., Mazzarella, L., and Quagliariello, E. (1970). *Biochim. Biophys. Acta* **205,** 136–147.

Lardy, H. A. and Ferguson, S. M. (1969). *Ann. Rev. Biochem.* **38,** 991–1034.

Lardy, H. A., Paetkau, V., and Walter, P. (1965). *Proc. Nat. Acad. Sci. U.S.* **53,** 1410–1415.

Lehninger, A. L., Carafoli, E., and Rossi, C. S. (1967). *Adv. Enzymo.* **29,** 259–320.

Leighton, F., Poole, B., Beaufay, H., Baudhuin, P., Coffey, J., Fowler, S., and de Duve, C. (1968). *J. Cell Biol.* **37,** 482–513.

Ling, A. -M. and Keech, D. B. (1966). *Enzymologia* **30,** 367–380.

Linn, T. C., Pelley, J. W., Pettit, F. H., Hucho, F., Randall, D. D., and Reed, L. J. (1972). *Arch. Biochem. Biophys.* **148,** 327–342.

Linn, T. C., Pettit, F. H., and Reed, L. J. (1969). *Proc. Nat. Acad. Sci. U.S.* **62,** 234–241.

Lopes-Cardoza, M., Vaartjes, W. J., and Van den Bergh, S. G. (1972). *FEBS Let.* **28,** 265–270.

Madappally, M. M. and Mistry, S. P. (1970). *Biochim. Biophys. Acta* **215,** 316–322.

McClure, W. R. and Lardy, H. A. (1971). *J. Biol. Chem.* **246,** 3591–3596.

McClure, W. R., Lardy, H. A., and Kneifel, H. P. (1971). *J. Biol. Chem.* **246,** 3569–3578.

Middleton, B. (1973) *Biochem. J.* **132,** 717–730.

Morikofer-Zwez, S., Kunin, A. S., and Walter, P. (1973). *J. Biol. Chem.* **248,** 7588–7594.

Moriyama, T. and Srere, P. A. (1971). *J. Biol. Chem.* **246,** 3217–3223.

Newsholme, E. A., Crabtree, B., Higgins, S. J., Thornton, S. D., and Start,C. (1972). *Biochem. J.* **128,** 89–97.

Newsholme, E. A. and Gevers, W. (1967). Vitam. Horm. **25,** 1–87.

Nordlie, R. C. and Lardy, H. A. (1963). *J. Biol. Chem.* **238,** 2259–2263.

Prinz, W., Schoner, W., Haag, U., and Seubert, W. (1966). *Biochem. Z.* **346,** 206–211.

Prinz, W. and Seubert, W. (1964). *Biochem. Biophys. Res. Commun.* **16,** 582–585.

Reed, L. J. (1969). In *Current Topics in Cellular Regulation,* Vol. 1 (B. L. Horecker and E. R. Stadtman, eds.), Academic Press, New York, pp. 233–251.

Reiner, J. M. (1959). *Behavior of Enzyme Systems,* Burgess, Minneapolis, Minn., pp. 54–63.

Rognstad, R. and Katz, J. (1966). *Proc. Nat. Acad. Sci. U.S.* **55,** 1148–1156.

Rolleston, F. S. (1972). In *Current Topics in Cellular Regulation,* Vol. 5 (B. L. Horecker and E. R. Stadtman, eds.), Academic Press, New York, pp. 47–75.

Rubin, M. M. and Changeux, J. -P. (1966). *J. Mol. Biol.* **21,** 265–274.

Salganicoff, L. and Koeppe, R. E. (1968). *J. Biol. Chem.* **243,** 3416–3420.

Sanadi, D. R. (1963). In *The Enzymes,* Vol. 7, 2nd ed. (P. D. Boyer, H. Lardy, and K. Myrback, eds.), Academic Press, New York, pp. 307–344.

Sanadi, D. R. (1969). In *Methods in Enzymology,* Vol. 13 (J. M. Lowenstein, ed.), Academic Press, New York, pp. 52–55.

Schaefer, P. C. and Veneziale, C. M. (1973). *Eur. J. Biochem.* **35,** 18–22.

Scholz, R. and Bucher, T. (1965). In *Control of Energy Metabolism* (B. Chance, R. W. Estabrook, and J. R. Williamson, eds.), Academic Press, New York, pp. 393–414.

Scrutton, M. C., Olmsted, M. R., and Utter, M. F. (1969). In *Methods in Enzymology,* Vol. 13 (J. M. Lowenstein, ed.), Academic Press, New York, pp. 235–249.

Scrutton, M. C. and Utter, M. F. (1965). *J. Biol. Chem.* **240,** 1–9.

Scrutton, M. C. and Utter, M. F. (1967). *J. Biol. Chem.* **242,** 1723–1735.

Scrutton, M. C. and Utter, M. F. (1968). *Ann. Rev. Biochem.* **37,** 249–302.

Scrutton, M. C. and White, M. D. (1972). *Biochem. Biophys. Res. Commun.* **48,** 85–93.

Scrutton, M. C. and White, M. D. (1973a). *J. Biol. Chem.* **248,** 5541–5544.

Scrutton, M. C. and White, M. D. (1973b). *Fed. Proc.* **32,** 575 Abs.

Shrago, E. and Lardy, H. A. (1966). *J. Biol. Chem.* **241,** 663–668.

Shug, A. L. and Shrago, E. (1973). *Biochem. Biophys. Res. Commun.* **53,** 659–665.

Siekevitz, P. (1952). *J. Biol. Chem.* **195,** 549–565.

Smith, C. M., Bryla, J., Damon, S., LaNoue, K., and Williamson, J. R. (1973). *Anal. Biochem.* **51,** 408–420.

Smith, C. M. and Williamson, J. R. (1971). *FEBS Let.* **18,** 35–38.

Söling, H. D. and Volkmann, B. (1973). *Anal. Biochem.* **52,** 305–310.

Sols, A. and Marco, R. (1970). In *Current Topics in Cellular Regulation,* Vol. 2 (B. L. Horecker and E. R. Stadtman, eds.), Academic Press, New York, pp. 227–273.

Spivey, H. O. and Hill, D. E. (1973). *Fed. Proc.* **32,** 660 Abs.

Srere, P. A. (1968). In *The Metabolic Roles of Citrate* (T. W. Goodwin, ed.), Academic Press, New York, pp. 11–21.

Srere, P. A. (1972a). In *Current Topics in Cellular Regulation,* Vol. 5 (B. L. Horecker and E. R. Stadtman, eds.), Academic Press, New York, pp. 229–283.

Srere, P. A. (1972b). In *Energy Metabolism and the Regulation of Metabolic Processes in*

Mitochondria (M. A. Mehlman and R. W. Hanson, eds.), Academic Press, New York, pp. 79–91.

Stadtman, E. R. (1966). *Adv. Enzymol.* **28,** 41–154.

Stucki, J. W., Branand, F., and Walter, P. (1972). *Eur. J. Biochem.* **27,** 181–191.

Sugiyama, T., Clinkenbeard, K., Moss, J., and Lane, M. D. (1972). *Biochem. Biophys. Res. Commun.* **48,** 255–261.

Tanaka, N., Koike, K., Hamada, M., Otsuka, K. -I., Suematsu, T., and Koike, M. (1972). *J. Biol. Chem.* **247,** 4043–4049.

Taylor, H., Nielsen, J., and Keech, D. B. (1969). *Biochem. Biophys. Res. Commun.* **37,** 723–728.

Taylor, P. H., Wallace, J. C., and Keech, D. B. (1971). *Biochim. Biophys. Acta* **237,** 179–191.

Toews, C. J., Lowy, C., and Ruderman, N. B. (1970). *J. Biol. Chem.* **245,** 818–824.

Utter, M. F. (1959). *Ann. N.Y. Acad. Sci.* **72,** 451–461.

Utter, M. F. (1969). *FEBS Symp.* **19,** 91–98.

Utter, M. F., Barden, R. E., and Taylor, B. L. *Adv. in Enzymology* (1975) **42,** 1–72.

Utter, M. F. and Fung, C. -H. (1971). In *Regulation of Gluconeogenesis* (H. -D. Söling and B. Willms, eds.), Academic Press, New York, pp. 1–10.

Utter, M. F. and Keech, D. B. (1963). *J. Biol. Chem.* **238,** 2603–2608.

Utter, M. F., Keech, D. B., and Scrutton, M. C. (1964). In *Advances in Enzyme Regulation,* Vol. 2 (G. Weber, ed.), Pergamon Press, New York, pp. 49–68.

Utter, M. F. and Scrutton, M. C. (1969). In *Current Topics in Cellular Regulation,* Vol. 1 (B. L. Horecker and E. R. Stadtman, eds.), Academic Press, New York, pp. 253–296.

Valentine, R. C., Wrigley, N. G., Scrutton, M. C., Irias, J. J., and Utter, M. F. (1966). *Biochemistry* **5,** 3111–3116.

Veloso, D., Guynn, R. W., Oskarsson, M., and Veech, R. L. (1973). *J. Biol. Chem.* **248,** 4811–4819.

Villar-Palasi, C. and Larner, J. (1961). *Arch. Biochem. Biophys.* **94,** 436–442.

Walter, P. and Anabitarte, M. (1971). *FEBS Lett.* **12,** 289–292.

Walter, P., Paetkau, V., and Lardy, H. A. (1966). *J. Biol. Chem.* **241,** 2523–2532.

West, D. W., Chase, J. F. A., and Tubbs, P. K. (1971). *Biochem. Biophys. Res. Commun.* **42,** 912–918.

Wieland, O., Evertz-Prusse, E., and Stukowski, B. (1968). *FEBS Let.* **2,** 26–28.

Wieland, O. H., Patzelt, C., and Loffler, G. (1972). *Eur. J. Biochem.* **26,** 426–433.

Williamson, D. H., Bates, M. W., and Krebs, H. A. (1968). *Biochem. J.* **108,** 353–361.

Williamson, J. R., Browning, E. T., and Olson, M. S. (1968a). In *Advances in Enzyme Regulation,* Vol. 6 (G. Weber, ed.), Pergamon Press, New York, pp. 67–100.

Williamson, J. R., Browning, E. T., Scholz, R., Kreisberg, R. A., and Fritz, I. B. (1968b). *Diabetes* **17,** 194–208.

Williamson, J. R., Browning, E. T., and Scholz, R. (1969a). *J. Biol. Chem.* **244,** 4607–4616.

Williamson, J. R., Browning, E. T., Thurman, R. G., and Scholz, R. (1969b). *J. Biol. Chem.* **244,** 5055–5064.

Williamson, J. R. and Corkey, B. E. (1969). In *Methods in Enzymology,* Vol. 13 (J. M. Lowenstein, ed.), Academic Press, New York, pp. 434–513.

Williamson, J. R., Rostand, S. G., and Peterson, M. J. (1970). *J. Biol. Chem.* **245,** 3242–3251.

Williamson, J. R., Smith, C. M., LaNoue, K. F., and Bryla, J. (1972). In *Energy Metabolism and the Regulation of Metabolic Processes in Mitochondria* (M. A. Mehlman and R. W. Hanson, eds.), Academic Press, New York, pp. 185–210.

Wimhurst, J. M. and Manchester, K. L. (1970). *Biochem. J.* **120,** 79–93.

Wise, E. M. and Ball, E. G. (1964). *Proc. Nat. Acad. Sci. U.S.* **52,** 1255–1263.

Yamazaki, R. K., Harano, Y., Utter, M. F., and Miller, M. (1976). To be published.

CHAPTER TWO

Hormonal Regulation of Phosphoenolpyruvate Carboxykinase (GTP) in Mammalian Tissues

SHIRLEY M. TILGHMAN, RICHARD W. HANSON, AND F. JOHN BALLARD

I. INTRODUCTION

In 1954 Utter and Kurahashi (1954) described an enzyme in chicken liver mitochondria that catalyzed the Mn^{2+} and IDP dependent carboxylation of phosphoenolpyruvate to oxaloacetate (Kurahashi *et al.*, 1959). The reaction was freely reversible. Subsequently the enzyme was called phosphoenolpyruvate carboxykinase (EC 4.1.1.32) and, since it was argued that guanine nucleotides were the physiological substrates, the reaction becomes:

$$\text{oxalacetate} + \text{GTP} \overset{Mn^{2+}}{\rightleftharpoons} \text{P-enolpyruvate} + \text{GDP} + CO_2$$

Because an excellent review of the properties and kinetic mechanism of the reaction has appeared recently (Utter and Kolenbrander, 1972), we are concerned in this chapter only with the hormonal regulation of the enzyme.

Phosphoenolpyruvate carboxykinase (PEP-CK) is present at high specific

Supported by grants AM-16009, AM-18034 and CA-12227 from the U.S. Public Health Service and by general support from the Samuel S. Fels Fund. R. W. H. was the recipient of a U.S. Public Health Service Career Development Award K04 AM-15365 activities in liver, kidney cortex, and white adipose tissue and at lower specific activities in lung, jejunum, and brain (Hanson and Garber, 1972). It was initially proposed that PEP-CK was involved in the fixation of CO_2 in the dicarboxylic acid shuttle, but subsequent studies indicated that its role was to catalyze the formation of phosphoenolpyruvate as a precursor for the synthesis of glucose in liver and kidney (Utter, 1969) and for the synthesis of α-glycerol phosphate in adipose tissue (Hanson and Garber, 1973). In mammals the enzyme exhibits a diverse pattern of regulation with respect to its intracellular distribution and its adaptability by a wide variety of hormones.

The intracellular distribution of PEP-CK is not constant but varies between species. Livers of all species studied to date contain both a cytosol form and a mitochondrial form of the enzyme, although in markedly different proportions. While the rat, hamster, and mouse contain almost solely cytosol enzyme, the chicken and rabbit have 95% of the enzyme in the mitochondria (Hanson and Garber, 1972). All other species, including man, have an approximately equal distribution between the cytosol and mitochondria. The two forms of PEP-CK have been shown to be different proteins on the basis of immunochemical studies in rat tissues (Ballard and Hanson, 1969) and chemical studies in the livers of humans (Diesterhaft *et al.*, 1971) and chickens (Ballard, 1971). Another important distinction, first recognized by Lardy and his co-workers (Shrago *et al.*, 1963; Nordlie *et al.*, 1965) is that only the cytosol enzyme is responsive to dietary and hormonal stimuli.

The participation of cytosol PEP-CK in gluconeogenesis was originally demonstrated by isotopic studies with rat liver (see Utter, 1969, for review). These observations were supported by evidence that dietary and endocrine treatment in rats, known to increase gluconeogenesis, also increased the activity of the cytosol enzyme (Shrago *et al.*, 1963). Finally, "crossover" studies showed that PEP-CK was the rate-limiting step in gluconeogenesis under certain conditions. In livers of some species such as chicken, pigeon, and rabbit, the evidence is convincing that the mitochondrial enzyme is also involved in gluconeogenesis (Hanson and Garber, 1972; Arinze *et al.*, 1973; Söling *et al.*, 1973). These studies, reviewed in detail by Söling and Kleineke in another chapter, represent the beginning of a quantitation of mitochondrial phosphoenolpyruvate formation for glucose synthesis in such species. However, some controversy has arisen over the relative contributions to phosphoenolpyruvate synthesis of the cytosol and mitochondrial enzymes in species that contain them in approximately equal proportions. There is now considerable evidence of a unique pattern of regulation of gluconeogenesis in these species. It has been established that mitochondrial PEP-CK can contribute a substantial fraction of the total phosphoenolpyruvate synthesized in gluconeogenesis, but the physiological interaction between the cytosol and mitochondrial enzyme remains to be fully elucidated.

The importance of PEP-CK in nongluconeogenic tissues has been most thoroughly studied in adipose tissue (Reshef *et al.*, 1970; Hanson and Garber, 1973). The activity of the cytosol enzyme in this tissue is increased during metabolic conditions in which there is an acceleration of free fatty acid (FFA) release (Reshef *et al.*, 1969a). It has been proposed that under such conditions PEP-CK participates in the synthesis of α-glycerol phosphate from pyruvate, thus allowing reesterification of FFA. A similar role has been proposed by Scholz (1972) for the enzyme in lung.

II. HORMONAL REGULATION OF PEP-CK

Interest in the adaptation of PEP-CK was originally stimulated by the enzyme's important physiological role in gluconeogenesis, but in addition, it can be argued that PEP-CK is an excellent model for studies of the mechanisms by which hormones regulate enzyme levels in mammalian tissues. Its activity is altered *in vivo* by administration of a wide variety of hormones including glucagon (Shrago *et al.*, 1963; Reshef and Hanson, 1972), norepinephrine (Reshef and Hanson, 1972), ACTH (Reshef and Hanson, 1972), glucocorticoids (Shrago *et al.*, 1963; Foster *et al.*, 1966a; Reshef *et al.*, 1970; Exton *et al.*, 1973b; Gunn *et al.*, 1975b), thyroxine (Bottger *et al.*, 1970; Nagai and Nakagawa, 1972), and insulin (Shrago *et al.*, 1963; Foster *et al.*, 1966a; Shrago *et al.*, 1967; Tilghman *et al.*, 1974). The response to each hormone is tissue specific (Table I). For example, glucocorticoids increase the activity of the kidney enzyme (Longshaw and Pogson, 1972), decrease the activity in white adipose tissue (Reshef *et al.*, 1969b; Reshef *et al.*, 1970), and have only a marginal effect on the activity of the liver enzyme (Shrago *et al.*, 1963; Reshef *et al.*, 1970). Likewise, insulin administration can decrease PEP-CK activity in liver and adipose tissue (Shrago *et al.*, 1967; Reshef *et al.*, 1970) but has no effect in the kidney (Kamm *et al.*, 1974). It is possible that such diverse hormonal responses from tissue to tissue will eventually be useful in identifying specific factors (e.g., receptors, binding proteins) required in target tissues for hormonal action. The hormonally induced changes in enzyme activity are of special importance, since no cofactors or prosthetic groups have been discovered.

From a technical point of view, PEP-CK is a relatively easy protein to quantitate, both enzymatically and immunochemically. It is present in large amounts in liver and kidney (2–30 μmole/min/g weight tissue), and it can be readily purified from liver (Ballard and Hanson, 1969). The relative

Table I. Factors that alter the levels of PEP-CK in rat tissues

Tissue	Hormone	Mechanism	References
I. Inducers			
Liver	Glucagon Epinephrine Norepinephrine	↑ Synthesis cyclic AMP	Wicks *et al.* (1972), Wicks and McKibbin (1972), Hanson *et al.* (1973), Gunn *et al.* (1975a)
	Glucocorticoids	↑ Synthesis in absence of insulin	Gunn *et al.* (1975a, 1975b)
	Tryptophan	↑ Synthesis; ↓ degradation	Ballard and Hopgood (1973)
	Thyroxine	Not known	
Kidney	Acidosis	↑ Synthesis	Iynedjian *et al.* (1975)
	Glucocorticoids	↑ Synthesis	Iynedjian *et al.* (1975)
Adipose tissue	Norepinephrine Epinephrine	↑ Synthesis	Meyahus and Reshef, unpublished observations
II. Deinducers			
Liver	Insulin	↓ Synthesis	Tilghman *et al.* (1974), Tilghman *et al.* (1975)
Kidney	Alkalosis	↓ Synthesis	Iynedjian *et al.* (1975)
Adipose tissue	Insulin	↓ Synthesis	Meyahus and Reshef, unpublished observations
	Glucocorticoids	↓ Synthesis	Meyahus and Reshef, unpublished observations

amount of PEP-CK synthesized in liver can be as high as 4% of cytosol protein (Hopgood *el al.*, 1973; Tilghman *et al.*, 1974) and in kidney cortex 9% of cytosol protein synthesis (Iynedjian *et al.*, 1975). On the basis of sodium dodecyl sulfate electrophoresis, PEP-CK is a single polypeptide with a molecular weight of about 85,000 (Fig. 1). These characteristics of the enzyme simplify considerably the problems in measuring its turnover. However, Bentle and Snoke (1974) have suggested that the enzyme is activated by a protein of high molecular weight that stabilizes the binding of iron to the enzyme protein. From these and other studies (Snoke *et al.*, 1971), it has been argued that Fe^{2+} rather than Mn^{2+} is the most likely activator of PEP-CK in the intact liver. Although further studies are required to clarify the role of activated protein, Fe^{2+}, and Mn^{2+} in the regulation of the enzyme, work in our own laboratory has shown that Mn^{2+} and not Fe^{2+} will protect the enzyme from heat denaturation or inactivation *in vitro* (Hopgood and Ballard, unpublished experiments). Except for these experiments on enzyme activation and stabilization, there have been no reports of any allosteric-type modifications of the enzyme, despite the accumulation of evidence that the rapid increase in hepatic gluconeogenesis caused by glucagon results in a metabolite "crossover" pattern consistent with regulation at PEP-CK (Exton and Park, 1969; Mallette *et al.*, 1969; Exton *et al.*, 1972; Exton *et al.*, 1973a; 1973b).

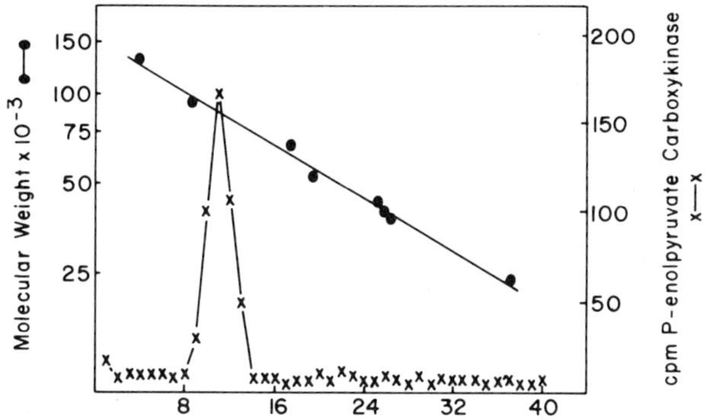

Figure 1. Molecular weight of P-enolpyruvate carboxykinase. Purified ^{3}H-PEP-CK was electrophoresed in 10% polyacrylamide gels in the presence of sodium dodecyl sulfate for 18 hours. At the same time, the following molecular markers were treated identically and run in separate gels (in descending order by molecular weight): β-galactosidase, phosphorylase *b*, serum albumin, immunoglobulin heavy chain, ovalbumin, creatine phosphokinase, lactate dehydrogenase, and immunoglobulin light chain. The gels were scanned at 280 nm using a Gilford linear transporter, and the relative mobility of each protein was calculated. (Solid points) represent a plot of molecular weight against mobility for each protein marker. The gel containing the ^{3}H-enzyme was extruded into 1 mm fractions and the radioactivity in each determined (crosses).

A. Hormonal Regulation of PEP-CK in Liver

Lardy and co-workers (Shrago *et al.*, 1963) were the first to demonstrate that an overnight fast produced a threefold increase in the activity of the enzyme, which could be reversed by refeeding a diet high in carbohydrate. The requirement for carbohydrate was specific; glucose, fructose, and glycerol were each effective in lowering PEP-CK activity, whereas galactose, 2-deoxyglucose, and ribose had no effect (Shrago *et al.*, 1967). Surprisingly, the refeeding of a high protein diet caused a further increase in activity rather than a decline. The fasting-refeeding cycle has now been thoroughly studied using immunological techniques that allow direct measurements of the rate of enzyme synthesis and degradation. Hopgood *et al.* (1973) demonstrated that the rate of PEP-CK synthesis, expressed as a percentage of total cytosol proteins, decreased fifteenfold when starved rats were refed (Fig. 2). When the time course of this decline was examined in more detail, it became apparent that the synthesis rate of the enzyme decreased after refeeding, with a half-time of 30 minutes (Tilghman *et al.*, 1974).

Alterations in the rate of enzyme synthesis is the major factor responsible for changes in hepatic PEP-CK levels. However, neither the specific hormones responsible, nor their mechanisms of action have been unequivocally resolved. Most of the work in this area has been carried out using intact animals, where a perturbation of one endocrine system often produces change in another, making it difficult to isolate the effect of a single mediator. Now that we have descriptions of several *in vitro* cell and organ culture systems that not only contain the enzyme but also respond to hormonal stimuli, it should be possible to overcome the problems inherent in working *in vivo*. Two systems in particular, the Reuber H-35 hepatoma (Barnett and Wicks, 1971) and the fetal liver organ culture (Wicks, 1969), have been used successfully to examine the hormonal control of PEP-CK. Nevertheless, it is always necessary to extrapolate back to the whole animal to ascertain whether observations noted with cultured cells are truly reflections of the physiological situation.

The importance of cyclic AMP-mediated hormones in the control of PEP-CK has been recognized for some time. Glucagon administered *in vivo* elevates the activity of the enzyme severalfold in 4 hours (Shrago *et al.*, 1963; Reshef and Hanson, 1972). The injection of a cyclic AMP derivative, dibutyryl cyclic AMP, produces a similar increase, suggesting that glucagon acts directly on the liver by elevating the levels of cyclic AMP (Reshef and Hanson, 1972; Wicks *et al.*, 1972). This was confirmed using the fetal organ culture system, where glucagon was able to induce PEP-CK directly as the result of an activation of adenyl cyclase (Wicks, 1969; 1971). The liver is also sensitive to the action of catecholamines, and several labo-

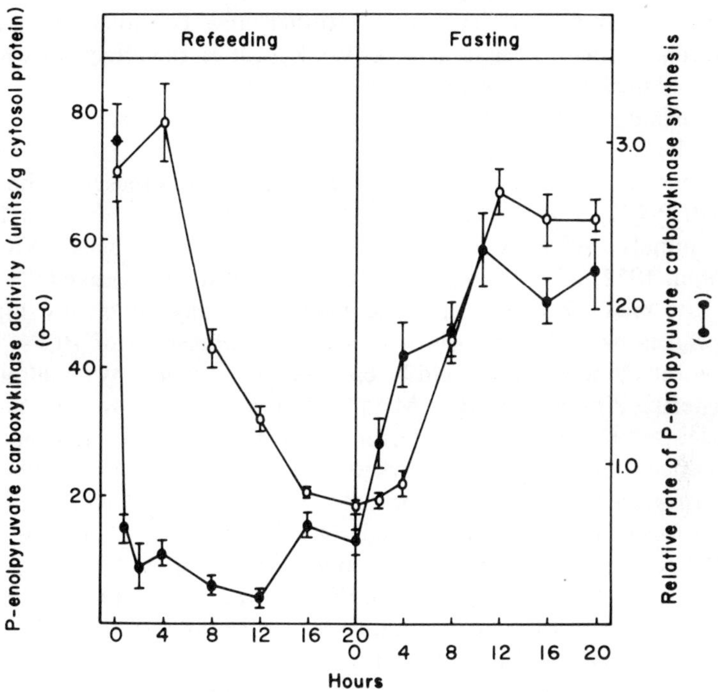

Figure 2. Changes in the activity and relative rate of P-enolpyruvate carboxykinase synthesis during a fasting-refeeding cycle. Rats were fasted 20 hours prior to the start of the experiment. At zero time, the animals were refed laboratory chow for 20 hours in the dark, and food was withdrawn again for 20 additional hours. Thirty minutes prior to killing at the indicated times, the rats were injected with 100 μCi ^{3}H-leucine. A liver cytosol fraction was prepared after homogenization of the liver in 0.25 M sucrose, and the activity of PEP-CK (expressed as units/g cytosol protein, O——O) determined in that fraction. To a volume of cytosol containing 180 mU of PEP-CK was added 250 mU of specific antibody to PEP-CK. After incubation of the reaction at 37°C for 15 minutes, the tube was left overnight at 0°C. The immunoprecipitate was collected by centrifugation, and the radioactivity was determined. A second control precipitation was performed on the supernatant of the first precipitation with the addition of 180 mU of unlabeled purified PEP-CK. The difference in radioactivity between the first and second precipitations was taken as the radioactivity in the enzyme. The incorporation of radioactivity into total cytosol proteins was determined after trichloroacetic acid precipitation of an aliquot of the liver cytosol. The rate of PEP-CK synthesis was then expressed as the radioactivity in PEP-CK relative to the radioactivity in total cytosol protein (●——●). *Source:* Hopgood *et al.* (1973).

ratories have reported increases in PEP-CK activity in response to epinephrine administration *in vivo* (Reshef and Hanson, 1972) and to isoproterenol in fetal liver in culture (Wicks, 1969), the effects of which are presumably mediated by cyclic AMP.

The mechanism by which cyclic AMP induces PEP-CK has been shown immunochemically to be the result of *de novo* synthesis of the enzyme. Wicks and his collaborators have reported fivefold increases in the rate at which PEP-CK is synthesized relative to other cytosol proteins in both intact animals (Wicks *et al.,* 1972) and Reuber H-35 cells (Wicks and McKibbin, 1972) after a single administration of dibutyryl cyclic AMP. In an effort to relate this effect of cyclic AMP to the physiologically important fasting-refeeding cycle, we showed that the suppression of PEP-CK synthesis by refeeding glucose could be completely overcome by simultaneously administering dibutyryl cyclic AMP (Tilghman *et al.,* 1974). This suggested that at least one aspect of the refeeding response was a fall in hepatic cyclic AMP concentrations. Recent work in our laboratory using Reuber H-35 cells has strengthened this proposal. When dibutyryl cyclic AMP is removed from the tissue culture medium after 18 hour incubation, the rate of PEP-CK synthesis in the cells declines rapidly, with a half-time of approximately 40 minutes (Fig. 3) (Tilghman *et al.,* 1975). Thus the response of hepatic PEP-CK to refeeding and to removal of dibutyryl cyclic AMP in Reuber H-35 cells is identical and strongly suggests that they share the same hormonal mediator and molecular mechanism.

Insulin is also involved in the control of PEP-CK. When rats were made insulin deficient by alloxan or mannoheptulose administration or by pancreatectomy, the activity of PEP-CK rose sharply (Shrago *et al.,* 1963, 1967; Foster *et al.,* 1966a). A subsequent injection of insulin was sufficient to return the activity to normal values. This effect of insulin on PEP-CK in diabetic rats was later shown to be the result of a decrease in the rate of synthesis of the enzyme (Tilghman *et al.,* 1974). The rate at which insulin decreased PEP-CK synthesis was identical to that observed when fasted rats were refed. This pointed to insulin as a possible mediator of the fasting-refeeding response, perhaps by lowering intracellular cyclic AMP levels. What could not be determined, however, was whether insulin acted directly on the liver or blocked the action or release of another hormone, such as glucagon. The indirect alternative was suggested by the observation that insulin was ineffective in the absence of carbohydrate. Thus the administration of insulin to either fasting normal or diabetic rats did not lower PEP-CK activity (Shrago *et al.,* 1967; Tilghman *et al.,* 1974). This could be explained in terms of a requirement for both glucose and insulin for a suppression of glucagon secretion (Muller *et al.,* 1971). Alternatively, if glucose or a metabolite of glucose is required for insulin action in the liver, the low

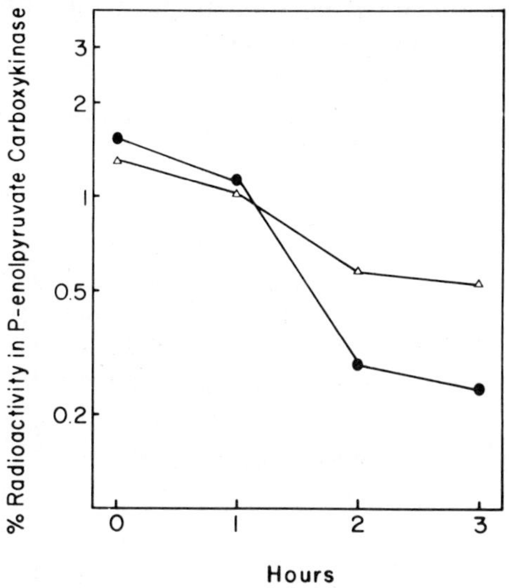

Figure 3. Deinduction of P-enolpyruvate carboxykinase in Reuber H-35 cells. Reuber H-35 cells were incubated for 18 hours in the absence of serum and presence of 0.5 mM dibutyryl cyclic AMP and 1 mM theophylline. At zero time the medium was either replaced with serum-free medium minus inducers (●——●); or glucagon-free insulin was added to a final concentration of 50 nM without changing the medium (△——△). Next 50 μCi of ^{3}H-leucine was added to the cells 15 minutes before harvesting at the time indicated. A cytosol fraction was prepared and the relative rate of PEP-CK synthesis determined as described in the legend to Fig. 2. *Source*: Tilghman *et al.* (1975).

glucokinase activity in the liver of fasting rats (Sharma *et al.*, 1963) might also explain these results. Using Reuber H-35 cells, we have shown that insulin, even in the absence of glucose, is able to prevent the induction of PEP-CK by dibutyryl cyclic AMP as well as cause a decrease in the rate of synthesis of the enzyme in cells previously induced with dibutyryl cyclic AMP (Fig. 3) (Tilghman *et al.*, 1975; Gunn *et al.*, 1975a). Thus although insulin may have a number of other sites of action *in vivo*, it may directly affect liver cells to suppress PEP-CK synthesis.

The role of the glucocorticoids in the regulation of PEP-CK is rather confusing (see Table II). Adrenalectomy of adult rats neither alters the activity of this enzyme nor prevents the response to fasting (Shrago *et al.*, 1963; Nordlie *et al.*, 1965; Reshef *et al.*, 1969b, 1970), which suggests that the glucocorticoids are not required for either maintenance or induction of PEP-CK. However, a single dose of hydrocortisone or a fluorinated glucocorticoid, triamcinolone, produced a transient 40% increase in enzyme activity (Foster *et al.*, 1966a). If the treatment was continued for several

Table II. Effect of glucocorticoids on the levels of PEP-CK in rat liver, kidney, and Reuber H-35 hepatoma cells

Tissue	Status	Hormone administered	Activity units/g tissue	Synthesis rate % radioactivity in PEP-CK
Liver	Fed	—	4.05 ±0.29	1.05 ±0.15
		Triamcinolone	4.57 ±0.28	0.28 ±0.06
		Hydrocortisone	4.80 ±0.44	0.42 ±0.04
	Diabetic	—	10.74 ±1.29	2.95 ±0.14
		Triamcinolone	13.84 ±1.37	4.02 ±0.19
Reuber H-35 hepatoma cells	—	—	46[a] ±3.2	0.53 ±0.08
		Dexamethasone	57[a] ±6	1.38 ±0.08
Kidney	Fed	—	3.93 ±0.26	1.31 ±0.18
		Triamcinolone	7.11 ±0.47	3.92 ±0.65
		Hydrocortisone	5.94 ±0.39	3.94 ±0.25

[a] Activities expressed as mU/mg cytosol protein. Animals were injected with triamcinolone (5 mg/100 g body weight) and hydrocortisone (5 mg/100 g body weight) and killed 6.5 hours later. Details on the assay of PEP-CK and the determination of its relative synthesis rate are presented in Fig. 2. and in Gunn *et al.* (1975b). In experiments with Reuber H-35 cells, 2×10^{-7} M dexamethasone was added to cells grown in Eagle's Minimal Essential Medium. After 3 hours ^{3}H-leucine was added to each Petri dish. The cells were collected and the activity and relative synthesis rate of PEP-CK determined as described in Gunn *et al.* (1975a).

days, PEP-CK increased approximately threefold (Shrago *et al.*, 1963). In studies conducted in our laboratory, however, triamcinolone actually depressed PEP-CK synthesis in liver after 24 hours, by stimulating insulin secretion (Gunn *et al.*, 1975b). This is in accord with the observation that only in diabetic-adrenalectomized animals was triamcinolone able to increase the activity of PEP-CK (Reshef *et al.*, 1969b). PEP-CK activity in fetal liver explants is increased by hydrocortisone, but the effect is small, especially when compared to the large increase in tyrosine aminotransferase activity under the same conditions (Wicks, 1969). This is pertinent because both enzymes were induced to the same extent by dibutyryl cyclic AMP (Wicks, 1969). In contrast, PEP-CK in Reuber H-35 cells is induced by another glucocorticoid derivative, dexamethasone, and to the same degree as tyrosine aminotransferase (Barnett and Wicks, 1971). The effect of dexamethasone in Reuber H-35 cells was due to an increase in *de novo* synthesis of the enzyme (Gunn *et al.*, 1975a), which supported other observations that the effect could be blocked by cycloheximide or actinomycin D (Barnett and Wicks, 1971; Wicks and McKibbin, 1972). Although these experiments with isolated systems have shown that PEP-CK is able to respond to the glucocorticoid hormones, the site at which the steroids participate in the *in vivo* regulation of the hepatic enzyme remains to be elucidated.

Thyroxine increases the activity of PEP-CK two- to threefold in both normal and thyroidectomized rats, and the effect is blocked by cycloheximide (Bottger *et al.,* 1970; Nagai and Nakagawa, 1972; Suda *et al.,* 1973). However, as in the case of adrenalectomy, thyroidectomy alone does not significantly alter basal enzyme levels, although it markedly reduces the activity of another gluconeogenic enzyme, pyruvate carboxylase (Bottger *et al.,* 1970; Nagai and Nakagawa, 1972; Suda *et al.,* 1973). In addition, thyroidectomy does not prevent the circadian rhythm of the enzyme that is the result of the nocturnal eating habits of rats (Suda *et al.,* 1973). We do not know the precise physiological conditions under which the thyroid gland becomes important in the regulation of PEP-CK or the mechanism by which thyroxine increases enzyme activity.

One rather unusual modulator of PEP-CK activity *in vivo* is tryptophan. Interest was focused on this amino acid as the result of studies suggesting that tryptophan inhibits gluconeogenesis *in vivo,* with the block occurring at the level of PEP-CK (Ray *et al.,* 1966; Snoke *et al* ., 1971). Yet when the livers of tryptophan-treated animals were removed and PEP-CK was assayed, the activity was actually increased (Foster *et al.,* 1966b). The time course of this increase was biphasic with only the slower second phase sensitive to inhibitors of protein synthesis. The apparent paradox was resolved by Veneziale *et al.* (1967), who showed that the initial rapid increase in PEP-CK was the result of activation of existing enzyme by divalent metal ions. The inhibition *in vivo* is explained by the formation of a ternary complex between the enzyme, Fe^{2+}, and quinolinate, a metabolite of tryptophan. This complex dissociates during the large dilution of the enzyme required for the assay, leaving the enzyme in a metal-activated form (Foster *et al.,* 1967). The metal activation occurs because Fe^{2+}, Co^{2+}, and Mn^{2+} stimulate PEP-CK to give a higher activity than found when Mg^{2+} is the only divalent metal ion included in the assay. In our work we did not find a tryptophan or Fe^{2+} stimulation of the enzyme, a result explained by our use of Mn^{2+} rather than Mg^{2+} in the routine enzyme assay. The second or puromycin-sensitive phase of the increase in enzyme activity is due to an increase in the *de novo* synthesis of PEP-CK, coupled with a decrease in its rate of degradation. Using a specific antibody, Ballard and Hopgood (1973) demonstrated a twofold increase in the relative synthesis rate of the enzyme in fed but not in starved rats after tryptophan injection. On the other hand, the effect on degradation was not dependent on the dietary status of the animal.

Similar effects of tryptophan on increasing the activity of other hepatic enzymes have been reported (Peraino *et al.*, 1965; Jost *et al;* 1968; Kenney and Flora, 1971), but only in the case of tryptophan oxygenase has the mechanism been ascribed to substrate-mediated stabilization of the protein (Schimke *et al.*, 1965). It is not clear at present whether these observations with high concentrations of tryptophan are physiologically important in modulating enzyme levels in the liver.

B. The Neonatal Development of Hepatic PEP-CK

1. Developmental Pattern of Hepatic Gluconeogenesis. During fetal life the rat is dependent on the maternal blood for maintenance of its blood glucose levels, a dependency reflected in a total absence of fetal gluconeogenesis. The amount of liver glycogen in the fetus rises during the last quarter of gestation, and glycogen accounts for 10% of the liver wet weight in the term fetus (Ballard, 1970). When the rat is born the maternal blood supply of glucose is cut off, resulting in a transient neonatal hypoglycemia. To compensate for this fall in blood glucose, the hepatic glycogen reserves are mobilized and gluconeogenesis is initiated (Fig. 4). As pointed out by Ballard (1970), the rate of glycogen mobilization is sufficient to provide most, if not all, of the glucose requirements of the young rat during the first 8 hours after birth. After this time the rate of gluconeogenesis is sufficient to maintain the blood glucose concentration. All the enzymes required for gluconeogenesis are present in fetal rat liver, with the exception of PEP-CK, which develops immediately after birth (Ballard and Hanson, 1967). The appearance of this enzyme completes the gluconeogenic pathway; thus the time course of development of gluconeogenesis and PEP-CK are identical (Fig. 4).

Other factors also contribute to the marked increase in hepatic gluconeogenesis observed at birth. Philippidis and Ballard (1970) showed that glucose synthesis in liver slices from fetal rats was negligible, even though PEP-CK had been induced *in utero* with glucagon. This was apparently due to the highly reduced cytosol NAD^+ to NADH ratio of approximately 40 in fetal rat liver. This ratio shifts to 700 immediately after birth, indicating that the relative anoxia characteristic of the fetal rat can markedly inhibit glucose synthesis.

The implications of this proposed hypoxia-normoxia regulation of gluconeogenesis are more important in species such as the sheep which have high activities of hepatic PEP-CK in the fetus. To investigate the control of gluconeogenesis in developing sheep, Warnes *et al.* (1974) cannulated the femoral blood vessels of lambs *in utero,* injected pulses of ^{14}C-labeled lactate into the fetus, and measured the formation of ^{14}C-glucose and ^{14}C-fructose at various times after the mother had recovered from the opera-

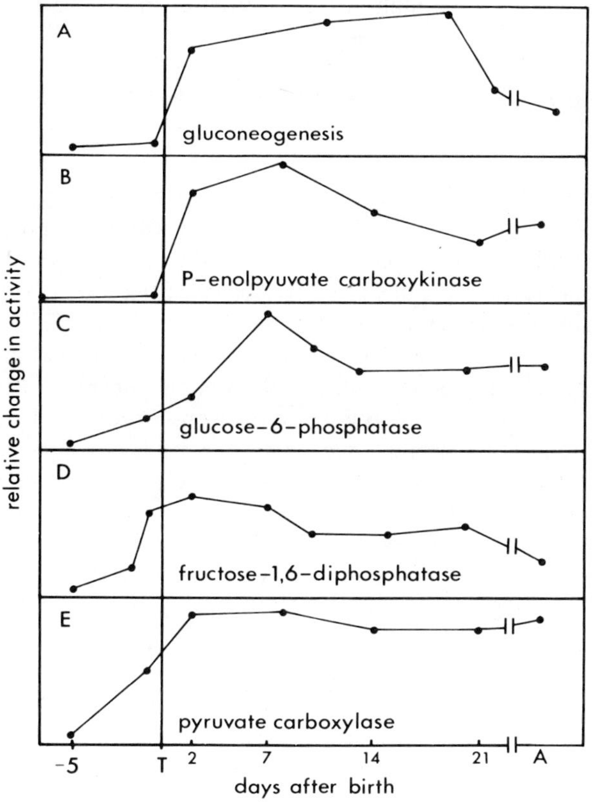

Figure 4. Changes in the relative activities of gluconeogenic enzymes in rat liver during development. The activities of four hepatic gluconeogenic enzymes were measured from 5 days prenatal to 30 days postnatal and in the livers of adult animals. The values are expressed as relative changes with the maximum activity measured over this period taken as 100%. Gluconeogenesis was determined by measuring the incorporation of lactate-2-^{14}C into glycogen by liver slices from animals of ages indicated. Values from Ballard and Oliver (1963) and Ballard and Hanson (1967).

tion. Complications caused by anesthesia were thus eliminated. These experiments showed a complete absence of gluconeogenesis in the fetal sheep, even though PEP-CK and all the other gluconeogenic enzymes were present at activities comparable to those found in the adult animal. However, there was a rapid rate of gluconeogenesis in neonatal lambs that could be detected as quickly as 3 minutes after birth. Gluconeogenesis could not be detected unless the lamb was breathing normally. These experiments show that although PEP-CK development is required for an active pathway of gluconeogenesis to occur, high enzyme activities must also be accompanied by oxygenation of the hepatic cells.

2. ***Hormonal Control of the Development of Rat Liver PEP-CK.*** Yeung and Oliver (1968a) suggested that the appearance of PEP-CK at birth involved *de novo* enzyme synthesis, since the premature delivery of fetal rats resulted in induction of the enzyme that could be blocked by the administration of puromycin or ethionine to the rats while still in the uterus. This conclusion was later confirmed by Philippidis *et al.* (1972) using immunochemical techniques that directly measured the rate of synthesis of the enzyme. During the period of rapid enzyme accumulation after delivery, the rate of synthesis of hepatic PEP-CK increased from negligible levels to 2.4% of total cytosol protein synthesis. It was also observed that during the initial period of increasing synthesis there was no degradation of the enzyme (Section III.B.1). Measurable degradation rates were not noted until 24 hours after delivery. The appearance of PEP-CK at birth can thus be attributed to two related conditions: rapid increase in the rate of enzyme synthesis and absence of its degradation.

Much effort has been directed toward defining the hormonal mediators responsible for the development of PEP-CK in the newborn rat. Yeung and Oliver (1968a) showed that by preventing the hypoglycemia accompanying birth by injecting glucose intraperitoneally, the induction of PEP-CK is forestalled. A number of other carbohydrates, including galactose, fructose, mannose, and glycerol, were also capable of suppressing PEP-CK appearance; ribose and 2-deoxyglucose were ineffective. Subsequent work suggested, however, that the decrease in blood glucose itself was not directly affecting PEP-CK but rather was stimulating a change in hormonal status of the neonate (see Hanson *et al.*, 1975, for a review).

When glucagon was administered to fetuses at birth, the activity of PEP-CK increased in a 5 hour period to twice the level reached in untreated newborns (Yeung and Oliver, 1968a). Glucose was unable to repress the enzyme under these conditions, despite its ability to prevent hypoglycemia. On the other hand, insulin given at birth prevented the induction of PEP-CK but at the same time prolonged the period of hypoglycemia (Yeung and Oliver, 1968b). Although hypoglycemia is probably the physiological trigger responsible for the appearance of the enzyme, its action is by way of an increase in glucagon and depression of insulin secretion. Support for this view comes from the observations of Girard *et al.* (1973), who measured both an increase in plasma glucagon and a marked decrease in plasma insulin within 20 minutes after detachment of the fetus from the placenta (Fig. 5).

The enhanced induction of PEP-CK by glucagon and the antagonistic ac-

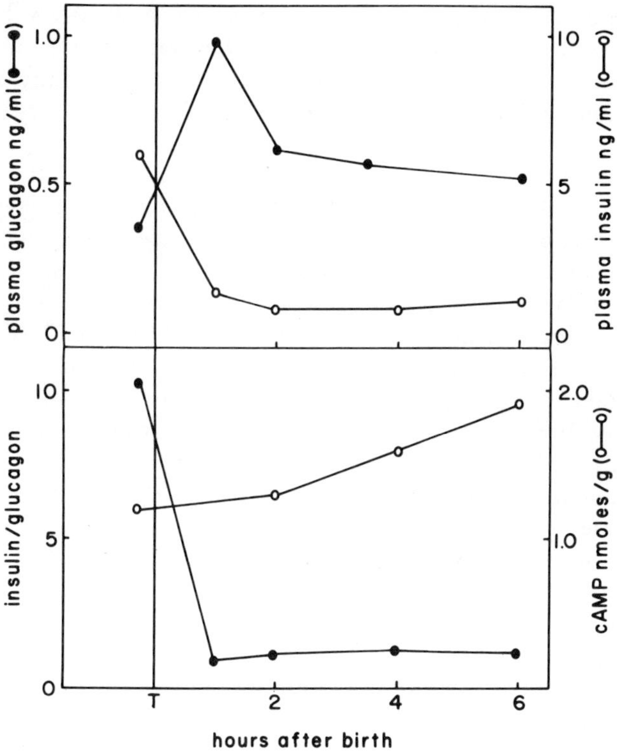

Figure 5. Changes in the concentrations of insulin, glucagon, and hepatic cyclic AMP at birth. Values for insulin and glucagon from Girard *et al.* (1973); cyclic AMP values from Butcher and Potter (1972).

tion of insulin points to a central role for cyclic AMP in the onset of enzyme synthesis at birth. Indeed, an increase in hepatic cyclic AMP has been measured in newborns (Butcher and Potter, 1972). Yeung and Oliver (1968a, 1968b) first reported that glucagon as well as two other cyclic AMP mediated hormones, epinephrine and norepinephrine, were effective in inducing PEP-CK *in utero*. More directly, it was discovered that cyclic AMP and its dibutyryl derivative produce large increases in enzyme activity in both fetuses and newborns (Yeung and Oliver, 1968b; Hanson *et al.*, 1973). It was demonstrated that the induction by glucagon or cyclic AMP *in utero* was the result of *de novo* enzyme synthesis, as measured immunochemically (Hanson *et al.*, 1973). Hanson *et al.* (1973) found that the ability of dibutyryl cyclic AMP to induce PEP-CK *in utero* appeared at 15 days of uterine life, and the extent of the response increased with the age of the fetuses up to 19 days.

The participation of insulin in the suppression of PEP-CK before birth has not been unequivocally demonstrated. However, given the marked drop in plasma insulin levels at birth (Girard *et al.*, 1973), the ability of insulin to prevent induction when administered immediately after delivery (Yeung and Oliver, 1968b), and our own observations that insulin can antagonize the action of dibutyryl cyclic AMP *in utero* (Hanson *et al.*, 1975), it is reasonable to assume that the decrease in insulin at birth in response to hypoglycemia is an important prerequisite for PEP-CK appearance.

The involvement of hormones other than insulin and glucagon in the development of PEP-CK at birth is less well understood. Yeung and Oliver (1968b) proposed that at least part of the cyclic AMP mediated increase in the enzyme at birth is due to the action of epinephrine, since an inhibitor of catecholamine action, ergotamine tartrate, was able to decrease the response of PEP-CK to premature delivery. It seems unlikely that glucocorticoids play a primary role in the development of the enzyme at birth, since the administration of triamcinolone *in utero* had no effect on PEP-CK activity under conditions where another hepatic enzyme, tyrosine aminotransferase, was induced by the steroid (Yeung *et al.*, 1967).

C. PEP-CK in Kidney Cortex

1. The role of the Enzyme in Renal Metabolism. The activity of PEP-CK in rat kidney cortex is approximately 10 μmole bicarbonate incorporated into oxaloacetate per minute per gram, a value somewhat higher than that found in liver. Recently Guder and Schmidt (1974) have demonstrated that the enzyme is present exclusively in the proximal convoluted tubule of the kidney cortex, implying that this segment of the kidney is involved in gluconeogenesis. It is clear from the studies of Krebs *et al.* (1963) that the kidney cortex can synthesize glucose from a variety of precursors; later studies with the isolated, perfused rat kidney demonstrated that the rate of renal gluconeogenesis per gram of tissue is similar to hepatic glucose synthesis (Bowman, 1970). In the human, during prolonged starvation, total renal gluconeogenesis is almost equal to total hepatic glucose synthesis (Owen *et al.*, 1969).

During starvation the concentration of plasma β-hydroxybutyrate and acetoacetate can increase approximately seventyfold, resulting in an enhanced rate of ketone body filtration by renal glomeruli. This increased ketonuria necessitates the excretion of ammonia to maintain urinary neutrality. The ammonia is derived from glutamine that has been mobilized from muscle protein and taken up by the kidney. During its metabolism in the kidney, glutamine is converted to glutamate by glutaminase I, and the glutamate is oxidatively deaminated to α-ketoglutarate by glutamate dehydrogenase. For the prolonged utilization of glutamine by the kidney this α-

ketoglutarate must be removed by subsequent reactions. Goodman *et al.*, (1966) reported that metabolic acidosis increased the rate of renal gluconeogenesis in the rat, and they linked the synthesis of glucose with the disposal of glutamine. In support of this suggestion Rosenzweig and Frascella (1968) and Alleyne and Scullard (1969) reported that rat kidney PEP-CK was markedly increased during acidosis induced by the administration of ammonium chloride. It has therefore been suggested that the rate of renal gluconeogenesis can control the rate of ammonia secretion by a mechanism related to the removal of common metabolic intermediates.

The validity of such mechanisms depends on their demonstration *in vivo*. A number of published investigations supports this pathway linking ammoniagenesis and gluconeogenesis (Pitts, 1972; Goodman, 1972). Many of these studies are based on work with slices of rat kidney cortex and employ high concentration of glutamine. In a carefully controlled study of glutamine metabolism in the intact, functioning kidney of the dog, Pitts *et al.* (1972) compared the renal conversion of glutamine to glucose and to CO_2 in dogs under conditions of chronic metabolic acidosis or alkalosis. Of the total glutamine extracted during acidosis, only one-fifth was converted to glucose, the rest was recovered as CO_2. It was therefore concluded that gluconeogenesis did not limit ammonia formation during acidosis in the kidney. The authors suggested (see also a review by Goodman, 1972) that PEP-CK could play a role in ammonia formation as part of a nongluconeogenic pathway. As Pitts (1972) states:

> It is possible to explain, in part, increased extraction and utilization of glutamine and increased excretion of ammonia in acidosis on the basis of an adaptive and rate limiting increase in tubular cell content of PEP-CK. This is different from saying that gluconeogenesis from oxalacetate is rate limiting for, on the basis of our experiments, it obviously is not. However, it is possible that CO_2 production from oxalacetate derived from glutamine occurs as a pathway involving conversion of oxalacetate to phosphoenolpyruvate and its re-entry into the Krebs cycle as acetyl CoA.

This pathway (Fig. 6) would not result in "futile cycling" of carbon, as may occur with lactate during hepatic gluconeogenesis (Friedmann *et al.*, 1971), since with glutamine as precursor, there is no requirement for the initial carboxylation of pyruvate by pyruvate carboxylase. Rognstad and Katz (1972) have shown an extensive recycling of carbon through PEP-CK and pyruvate kinase in rat kidney cortex slices. It therefore seems likely that PEP-CK plays an important role in kidney as part of a pathway that oxidizes glutamine carbon and at the same time supports the generation of ammonia during periods of acidosis.

2. ***Hormonal Regulation of Renal PEP-CK.*** Although studies on the hormonal regulation of PEP-CK in kidney are less extensive than those on

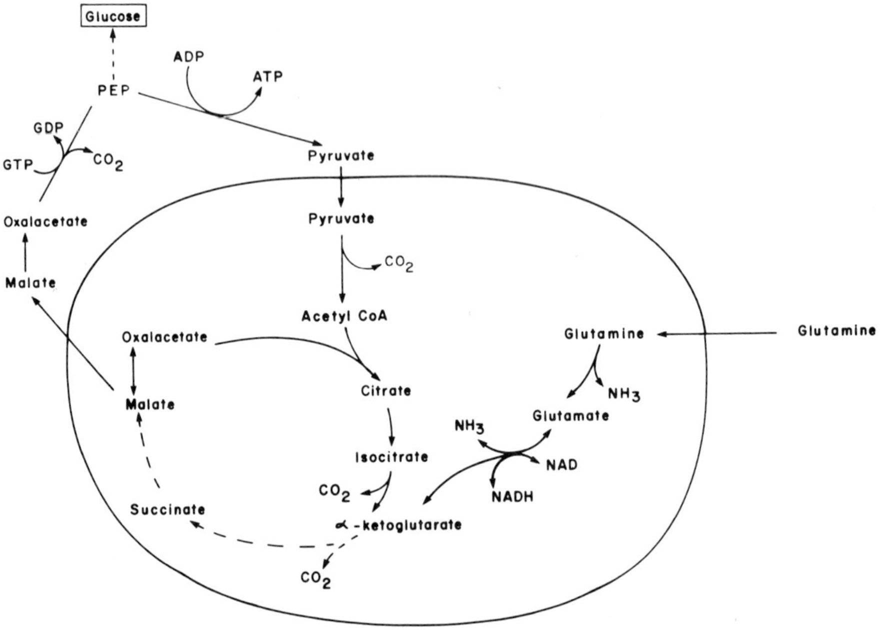

Figure 6. Pathway for glutamine metabolism in rat kidney cortex. The circle represents the mitochondria; the illustration emphasizes the recycling of glutamine carbon as pyruvate by way of PEP-CK and pyruvate kinase.

the liver enzyme, three factors which increase cytosol enzyme activity in the rat have been identified: starvation (Iynedjian and Peters, 1974), glucocorticoid administration (Longshaw and Pogson, 1972), and acidosis induced by ammonium chloride (Rosenzweig and Frascella, 1968; Alleyne and Scullard, 1969; Kamm *et al.*, 1974). Refeeding starved rats and $NaHCO_3$ treatment of acidotic animals reduces PEP-CK activity, but the rates of deinduction are less rapid than observed in liver during refeeding (Tilghman *et al.*, 1974).

The acidosis-mediated increase in renal PEP-CK activity was first reported by Rosenzweig and Frascella (1968). Subsequently a number of groups have investigated the mechanism underlying the increase in enzyme activity after NH_4Cl administration but with conflicting results. Since neither ithionine nor actinomycin D could prevent the induction of PEP-CK by acidosis in their studies, Flores and Alleyne (1971) concluded that protein synthesis was not involved. In an extension of this work, Longshaw *et al.* (1972) measured the rate of loss of enzyme activity after $NaHCO_3$ administration to acidotic rats and suggested that the increase in renal PEP-CK was due to a decrease in the degradation rate of the protein. The authors also reported an increase in the molecular weight from 83,000 to 128,000 after NH_4CL administration. This finding is difficult to interpret because the enzyme, although large, is not known to have subunits or to be allosterically activated. It should be noted, however, that the kidney enzyme has not been purified to homogeneity nor studied as extensively as the hepatic enzyme. It is known that renal PEP-CK is immunichemically similar to the liver form of the enzyme (Longshaw and Pagson, 1972; Iynedjian *et al.*, 1975), and it has a similar molecular weight, as determined by sodium dodecyl sulfate gel electrophoresis (Iynedjian *et al.*, 1975).

Recently in our laboratory (Iynedjian *et al.*, 1975) we have studied the effects of acidosis on the rate of PEP-CK synthesis in rat kidney (Fig. 7). We found that acidosis increased the rate of enzyme synthesis and did not alter the rate of degradation. These immunochemical experiments showed that the rate of synthesis increased from 2% in fed animals to 5%, 5 hours after NH_4Cl administration. The half-life of the enzyme was approximately 14 hours in kidneys of both normal and acidotic animals. Treatment of acidotic

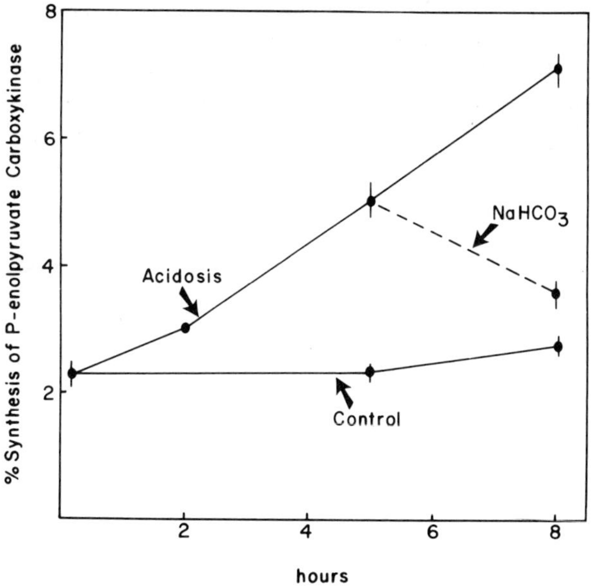

Figure 7. The effects of acidosis and alkalosis on renal P-enolpyruvate carboxykinase synthesis. Rats were administered either NH_4Cl (10 mmole/kg body weight) at zero time (acidosis) or the same amount of NaCl (control). One group of NH_4Cl-treated rats was given $NaHCO_3$ (15 mmole/kg body weight) 5 hours later ($NaHCO_3$). Thirty minutes prior to killing at the indicated times, rats were administered 200 μCi ^{3}H-leucine. The kidney cortex was removed, and a cytosol fraction was prepared and used to determine the relative rate of PEP-CK synthesis as described in the legend to Fig. 2. *Source*: Iynedjian *et al.* (1975).

animals with $NaHCO_3$ decreased renal PEP-CK activity by decreasing its rate of synthesis, with a half-time of approximately 3 hours.

It is improbable that a change in the acid-base status of the animal is *itself* the cellular modulator for the increased rate of enzyme synthesis. Acidosis does not act by stimulating the secretion of glucocorticoids by the adrenal cortex, since the response of adrenalectomized rats to NH_4Cl acidosis resembles that of intact animals (Iynedjian *et al.*, 1975). It is possible that a change in the acid-base status results in the secretion of another hormone, which then triggers the increased synthesis of PEP-CK. Thus $NaHCO_3$ could proceed by shutting off the secretion of the stimulatory hormone. The rather slow development of the effects of both acidosis and alkalosis on enzyme synthesis suggest, but do not prove, an extrarenal involvement.

Renal PEP-CK undergoes a fasting-refeeding cycle similar to the liver enzyme but exhibits a different time course (Iynedjian *et al.*, 1975). Whereas the liver enzyme reaches a peak activity after 24 hours of starvation, the enzyme in kidney increases in activity for up to 36 hours and maintains that level for up to 72 hours. Refeeding glucose to starved rats results in a very slow decline in the activity of the renal enzyme when compared to the rapid response of the liver enzyme. Both the increase and decrease in enzyme activity during this cycle have been shown to be the result of changes in the *de novo* synthesis rate of the enzyme (Iynedjian *et al.*, 1975).

It is evident that the pancreatic hormones insulin and glucagon, which are intimately involved in the regulation of the fasting-refeeding cycle in liver, are not mediating the rise in enzyme activity in the kidney. Insulin depletion caused by alloxan treatment increases the activity and rate of synthesis of kidney PEP-CK, but $NaHCO_3$ is more effective than insulin in subsequently reducing the activity to normal levels (Iynedjian *et al.*, 1975). Thus the effect of diabetes is most probably due to changes in the acid-base balance and is not directly the result of a decrease in circulating insulin concentrations. The renal enzyme also appears to be insensitive to changes in the concentration of intracellular cyclic AMP. The injection of dibutyryl cyclic AMP and theophylline into normal animals elevates the synthesis rate of PEP-CK, but this effect is abolished in adrenalectomized rats (Iynedjian *et al.*, 1975). The adrenosteroidogenic effect of cyclic AMP (Bieck *et al.*, 1969), therefore, explains the rise in enzyme activity in intact animals. In addition, parathyroid hormone, which is known to increase cyclic AMP levels in the kidney, has no effect on PEP-CK activity (Iynedjian *et al.*, 1975). On the basis of this work we conclude that the elevation in renal PEP-CK activity upon starvation is due primarily to changes in the acid-base balance and not to the intervention of the pancreatic hormones.

The inductive effect of glucocorticoids have been examined by several groups of investigators. The glucocorticoid-mediated increase in renal PEP-CK has been shown in our laboratory to be due to a two- to threefold increase in the *de novo* rate of enzyme synthesis (Table II) (Iynedjian *et al.*, 1975). The induction of the renal enzyme by glucocorticoids is first observed after 3 hours after and at 12 hours reaches a maximal rate of enzyme synthesis that resembles the effect of the adrenal steroids on several liver enzymes such as tyrosine aminotransferase. Longshaw and Pogson (1972) demonstrated that simultaneous administration of triamcinolone and NH_4Cl produced an additive effect on enzyme activity. In addition, adrenalectomy does not diminish the effect of acidosis, as noted earlier.

D. PEP-CK in White Adipose Tissue

1. The Role of the Enzyme in Adipose Tissue. Although adipose tissue has approximately a tenfold lower activity of cytosol PEP-CK than liver or kidney, when activities are compared on a wet weight basis, the specific activities per unit of protein are similar. The enzyme is believed to participate in the synthesis of α-glycerophosphate from compounds such as lactate, pyruvate, and especially amino acids, by way of an abbreviated form of gluconeogenesis that we termed "glyceroneogenesis" (Ballard *et al.*, 1967, Reshef *et al.*, 1970). This suggestion was based on labeling studies indicating that twice as much carbon-2 as carbon-1 of pyruvate-^{14}C was incorporated into glyceride glycerol by adipose tissue. When the glycerol was chemically degraded, more than 90% of the pyruvate carbon-1 and 43 to 49% of the pyruvate carbon-2 incorporated were in the α-carbons (Ballard *et al.*, 1967). These results, which have been extended by White *et al.* (1968), indicate that glyceride-glycerol can be synthesized from 3-carbon compounds such as pyruvate and that the dicarboxylic acid shuttle involving pyruvate carboxylase and PEP-CK is operative in adipose tissue.

The physiological role of glyceroneogenesis in adipose tissue may be related to the reesterification of FFA during starvation, using substrates other than glucose as a carbon source. Glyceroneogenesis in adipose tissue increases during starvation (Ballard *et al.*, 1967) in direct relationship to the increased activity of PEP-CK. There is a substantial rate of reesterification of FFA in epididymal fat pads from starved rats (Reshef *et al.*, 1970), which can only be accounted for on the basis of α-glycerolphosphate formation by way of glyceroneogenesis. Furthermore, low concentrations of FFA substantially increase the rate of pyruvate conversion to glyceride glycerol, especially in adipose tissue from fasted rats (Reshef and Shapiro, 1970). Since FFA levels do indeed increase in adipose tissue of fasted rats (Rodbell, 1965) there will be a synergistic response caused by elevated PEP-

CK activation and the FFA-mediated stimulation. Glyceroneogenesis, therefore, acts as a repressing force on all-out FFA release from adipose tissue during periods of starvation, and it is antiketogenic.

2. ***Hormonal Regulation of Adipose Tissue PEP-CK.*** Adipose tissue PEP-CK is increased in activity by starvation (Ballard *et al.*, 1967, Reshef *et al.*, 1969b), epinephrine, norepinephrine, and ACTH (Reshef and Hanson, 1972); it is decreased by refeeding (Ballard *et al.*, 1967, Reshef *et al.*, 1969b), glucocorticoids (Reshef *et al.*, 1969b, 1970; Gorin *et al.*, 1969), and the administration of nicotinamide (Reshef and Hanson, 1972). Hopgood *et al.* (1973) have shown that the three- to fourfold increase in the adipose tissue enzyme activity during starvation is due to an increased rate of enzyme synthesis. Upon refeeding, the synthesis rate of PEP-CK decreases, with a half-time of 8 hours. This is much slower than the decrease in enzyme synthesis in livers from the same animals, which occurred with a half-time of less than 1 hour (Tilghman *et al.*, 1974). The rate of degradation of adipose tissue PEP-CK, determined immunochemically, did not change significantly during a fasting-refeeding cycle (Hopgood *et al.*, 1973).

These findings indicate different sensitivities of PEP-CK from adipose tissue and liver to the hormonal stimuli mediating the refeeding response. Although *in vitro* systems, using adipocytes, have not been investigated to determine direct hormonal interactions on enzyme turnover, it is probable that the fasting-refeeding cycle is mediated by insulin and cyclic AMP as noted with liver (Tilghman *et al.*, 1974; Tilghman *et al.*, 1975). The adipose tissue enzyme is insulin sensitive as shown by the three- to fourfold increase in its activity after alloxan treatment, whereas insulin administration reduces the activity to basal levels (Gorin *et al.*, 1969; Reshef *et al.*, 1970). It is possible that the more rapid changes in the synthesis rate of the enzyme in liver are due to the greater insulin concentrations in portal versus peripheral blood. It is also possible that intracellular cyclic AMP concentrations are not as rapidly increased or decreased in adipose tissue. Studies by Reshef and Hanson (1972) demonstrated that two cyclic AMP mediated hormones, ACTH and norepinephrine, increased the activity of the adipose tissue enzyme *in vivo,* whereas administration of either nicotinic acid or nicotinamide decreased the activity. The effect of nicotinamide, reported to cause a decrease in cyclic AMP concentrations in adipose tissue, was greater in starved animals, which implies that cyclic AMP levels were elevated under these conditions. However, neither dibutyryl cyclic AMP nor glucagon produced, on injection, an increase in enzyme activity (Reshef and Hanson, 1972). Until a satisfactory isolated system is developed for studying hormonal inductions of adipose tissue PEP-CK, it is not possible to state categorically whether cyclic AMP is involved in the regulation of the fasting-refeeding response.

A depressive effect of glucocorticoids on the activity of the adipose tissue enzyme has been demonstrated by several groups (Gorin *et al.*, 1969; Reshef *et al.*, 1969b, 1970). Adrenalectomy increases the PEP-CK activity three- to fourfold, an effect that can be overcome by administration of triamcinolone (Reshef *et al.*, 1969b). This effect of adrenalectomy in adipose tissue is in marked contrast to the lack of its effect on the liver and kidney enzymes, and implies that the adipose tissue is under continual modulation by the adrenal steroids. Adrenalectomized rats exhibit a normal fasting-refeeding cycle with respect to PEP-CK activity suggesting that the glucocorticoid suppression is distinct from any insulin–cyclic AMP effect (Reshef *et al.*, 1969b). In addition, adrenalectomy of alloxan-diabectic animals produces a further increase in enzyme activity over that caused by insulin deficiency alone (Reshef *et al.*, 1970). The eight- to tenfold elevation in enzyme activity in diabetic-adrenalectomized rats can only be reduced to basal levels by simultaneous treatment with insulin and glucocorticoid, giving additional evidence that the two hormones are affecting the enzyme by separate mechanisms.

It has been suggested that the decrease in PEP-CK activity and synthesis rate found in livers of glucocorticoid-treated, intact animals is the result of a glucocorticoid-mediated increase in insulin secretion (Gunn *et al.*, 1975b). Quite clearly, a different mechanism must be proposed for the depressive effect of glucocorticoids in adipose tissue, althought it remains to be established whether the glucocorticoids act directly on adipose tissue. The observation that triamcinolone suppresses PEP-CK activity in both hypophysectomized and hypophysectomized-adrenalectomized animals rules out the pituitary as a possible site of glucocorticoid mediation (Tilghman, unpublished observations). Whatever the mechanism of glucocorticoid action, adipose tissue PEP-CK is probably being continually modulated by the circulating steroid hormones.

III. MECHANISMS OF PEP-CK ADAPTATION

A. Changes in the Synthesis Rate of PEP-CK

In the preceding section we showed that hormones alter the activity of PEP-CK by changing its rate of synthesis. Because of the metabolic importance of this enzyme, it is of primary interest to determine the molecular mechanisms that control the amount of PEP-CK synthesized in the cell. This enzyme may also serve as a model for studies of the effects of hormones on gene transcription, mRNA processing from the nucleus to the cytoplasm, and the rate of mRNA translation. The adaptive responses of hepatic PEP-CK are very similar to other inducible hepatic enzymes. For example, the stimulatory effects of glucagon and cyclic AMP and their antagonism by insulin are also observed for glucose-6-phosphatase (Langdon and Weakley, 1955), ornithine transaminase (Peraino and Pitot, 1964; Pitot and Peraino, 1964), and serine dehydratase (Freedland and Avery, 1964; Peraino *et al.*, 1966, Jost *et al.*, 1968). In the case of serine dehydratase, the rapid effect of feeding glucose to starved rats to decrease the rate of enzyme synthesis is similar to that found for PEP-CK (Jost *et al.*, 1968). Barnett and Wicks (1971) have shown a remarkable similarity in the response to glucocorticoids of PEP-CK and tyrosine aminotransferase in Reuber H-35 cells. From these observations, it can be argued that an understanding of PEP-CK regulation at the molecular level will provide information that can be applied to studies of other adaptable enzymes.

In the following sections we consider the mechanism of action of the two most thoroughly studied effectors of PEP-CK, cyclic AMP and glucocorticoids. Until very recently, studies in the area of enzyme induction have been limited by the technical difficulties inherent in quantitating intracellular levels of specific mRNA. Thus even the most basic question of whether a hormone is affecting the concentration of a specific mRNA or the rate of translation of a stable mRNA pool could not be answered, except indirectly with the use of general inhibitors of transcription, such as actinomycin D. Inhibitor studies have provided important clues to the site of action of hormones on PEP-CK synthesis, but in some instances their use has posed more problems than have been answered, as we discuss later. With the advent of techniques designed to measure intracellular mRNA, it should now be possible to determine directly the mechanisms by which hormones change the levels of PEP-CK.

1. The Mechanism of the Cyclic AMP Mediated Induction of Hepatic PEP-CK. The mechanism of the cyclic AMP mediated induction of hepatic PEP-CK has been studied using a number of systems. Actinomycin D, which is an inhibitor of total RNA transcription when used at the appropriate concentration, completely blocked the inductive effect of dibutyryl cyclic AMP, glucagon, or isoproterenol on enzyme activity in fetal organ culture (Wicks, 1969). Likewise, Yeung and Oliver (1968b) and Hanson *et al.* (1973) observed that injection of actinomycin D, together with either dibutyryl cyclic AMP or glucagon into fetal rats, also blocked the subsequent induction of enzyme activity. However, the effect of actinomycin D was time dependent (Hanson *et al.*, 1973). If the drug was given 1 hour after either inducer, there was no observable inhibition of the increase in enzyme activity. This is in contrast to the effect of actinomycin D on the increase in PEP-CK activity after premature delivery, an increase presumably mediated by cyclic AMP. In this instance, not only does the administration of actinomycin D just prior to delivery block the appearance of the enzyme, it is able to prevent further increases in enzyme activity when given up to 5 hours after delivery (Yeung and Oliver, 1968b). The likeliest explanation for the discrepancy is the transient effect of externally administered dibutyryl cyclic AMP on enzyme activity (Hanson *et al.*, 1973) compared to the continual exposure to elevated intracellular cyclic AMP after birth (Butcher and Potter, 1972). These studies suggest that DNA-dependent RNA synthesis is required for the initial induction by cyclic AMP in livers of fetal and newborn rats, the simplest explanation being the requirement for PEP-CK mRNA synthesis.

However, quite different results were obtained in studies on the enzyme in Reuber H-35 cells by Wicks and McKibbin (1972) and by Gunn *et al.* (1975a). The rapid increase in PEP-CK activity and synthesis produced by the addition of dibutyryl cyclic AMP to the culture media is *not* prevented by concentrations of actinomycin D that are sufficient to inhibit by 90% the synthesis of poly-(A)-containing RNA (Gunn *et al.*, 1975a). Since poly-(A)-containing RNA is mostly mRNA, this result implies that mRNA synthesis is not required for the cyclic AMP effect in Reuber cells. These results were confirmed using a second transcriptional inhibitor, cordycepin, which preferentially inhibits the polyadenylation of mRNA (Penman *et al.*, 1970; Darnell *et al.*, 1971; Singer and Penman, 1972). At concentrations that produced maximal inhibition of ^{3}H-orotic acid incorporation into mRNA, the induction of the enzyme by dibutyryl cyclic AMP was not blocked (Gunn *et al.*, 1975a).

The effect of these inhibitors exhibited time and concentration dependencies (Gunn *et al.*, 1975a). At longer times after the addition of actinomycin D to the medium, the ability of dibutyryl cyclic AMP to maintain elevated activities of the enzyme was decreased. Although other explanations are possible, it is likely that the decline in the rate of enzyme synthesis after long exposure to actinomycin D or cordycepin represents the inhibition of the normal replenishment of a relatively labile PEP-CK mRNA pool. The estimated half-life of the enzyme mRNA is 4 to 8 hours (Tilghman *et al.*, 1975), which is considerably shorter than the half-life of average mRNA. When concentrations of inhibitors were used that were greater than those required for the maximal inhibitory effect on mRNA synthesis, the induction of PEP-CK by dibutyryl cyclic AMP was completely blocked (Gunn *et al.*, 1975a). However, at these concentrations the inhibitors were also affecting the rate of translation of general proteins (Gunn *et al.*, 1975a). It is reasonable to assume that translational rather than transcriptional effects at high concentrations of inhibitors were responsible for the prevention of induction of the enzyme. Otherwise it is necessary to postulate that the PEP-CK mRNA has a sensitivity to these drugs different from that of 75 to 95% of all cellular mRNA. The importance of these rather esoteric observations is the warning they offer regarding the use of metabolic inhibitors. The time course of the experiment as well as the concentration of drug used can radically affect the interpretation of the results.

The experiments using Reuber H-35 cells imply a posttranscriptional site of regulation for the cyclic AMP-mediated induction of PEP-CK and are in contrast to the results from fetal rat liver. There are several possible reasons for this discrepancy. One trivial explanation is that the concentrations of actinomycin D used in the fetal liver experiments were greater than those required to inhibit transcription and may have interfered with translation. Unfortunately concentration-dependent effects of the drug on the fetus were not examined, and it is not possible to eliminate such an interaction as the cause of the inhibition. Another more biologically significant explanation for the difference between the fetal liver and Reuber H-35 cells is a different site of action for cyclic AMP in developing liver. As discussed in a previous section, the level of PEP-CK in fetal liver is very low, and its appearance is observed only after birth. It may be that the *initial* development of enzyme activity requires transcription of PEP-CK mRNA which is stimulated by cyclic AMP. Once adult levels of the template accumulate in the cells, the action of cyclic AMP shifts to a posttranscriptional site. It is also possible that there is a differential stability between translated and untranslated mRNA. Accordingly, the fetal liver may be synthesizing sufficient quantitities of PEP-CK mRNA, but in the absence of a minimal concentration of cyclic AMP, the mRNA is not being translated. Under this scheme, the mRNA unattached to ribosomes or attached to smaller polysomes would be more rapidly degraded, and thereby the inhibition of cyclic AMP action by actinomycin D would reflect a smaller mRNA pool as the result of increased lability and not an absence of transcription. Very little information is available on the control of degradation of specific mRNA, but a recent report by Palmiter and Carey (1974) demonstrated a differential stability of ovalbumin mRNA in oviducts from estrogen stimulated and estrogen withdrawn chicks. In additions, Kenney *et al.* (1972) reported that after treatment of Reuber H-35 cells with cycloheximide, the recovery of tyrosine aminotransferase synthesis after washout was much more rapid than after treatment and washout of puromycin. Assuming the maintenance of polysome structure during cycloheximide treatment, in contrast to the polysome disaggregation in the presence of puromycin, these results could be interpreted as indicating increased lability of released mRNA after puromycin treatment, as compared to polysome-stabilized mRNA in the presence of cycloheximide.

2. The Mechanism of the Insulin-Cyclic AMP Antagonism on Hepatic PEP-CK. The administration of insulin to diabetic rats (Tilghman *et al.*,

1974) or the refeeding of fasted animals (Hopgood *et al.*, 1973; Tilghman *et al.*, 1974) resulted in a rapid decline in PEP-CK synthesis, with a half-life of 30 to 40 minutes. The observation that an equally rapid decrease in enzyme synthesis could be effected in Reuber H-35 cells by washing out dibutyryl cyclic AMP from the medium (Tilghman *et al.*, 1975) led us to propose that at least one action of insulin on the enzyme *in vivo* is to lower the concentration of hepatic intracellular cyclic AMP. Whether insulin has an alternate mechanism of action in addition to a cyclic AMP antagonism has not been established. Nevertheless, the deinduction of PEP-CK synthesis *in vivo* and in Reuber H-35 cells was studied in an effort to understand the mechanism whereby enzyme synthesis could be rapidly decreased. It was proposed that if this decrease in enzyme synthesis occurred as the result of inhibition at the level of transcription of PEP-CK mRNA, the remaining mRNA must have a half-life not exceeding 40 minutes. On the other hand, inhibition of enzyme synthesis at translation would not necessarily require a short mRNA half-life.

To distinguish between these two possibilities, we estimated the half-life of the PEP-CK template in fasted rats by administering either actinomycin D or cordycepin to prevent synthesis of mRNA, and following the decrease in the rate of enzyme synthesis with time. It was assumed that the inhibitors were affecting transcription only and that the rate of decline of enzyme synthesis reflected the rate of degradation of the PEP-CK mRNA pool. The estimated half-life of the enzyme template using actinomycin D was approximately 4 hours, but a much shorter half-life of 40 to 60 minutes was found using cordycepin (Fig. 8) (Tilghman *et al.*, 1974). This difference in the estimates could not be readily explained, however, anomalous effects of these two inhibitors have been noted in other systems (Butcher *et al.*, 1972; Grayson and Berry, 1973). On the basis of an inhibition of total cytosol protein synthesis in the actinomycin D-injected rats, coupled with an apparent prevention of deinduction of PEP-CK synthesis in the presence of this drug, it was suggested that actinomycin D causes an apparent "stabilization" of the enzyme template, perhaps related to the observed effects on translation. Cordycepin neither inhibited total protein synthesis nor interfered with the deinduction of the enzyme in refed rats (Tilghman *et al.*, 1974). Because of the very short half-life estimated using cordyceptin, a block in PEP-CK mRNA synthesis could not be ruled out as the mechanism responsible for the rapid fall in enzyme synthesis.

The deinduction of PEP-CK that occurs as the result of removing dibutyryl cyclic AMP from the medium of previously induced Reuber H-35 cells (Tilghman *et al.*, 1975) represented an excellent model system in which to investigate the underlying mechanism. The time course of the decline in enzyme synthesis in these cells is remarkably similar to that observed after

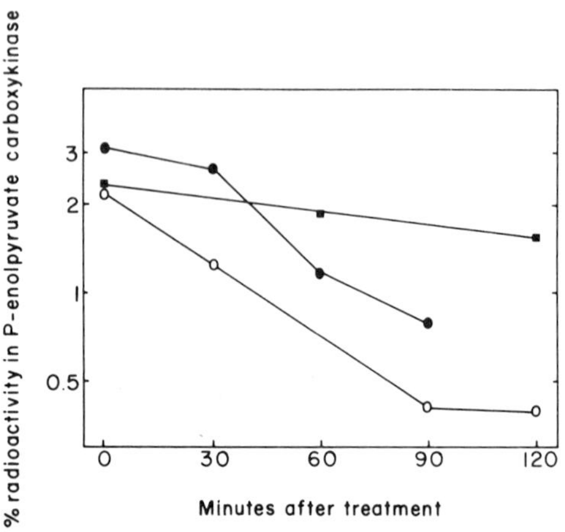

Figure 8. Comparison between the effects of refeeding, actinomycin D, and cordycepin on P-enolpyruvate carboxykinase synthesis. Rats were fasted 20 hours, then either refed laboratory chow (●——●), or injected with actinomycin D (■——■; 100 μg/100 g body weight) or cordycepin (○——○; 1.5 mg/100 g body weight). The relative rate of PEP-CK synthesis was determined at the indicated times as described in the legend to Fig. 2. *Source:* Tilghman *et al.* (1974).

refeeding fasted rats. Thus the relationship between enzyme template stability and deinduction could be examined under conditions where extrahepatic factors were not involved.

Estimates of the PEP-CK template half-life in dibutyryl cyclic AMP induced cells yielded different values with actinomycin D and cordycepin—approximately 8 and 4 hours, respectively. Under no conditions did a block in transcription produce a decrease in enzyme synthesis as rapid as that resulting from the removal of dibutyryl cyclic AMP from the medium. Thus it was tentatively concluded that the enzyme template was not sufficiently short-lived to account for the decrease in enzyme synthesis. In support of this it was shown that 3 hours after the removal of dibutyryl cyclic AMP, when the rate of enzyme synthesis had fallen to 10% of the initial rate, the readdition of the cyclic nucleotide to the medium, even in the presence of actinomycin D, caused a rapid restimulation of enzyme synthesis (Tilghman *et al.*, 1975). This experiment implied the presence of specific mRNA for PEP-CK in the deinduced cells that either was not being initiated or was being translated at a decreased rate in the absence of dibutyryl cyclic AMP.

Unfortunately, attempts to demonstrate a translational effect of cyclic AMP *in vitro* on PEP-CK synthesis have been unsuccessful. In a recent publication, Ballard *et al.* (1974b) demonstrated that the enzyme is preferentially synthesized on free rather than membrane-bound polysomes in livers from fasted rats. Postmitochondrial supernatants (PMS) and isolated polysomes from livers of refed rats synthesize a decreased proportion of the enzyme, at about the same levels as observed *in vivo* (Table III). Neither the addition of supernatant factors from the liver of fasted rats nor the addition of cyclic AMP or dibutyryl cyclic AMP stimulated the synthesis of the enzyme on these polysomes (Ballard *et al.*, 1974b). However, since homologous polysome systems exhibit only a limited degree of new initiation, it is unlikely that an effect of cyclic AMP at the initiation step would be detected.

Other investigators have studied the mechanism of the cyclic nucleotide effect on inducible enzymes *in vitro*. Chuah and Oliver (1971, 1972) incubated a microsomal fraction from neonatal rat liver with cyclic AMP and an inhibitor of protein synthesis and showed a threefold elevation of

Table III. Synthesis of PEP-CK *in Vivo* and on homologous polysomes *in Vitro*[a]

Dietary state		dpm × 10^{-3} incorporated into		
Starved rat	Refed rat	Released protein	Enzyme	% Radioactivity in PEP-CK
In vivo	—	45.2 ±3.9	1.17 ±0.26	2.58 ±0.42
—	*In vivo*	32.2 ±3.2	0.19 ±0.04	0.59 ±0.11
PMS	—	2320	40.3	1.74
—	PMS	5050	13.7	0.27
Cytosol + polysomes	—	2710	45.0	1.66
—	Cytosol + polysomes	4740	10.8	0.23
Cytosol	Polysomes	2400	7.3	0.30
Polysomes	Cytosol	3730	61.9	1.66

[a] The *in vivo* experiments are from rats either starved for 20 hours or starved for 20 hours and refed for 2 hours. The procedures for the determination of the relative rates of PEP-CK synthesis are given in Fig. 2. Cytosol and total polysomes were prepared from postmitochondrial supernatants (PMS) of livers from either starved or 2-hours-refed rats and were mixed in different combinations prior to the measurement of protein and PEP-CK synthesis. Incorporated radioactivity is expressed as dpm × 10^{-3}/ml of postmitochondrial supernatant or the polysomes derived from that amount. An equivalent amount of cytosol was added to that present in the postmitochondrial supernatant.

Source: Data from Ballard *et al.* (1974b).

tyrosine aminotransferase activity. In addition to cyclic AMP, a cyclic AMP binding protein was required for the appearance of enhanced enzyme activity. The enzyme released from membrane-bound polysomes exhibited the usual molecular weight for tyrosine aminotransferase, thus implicating termination as the site of cyclic AMP action. However, the release of nascent tyrosine aminotransferase from microsomes of adult or fetal liver was not observed (Chuah and Oliver, 1971). It was suggested by the authors that the termination step is rate limiting in newborns, whereas in earlier or later development, other steps in the synthesis of the enzyme become limiting. Chuah and Oliver (1971) were also unable to observe an effect of cyclic AMP on PEP-CK release in newborns, even though both this enzyme and tyrosine aminotransferase have identical responses to cyclic AMP in the fetus and both are induced at birth. However, since an increase in activity was the sole criterion for enzyme release, a small effect may have gone unnoticed. Khairallah and Pitot (1967), reported a similar effect of cyclic AMP on serine dehydratase synthesis on polysomes *in vitro*, but further work describing the system has not appeared.

We have attempted to determine whether the lack of PEP-CK synthesis in postmitochondrial supernatants from livers of refed rats is due to a block at termination. These experiments were performed using puromycin to release nascent peptides, followed by the immunochemical isolation of nascent PEP-CK. We found neither a significant release of completed PEP-CK molecules nor a large amount of incomplete enzyme molecules. Although these results could be caused by a cessation of protein synthesis with the peptidyl tRNA occupying the amino acid site on the ribosome (i.e., a block in elongation), it seems unlikely that any specific block occurs at termination. Control of PEP-CK synthesis is therefore apparently different from that shown for tyrosine aminotransferase.

3. ***The Mechanism of Glucocorticoid Action.*** There are relatively few studies on the mechanism of action of the glucocorticoids on PEP-CK synthesis, mainly because in liver, the tissue most studied, the enzyme responds marginally to glucocorticoid stimulation and only in the absence of insulin (Reshef *et al.*, 1969b; Exton *et al.*, 1973b; Gunn *et al.*, 1975b). The hepatic enzyme in the intact fetus (Yeung *et al.*, 1967; Hanson *et al.*, 1975) and in fetal organ culture (Wicks, 1969) is insensitive to steroid hormones. Reuber H-35 cells constitute the only system in which an induction of PEP-CK by glucocorticoids has been observed in tissue culture. Wicks and McKibbin (1972) showed that the twofold increase in PEP-CK activity observed after addition of dexamethasone to Reuber H-35 cells for 3 hours could be completely blocked by simultaneous addition of actinomycin D. The inhibition of the dexamethasone induction of the enzyme by actinomycin D contrasted to the lack of effect of the drug on the dibutyryl cyclic AMP mediated induction. The authors proposed that dexamethasone was acting to increase the transcription of PEP-CK mRNA.

A transcriptional effect of glucocorticoid was postulated for several other inducible hepatic enzymes, notably tyrosine aminotransferase (Kenney, 1970; Schimke and Doyle, 1970; Gelehrter, 1973). This proposal is largely based on the use of inhibitors of transcription that completely block the effect of adrenal steroids on both enzyme activity and synthesis (Lee *et al.*, 1970; Kenney *et al.*, 1973). In only one example—rat liver tryptophan oxygenase—has it been shown unequivocally that the glucocorticoid-mediated induction was the result of an increase in specific mRNA. In that study, Schutz *et al.* (1973) isolated cellular RNA from livers of control and hydrocortisone-treated rats and demonstrated, using the Krebs II ascites cell-free protein synthesizing system, an increase in tryptophan oxygenase mRNA after steroid treatment.

The most promising tissue in which to study the mechanism of the glucocorticoid induction of PEP-CK is the kidney, since the response of PEP-CK to adrenal hormones is greatest in this tissue. Several groups have measured a significant increase in renal enzyme activity after triamcinolone treatment and have shown that the increase can be completely prevented by actinomycin D (Flores and Alleyne, 1971; Longshaw and Pogson, 1972). Iynedjian *et al.* (1975) have demonstrated that the two- to threefold elevation in the rate of enzyme synthesis during acidosis is blocked by either actinomycin D or cordycepin.

4. Approaches to the Quantitation of Specific mRNA. The pioneering work on the quantitation of the mRNA for the globin chains (Evans and Lingrel, 1969; Ross *et al.*, 1972; Terada *et al.*, 1972) and its extension to a wide variety of eucaryotic templates (Brawerman, 1974) have opened up a new area of investigation into the mechanism of hormone action on specific protein synthesis. Using heterologous protein synthesizing systems, such as those developed from rabbit reticulocytes (Lockard and Lingrel, 1969; Stavnezer and Huang, 1971; Rhoads *et al.*, 1971), Krebs II ascites cells (Matthews and Korner, 1970; Metafora *et al.*, 1972), and wheat germ embryo (Roberts and Paterson, 1973), it is now possible to estimate the amount of a specific mRNA after various hormonal treatments. RNA isolated from the tissue under study is added to a heterologous protein synthesizing system, and the amount of *specific* protein synthesized is used as a measure of the amount of *specific* mRNA added. The system serves as a bioassay for mRNA, with the assumption that the amount of specific protein synthesized is limited only by the amount of added mRNA.

However, the technical difficulties associated with isolating intact mRNA, developing a satisfactory cell-free protein synthesizing system, and quantitating the protein product should not be underestimated. With the notable exceptions of tryptophan oxygenase (Shutz *et al.*, 1973) and lysozyme (Palmiter, 1974), the proteins thus far studied using these techniques have been in relatively high concentrations in the cell. For example, one of the most thoroughly investigated hormonal systems is the estrogen-mediated induction of ovalbumin in the developing chick oviduct (Rhoads *et al.*, 1971; Rosenfeld *et al.*, 1972; Palmiter *et al.*, 1972: Chan *et al.*, 1973; Palmiter, 1973). After estrogen stimulation, the synthesis of ovalbumin comprises approximately 60% of total protein synthesis in the cell. We have recently used polysomal RNA isolated from fasted rat liver to direct the synthesis of PEP-CK in a cell-free system derived from rabbit reticulocytes. The product was isolated with a specific antibody, and was found to co-migrate on SDS-acrylamide gel electrophoresis with labeled PEP-CK from rat liver (Fig. 9). When polysomal RNA was fractionated according to size on sucrose gradients, the PEP-CK mRNA activity migrated between the 28S and 18S ribosomal subunits, which is expected from the molecular weight of the protein (85,000) (Fig. 10). It is hoped that this bioassay will prove useful in determining the effects of hormones on tissue PEP-CK mRNA concentrations.

B. Regulation of PEP-CK Degradation

Degradation of hepatic PEP-CK has been measured by three different methods: (1) loss of enzymatic activity after the removal of an inducing stimulus (Reshef *et al.*, 1969b; Treadow and Khairallah, 1972); (2) loss of enzymatic activity following the inhibition of enzyme synthesis with cycloheximide or other inhibitors of general protein synthesis (Treadow and Khairallah, 1972); (3) loss of radioactivity from the enzyme protein pool after pulse labeling with an amino acid precursor (Philippidis *et al.*, 1972; Ballard and Hopgood, 1973; Hopgood *et al.*, 1973). All these methods have some disadvantages, but the difficulties are especially serious in methods 1 and 2, which measure only the loss of catalytic activity. Thus a calculation of enzyme half-life after the removal of an inducing stimulus presupposes that the subsequent rate of enzyme synthesis is reduced immediately to zero and that the degradation rate is the same in the presence of the inducer and in its absence. These restrictions are clearly unreasonable, and at best an approximate rate of enzyme degradation can be calculated. Likewise, the use of protein synthesis inhibitors to block the production of new enzyme molecules has the unfortunate disadvantage that the most effective inhibitor, cycloheximide, has been shown to depress protein degradation as well as protein synthesis (Hershko and Tomkins, 1971; Ballard and Hopgood,

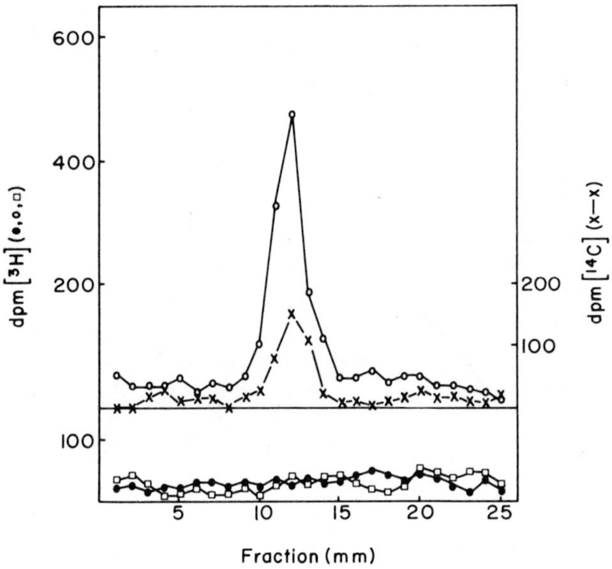

Figure 9. Sodium dodecyl sulfate polyacrylamide gel electrophoresis of P-enolpyruvate carboxykinase synthesized in reticulocyte lysates. Polysomal RNA from fasted rat liver or guinea pig liver was isolated by discontinuous sucrose gradient centrifugation followed by dissociation with SDS and centrifugation in 12 ml 5 to 20% sucrose gradients. The RNA migrating between 28S and 18S was collected and precipitated with ethanol. The RNA was added to a reticulocyte lysate system of 0.5 ml total volume. After 1 hour at 26°C, 0.3 ml of lysate was removed for immunoprecipitation of PEP-CK. The washed immunoprecipitates were dissociated and electrophoresed on SDS polyacrylamide gels: (●——●) incubation with no liver mRNA; (□——□) 20 μg rat spleen RNA; (○——○) 20 μg rat liver RNA; (×——×) ^{14}C-PEP-CK marker prepared from fasted rat liver.

1973). Therefore the stability of PEP-CK determined by this method will be overestimated.

The radioactivity lost from the PEP-CK pool has been measured by using a specific antibody to provide a rapid and convenient procedure for isolating the enzyme from other cytosol proteins. This method has fewer problems than the measurement of enzyme activity for determining rates of protein degradation, provided the antibody chosen is specific for PEP-CK. In the studies described below, we have established this specificity by showing that the radioactivity precipitated when the antibody is added to a cytosol fraction from rat liver moves as a single peak during sodium dodecyl sulfate acrylamide gel electrophoresis. This peak coincides exactly with pure PEP-CK added as an internal standard (Hopgood *et al.,* 1973). However, a further assumption is necessary before precise measurement of protein degradation can be made. This concerns the reutilization of isotope derived from the

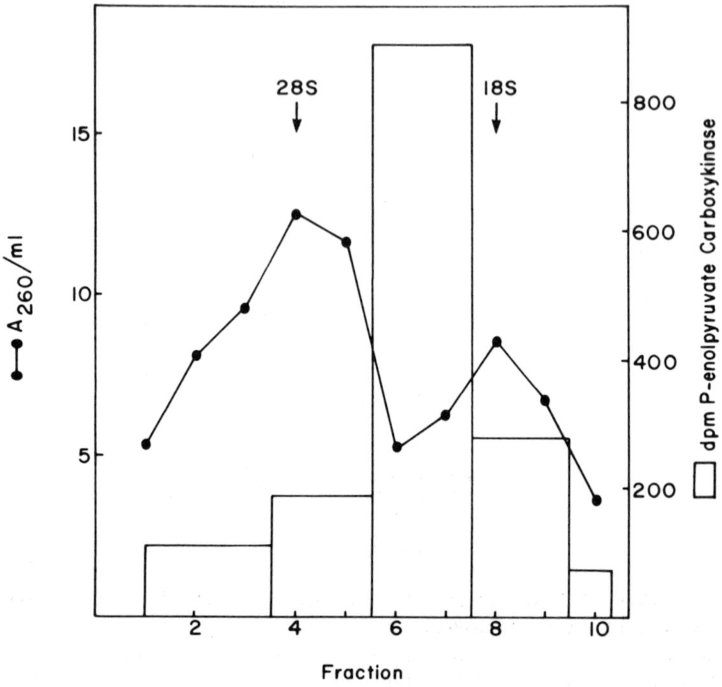

Figure 10. Sucrose gradient centrifugation of P-enolpyruvate carboxykinase mRNA activity. Polysomes were prepared from fasted rat liver and dissociated with SDS. The RNA was centrifuged in 12 ml 5 to 20% sucrose density gradients, and 1 ml fractions were collected. The RNA from six gradients was pooled into five fractions, precipitated, and collected by centrifugation. Twenty micrograms of each fraction was added to reticulocyte lysates, total volume 0.5 ml, and the ^{3}H-PEP-CK was synthesized at 26°C for 1 hour isolated by immunoprecipitation and quantitated on SDS-polyacrylamide gels. The dpm in PEP-CK (bars) represents the amount of enzyme synthesized in 0.5 ml reaction volume: (●——●) A_{260} per ml for a representative gradient.

breakdown of radioactive proteins and is distinct from the labeling of PEP-CK during an initial pulse period. In the studies on PEP-CK degradation we have routinely used a "chase" of nonradioactive leucine to dilute out any radioactive leucine remaining from the initial injection or derived from the breakdown of radioactive proteins. Even this "chase" technique may not be completely satisfactory in experiments with intact animals, however, since the intracellular, radioactive leucine may not equilibrate with the artificially large extracellular pool. The problem is less serious for measurements of PEP-CK degradation in Reuber H-35 cells, where the residual radioactive leucine can be removed by changing the culture medium to one with a high concentration of nonradioactive leucine.

1. Degradation of PEP-CK in Rat Tissues. In previous publications we have measured rates of PEP-CK degradation in liver (Philippidis *et al.*, 1972; Ballard and Hopgood, 1973; Hopgood *et al.*, 1973), epididymal adipose tissue (Hopgood *et al.*, 1973), and kidney of adult rats (Iynedjian *et al.*, 1975) under various nutritional and hormonal conditions, and we have expressed the loss of radioactivity in the PEP-CK pool relative to the radioactivity in total tissue proteins. Although this method reduces variability by correcting for a lack of precision in the injection or supply of isotopic precursor to the tissue in question, it is essential that the rate of PEP-CK degradation be rapid relative to the turnover of total tissue proteins. To avoid this restriction, PEP-CK degradation rate constants have been calculated on the basis of total radioactivity in hepatic PEP-CK per animal. Except for experiments with fetal rat liver during development, the animals used were all 7 weeks old and weighed between 180 and 200 g. This method is not affected by any changes in liver weight that might occur after the isotope pulse, such as would occur during fasting or refeeding. The degradation rate constants (K_d) reported in Table IV have been derived from the expression for a change in enzyme amount:

$$\frac{dE}{dt} = K_s - K_d E$$

where t is time in hours and E is the total enzyme radioactivity. The rate of synthesis of radioactive PEP-CK (K_s) is taken as zero, since it is assumed that isotope reincorporation has been prevented by the leucine chase.

Table IV reveals that liver PEP-CK is degraded with a half-time of approximately 6 hours in fed or fasted rats, and this degradation rate is not modified by either dibutyryl cyclic AMP or triamcinolone. Furthermore, previous studies have shown that the rapid deinduction caused by refeeding fasted animals a high carbohydrate diet does not result in any change in the degradation rate (Hopgood *et al.*, 1973). However, the injection of tryptophan into fed or fasted rats depresses markedly the rate of PEP-CK degradation and also increases the rate of enzyme synthesis (Ballard and Hopgood, 1973). The mechanism of these tryptophan effects is not understood, but they may be produced by compensatory changes in response to a sequestering of preexisting PEP-CK molecules by tryptophan metabolites (Snoke *et al.*, 1971). Alternatively, the effects may be related to the inhibition of general protein degradation that is caused by an excess of certain amino acids (Hershko and Tomkins, 1971; Woodside and Mortimore, 1972).

The appearance of hepatic PEP-CK soon after the birth of infants rats is accompanied by a period during which the enzyme is not degraded at a significant rate (Philippidis *et al.*, 1972) (Table IV). This was also true for

Table IV. Degradation of PEP-CK in rat liver[a]

Animal and treatment	K_d (hour^{-1})	Half-life (hours)
Adult, fed	0.113 ±0.033	6.1
fed + tryptophan (150 mg each 4 hours)	0.028 ±0.023	25
Adult, fasted	0.101 ±0.017	6.9
fasted + tryptophan (150 mg each 4 hours)	0.034 ±0.021	20
fasted + cycloheximide (200 μg)	0.077 ±0.021	9.0
Adult, fasted	0.112 ±0.023	6.2
fasted + dibutyryl cAMP (2 mg each hour)	0.108 ±0.023	6.4
Adult, fasted	0.100 ±0.009	6.9
fasted + triamcinolone (8 mg)	0.094 ±0.010	7.4
Newborn	0.014 ±0.011	50
1 day old	0.034 ±0.003	20
2 days old	0.066 ±0.008	11
3 days old	0.098 ±0.027	7.1

[a] Degradation rate constants (K_d) have been calculated from the rate of loss of radioactivity from the PEP-CK pool after the animals had been injected with ^{3}H-leucine. Variability is indicated by standard error (see Philippidis *et al.* 1972; Ballard and Hopgood, 1973; Gunn *et al.*, 1975b). All injections (expressed as dose per 100 g body weight) were given intraperitoneally.

PEP-CK induced *in utero* by injection of dibutyryl cyclic AMP (Hanson *et al.*, 1973). The implication that proteins are not degraded during the time of initial induction has been shown for one other enzyme, chicken liver malic enzyme (Silpananta and Goodridge, 1972). We will not know whether this finding is of general importance until measurements have been made of the degradation of other enzymes during differentiation, but the situation is comparable to induction in bacteria where no degradation occurs (Mandelstam, 1960). The initial appearance of PEP-CK and the concomitant lack of degradation contrasts sharply with the "secondary" induction that can occur in tissues of adult rats in response to fasting or diabetes (Table IV). We must be cautious in proposing a qualitative difference between "primary" induction during differentiation and a secondary stimulus, because a substantial reincorporation of isotope may occur during any induction period. Since cycloheximide has been shown to inhibit PEP-CK degradation (Ballard and Hopgood, 1973), (Table IV) we cannot usefully employ such inhibitors of protein synthesis to prevent isotope reincorporations. We cannot, therefore, make an unequivocal statement that PEP-CK degradation is absent during the neonatal induction period. We do know that PEP-CK from these newborn rats is rapidly inactivated when homogenates are incubated *in vitro*; therefore the capacity to degrade the enzyme is certainly present in the neonate (Ballard *et al.*, 1973). Presumably the enzyme *in vivo* is present in a form that is protected from degradation, a situation that is not maintained after tissue disruption.

The method used to calculate the degradation constant for hepatic PEP-CK (Table IV) is not applicable to epididymal adipose tissue because very large variability is found in the total isotope incorporation. However, a rate of PEP-CK degradation can be calculated relative to the total protein radioactivity. By this method we find a half-time of degradation of 20 hours in fasted rats and 12 hours during refeeding. These values offer only a rough approximation of the rate of PEP-CK degradation, since reincorporation of isotope may be a serious problem in adipose tissue.

PEP-CK in kidney cortex is degraded with a relative half-life of 13 hours, although the enzyme *appeared* to be considerably more stable during ammonium chloride induction (Iynedjian *et al.*, 1975). In kidney, however, cycloheximide increased the apparent degradation rate when the drug was given at the same time as the ammonium chloride stimulus (Iynedjian *et al.*, 1975). It seems that induction was accompanied by an increase in reincorporation of radioactive leucine into PEP-CK. It is pertinent to note that this effect of cycloheximide is opposite to that found in liver where an *inhibition* of PEP-CK degradation was measured (Table IV).

2. Degradation of PEP-CK in Reuber H-35 Cells. We have used Reuber H-35 cells for studies on PEP-CK degradation since, unlike intact animals, there is no doubt that a drug or possible modulator is acting directly on the cells rather than by an indirect mechanism. Techniques comparable to those employed for liver may be used to measure PEP-CK degradation in Reuber H-35 cells because the enzymes from the two sources are immunochemically indistinguishable (Gunn *et al.*, 1975a; Knowles *et al.*, 1975). The average rate of PEP-CK degradation in Reuber H-35 cells is similar to that found in liver (cf. Tables IV, V). However, the enzyme is degraded somewhat more slowly in uninduced cells than it is during induction produced by a mixture of 0.5 mM dibutyryl cyclic AMP, 1 mM theophylline, and 5 μM dexamethasone, or following the removal of these inducers (deinduction). It is unlikely that these differences are artifacts caused by altered rates of isotope reincorporation, since it has been shown that the addition of 2 mM leucine as a "chase" is effective in preventing leucine incorporation into PEP-CK (J. M. Gunn, unpublished experiments).

The protein synthesis inhibitors cycloheximide and puromycin reduce

Table V. Degradation of PEP-CK in Reuber H-35 cells[a]

Treatment	K_d (hour^{-1})	Half-life (hours)
Uninduced	0.087 ±0.024	8.0
Induced	0.118 ±0.030	5.9
Deinduced	0.142 ±0.023	4.9
Deinduced	0.156 ±0.013	4.4
Cordycepin (4 × 10^{-5} M)	0.156 ±0.012	4.4
Actinomycin D (3 × 10^{-7} M)	0.129 ±0.003	5.4
Cycloheximide (10^{-5} M)	0.093 ±0.011	7.5
Puromycin (1 mM)	0.092 ±0.017	7.5
Deinduced	0.134 ±0.013	5.2
TPCK (1.5 × 10^{-4} M)	0.069 ±0.015	10
TLCK (2.5 × 10^{-4} M)	0.024 ±0.040	29
Deinduced, arginyl PEP-CK	0.146 ±0.019	4.7
Canavanyl PEP-CK	0.328 ±0.045	2.1
Tryptophan PEP-CK	0.159 ±0.026	4.4
6-Fluorotryptophan PEP-CK	0.247 ±0.025	2.8

[a] Methods for the measurement of enzyme degradation in Reuber H-35 cells have been reported previously (Knowles *et al.*, 1975). Degradation rate constants have been calculated from the rate of loss of radioactivity from the enzyme pool after a pulse with ^{3}H-leucine. Variability is indicated by standard error.

PEP-CK degradation (Table V), a result comparable to the effect of cycloheximide on PEP-CK degradation in liver (Table IV). Neither actinomycin D nor cordycepin, the two inhibitors of RNA synthesis tested, had any marked effect on PEP-CK degradation (Table V). Studies with other proteins have revealed that inhibitors of both protein synthesis and RNA synthesis can reduce protein degradation (Hershko and Tomkins, 1971). To explain these paradoxical effects, Tomkins has suggested that protein degradation is one of several cellular processes regulated coordinately by a "pleiotypic response" (Hershko *et al.*, 1971; Kram and Tomkins, 1973).

Both the trypsin inhibitor tosyllysylchloromethylketone (TLCK) and the chymotrypsin inhibitor tosylphenylalanylchloromethylketone (TPCK) inhibit PEP-CK degradation in Reuber H-35 cells (Table V). Although these inhibitors are substrate analogs for the respective proteases and are incorporated to form inactive dead-end complexes, TPCK and TLCK have the capacity to react with many functional groups on proteins; thus we cannot conclude that PEP-CK is degraded in the cells by enzymes similar to trypsin or chymotrypsin.

Reuber H-35 cells require both arginine and tryptophan for growth and protein synthesis. When these amino acids are replaced in the growth medium by canavanine and 6-fluorotryptophan, respectively, the PEP-CK synthesized will differ from the normal enzyme because it contains these analogs. We have shown that both types of aberrant PEP-CK are heat labile, have lower catalytic activities as compared to antibody reactivities (Knowles *et al.*, 1975), and are rapidly degraded by the cells (Table V). The degradation of aberrant PEP-CK and aberrant total proteins labeled in the presence of these analogs is not subject to control by insulin and serum (Knowles and Ballard, unpublished experiments). Consequently, we suggest that the cells contain a unique or an unregulated pathway for the removal of such proteins.

*3. **Regulation of PEP-CK Inactivation and Degradation** in* **Vitro.** PEP-CK degradation has been measured *in vitro* by following the radioactive products formed when pure ^{3}H-labeled enzyme is added to liver extracts (Ballard *et al.*, 1974a). These studies (Fig. 11) show a loss of catalytic activity, then a disappearance of immunological reactivity, and finally a

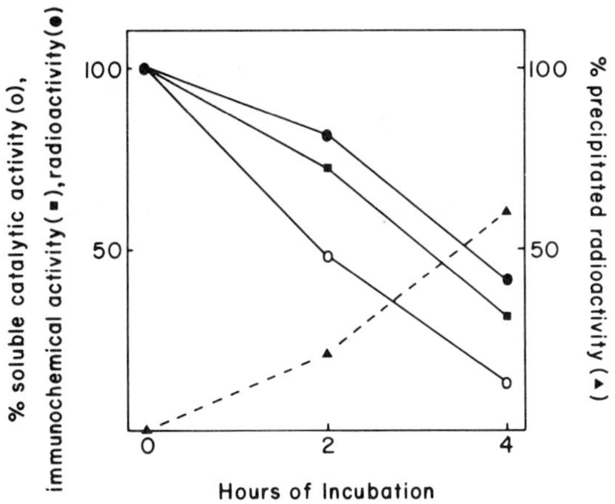

Figure 11. Inactivation and degradation of ^{3}H-PEP-CK in vitro. Purified ^{3}H-PEP-CK was added to unlabeled liver extracts and incubated at 37°C for the times indicated. The loss of PEP-CK soluble catalytic activity (○), immunochemical activity (■) and radioactivity (●) was followed. The percentage of ^{3}H-PEP-CK recovered in the pellet after centrifugation at 100,000 *g* is represented by ▲. *Source:* Values from Ballard *et al.* (1974a).

loss of solubility of the radioactive enzyme. There was no evidence of proteolytic cleavage in these experiments. However, if radioactive PEP-CK is incubated for several hours with an enriched liver microsome fraction, part of the radioactivity that is precipitated has a lower molecular weight than does the native PEP-CK. Further degradation of the protein to trichloroacetic acid soluble fragments has not been detected with this system.

Additional experiments on PEP-CK inactivation *in vitro* have been carried out in order to characterize the reactions involved (Table VI). The inactivation process is clearly localized in the microsomal fraction of liver and leads to the inactivated enzyme being bound to the microsomes (i.e., "precipitated" enzyme). Inactivation is stimulated by 5 mM L-cysteine or L-cystine and is strongly inhibited by $MnCl_2$ concentrations as low as 2×10^{-4} M. It is unlikely that contaminating lysosomes are responsible for PEP-CK inactivation, since the microsomes used were freeze dried and then purified by sucrose gradient centrifugation. The capacity to inactivate PEP-CK was clearly separated by these procedures from the lysosomal marker enzymes acid phosphatase and β-glucuronidase and was co-purified with glucose-6-phosphatase (Ballard and Hopgood, unpublished experiments). Furthermore, neither intact nor disrupted lysosomes were able to inactivate PEP-CK at neutral pH (at acid pH, PEP-CK is unstable with or without lysosomes).

Although the inactivation of PEP-CK by trypsin was prevented by the proteinase inhibitors in rat serum, the microsomal inactivation of the enzyme was unaffected by these inhibitors. A similar difference between tryptic and microsomal inactivation was found with $MnCl_2$. This salt inhibited PEP-CK inactivation by microsomes but had no effect on tryptic

Table VI. Properties of the PEP-CK inactivation reaction as measured in liver extracts[a]

1. Inactivation occurs at neutral pH and is located in the microsome fraction.
2. Inactivated PEP-CK is bound to microsomes.
3. Inactivation is stimulated by cysteine or cystine.
4. Inactivation is prevented by $MnCl_2$ but not by $MgCl_2$ or $FeCl_2$.
5. Inactivation of ^{3}H-labeled PEP-CK does not lead to proteolytic cleavage fragments.
6. Serum proteolytic inhibitors do not alter the rate of PEP-CK inactivation by microsomes.
7. Neither intact nor disrupted lysosomes inactivate PEP-CK at neutral pH.

[a] For details see Ballard *et al.* (1974a).

Table VII. Proposed sequence for PEP-CK degradation

Stable enzyme (e.g., in liver of newborn rat)
⇅
Unstable enzyme
↓
Denatured enzyme (loss of catalytic and immunochemical activity)
↓
Precipitated enzyme (attached to microsomes)
↓
Proteolytic cleavage fragments
↓
Amino acids

inactivation, indicating that the two types of PEP-CK inactivation are unrelated.

***4. Relationships Between PEP-CK Inactivation* in Vitro *and Degradation* in Vivo.** We have attempted to follow the degradation of PEP-CK *in vivo* by searching for intermediates between native enzyme and the final amino acid products. Under normal conditions immunochemically active but catalytically inactive enzyme cannot be detected in tissues or Reuber H-35 cells, thus the sequence proposed from the *in vitro* studies cannot be tested. From recent experiments with Reuber H-35 cells, however, we know that the proteolytic inhibitor TPCK inhibits the loss of immunochemically active PEP-CK without affecting the inactivation rate of the enzyme. Consequently, there is a buildup of catalytically inactive but immunochemically reactive enzyme in the cells. We interpret this finding as evidence for the sequence of reactions shown in Table VII and originally proposed from inactivation experiments *in vitro* (Ballard *et al.,* 1974a). Thus enzyme inactivation, the normal rate-limiting step for the whole pathway of PEP-CK degradation, is no longer rate limiting when the proteolytic part of the sequence is inhibited by TPCK. It is hoped that a further combination of *in vivo* and *in vitro* experiments will elaborate the proposed sequence of enzyme degradation.

REFERENCES

Alleyne, G. A. O. and Scullard, G. H. (1969). *J. Clin. Invest.* **48,** 364.

Arinze, I. J., Garber, A. J., and Hanson, R. W. (1973). *J. Biol. Chem.* **248,** 2266.

Ballard, F. J. (1970). In *Proceedings of the VII Congress of the International Diabetes Federation,* Excerpta Medica International Congress Series No. 231, pp. 592 ff.

Ballard, F. J. (1971). *Biochim. Biophys. Acta* **242,** 470.

Ballard, F. J. and Hanson, R. W. (1967). *Biochem. J.* **104,** 866.

Ballard, F. J. and Hanson, R. W. (1969). *J. Biol. Chem.* **244,** 5625.

Ballard, F. J. and Hopgood, M. F. (1973). *Biochem. J.* **136,** 259.

Ballard, F. J. and Oliver, I. T. (1963). *Biochim. Biophys. Acta* **71,** 578.

Ballard, F. J., Hanson, R. W., and Leveille, G. A. (1967). *J. Biol. Chem.* **242,** 2746.

Ballard, F. J., Hopgood, M. F., Reshef, L., and Hanson, R. W. (1973). In *The Biochemistry of Gene Expression in Higher Organisms* (J. K. Pollak and J. W. Lees, eds.), Australia and New Zealand Book Co., Sydney, pp. 274 ff.

Ballard, F. J., Hopgood, M. F., Reshef, L., and Hanson, R. W. (1974a). *Biochem. J.* **140,** 531.

Ballard, F. J., Hopgood, M. F., Reshef, L., Tilghman, S. M., and Hanson, R. W. (1974b). *Biochem. J.* **144,** 199.

Barnett, C. A. and Wicks, W. D. (1971). *J. Biol. Chem.* **246,** 7201.

Bentle, L. A. and Snoke, R. E. (1974). *Fed. Proc.* **33,** 1272.

Bieck, P., Stock, K., and Westerman, E. (1969). *Naunyn-Schmiedebergs Arch. Pharmakol. Exp. Pathol.* **263,** 387.

Bottger, I., Kriegel, H., and Wieland, O. (1970). *Eur. J. Biochem.* **13,** 253.

Bowman, R. H. (1970). *J. Biol. Chem.* **245,** 1604.

Brawerman, G. (1974). *Ann. Rev. Biochem.* **43,** 621.

Butcher, F. R. and Potter, V. R. (1972). *Cancer Res.* **32,** 2141.

Butcher, F. R., Bushnell, D. E., Becker, J. E., and Potter, V. R. (1972). *Exp. Cell Res.* **74,** 115.

Chan, L., Means, A. R., and O'Malley, B. W. (1973). *Proc. Natl. Acad. Sci. U.S.* **70,** 1870.

Chuah, C.-C. and Oliver, I. T. (1971). *Biochemistry* **10,** 2990.

Chuah, C.-C. and Oliver, I. T. (1972). *Biochemistry* **11,** 2547.

Darnell, J. E., Philipson, L., Wall, R., and Adesnik, M. (1971). *Science* **174,** 507.

Diesterhaft, M., Shrago, E., and Sallach, H. J. (1971). *Biochem. Med.* **5,** 297.

Evans, M. J. and Lingrel, J. B. (1969). *Biochemistry* **8,** 3000.

Exton, J. H. and Park, C. R. (1969). *J. Biol. Chem.* **244,** 1424.

Exton, J. H., Corbin, J. G., and Harper, S. C. (1972). *J. Biol. Chem.* **247,** 4996.

Exton, J. H., Harper, S. C., Tucker, A. L., and Ho, R.-J. (1973a). *Biochim. Biophys. Acta* **329,** 23.

Exton, J. H., Harper, S. C., Tucker, A. L., Flagg, J. L., and Park, C. R. (1973b). *Biochim. Biophys. Acta* **329,** 41.

Flores, H. and Alleyne, G. A. O. (1971). *Biochem. J.* **123,** 35.

Foster, D. O., Ray, P. D., and Lardy, H. A. (1966a). *Biochemistry* **5,** 555.

Foster, D. O., Ray, P. D., and Lardy, H. A. (1966b). *Biochemistry* **5,** 563.

Foster, D. O., Lardy, H. A., Ray, P. D., and Johnston, J. (1967). *Biochemistry* **6,** 2120.

Freedland, R. A. and Avery, E. H. (1964). *J. Biol. Chem.* **239,** 3357.

Friedmann, B., Goodman, E. H., Saunders, H. L., Kostos, V., and Weinhouse, S. (1971). *Arch. Biochem. Biophys.* **143,** 566.

Gelehrter, T. D. (1973). *Metabolism* **22,** 85.

Girard, J. R., Cuendet, G. S., Marliss, E. B., Kervran, A., Rieutort, M., and Assan, R. (1973). *J. Clin. Invest.* **52,** 3190.

Goodman, A. D. (1972). *Isr. J. Med. Sci.* **8,** 285.

Goodman, A. D., Fuisz, R. E., and Cahill, G. F., Jr. (1966). *J. Clin. Invest.* **45,** 612.

Gorin, E., Tal-Or, Z., and Shafrir, E. (1969). *Eur. J. Biochem.* **8,** 370.

Grayson, S. and Berry, S. J. (1973). *Science* **180,** 1071.

Guder, W. G. and Schmidt, U. (1974). *Hoppe-Seyler's Z. Physiol. Chem.* **355,** 273.

Gunn, J. M., Tilghman, S. M., Hanson, R. W., Reshef, L., and Ballard, F. J. (1975a). *Biochemistry,* **14,** 2350.

Gunn, J. M., Hanson, R. W., Meyuhas, O., Reshef, L. and Ballard, F. J. (1975b). *Biochem. J.* **150,** 195.

Hanson, R. W. and Garber, A. J. (1972). *Am. J. Clin. Nutr.* **25,** 1010.

Hanson, R. W. and Garber, A. J. (1973). *Am. J. Clin. Nutr.* **26,** 55.

Hanson, R. W., Fisher, L., Ballard, F. J., and Reshef, L. (1973). *Enzyme* **15,** 97.

Hanson, R. W., Reshef, L., and Ballard, F. J. (1975). *Fed. Proc.* **34,** 166.

Hershko, A. and Tomkins, G. M. (1971). *J. Biol. Chem.* **246,** 710.

Hershko, A., Mamont, P., Shields, R., and Tomkins, G. M. (1971). *Nature New Biol.* **232,** 206.

Hopgood, M. F., Ballard, F. J., Reshef, L., and Hanson, R. W. (1973). *Biochem. J.* **134,** 445.

Iynedjian, P. B. and Peters, G. (1974). *Am. J. Physiol.* **226,** 1281.

Iynedjian, P. B., Ballard, F. J., and Hanson, R. W. (1975). *J. Biol. Chem.* **250,** 5596.

Jost, J.-P., Khairallah, E. A., and Pitot, H. C. (1968). *J. Biol. Chem.* **243,** 3057.

Kamm, D. E., Strope, G. L., and Kuchmy, B. L. (1974). *Metabolism* **23,** 1073.

Kenney, F. T. (1970), In *Mammalian Protein Metabolism,* Vol. IV (H. N. Munro, ed.), 131, Academic Press, New York, p. 131.

Kenney, F. T. and Flora, R. M. (1961). *J. Biol. Chem.* **236,** 2699.

Kenney, F. T., Lee, K.-T., and Stiles, C. D. (1972). *Acta Endocrinol.* **168** (Suppl.) 369.

Kenney, F. T., Lee, K.-L., Stiles, C., and Fritz, J. E. (1973). *Nature New Biol.* **246,** 208.

Khairallah, E. A. and Pitot, H. C. (1967). *Biochem. Biophys. Res. Commun.* **29,** 269.

Knowles, S. E., Gunn, J. M., Reshef, L., Hanson, R. W., and Ballard, F. J. (1975). *Biochem. J.* **146,** 585.

Kram, R. and Tomkins, G. M. (1973). *Proc. Nat. Acad. Sci. U.S.* **70,** 1659.

Krebs, H. A., Bennett, D. A. H., deGasquet, P., Gascoyne, T., and Yoshida, T. (1963). *Biochem. J.* **86,** 22.

Kurahashi, K., Pennington, R. J., and Utter, M. F. (1959). *J. Biol. Chem.* **226,** 1059.

Langdon, R. G. and Weakley, D. R. (1955). *J. Biol. Chem.* **214,** 167.

Lee, K.-L., Reel, J. R., and Kenney, F. T. (1970). *J. Biol. Chem.* **245,** 5806.

Lockard, R. E. and Lingrel, J. B. (1969). *Biochem. Biophys. Res. Commun.* **37,** 204.

Longshaw, I. D. and Pogson, C. I. (1972). *J. Clin. Invest.* **51,** 2277.

Longshaw, I. D., Alleyne, G. A. O., and Pogson, C. I. (1972). *J. Clin. Invest.* **51,** 2284.

Mallette, L. E., Exton, J. H., and Park, C. R. (1969). *J. Biol. Chem.* **244,** 5713.

Mandelstam, J. (1960). *Bacteriol. Rev.* **24,** 289.

Matthews, M. B. and Korner, A. (1970). *Eur. J. Biochem.* **17,** 328.

Metafora, S., Terada, M., Dow, L. W., Marks, P. A., and Bank, A. (1972). *Proc. Nat. Acad. Sci. U.S.* **69,** 1299.

Muller, W. A., Faloona, G. R., and Unger, R. H. (1971). *J. Clin. Invest.* **50,** 1992.

Nagai, K. and Nakagawa, H. (1972). *J. Biochem.* **71.** 125.

Nordlie, R. C., Varricchio, F. E., and Holten, D. D. (1965). *Biochem. Biophys. Acta* **97,** 214.

Owen, O. E., Felig, P., Morean, A. R., Wahren, J., and Cahill, G. F., Jr. (1969). *J. Clin. Invest.* **48,** 574.

Palmiter, R. D. (1973). *J. Biol. Chem.* **248,** 8260.

Palmiter, R. D. (1974). *J. Biol. Chem.* **249,** 6779.

Palmiter, R. D. and Carey, N. H. (1974). *Proc. Nat. Acad. Sci. U.S.* **71,** 2357.

Palmiter, R. D., Palacios, R., and Schimke, R. T. (1972). *J. Biol. Chem.* **247,** 3296.

Penman, S., Rosbash, M., and Penman, M. (1970). *Proc. Nat. Acad. Sci. U.S.* **67,** 1878.

Peraino, C. and Pitot, H. C. (1964). *J. Biol. Chem.* **239,** 4308.

Peraino, C., Blake, R. L., and Pitot, H. C. (1965). *J. Biol. Chem.* **240,** 3039.

Peraino, C., Lamar, C., and Pitot, H. C. (1966). *Adv. Enzyme Regue.* **4,** 199.

Philippidis, H. and Ballard, F. J. (1970). *Biochem. J.* **120,** 385.

Philippidis, H., Hanson, R. W., Reshef, L., Hopgood, M. F., and Ballard, F. J. (1972). *Biochem. J.* **126,** 1127.

Pitot, H. C. and Peraino, C. (1964). *J. Biol. Chem.* **239,** 1783.

Pitts, R. F. (1972). *Kidney Int.* **1,** 297.

Pitts, R. F., Pilkington, L. A., McLeod, M. B., and Leal-Pinto, E. (1972). *J. Clin. Invest.* **51,** 557.

Ray, P. D., Foster, D. O., and Lardy, H. A. (1966). *J. Biol. Chem.* **241,** 3094.

Reshef, L. and Hanson, R. W. (1972). *Biochem. J.* **127,** 809.

Reshef, L. and Shapiro, B. (1970). In *Adipose Tissue* (B. Jeanrenaud and Hepp, D., eds.), Georg Thieme Verlag, Stuttgart, and Academic Press, New York and London, p. 136.

Reshef, L., Hanson, R. W., and Ballard, F. J. (1969a). *J. Biol. Chem.* **244,** 1994.

Reshef, L., Ballard, F. J., and Hanson, R. W. (1969b). *J. Biol. Chem.* **244,** 5577.

Reshef, L., Hanson, R. W., and Ballard, F. J. (1970). *J. Biol. Chem.* **245,** 5979.

Rhoads, R. E., McKnight, G. S., and Schimke, R. T. (1971). *J. Biol. Chem.* **246,** 7407.

Roberts, B. E. and Paterson, B. M. (1973). *Proc. Nat. Acad. Sci. U.S.* **70,** 2330.

Rodbell, M. (1965). *Ann. N.Y. Acad. Sci.* **131,** 302.

Rognstad, R. and Katz, J. (1972). *J. Biol. Chem.* **247,** 6047.

Rosenfeld, G. C., Comstock, J. P., Means, A. R., and O'Malley, B. W. (1972). *Biochem. Biophys. Res. Commun.* **46,** 1695.

Rosenzweig, S. and Frascella, D. (1968). *Bull. N.J. Acad. Sci.* **13,** 17.

Ross, J., Ikawa, Y., and Leder, P. (1972). *Proc. Nat. Acad. Sci. U.S.* **69,** 3620.

Schimke, R. T. and Doyle, D. (1970). *Ann. Rev. Biochem.* **39,** 929.

Schimke, R. T., Sweeney, E. W., and Berlin, C. M. (1965). *J. Biol. Chem.* **240,** 322.

Scholz, R. W. (1972). *Biochem. J.* **126,** 1219.

Schutz, G., Beato, M., and Feigelson, P. (1973). *Proc. Nat. Acad. Sci. U.S.* **70,** 1218.

Sharma, C., Manjeshwar, R., and Weinhouse, S. (1963). *J. Biol. Chem.* **238,** 3840.

Shrago, E., Lardy, H. A., Nordlie, R. C., and Foster, D. O. (1963). *J. Biol. Chem.* **238,** 3188.

Shrago, E., Young, J. W., and Lardy, H. A. (1967). *Science* **158,** 1572.

Silpananta, P. and Goodridge, A. G. (1971). *J. Biol. Chem.* **246,** 5754.

Singer, R. H. and Penman, S. (1972). *Nature* **240,** 100.

Snoke, R. E., Johnston, J. B., and Lardy, H. A. (1971). *Eur. J. Biochem.* **24,** 342.

Söling, H.-D., Kleineke, J., Willms, B., Janson, G., and Kuhn, A. (1973). *Eur. J. Biochem.* **37,** 233.

Stavnezer, J. and Huang, R. C. C. (1971). *Nature* **230,** 172.

Suda, M., Nagai, K., and Nakagawa, H. (1973). *J. Biochem.* **73,** 727.

Terada, M., Cantor, L., Metafora, S., Rifkind, R. A., Bank, A., and Marks, P. A. (1972). *Proc. Nat. Acad. Sci. U.S.* **69,** 3575.

Tilghman, S. M., Hanson, R. W., Reshef, L., Hopgood, M. F., and Ballard, F. J. (1974). *Proc. Nat. Acad. Sci. U.S.* **71,** 1304.

Tilghman, S. M., Gunn, J. M., Fisher, L., Hanson, R. W., Reshef, L., and Ballard, F. J. (1975). *J. Biol. Chem.* **250,** 3322.

Treadow, B. R. and Khairallah, E. A. (1972). *Nature New Biol.* **239,** 131.

Utter, M. F. (1969). In *Citric Acid Cycle* (J. M. Lowenstein, ed.) Dekker, New York and London, pp. 249 ff.

Utter, M. F. and Kolenbrander, H. M. (1972). In *The Enzymes,* Vol. VI (P. D. Boyer, ed.) Academic Press, New York and London, pp. 117 ff.

Utter, M. F. and Kurahashi, K. (1954). *J. Biol. Chem.* **207,** 787.

Veneziale, C. M., Walter, P., Kneer, N., and Lardy, H. A. (1967). *Biochemistry* **6,** 2129.

Warnes, D., Ballard, F. J., and Seamark, R. F. (1974). *J. Reprod. Fert.* **36,** 471.

White, L. W., Williams, H. R., and Landau, B. R. (1968). *Arch. Biochem. Biophys.* **126,** 552.

Wicks, W. D. (1969). *J. Biol. Chem.* **244,** 3941.

Wicks, W. D. (1971). *J. Biol. Chem.* **246,** 217.

Wicks, W. D. and McKibbin, J. B. (1972). *Biochem. Biophys. Res. Commun.* **48,** 205.

Wicks, W. D., Lewis, W., and McKibbin, J. B. (1972). *Biochim. Biophys. Acta* **264,** 177.

Woodside, K. H. and Mortimore, G. E. (1972). *J. Biol. Chem.* **247,** 6474.

Yeung, D. and Oliver, I. T. (1968a). *Biochem. J.* **108,** 325.

Yeung, D. and Oliver, I. T. (1968b). *Biochemistry* **7,** 3231.

Yeung, D., Stanley, R. S., and Oliver, I. T. (1967). *Biochem. J.* **105,** 1219.

CHAPTER THREE

Glucose-6-Phosphatase-Phosphotransferase: Roles and Regulation in Relation to Gluconeogenesis

ROBERT C. NORDLIE

I. INTRODUCTION

The liver plays a key, central role in maintaining blood glucose levels with relatively narrow, constant limits in health and disease and in dietary excess and starvation, as well as in the resetting of these limits as circumstances demand. It possesses the capacity to generate and release glucose into the blood in times of need and to remove this hexose from the blood and store it, as polymeric glycogen, against future requirements when an excess of blood glucose is present.

Paramount in the homeostatic role of the liver in body carbohydrate metabolism are its capacity both (*a*) to phosphorylate free glucose and (*b*) to hydrolyze glucose-6-phosphate that has been generated in either glycogenolytic or gluconeogenic processes. Indeed, the controlled balance between these two hepatic processes—glucose-6-phosphate formation and its hydrolysis—is believed to constitute a fulcrum around which the regulation of blood glucose levels is centered (Cahill *et al.*, 1959).

The multifunctional enzyme "glucose-6-phosphatase" (D-glucose-6-phosphate phosphohydrolase, EC 3.1.3.9) appears to play important roles in both these key processes.

In this chapter we review briefly characteristics of both hydrolytic and more recently characterized synthetic activities of glucose-6-phosphatase, consider in some detail the central role of the liver in glucose homeostasis, discuss the roles and regulation of hydrolytic and synthetic activities of hepatic glucose-6-phosphatase as they relate to the control of blood sugar levels, and finally attempt to incorporate hydrolytic and synthetic activities of glucose-6-phosphatase of kidney, small intestine, pancreas, and certain other organs, along with such activities of liver, in an integrated mechanism for overall control of body glucose metabolism. Emphasis is placed on such control mechanisms regulating blood glucose levels under conditions of accelerated gluconeogenesis.

II. GLUCOSE-6-PHOSPHATE PHOSPHOHYDROLASE

A. Historical

The enzyme "glucose-6-phosphatase" (reaction 1) occupies a unique metabolic position, catalyzing the terminal reaction not only in gluconeogenesis but in glycogenolysis as well[1] (Fig. 1). Early studies of the nature and physiological roles of this interesting biological catalyst related to its function in the latter process; however, more recent studies suggest highly interesting implications with respect to its role as a controlling factor in the gluconeogenic process as well.

$$\text{D-glucose-6-phosphate} + H_2O \rightarrow \text{D-glucose} + P_i \qquad (1)$$

[1] And, indeed, recent studies indicating a potent synthetic as well as hydrolytic function for this enzyme may serve to implicate it in glycogenesis and possibly hexose monophosphate shunt and glycolytic process as well, under certain conditions (see detailed discussion in Sections IV and V).

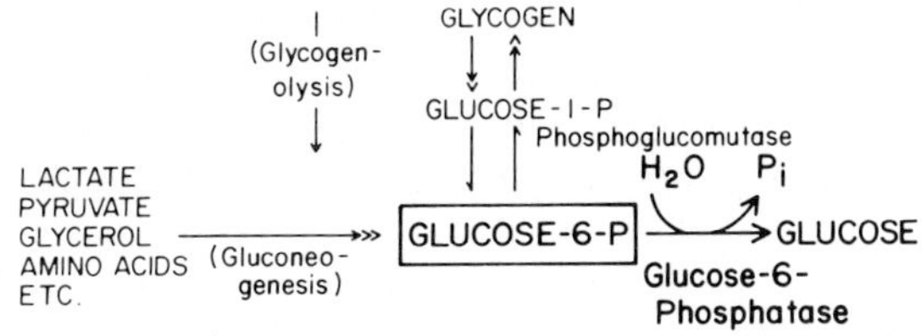

Figure 1. The unique, central position of glucose-6-P as a metabolite common both to gluconeogenic and glycogenolytic processes.

The involvement of a hexose phosphatase as an obligatory part of the catalytic machinery converting liver glycogen to blood glucose was first suggested by the Coris (Cori and Cori, 1938) and by Ostern *et al.* (1939). Because of the presence of potent phosphoglucomutase activity in their preparations, these workers were unable to determine whether the primary substrate for this hydrolysis reaction was glucose-6-P or glucose-1-P.

Fantl and Rome (1945) in Australia first demonstrated the existence in rabbit liver homogenates of a hydrolase distinct from nonspecific acid and alkaline phosphatases; on the basis of its apparent substrate specificity, they named the substance "glucose-6-phosphatase."

Broh-Kahn and Mirsky (1948) then showed that "during the course of glycogenolysis in the liver, glucose-1-P undergoes a preliminary conversion to glucose-6-P prior to dephosphorylation"; thus for the first time the key role for glucose-6-phosphatase in the liberation of glucose from the liver to the blood was directly established.

Subsequently Marjorie Swanson (1949, 1950) in the United States and Christian deDuve and co-workers (deDuve and Beaufay, 1951; deDuve *et al.*, 1949a, 1949b; Beaufay and deDuve, 1954a, 1954b; Beaufay *et al.*, 1954; Hers and deDuve, 1950; Hers *et al.*, 1951) in Belgium successfully separated this enzymic activity from phosphoglucomutase, distinguished it from nonspecific acid and alkaline phosphatases, and characterized catalytic and biological properties of this enzyme. Many of their specific observations relating to the characteristics of hydrolytic activity of this enzyme, as well as the results of more recent observations of others relating to synthetic activities of this enzyme (Section IV), are described in detail in this chapter and in earlier reviews (Nordlie, 1974, 1971, 1969, 1968; Ashmore and Weber, 1959; Byrne, 1961; Nordlie and Arion, 1966; Swanson, 1955; Cohn *et al.*, 1969).

B. Properties Characteristic of Glucose-6-P Phosphohydrolase

Specific glucose-6-P phosphohydrolase, catalyzing reaction 1, has been found classically in endoplasmic reticulum of the gluconeogenic organs liver and kidney (Hers *et al.*, 1951). More recent studies indicate its presence also in outer nuclear membrane (Kashnig and Kasper, 1969; Kanamura, 1971; Leskes *et al.*, 1971; Gunderson and Nordlie, 1973, 1975) and possibly also in plasma membrane (Emmelot and Bos, 1969, 1970; Nordlie and Jorgenson, 1975) and mitochondrial membrane (Nordlie and Gunderson, 1973b). In addition to liver and kidney, this enzyme has been shown to be present in mucosa of the small intestine (Lygre and Nordlie, 1968, 1969a, 1969b), in β-cells of the pancreas (Scott and Jones, 1970; Colilla and Nordlie, 1973b, 1975), and in adrenals, testes, brain, spleen, and lung (Colilla and Nordlie, 1973b, 1975). A variety of other tissues also display the ability to hydrolyze glucose-6-P at acid pH, but in most such cases activity appears to be due to nonspecific acid (or alkaline) phosphatases (see Morton, 1955; Anderson and Nordlie, 1967).

The enzyme is rather detergent sensitive (Beaufay and deDuve, 1954b; Ashmore and Nesbett, 1955; Snoke and Nordlie, 1967), exhibits maximal activity near pH 6 to 6.5 (Swanson, 1950; deDuve *et al.*, 1949b; Nordlie and Arion, 1964a), is not activated by supplemental Mg^{2+} (Nordlie and Arion, 1965; Johnson *et al.*, 1974) but appears to be a metalloenzyme (Nordlie and Johns, 1967, 1968), hydrolyzes a variety of sugar and polyol phosphate esters all at rates less than that for glucose-6-P (Beaufay and deDuve, 1954a; Crane, 1955; Maley and Lardy, 1956; Nordlie, 1971), and exhibits a K_m for the latter substrate of approximately 1 to 2 mM at pH 7 (see, e.g., Segal and Washko, 1959; Lueck *et al.*, 1972).

The enzyme also vigorously hydrolyzes such other compounds as carbamyl-P, PP_i, ATP and other nucleoside tri- and diphosphates, and phosphoenolpyruvate in a highly pH-dependent fashion, as described in more detail in Section IV.

The enzyme appears to be tightly membrane bound (Ernster *et al.*, 1962), requiring specific phospholipids for activity (Garland and Cori, 1972) and appearing to be stabilized by certain phospholipids (Cater *et al.*, 1970); but it is generally constrained by phospholipids (Pollak *et al.*, 1971; Zakim, 1970; Snoke and Nordlie, 1972) under physiological conditions. The liver enzyme has not been extensively purified (Nordlie, 1971). Purification of the brain enzyme to homogeneity has been described in a recent abstract (Anchors and Karnovsky, 1974).

C. The Classical Role of Glucose-6-P Phosphohydrolase in Gluconeogenesis

The metabolic importance of hydrolytic activity of glucose-6-phosphatase has been well established by the Coris (Cori, 1952–1953; Cori and Cori, 1952) and others. The congenital absence of this activity in humans (von

Gierke's disease; Glycogen Storage Disease Type I) is characterized by huge, pathological accumulations of hepatic glycogen, marked hypoglycemia in the absence of frequent glucose ingestion, and early death. Similar observations more recently have been made by C. Cori and coworkers (Erickson *et al.*, 1968), who noted phenomena resembling those just named in irradiation-induced mutant mice whose offspring were unable to form this enzyme. An essential role for hydrolytic activity of the enzyme in catalyzing the terminal reaction in both hepatic glycogenolytic and gluconeogenic processes (Fig. 1) is thus established.

Studies carried out with increasing intensity during the past 15 to 20 years have directed attention to three principal metabolic sites for the control of the gluconeogenic process. These three sites (Fig. 2) are concerned with the interconversions of (*a*) pyruvate and phosphoenolpyruvate, (*b*) fructose-1,6-diphosphate and fructose-6-P, and (*c*) glucose-6-P and glucose; all involve alternative enzymes or sets of enzymes responsible for metabolic flow in the direction of gluconeogenesis versus glycolysis. Hence discriminant control of the direction and intensity of net metabolic flux is permitted, and rapid response to meet the organism's immediate needs may be effected.

Site I, involving the interconversion of pyruvate and phosphoenolpyruvate through the combined action of pyruvate carboxylase plus phosphoenolpyruvate carboxykinase working versus pyruvate kinase, is presently believed to be the principal site of control of hepatic gluconeogenesis (see, e.g., Exton *et al.*, 1970; Williamson, 1967; Lardy, 1965), whereas sites II (fructose diphosphatase working vs. phosphofructokinase) and III (glucose-6-P phosphohydrolase functioning in apposition to the combined action of glucokinase plus hexokinase) may be less significant (Williamson, 1967; but see Exton *et al.*, 1973a, 1973b; also see *addendum*, p. 146).

Nonetheless, the metabolic control at the glucose-6-phosphatase level is extremely important when considered in relation to the whole picture of energy metabolism in the animal. Let us define *gluconeogenesis* as Lardy (1965) has: ". . . the synthesis of glucose and other hexose derivatives from molecules containing 5 carbon atoms or less. . . ." This allows us to consider the process as proceeding in two somewhat distinct phases—(*a*) the formation of hexose derivatives (fructose-1,6-P_2, fructose-6-P,glucose-6-P) from such precursors as lactate, glycerol, and amino acids, and (*b*) the further metabolism of the ultimate such hexose derivative, glucose-6-P, once it is formed (Fig. 3).

Thus this terminal enzyme in the gluconeogenic process, glucose-6-phosphatase, is a significant factor in determining the ultimate metabolic dispensation of gluconeogenically (as well as glycogenolytically) generated glu-

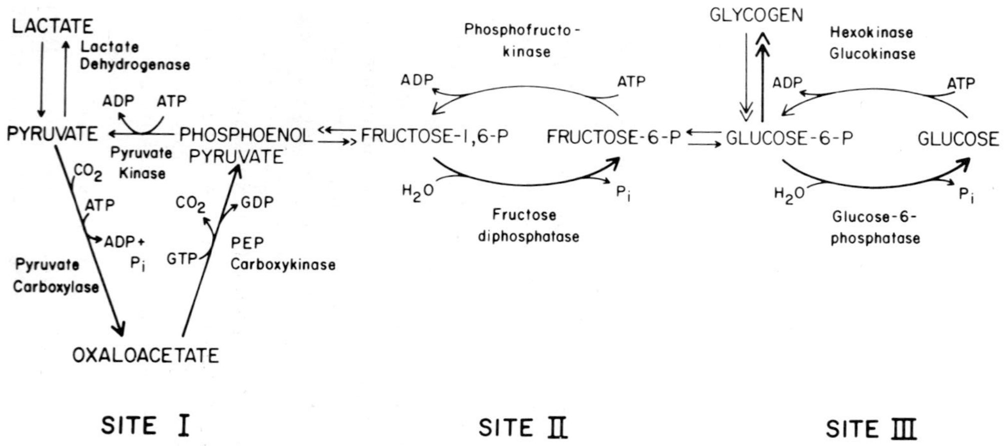

Figure 2. The three key sites for differential metabolic control of hepatic gluconeogenesis and glycolysis. At each of sites I, II, and III, unidirectional reactions (see arrows) important for gluconeogenesis (read from left to right) and for opposing glycolysis (read from right to left) are catalyzed by separate enzymes (or sets of enzymes). Glucose-6-P phosphohydrolase is the important such gluconeogenic catalyst at site III.

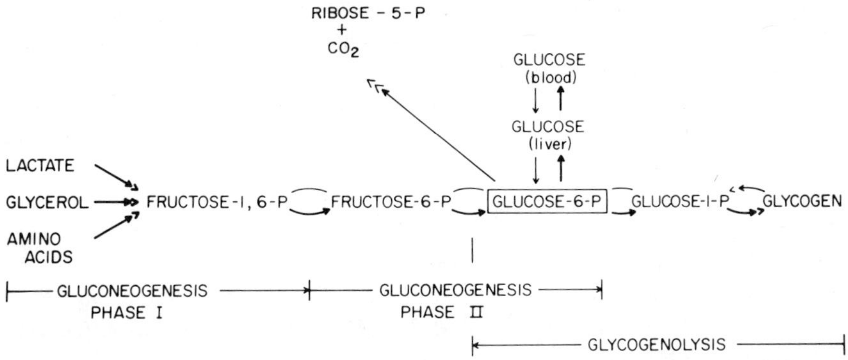

Figure 3. The two phases of gluconeogenesis: generation of hexose derivatives from lactate, amino acids, glycerol, and other substances (phase I) and conversion of such hexose derivatives to glucose (phase II). Participation of glucose-6-P phosphohydrolase in phase II overlaps its role in the glycogenolytic process.

cose-6-P. Such glucose-6-P alternatively may be converted to glycogen, may be oxidized by way of the hexose monophosphate shunt, may be involved in complex polysaccharide synthesis, or may be hydrolyzed to liberate free glucose from the hepatocyte to the blood with the ultimate objective of supplying the energy requirements of peripheral tissues.

Under conditions of stimulated gluconeogenesis, the last-mentioned process is, of course, markedly accelerated. However, glucose-6-P hydrolysis does not proceed totally uncontrolled, even under such conditions of markedly accelerated gluconeogenesis. This is made abundantly clear by the experimental observation (Friedmann *et al.*, 1967) that in diabetes, for example, where vigorous gluconeogenesis is the rule and glucose-6-phosphatase levels are markedly elevated, liver glycogen levels remain significantly high. Also important in this respect is the observed increase in the liver glycogen pool following glucocorticoid therapy (Friedmann *et al.*, 1967).

Furthermore, in both situations, although levels of the various key gluconeogenic enzymes (pyruvate carboxylase, phosphoenolpyruvate carboxykinase, fructose-1,6-diphosphatase, and glucose-6-phosphatase) all increase significantly and blood sugar levels are concomitantly elevated, such blood glucose levels generally reach very definite plateau values within newly defined limits (Cahill *et al.*, 1959) rather than continuing upward unconstrained (Nordlie, 1968, 1969).

Some of the strongest experimental support for control of site III[2] comes

[2] This chapter was submitted in October 1973. Since that time some highly relevant studies relating to glucose-6-P/glucose as a regulatory site have appeared. See Addendum for details.

from recent work by Exton and collaborators (Exton, 1973; Exton *et al.*, 1973a, 1973b). In studies of gluconeogenesis from lactate in isolated, perfused liver, these investigators have seen clear-cut "crossovers" between glucose-6-P and glucose induced by alloxan diabetes and by insulin administration to diabetic animals (Exton *et al.*, 1973a). They also have observed less pronounced *suggestions* of "crossovers" at this site in adrenalectomized-diabetic rats (Exton *et al.*, 1973b), in glucocorticoid-treated adrenalectomized-diabetic rats (Exton *et al.*, 1973b), and in cortisol-treated adrenalectomized, fasted rats (Exton, 1973).

Clearly, control may be exerted not only at earlier stages of gluconeogenesis per se (sites I and II), but at the stage of glucose-6-P hydrolysis (site III) as well.

Glucose-6-phosphatase is thus unique in a number of respects among enzymes classified as "gluconeogenic." First, as just pointed out, it plays a key role not only in the gluconeogenic process but in a second major hepatic function—glycogenolysis—as well. In addition, this enzyme also is a "maverick" in that (*a*) it alone of the complement of such enzymes is a part of cellular membranous structures (endoplasmic reticulum, nuclear membrane, and possibly mitochondrial and plasma membranes; see Section IV.C.2), and (*b*) it is multifunctional, exhibiting potent synthetic and hydrolytic capabilities (see Section IV.A).

III. GLUCOSE-6-P PHOSPHOHYDROLASE AND THE ROLE OF THE LIVER IN THE REGULATION OF BLOOD GLUCOSE LEVELS

The potent capacity of the liver for gluconeogenesis is only one feature of this metabolically prominent organ's more general role in the maintenance of carbohydrate homeostasis of the entire vertebrate or invertebrate organism. The liver possesses the ability to remove glucose from the blood and to generate this energy source from other metabolites and release it into the blood. Glucose-6-phosphatase is a crucial factor in the liver's ability to maintain carbohydrate homeostasis under a variety of conditions, and it is within this context that we wish to consider this enzyme in much of the remainder of the chapter.

In their review of "blood glucose and the liver," Cahill *et al.* (1959) have elegantly defined a variety of factors including, prominently, glucose-6-phosphatase action that appear, based on information available through 1959, to contribute to the liver's capacity for finely regulated glucose uptake, generation, storage (as glycogen), and release as dictated by the immediate metabolic needs of the organism under a variety of nutritional and hormonal stresses.

The following factors all are relevant in determining the concentration of glucose in the blood at a given time: (*a*) amount of dietary carbohydrate consumed, (*b*) rate of intestinal absorption of dietary carbohydrate, (*c*) rate of utilization of glucose by peripheral tissues, (*d*) rate of (possible) loss of this hexose through the kidney, (*e*) rate of removal of glucose by the liver (mainly deposited as glycogen or converted to fatty acids), (*f*) rate and extent of release of glucose from the liver glycogen pool, and (*g*) rate of gluconeogenesis in liver (and kidney[3]).

The importance of the liver in the animal's overall carbohydrate economy is quite apparent from an examination of the foregoing list. Cahill *et al.* (1958a, 1959) have suggested that all the following variables are relevant to the control of the direction and net rate of flux of glucose between the blood and the liver cell: (*a*) the activity of glucokinase,[4] (*b*) the level of hepatic glucose-6-phosphatase activity, (*c*) the size of the glucose-6-P pool in the liver cell, (*d*) the concentration of circulating blood glucose, and possibly also (*e*) the intracellular accessibility of glucose and glucose-6-P to glucokinase and glucose-6-phosphatase, respectively, (*f*) availability of ATP, and (*g*) the presence of the proper ionic environment.

It is also highly relevant to terms of the arguments to follow that permeability of glucose to the hepatic cell appears not to be a significant controlling factor, the hexose moving freely between the liver cell and blood (see Cahill *et al.*, 1958a, but see also Steele, 1966).

Cahill *et al.* (1959) cited as two sets of observations as essential to their hypothesis of control of blood glucose at the hepatic level: (*a*) Soskin *et al.* (1938) found that at a glucose concentration of 150 mg %, the rate of removal of blood glucose by liver just equaled hepatic glucose output in the dog, and (*b*) Cahill *et al.* (1958) demonstrated with rats that above 150 mg % plasma glucose, glucose levels in the plasma exceeded those of liver, at the 150 mg % level plasma glucose just equaled that in the liver cell, and below 150 mg % plasma glucose, liver cell glucose concentration was greater than that in the blood. Thus the ability of the liver either to supply blood glucose when the latter is less than 150 mg % or to remove this hexose from the blood when levels exceed this figure, appears to be established.

These authors theorize that the ability of the liver to remove (and store as

[3] Although kidney, on an *activity per gram of wet tissue* basis, equals or even exceeds the liver's capacity for gluconeogenesis, its contribution to total gluconeogenesis in the body is relatively small because of the kidney's comparatively small mass (Newsholme and Gevers, 1967).

[4] The term "glucokinase" as employed by Cahill *et al.* (1959) relates to the ability of liver to phosphorylate glucose as determined in studies involving liver slices. Specific, inducible, insulin-requiring liver glucokinase (diPietro *et al.*, 1962; Viñuela *et al.*, 1963; Walker and Parry, 1966) had not been characterized when their review was prepared.

glycogen) blood glucose in times of excess, and to supply glucose to the blood in times of need, may be explained on the basis of the relative rates of two opposing hepatic enzymic processes—glucose phosphorylation, the prerequisite initial step in hepatic liver uptake, and glucose-6-P hydrolysis, the terminal step in hepatic glycogenolysis and gluconeogenesis. Controlled "futile cycling," in today's terms, was thus anticipated at the glucose-6-P/glucose site as early as 1959. Glucose-6-P generated is effectively "captured" within the liver cell, since the charged phosphate ester cannot pass through the cell wall, and release of the sugar from this cell to the blood must await removal of phosphate through phosphatase action.

Key enzymes in these two facets of the overall control process (described in detail in Fig. 4) are, respectively, glucokinase (plus hexokinase) for glucose phosphorylation, and glucose-6-phosphatase for glucose-6-P hydrolysis. At a blood glucose level of 150 mg % (8.3 mM), effective physiological rates of these two enzymic processes are postulated to be equal, and no net flux of glucose between hepatocyte and blood ensues. Above this glucose concentration, it is proposed that the rate of kinase activity exceeds that of hydrolysis, and net hepatic glucose uptake ensues as a consequence; in contrast, with lower levels of blood glucose (< 50 mg %), the rate of glucose-6-P hydrolysis exceeds that of hepatic glucose kinase, and there is a net release of glucose from hepatic cell to blood.

The authors attempt to demonstrate the validity of their hypothesized homeostatic mechanism by showing its general applicability under a variety of types of metabolic stress. Included are the postabsorptive state, fasting,

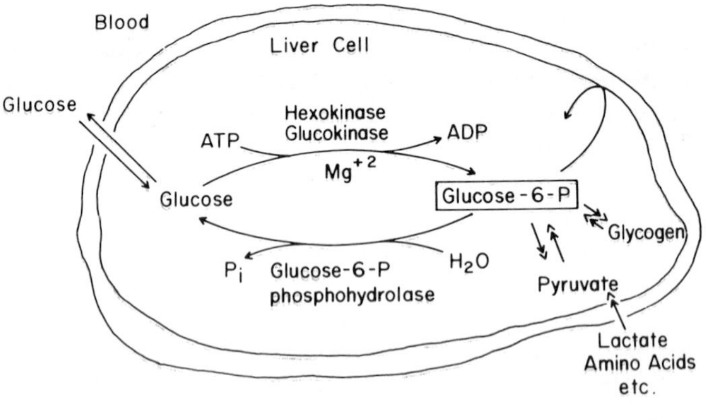

Figure 4. Role of the liver in blood and body glucose homeostasis, based on the proposal of Cahill et al. (1959). The relative rates of hepatic glucose phosphorylation by way of kinase action, and of glucose-6-P hydrolysis by glucose-6-P phosphohydrolase activity, as postulated to determine the direction and net rate of flux of glucose between blood and the liver cell.

untreated diabetes, adrenalectomy of normal and diabetic animals, insulin and glucocorticoid therapy, glycogen storage disease, and hepatomas. Some of these patterns of response are now discussed in detail.

A. Induced Changes in Levels of Glucose-6-P Phosphohydrolase, *In Vivo*

To define as directly as possible the role of hepatic glucose-6-phosphatase in the overall regulation of carbohydrate metabolism, many workers have measured changes in levels of this enzyme in response to hormonal, nutritional, or other stresses or conditions known either to increase or to decrease profoundly the organism's need for glucose. According to the philosophy of such an experimental approach,[5] an increase in the level of glucose-6-phosphatase might be expected in response to those stimuli producing an increased production of glucose by the liver; conversely, a decrease in level of this enzyme might be expected in response to stimuli reducing the requirement for hepatic glucose production, if indeed glucose-6-phosphatase does have a central regulatory role. Conversely, if this enzyme does not constitute a significant site of control, no responses would be anticipated.

Such direct correlations have been noted under many circumstances. For example, glucose-6-phosphatase levels increase significantly in response to experimentally induced diabetes (see, e.g., Weber and Cantero, 1957; Ashmore and Weber, 1959; Cahill *et al.*, 1959; Nordlie and Snoke, 1967), fasting (Ashmore *et al.*, 1954; Nordlie *et al.*, 1968a), and high-protein, low-carbohydrate diets (Freedland and Harper, 1958; Ashmore and Weber, 1959)—all situations or conditions dictating increased hepatic glucose production by way of accelerated gluconeogenesis and glycogenolysis.

Response of enzymic activity to glucocorticoid administration contrasts with those of experimental diabetes or acute fasting. Latency of microsomal activities increases in diabetes but is diminished markedly following glucocorticoid therapy. It appears that induction of new enzyme synthesis may be involved in responses to the former two stimuli (Nordlie and Snoke, 1967; Nordlie *et al.*, 1968b; Proctor and Nordlie, 1972), whereas the response to glucocorticoid administration may involve principally or exclusively an activation of existing enzyme molecules rather than induced protein synthesis (Nordlie *et al.*, 1968b; Nordlie and Snoke, 1967; Nordlie *et al.*, 1965; Arion and Nordlie, 1967).

Responses of glucose-6-phosphatase levels of liver to a variety of hormonal and nutritional manipulations, drug administration, and other

[5] Inherent in such an approach, included here for historical perspective, was the assumption that all regulatory enzymes are inducible and/or repressible, an assumption that is no longer tenable.

factors or conditions perturbing normal blood sugar levels and hepatic glucose production, as observed by a variety of workers, are summarized in Table I. Here the author has attempted to categorize responses of enzymic levels to a wide variety of stimuli; the documentation of each noted response is not exhaustive, however. This subject is also considered in great detail in the excellent earlier review by Ashmore and Weber (1959). For illustrative purposes, a number of interesting correlations apparent from information in Table I are amplified in the following paragraphs.

The essential metabolic role of glucose-6-P phosphohydrolase is made dramatically apparent in von Gierke's disease, where the enzyme is congenitally absent, as pointed out earlier.

Two additional interesting comparisons illustrating variations in levels of enzymic activity correlative with metabolic requirements involve (*a*) comparative embryological considerations and (*b*) a consideration of hepatomas contrasted with normal livers.

Differences in the need for carbohydrate production in the developing avian embryo as compared with the mammalian embryo are dramatically apparent from comparative developmental studies in which glucose-6-phosphatase levels were measured. In the avian embryo, which develops independently of the mother, glucose production in early stages is essential, and glucose-6-phosphatase activity is detectable very early during the incubation of domestic chicken eggs (Bot *et al.*, 1960; Kilsheimer *et al.*, 1960). In contrast, the rat embryo utilizes glucose from the mother's blood and normally does not manifest glucose-6-phosphatase activity until immediately postpartum (Bot and Vereb, 1966; Greengard, 1969a, 1969b; Herrman and Nordlie, 1973). The potential to manifest such activity is present in the embryo, however, and administration of dibutyryl cyclic AMP to the fetus, *in utero,* rapidly leads to development of demonstrable glucose-6-phosphatase activity (Greengard, 1969a, 1969b). In contrast with glucose-6-phosphatase-phosphotransferase, which normally appears at birth in rats, glucokinase is not demonstrable until approximately 16 days postpartum (Walker, 1963).

Greengard (1969a, 1969b) has concluded from the results of her embryological studies that 3′,5′-cyclic AMP acts as a "second messenger" involved in "turning on" glucose-6-phosphatase activity in rats at the time of birth, glucagon or thyroxine acting as the initial triggering agents.

In well-developed hepatomas, glucose-6-phosphatase is absent (Weber and Cantero, 1955), consistent with the reversion to a purely glycolytic state as the tumor develops.

Other correlations also are readily apparent from information in Table I, and are not belabored further.

Table I. Responses of hydrolytic and synthetic activities of hepatic glucose-6-phosphatase-phosphotransferase to hormonal and dietary manipulations and other factors

		Some key references[a]	
Treatment, manipulation, or condition	Response noted	Glucose-6-P phospho-hydrolase	Phospho-transferase activities
Hormonal manipulations			
Diabetes, natural or experimental	Increase	*b–h*	*d–h*
Insulin administration to diabetic or normal subjects	Decrease	*b–h*	*d–h*
Adrenalectomy	Slight decrease	*b–d, i*	*d, i*
Adrenalectomy of diabetic animal	Decreased to normal	*b–d*	*b*
Glucocorticoid administration, adrenalectomized or normal animals	Apparent[k] increase	*b–d, f, g, i, j*	*d, f, g, i, j*
Glucocorticoid administration to diabetic animal	Apparent[k] increase superimposed on diabetic response	*h*	*h*
Growth hormone administration	Increase or no change in normal animals; decrease in hypophysectomized rats	*b, c, l*	
Hypophysectomy	Decrease	*b, d, l*	
ACTH administration	Increase	*b, c*	
Glucagon administration	Increase	*b, c*	
Thyroxine administration	Increase	*b, c*	
Dietary manipulations			
Acute fasting	Increase	*b–d, l–o*	*d, l, m, p*[q]
High-maltose diet	Decrease	*r*	*r*
High-fructose diet	Increase	*b, c, r*	*r*
High-galactose diet	Increase	*b, c*	
High-fat diet	Increase	*b, c*	
High-protein diet	Increase	*b, c*	

Table I. *Continued*

Treatment, manipulation, or condition	Response noted	Some key references[a] Glucose-6-P phospho-hydrolase	Phospho-transferase activities
Other situations or conditions			
Developing fetus	Appears very early in chick embryo; detectable at 18th day in rat fetus	*b, c, r–u*	*r–u*
Newborn	Marked increase at birth (rat)	*r–u*	*r, u*
Congenital aberration; normally inherited (von Gierke's disease) or radiation induced	Markedly decreased or absent	*b, c, v, w, x*	*y*[z]
Hepatomas	Decreased in minimal deviation tumor; absent in well-advanced tumors	*b, c*	
Induced carcinogenesis	Decreased	*b, c*	
Ethanol administration	Increase	*aa*	
Phenobarbital administration	Decrease	*o*	*o*
Hypoglycemic agent administration	Decrease	*b, c, bb*	*bb*

[a] Representative references only are presented for illustrative purposes; no attempt is intended to imply citation of the voluminous literature relating to responses to certain factors, diabetes, and insulin or glucocorticoid administration, for example.
[b] Ashmore and Weber (1959). This review contains many relevant primary references.
[c] Cahill *et al.* (1959). This review contains additional primary references.
[d] Fisher and Stetten (1966).
[e] Nordlie and Arion (1965).
[f] Nordlie *et al.* (1968b).
[g] Nordlie and Snoke (1967).

B. Some Limitations of the Classical Theory of Cahill *et al.*

The generalized direct correlation in responses of glucose-6-phosphatase levels to conditions altering the body's demand for glucose, as postulated by Cahill *et al.*, certainly appears to implicate this enzyme as a focal control point in the regulation of hepatic carbohydrate metabolism. This mechanism of control involving glucose-6-phosphatase, as summarized earlier and in Fig. 4 and as set forth in detail by Cahill *et al.* (1959) based on information available up to 1959 is elegant in its simplicity and impressive in its ability to correlatively rationalize a large and divergent body

---

[h] Proctor and Nordlie (1972).
[i] Nordlie *et al.* (1965).
[j] Arion and Nordlie (1967).
[k] The principal effect of glucocorticoid administration appears to be *activation* of existing enzyme rather than induction of new enzyme protein biosynthesis (Nordlie *et al.*, 1965, 1968b; Nordlie and Snoke, 1967; Proctor and Nordlie, 1972; Arion and Nordlie, 1967). Such activation by glucocorticoids has been observed even in the presence of inhibitors of protein synthesis (Arion and Nordlie, 1967) and in diabetic (Proctor and Nordlie, 1972) as well as normal (Nordlie *et al.*, 1965, 1968b; Nordlie and Snoke, 1967) animals.
[l] Collipp (1967).
[m] Nordlie *et al.* (1968a).
[n] Pandhi and Baum (1967).
[o] Pandhi and Baum (1970).
[p] Arion and Nordlie (1965).
[q] The response to fasting is complex, only that of the hydrolase being statistically significant in the absence of detergent activation of enzyme preparations (Nordlie *et al.*, 1968a; Arion and Nordlie, 1965). Both hydrolase and transferase respond parallelly, based on assays with detergent-activated preparations, however (Nordlie *et al.*, 1968a; Arion and Nordlie, 1965).
[r] Bot and Vereb (1966).
[s] Greengard (1969a).
[t] Greengard (1969b).
[u] Herrman and Nordlie (1973).
[v] Cori (1952–1953).
[w] Cori and Cori (1952).
[x] Erickson *et al.* (1968).
[y] Illingworth and Cori (1965).
[z] Transferase has been studied only in livers of von Gierke's patients (Illingworth and Cori, 1965).
[aa] Nelson *et al.* (1970).
[bb] Jorgenson and Nordlie (1973).

of experimental observations. Nonetheless, it appears to be incomplete in several respects.

1. ***Sluggishness of Response.*** Cahill *et al.* (1959) pointed out that time-response studies often indicate that changes in blood glucose levels and in vigor of gluconeogenesis precede, often by hours, demonstrable changes in liver glucose-6-phosphatase levels. For example, blood glucose levels return promptly to normal following insulin administration to diabetic animals, whereas diminution in hepatic glucose-6-phosphatase levels (as well as reduction in rate of gluconeogenesis and elevation of glucokinase) begin several hours after such treatment and may take 24 hours to return to "normal" values (Fig. 5*a*, taken from the work of Ashmore *et al.*, 1959, 1956, and Renold *et al.*, 1955).[6] A second example of "sluggishness" of response of measured glucose-6-phosphatase levels is seen in glucocorticoid therapy (Fig. 5*b*), where the gluconeogenic rate increases rapidly, blood glucose levels rise concomitantly, glucokinase activity remains unchanged, and apparent (see Section III.A) glucose-6-phosphatase levels rise slowly and only beginning several hours later (Ashmore *et al.*, 1956, 1959). Other factors (e.g., rapidly increased uptake by muscle cells of glucose following insulin administration, and changes induced by glucocorticoids in levels of other gluconeogenic enzymes) undoubtedly are involved, but it appears reasonable to suggest that there may also be operative alterations in manifest glucose-6-phosphatase activity that are much more immediate in effect than relatively slow induction (or repression) of glucose-6-phosphatase synthesis. Such rapid response might supplement the less responsive induction/repression mechanisms; the former could provide for "emergency" immediate alterations in functional activity, while the latter would allow for slower but more permanent readjustments in response to altered demands of the body for carbohydrate.

This need for additional metabolite (or other) control of existing levels of glucose-6-phosphatase is also apparent from a second, somewhat different point of view, as considered in detail in Sections III.B.2 and III.B.3.

2. ***Insufficiency of General Glucose Phosphorylative Capacity.*** A second problem associated with Cahill's overall theory of hepatic glucose control has become obvious more recently, it relates to the apparent deficiency in hepatic glucose phosphorylative capacity by the presently established liver kinases (glucokinase and hexokinase) under certain conditions.

Friedmann *et al.* (1967) studied the rate and ultimate extent of hepatic

[6] The main factor in decreasing blood glucose concentration here may be increased uptake by peripheral tissues (and liver), *not* decreased gluconeogenic reaction of glucose-6-phosphatase.

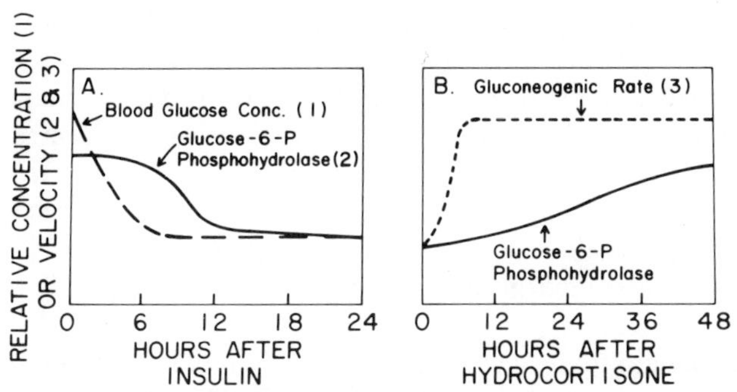

Figure 5. Time course of responses of apparent hepatic glucose-6-P phosphohydrolase activity levels (*a* and *b*), blood glucose levels (*a*), and hepatic gluconeogenic rate (*b*) to hormonal manipulations, *in vivo*. (*a*) Insulin was administered to diabetic rats at zero hours; (*b*) cortisone was injected at zero time to adrenalectomized, diabetic rats. *Source:* Cahill *et al.* (1959), by permission of Dun-Donnelley Publishing Co.

glycogen synthesis from high doses of orally administered glucose in fasted normal, diabetic, adrenalectomized, and cortisol-treated rats. They observed, for example, that the rate of glycogenesis in diabetic animals that had initially been depleted of liver glycogen by glucagon treatment was identical to the rate noted with normal animals. On the basis of this and other observations, these authors conclude that "although the high dependence of liver glycogen synthesis on the level of glucose administration is compatible with a rate-determining role of hepatic glucokinase, its activity is too low in fasted and diabetic rats to account for the rate of glycogen synthesis under these conditions." They suggest that some other synthetic activity for glucose phosphorylation, possibly PP_i-glucose phosphotransferase activity of glucose-6-phosphatase (see Sections IV and V) may be involved as an auxiliary mechanism for hepatic glucose phosphorylation.

McCraw *et al.* (1967) and McCraw (1968) have noted a marked dependence of glucose uptake on perfusate glucose concentration over a wide range (50–600 or 1000 mg %) in the isolated, perfused liver from the diabetic and fasted rat. These investigators have postulated the involvement of PP_i-glucose phosphotransferase to explain the occurrence of the phenomena in the virtual absence of glucokinase.

In quite recent studies, Brunengraber *et al.* (1973) have demonstrated a relationship between perfusate glucose concentration and rate of fatty acid synthesis in isolated, perfused livers from both fed and fasted rats that appears to be inconsistent with known properties of glucokinase. A direct, nearly linear correlation was noted over the entire range of glucose concentrations studied—4.5 to 25 mM.

Similar unsaturability with respect to glucose (5–50 mM) was observed in liver slice studies when glycogen formation was studied by Cahill *et al.* (1958b); in direct contrast, the system for CO_2 production in such preparations was saturated at relatively low (15 mM) glucose concentrations.

Despite their demonstration of the absence of glucokinase from both fetal and adult sheep liver, Ballard and Oliver (1964) have observed a direct dependence on varied glucose concentrations, between 20 and 200 mM, in the rate of ^{14}C-glucose incorporation into glycogen in liver slices from adult sheep.

Levels of glucokinase also appear to be very low or absent in livers of certain other species under normal conditions[7] (see Herrman and Nordlie, 1972). Birds, for example, normally exhibit blood glucose levels between 200 and 350 mg % (Sturkie, 1965), and the enzyme is absent from kidney (Sharma *et al.*, 1963).

Because no one has yet been able to demonstrate the general distribution of the enzyme in many relevant tissues, we conclude that an all-encompassing metabolic regulatory role of glucokinase in the control of blood glucose levels is questionable, if not contraindicated.

3. ***A Quantitative Analysis.*** Following the arguments of Cahill *et al.* (1959), the rate of glucose-6-P hydrolysis should exceed that of glucose phosphorylation at lower glucose levels (i.e., concentrations reflective of hypoglycemia, *in vivo*), whereas the rate of phosphorylation should exceed that of hydrolysis at higher glucose levels consistent with the liver's role in net glucose uptake in hyperglycemic, otherwise normal rats. At some critical, intermediate level of glucose, the rate of phosphorylation should just equal that of glucose-6-P hydrolysis, theoretical plots for these two processes should intersect, and the situation should reflect the state involving no net flux of glucose between liver and blood, *in vivo*. Hypothetical plots (Fig. 6) of rates of glucose phosphorylation and of glucose-6-P hydrolysis considered as a function of varied glucose concentrations illustrate these concepts, which are tested with actual experimental data in following paragraphs.

Quantitative demonstrations of the need for both additional control mechanisms for the regulation of glucose-6-P phosphohydrolase under ambient conditions and additional glucose phosphorylating mechanisms supplementary to hepatic hexokinase and glucokinase are made in Fig. 7.

[7] It may, of course, be possible that because of the enzyme's instability (Walker, 1963), because of the presence of natural inhibitors, or for other unknown reasons, significant levels of existing enzyme have not been detected in certain cases. Whatever the explanation, the generalized distribution of hepatic glucokinase has not been established as it has for synthetic activities of glucose-6-phosphatase (Herrman and Nordlie, 1972).

Figure 6. A theoretical consideration of the relative rates of hepatic glucose phosphorylation and glucose-6-P hydrolysis as affected by varying levels of glucose. At the concentration of glucose indicated by the arrow, the two rates are equal, and no net flux of glucose between liver and blood occurs. To the right of the arrow, the rate of glucose phosphorylation exceeds that of glucose-6-P hydrolysis, and net hepatic uptake of glucose is observed. Under relatively hypoglycemic conditions (to the left of the arrow), glucose flows from the liver to the blood concomitant with the inequality, *hepatic glucose phosphorylation rate < hepatic glucose-6-P phosphohydrolase rate.*

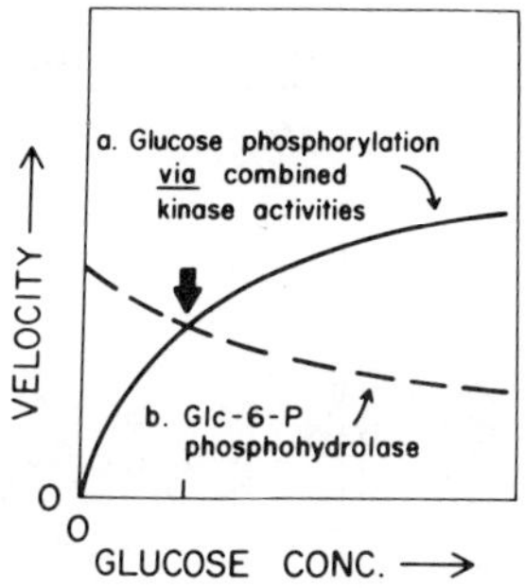

For illustrative purposes, we examine normal rat liver (Fig. 7*a*), diabetic rat liver (Fig. 7*b*), avian liver (Fig. 7*c*), and beef liver (Fig. 7*d*).

These specimens chosen for Fig. 7 represent an attempt by the author to assess formally and quantitatively, on the basis of currently available kinetic data, the concepts proposed by Cahill *et al.* (1959) relating to the control of blood glucose at the hepatic level through the apposing action of liver glucose kinase activity and glucose-6-phosphatase (Fig. 6). In all cases, rates of glucose-6-P hydrolysis and of glucose phosphorylation by both hexokinase and glucokinase have been calculated as a function of varied glucose concentrations (axes of abscissas in Fig. 7). These theoretical plots are based on certain basic assumptions listed specifically by the author (Nordlie, 1974) elsewhere. Hepatic glucose-6-P concentrations, based on literature values, are assumed to be 0.13 mM for normal animals (Young, 1966; Vianna and Nordlie, 1969; Nordlie *et al.*, 1968c) and 0.08 mM for diabetic rats (Sacks, 1953; Young, 1966; Nordlie *et al.*, 1968c); carbamyl-P was taken as 0.2 mM in all cases (Raijman, 1974; see also Herrman *et al.*, 1971). Kinetic parameters are taken from the literature (Lueck *et al.*, 1972; Herrman and Nordlie, 1972, 1973; Colilla and Nordlie, 1975), as are activity levels of glucose-6-phosphatase (Herrman and Nordlie, 1972), glucokinase (see Herrman and Nordlie, 1973; Nordlie, 1974), and hexokinase (see Herrman and Nordlie, 1972; Nordlie, 1974) measured under conditions of constant substrate concentration. Rate equations for glucose-6-P hydrolysis in the presence and absence of glucose (Lueck *et al.*, 1972; Hass and Byrne, 1960), for hexokinase (Herrman and Nordlie, 1973), and for glucokinase (Herrman and Nordlie, 1973) are from the literature or are based on unpublished kinetic studies of Dr. John Herrman in this laboratory when essential information was unavailable. Kinetic parameters, rate equations, assumed metabolite concentrations, other basic assumptions, and specific details regarding the actual calculations are given elsewhere (Nordlie, 1974). Observed activity levels at constant substrate concentrations are as in Tables II (see also Section V.A).

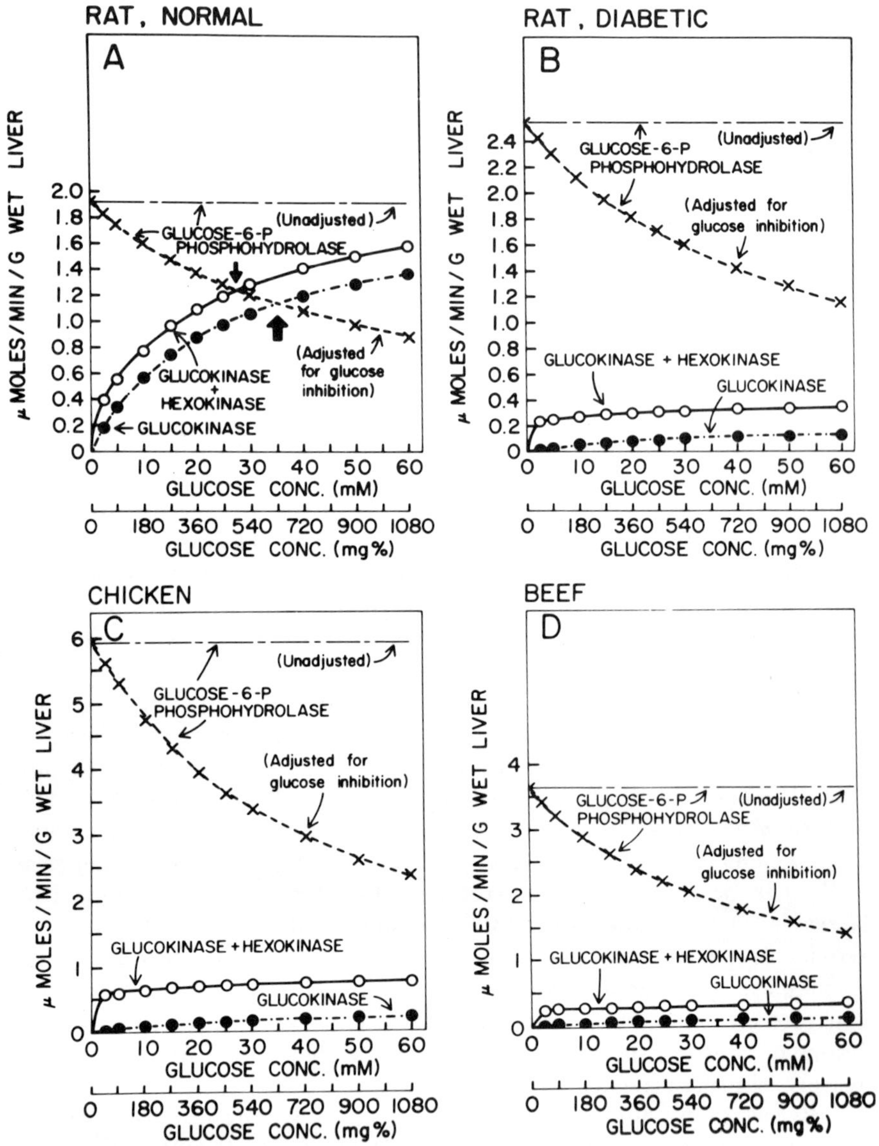

Figure 7. Comparative rates of hepatic glucose-6-phosphohydrolase, glucokinase, and hexokinase activities, calculated for presumed physiological substrate levels and the indicated, varied concentrations of glucose. Further details are given in the text and elsewhere (Nordlie, 1974).

Table II. Activity levels of hydrolytic and synthetic activities of multifunctional hepatic glucose-6-phosphatase-phosphotransferase in certain selected species[a]

Species	Activity levels[b]		
	Glucose-6-P phosphohydrolase	Carbamyl-P:glucose phospho-transferase	PP_i:glucose phospho-transferase
Mammals			
Beef	26	31	21
Deer	23	31	10
Dog	20	16	—
Cat	14	16	9
Sheep	10	15	—
Mouse	13	11	10
Rat	14	10	9
Guinea pig	9	9	8
Rabbit	8	8	6
Man[c]	6	4	6
Birds			
Chicken	29	25	21
Pigeon	20	19	16
Duck	9	9	6
Reptiles			
Garter snake	8	8	8
Turtle	3	4	3
Amphibians			
Grass frog	3	5	3
Mud puppy	1	3	0.7
Bull frog	2	2	1
Fish			
Bullhead	6	6	5
Arthropod			
Crayfish	0.4	0.8	0.8

[a] Glucose-6-P phosphohydrolase and carbamyl-P:glucose phosphotransferase values are from Herrman and Nordlie (1972); PP_i:glucose phosphotransferase activity values are from unpublished work of John Herrman and the author (Herrman and Nordlie, 1973). The former two activities were assayed at pH 7 and the latter at pH 6. Phosphoryl substrate concentration was 5 mM; glucose was 100 mM in phosphotransferase assay mixtures. Liver homogenates, supplemented to 0.2%, w/v, with deoxycholate, served as enzyme source. Bovine serum albumin, 10 mg/1.5 ml assay mixture, also was present.

[b] Activity levels are presented as micromoles per minute per gram of wet liver.

[c] Human liver autopsy samples were employed; deoxycholate in this case was 0.1%, w/v (see Herrman *et al.*, 1971).

In Fig. 7*a* (normal rat), an intersection of plots for glucose-6-P hydrolysis and glucose phosphorylation is indeed noted; however, such a crossover is obtained with glucose levels approximately 500 mg % (glucokinase plus hexokinase) or 630 mg % (glucokinase, only) rather than 150 mg % as predicted from the *in vivo* observations of Cahill *et al.* (1958, 1959) and Soskin *et al.* (1938) described previously. Close scrutiny of Fig. 7*a* suggests that a "null point" in the vicinity of a glucose concentration of approximately 150 mg % could be obtained either (*a*) by lowering the rate of glucose-6-P hydrolysis, (*b*) by increasing the total rate of glucose phosphorylation, or (*c*) both by (*a*) and (*b*).

The need for additional constraints on glucose-6-P hydrolysis and for supplementary mechanisms of hepatic glucose phosphorylation are even more dramatically and abundantly clear in the situations described in Figs. 7*b* to 7*d*. In diabetic rat liver (Fig. 7*b*) and also that of birds (typified in Fig. 7*c* by the domestic chicken) and beef (Fig. 7*d*), glucose-6-P phosphohydrolase levels significantly exceed those of normal rat liver, and glucokinase is comparatively very low (Wallace and Newsholme, 1967; Herrman and Nordlie, 1972). In both the latter two cases, no crossover of plots describing glucose-6-P phosphohydrolase and glucose phosphorylation rates is noted, even with glucose levels far in excess of 1080 mg %, the highest glucose level considered. Thus it would appear either that (*a*) regulation of the rate and direction of flux of glucose between liver and blood is lost with the development of the diabetic state and is never present in normal birds, or (*b*) additional factors for constraint of glucose-6-P hydrolysis and for supplementary glucose phosphorylation complementing hexokinase plus glucokinase must exist.

The absence of such control in one species (chicken) and presence normally in another (rat) is philosophically unattractive. The loss of all control at this level in the diabetic animal appears to be contradicted by the observation (Cahill *et al.*, 1959) that ". . . in all but the most severe states of insulin deficiency a new equilibrium with essentially normal rates of peripheral glucose assimilation—albeit obtained at much higher blood glucose concentrations—is reached" (the higher-than-normal plateau levels of blood sugar being accomplished through accelerated liver glucose production). Thus it appears reasonable to the present author that if, as seems likely, control of hepatic glucose flux is generally existent and involves fundamentally the basic principles outlined previously by Cahill *et al.* (1959), then we will eventually find additional mechanisms (*a*) providing for very rapid responses of glucose-6-phosphatase preceding those of enzyme induction or repression, (*b*) accounting for additional continuous, but highly flexible constraints on glucose-6-phosphatase activity, and (*c*) providing for hepatic glucose phosphorylation supplemental to that which may be accomplished by hexokinase plus glucokinase.

IV. SYNTHETIC ACTIVITIES OF GLUCOSE-6-PHOSPHATASE

Newly elucidated characteristics of classical glucose-6-phosphatase—potent glucose phosphorylating ability and a variety of associated mechanisms for control of both hydrolytic and synthetic activities—discovered in the laboratories of the author, of M. Stetten, and of others, seem to satisfy well all the additional requirements for glucose regulation at the hepatic level just outlined.[8] They appear to this author to nicely complement and in no way to detract from the original elegant basic hypothesis of Cahill *et al.* (1959) or the more recently postulated regulatory roles for hepatic glucokinase by Weinhouse (diPietro *et al.*, 1962; Sharma *et al.*, 1963; Friedmann *et al.*, 1967), Sols (Viñuela *et al.*, 1963; Salas *et al.*, 1963), Walker (1963; Walker and Parry, 1966) and co-workers.

We now set the following objectives: (*a*) to list additional newly characterized synthetic and hydrolytic activities of classical glucose-6-phosphatase, (*b*) to summarize briefly relevant catalytic and biological properties of the various activities of this enzyme, (*c*) to consider mechanisms for constraint of glucose-6-P hydrolysis based on recent experimental observations, and (*d*) to examine the possibility that newly discovered phosphotransferase activities of this complex catalyst may serve as the needed supplementary system for glucose phosphorylation.

A. The Multifunctional Nature of Glucose-6-Phosphatase[9]

A variety of studies carried out by several groups of workers beginning in 1963 (Nordlie and Arion, 1964a, 1964b; Arion and Nordlie, 1964; Stetten,

[8] Properties of the multifunctional enzyme have been extensively reviewed elsewhere (Nordlie, 1968, 1969, 1971, 1974; Nordlie and Jorgenson, 1975).

[9] Krebs (1954) has indicated reasons why he believes different enzymes are needed for controlled glucose-6-P synthesis and hydrolysis within the cell. We believe that multifunctional glucose-6-phosphatase-phosphotransferase may well play both such physiological roles under appropriate conditions. First, the enzyme appears to be under activity-discriminant control, as outlined in detail in this chapter and elsewhere (Nordlie, 1968, 1969, 1971, 1974; Nordlie *et al.*, 1967; Lygre and Nordlie, 1969a); thus the catalyst may function either predominantly as a hydrolase or as a synthetase, as conditions dictate. Second, both hydrolytic and synthetic capabilities of the enzyme are thermodynamically highly favored. Both hydrolytic and synthetic reactions described in this chapter are characterized by quite negative free energy changes under both standard and "physiological" conditions.

riety of applied criteria, all the following hydrolytic and synthetic reactions
1964a, 1964b; Stetten and Taft, 1964) support the multifunctional nature of
glucose-6-phosphatase (see Nordlie, 1968, 1969, 1971). According to a va-
(Equations 1–15) appear to be catalyzed by this enzyme:

$$\text{glucose-6-P} + H_2O \rightarrow \text{glucose} + P_i \qquad (1)$$

$$\text{glucose-6-P} + \text{glucose*} \rightleftarrows \text{glucose*-6-P} + \text{glucose} \qquad (2)$$

$$\text{glucose-6-P} + \text{fructose} \rightleftarrows \text{fructose-6-P} + \text{glucose} \qquad (3)$$

$$\text{mannose-6-P} + \text{glucose} \rightleftarrows \text{glucose-6-P} + \text{mannose} \qquad (4)$$

$$PP_i + \text{glucose} \rightarrow \text{glucose-6-P} + P_i \qquad (5)$$

$$PP_i + H_2O \rightarrow 2P_i \qquad (6)$$

$$\begin{array}{l} \text{nucleoside-5}'\text{-triphosphate or} \\ \text{nucleoside-5}'\text{-diphosphate} \\ + \text{glucose} \end{array} \rightarrow \begin{array}{l} \text{glucose-6-P} + \text{nucleoside-5}'\text{-} \\ \text{di- or monophosphate} \end{array} \qquad (7)$$

$$\begin{array}{l} \text{nucleoside-5}'\text{-triphosphate or} \\ \text{diphosphate} + H_2O \end{array} \rightarrow \begin{array}{l} P_i + \text{nucleoside-5}'\text{-di- or} \\ \text{monophosphate} \end{array} \qquad (8)$$

$$\text{carbamyl-P} + \text{glucose} \rightarrow \text{glucose-6-P} + NH_3 + CO_2 \qquad (9)$$

$$\text{carbamyl-P} + H_2O \rightarrow P_i + NH_3 + CO_2 \qquad (10)$$

$$\text{phosphoenolpyruvate} + \text{glucose} \rightarrow \text{glucose-6-P} + \text{pyruvate} \qquad (11)$$

$$\text{phosphoenolpyruvate} + H_2O \rightarrow P_i + \text{pyruvate} \qquad (12)$$

$$\text{phosphoramidate} + \text{glucose} \rightarrow \text{glucose-6-P} + NH_3 \qquad (13)$$

$$PP_i + \text{glycerol} \rightarrow \text{glycerol-1-P} + P_i \qquad (14)$$

$$\text{1,3-diphosphoglycerate} + \text{glucose} \rightarrow \text{glucose-6-P} + \text{3-phosphoglycerate} \qquad (15)$$

Catalysis of reaction 2 by liver microsomal glucose-6-phosphatase was first demonstrated by Hass and Byrne (1960) and by H. Segal (1959); the former workers also showed catalysis of phosphotransferase reaction 3, as well. Rafter's (1958, 1960) demonstration of PP_i-glucose phosphotransferase (reaction 5) and PP_i phosphohydrolase (reaction 6) by mouse liver mitochondrial preparations was followed by our elucidation of similar activities in rat liver mitochondrial (Nordlie and Lardy, 1961) and microsomal (Nordlie and Gehring, 1963) preparations. In 1964 Nordlie and Arion (1964a, 1964b) and M. Stetten (1964a, 1964b) independently established for the first time the probable identity of these activities with classical rat liver microsomal glucose-6-phosphatase. More recent studies in the author's laboratory have implicated mannose-6-P (reaction 4; Arion and Nordlie, 1964), nucleoside-5′-tri- and diphosphates (reactions 7 and 8, Nordlie and Arion, 1965; Hanson *et al.*, 1970), phosphoenolpyruvate (reactions 11 and 12; Hanson *et al.*, 1970), and most recently and perhaps most importantly (see Nordlie, 1971) carbamyl-P (reactions 9 and 10; Lueck and Nordlie, 1970, 1972; Herrman *et al.*, 1971; Lueck *et al.*, 1972) as additional phosphoryl substrates for both hydrolase and phosphotransferase reactions catalyzed by this enzyme. The involvement of phosphoramidate (reaction 13) was first demonstrated by Parvin and Smith (1969), while Stetten and Rounbehler (1968) have characterized glycerol phosphorylation (reaction 14) by this enzyme and Stetten and Kehoe (1971) have described similar activity with ribitol. Stetten and Goldsmith (1974) discovered activity with 1,3-diphosphoglycerate (reaction 15). Of the nucleoside tri- and diphosphates, the cytidine nucleotides are the most effective substrates (Nordlie and Arion, 1965); activity also has been demonstrated with adenine, guanosine, and inosine nucleotides (Nordlie and Arion, 1965).

The principal reactions involving phosphoryl acceptor specificity thus far characterized are listed in Section IV.A. More than 30 hexoses, hexose derivatives, heptoses, heptuloses, pentoses, trioses, and polyols have been shown by Stetten (1965) to function with at least some degree of efficiency as phosphoryl acceptors; inhibition studies by the present author and colleagues (Lygre and Nordlie, 1969a; Nordlie and Arion, 1965; Arion and Nordlie, 1964) and by Hass and Byrne (1960) also serve to implicate the interaction of a variety of such compounds with the enzyme. On the basis of their own inhibition studies and a consideration of the results of specificity studies of Stetten (1965), Lygre and Nordlie (1969a) have concluded that structural and configurational features indicated in Fig. 8 are essential for the efficient binding of phosphoryl acceptors to the enzyme.

The rather voluminous evidence on which the functional identity of the variety of activities just described is based has been treated in detail by the author in earlier reviews (Nordlie, 1968, 1969, 1971, 1974) and is not repeated here.

B. Relative Magnitude of Synthetic Activities of Glucose-6-Phosphatase

Unlike nonspecific acid and alkaline phosphatases, which often exhibit relatively minor phosphotransferase activity (constituting but a fraction of a percent of hydrolytic capacity, and then only at molar levels of phosphoryl acceptor: Morton, 1955; Anderson and Nordlie, 1967), glucose-6-phosphatase exhibits synthetic activity that actually may exceed by a significant

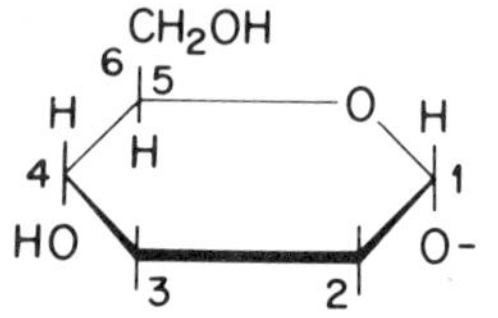

Figure 8. Haworth formula depicting structural and configurational features essential for effective binding of aldohexoses to multifunctional glucose-6-phosphatase-phosphotransferase. *Source:* Lygre and Nordlie (1969), by permission of Elsevier Scientific Publishing Co.

margin the enzyme's maximal catalytic capacity for hydrolysis of glucose-6-P or other phosphate substrates (Nordlie and Arion, 1964a; Arion and Nordlie, 1964; Lueck *et al.*, 1972). This property is of major importance—indeed, glucose is the preferred substrate over water as phosphoryl group acceptor, as indicated, for example, by the observations (*a*) that V_{max} for phosphotransferase at pH 7 exceeds that of phosphohydrolase by more than 50% (Lueck *et al.*, 1972) and (*b*) that transferase activity with 100 mM glucose approximates hydrolase activity in the presence of 55.5 M H_2O (see Table II). Furthermore, levels of transferase activities of glucose-6-phosphatase are significant even at physiological glucose concentrations (see Section V). Thus such synthetase activities as those described in Eqs. 2 to 5, 7, 9, 11, 13, 14 and 15—especially Eqs. 5, 7, and 9—certainly appear to be much more than mere experimental curiosities and may be of considerable physiological significance, as we point out in detail in later sections. Relative levels of glucose-6-P phosphohydrolase, carbamyl-P: glucose phosphotransferase, and PP_i-glucose phosphotransferase activities at constant substrate concentration, obtained with liver preparations from a variety of vertebrate species, are presented in Table II to illustrate this point.

C. Distribution and Location of the Enzyme

*1. **Phylogenetic and Organ Distribution.*** Both hydrolase and phosphotransferase activities of the liver enzyme are widely and generally distributed in higher organisms in nature. Herrman and Nordlie (1972) recently demonstrated the presence of hepatic glucose-6-P phosphohydrolase and carbamyl-P: glucose phosphotransferase in all of 21 species of vertebrates studied, including 10 species of mammals, 4 of birds, 2 of reptiles, 3 amphibia, 1 fish, and an arthropod (see Table II). In contrast, the enzyme has not been detected in plants, bacteria, yeasts, or ciliates (Herrman and Nordlie, 1972) and thus appears to be confined to higher, multicellular members of the animal kingdom. [It should be pointed out, however, that enzymic activities capable of hydrolyzing glucose-6-P at acid pH seem to be universally present but probably involve nonspecific phosphatase action (see Nordlie, 1971)].

The presence of multifunctional glucose-6-phosphatase in mammalian liver, kidney, small intestine, and pancreas appears to be well established (see Nordlie, 1971). A variety of reports of the capacity of many other organs and tissues for glucose-6-P hydrolysis have been summarized in tabular form by Nordlie (1971). In many of these cases, it is not certain whether specific glucose-6-phosphatase, as considered in this chapter, is the responsible enzyme. Recent studies by Dr. William Colilla in our laboratory (Colilla and Nordlie, 1973b, 1975) do establish the presence of carbamyl-P:glucose phosphotransferase and glucose-6-P phosphohydrolase in beef adrenal gland and pancreas and rat adrenals, testes, and brain (and possibly lung and spleen), indicating that this multifunctional catalyst is more widely distributed in the body than has generally been suspected. Interesting recent histochemical studies of Rosen (1970a, 1970b) rather strongly suggest that glucose-6-phosphatase may be present in cerebellum (1970a) and in various cells of the reticuloendothelial system (1970b).

Liver and kidney are the significant gluconeogenic tissues (Newsholme and Gevers, 1967), and the physiological gluconeogenic role of glucose-6-phosphatase in these organs is well established. However, gluconeogenesis is of no significance in other organs in which the enzyme definitely is present [e.g., the absence of other key gluconeogenic enzymes from mucosal cells of small intestine has been demonstrated (Utter, 1959, 1961; Utter and Keech, 1963; Lygre and Nordlie, 1968)]. Apparently, then, glucose-6-phosphatase must have additional significant physiological functions. A role in glucose transport has been suggested from time to time, both with respect to originally characterized phosphohydrolase activity (Drabkin, 1948; Kalckar, 1937; Lundsgaard, 1933; Ernster *et al.*, 1962; deDuve, 1962) and in view of observed phlorizin-induced inhibition of both hydrolase and more recently characterized phosphotransferase activities (Nordlie and Soodsma, 1966; Soodsma *et al.*, 1967; Lygre and Nordlie, 1969a; Nordlie, 1971; Zerr and Novoa, 1968). The more generalized distribution of this multifunctional enzyme in various tissues and organs, and in various membranes of and within the cell (see below), which now is becoming increasingly apparent, seems to be consistent with this hypothesis, as does the considerable correspondence in properties of this multifunctional enzyme system and the bacterial "PEP phosphotransferase system" of Roseman, Kundig, and associates (see Roseman, 1969). This subject is considered in more detailed fashion elsewhere (Nordlie, 1971, 1974) and in Section VI.

2. ***Location of the Enzyme Within the Cell.*** Glucose-6-phosphatase has long been believed to be an enzyme characteristic of the endoplasmic reticulum (Hers *et al.*, 1951; see Nordlie, 1971). Indeed, this catalyst traditionally has been employed as a "marker enzyme" characteristic of the microsomal fraction obtained through fragmentation of the cell's endoplasmic reticulum accomplished by homogenization of tissue followed by differential centrifugation.

Recently several histochemical and biochemical studies have established the presence of glucose-6-P phosphohydrolase activity in hepatic nuclear membrane (Leskes *et al.*, 1971; Kanamura, 1971; Kashnig and Kasper, 1969; Berezney *et al.*, 1972). Studies carried out by Hans Gunderson in our laboratory demonstrate that this enzyme of nuclear membrane possesses a multiplicity of functions, like its counterpart from endoplasmic reticulum, thus proving the presence of classical glucose-6-phosphatase (not just non-specific phosphatase acting on glucose-6-P) in nuclear membrane (Gunderson and Nordlie, 1973, 1975; Nordlie and Gunderson, 1973b, 1973c).

The possible presence of glucose-6-phosphatase displaying phosphotransferase activity in plasma membrane is supported by the findings of Emmelot and Bos (1969, 1970) and of Gunderson and Nordlie [see Nordlie and Jorgenson (1975)]. Recent studies in our laboratory, which indicate the retention of plateau levels of multifunctional glucose-6-phosphatase-phosphotransferase in liver mitochondria even following as many as 11 washings, suggest that this enzyme is present in the outer mitochondrial membrane (Nordlie and Gunderson, 1973a, 1973b; Nordlie and Jorgenson, 1975). It must be pointed out that most histochemical studies to date have failed to indicate the presence of this enzyme in either plasma membrane (but see Pollak and Shorey, 1968) or mitochondria, and contamination by microsomes cannot be entirely ruled out. However, because certain techniques must be used in histochemical studies: [the fixative glutaraldehyde inhibits glucose-6-phosphatase (Kanamura, 1973)], lead, a potent inhibitor of glucose-6-phosphatase activity, must be included along with substrate during incubation with tissue slices (see, e.g., Rosen 1970a, 1970b; Kanamura, 1971). Detergent use is precluded, since it would destroy structural integrity. Thus it appears to this author that negative results in such studies are not necessarily conclusive.

D. Kinetic Mechanism

Figure 9 depicts the simplest kinetic mechanism fully consistent with a large body of kinetic observations of various activities of glucose-6-phosphatase made by Hass and Byrne (1960), Segal (1959), and Nordlie and co-workers (Arion and Nordlie, 1964; Lueck *et al.*, 1972) over the past 12 years. This kinetic mechanism encompasses both hydrolytic and synthetic activities and also rationalizes a variety of inhibitions that have been observed.

Basically, this most elemental of kinetic mechanisms involves compulsory sequences of enzyme-substrate interactions proceeding through (*a*) initial formation of binary enzyme-phosphoryl substrate complexes, (*b*) dissociation leading to formation of phosphoryl-enzyme intermediate, which then may (*c*) transfer the phosphoryl group alternatively to water (phosphohydrolase) or glucose or other alternate acceptor (phosphotransferase).

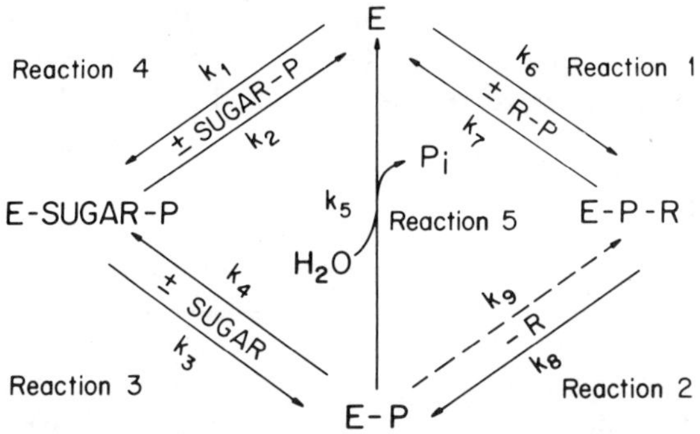

Figure 9. Kinetic mechanism encompassing phosphotransferase and phosphohydrolase activities of multifunctional glucose-6-phosphatase. Sugar-P represents glucose-6-P, RP indicates generalized phosphoryl substrate other than glucose-6-P, and E depicts enzyme. Further details are given in the text. *Source:* Modified from Arion and Nordlie (1964), by permission of American Society of Biological Chemists, Inc.

Hydrolysis of glucose-6-P, for example, proceeds through a sequence involving reactions 4 + 3 + 5 (Fig. 9); hydrolysis of generalized phosphoryl substrate RP (carbamyl-P, PP_i, mannose-6-P, etc.) analogously is described by combined reactions 1 + 2 + 5 (Fig. 9); and the transfer of a phosphoryl group from RP to glucose proceeds as indicated in reactions 1 + 2 + 3 + 4 (Fig. 9).

It is to be emphasized that a mechanistic scheme such as that in Fig. 9 indicates only steps of *kinetic* significance; the actual series of chemical transformations involved may be much more complex. It is not inconceivable, for example, that a multienzyme complex rather than a single enzymic site on a single protein enzyme molecule is involved, although data to this effect are totally lacking.

Various features of the kinetic mechanism in Fig. 9 are considered with respect to the regulation of both hydrolytic and synthetic activities, and relative to possible physiological functions of synthetic activities, in Sections V and VI.

E. Catalytic Properties

A tabulation of catalytic properties of transferase and additional hydrolase activities of glucose-6-phosphatase is complicated by the complex kinetics of the system (see Lueck *et al.*, 1972) and by differential, phosphate-substrate-discriminant latency of the enzyme as obtained generally in isolated liver or kidney microsomal preparations (Snoke and Nordlie, 1967; Nordlie and Arion, 1965; Arion *et al.*, 1972; Soodsma and Nordlie, 1969).

1. Complexity of Two-Substrate Kinetics. A thorough kinetic analysis of the complex glucose-6-phosphatase system has been presented by Lueck *et al.* (1972), and the interested reader is directed to it for detailed considerations. Basically, the apparent K_m values for phosphoryl donor substrates are directly dependent on the concentration of phosphoryl acceptor glucose, and vice versa. Such apparent K_m values for phosphoryl donors, extrapolated for infinite glucose concentration, are equal numerically to K_m values for the same phosphate compounds functioning as substrates in phosphohydrolase reactions of the enzyme, and also equal K_i values for the same phosphoryl compounds functioning as inhibitors of both phosphohydrolase and phosphotransferase activities competitive with other phosphoryl substrates. Thus both minimal apparent K_m values (extrapolated to zero concentration of second substrate) and maximal apparent K_m values (extrapolated for infinite concentrations of second substrate) are included in Table III.

2. Relationship of Membrane Integrity and Relative Enzymic Latency. Although previous studies with microsomes have failed to indicate whether differential latency of various activities of glucose-6-phosphatase are truly reflective of intrinsic catalytic properties of the enzyme *in vivo*, or whether they arise as an artifact of manufacture and isolation of fragments of endoplasmic reticulum (microsomes), very recent studies rather strongly indicate that the latter situation may obtain.

In contrast with endoplasmic reticulum, which must be fragmented and microsomes manufactured before isolation is possible, nuclei with membranes intact may be isolated directly (Berezney *et al.*, 1972; Blobel and Potter, 1966; Gunderson and Nordlie, 1975). And in such intact membranes of this cellular organelle isolated from avian and mammalian liver, synthetic and hydrolytic activities of glucose-6-phosphatase that are both present are totally or nearly totally (more than 90%), manifest without the need for detergent supplementation or other form of activation (Gunderson and Nordlie, 1973, 1975). Of further interest, disruption and isolation of such membranes from avian hepatic nuclei results in the acquisition of latency of both hydrolytic and synthetic activities of glucose-6-phosphatase reminiscent of that observed with untreated microsomal preparations (Gunderson and Nordlie, 1973, 1975).

Clearly, membrane integrity and catalytic function are closely related. And it is strongly implied by these observations that both transferase and hydrolase activities of intact endoplasmic reticulum *in situ* (in marked contrast with isolated fragments thereof) are to an extensive degree manifest rather than latent (Gunderson and Nordlie, 1973, 1975).

3. K_m', V_{max}, pH Optima, and K_i Values. Minimal and maximal apparent Michaelis constant (K_m), apparent V_{max}, and pH optima values are accordingly presented in Table III under two categories—(*a*) as noted with isolated but otherwise untreated microsomal preparations, and (*b*) as observed with maximally activated microsomal preparations, partially purified preparations of the microsomal enzyme, or with intact avian hepatic nuclei. It is of particular interest that such parameters for all three enzyme preparations in category *b* are in each instance essentially equal; quite probably they reflect the ultimate intrinsic catalytic characteristics of this complex enzyme.

Inhibitions of both synthetase and hydrolase activities of the enzyme by a variety of compounds have been demonstrated. Table III includes K_i values for some well-characterized inhibitors, along with relevant literature references and details relating to modes of inhibition. The possible physiological relevance of the effects of some of these compounds are considered in Section V.B.

The effects of reaction pH on activity values are highly complex and are phosphate-substrate discriminant (Nordlie *et al.*, 1971; Lueck and Nordlie, 1970; Nordlie, 1971, 1974). Activity-pH profiles for a representative group of activities are presented in Fig. 10*a*. In Fig. 10*b* the ability of phos-

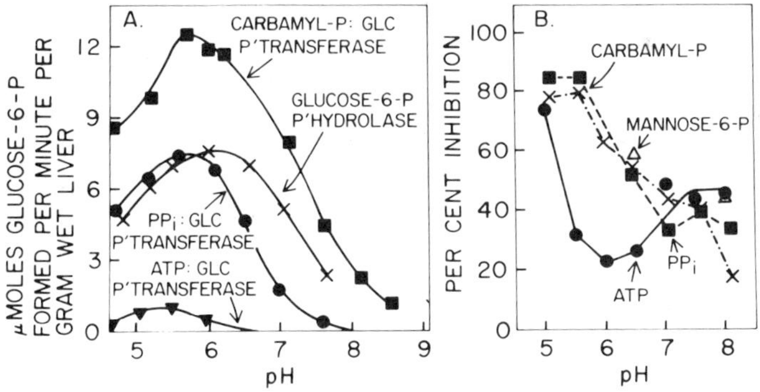

Figure 10. Contrasting, discriminant effects of pH on the ability of various phosphate compounds to function as substrates (*a*), and as inhibitors of glucose-6-P phosphohydrolase activity (*b*). (*a*) Activities were measured at the various indicated pH values with 5 mM phosphate substrate and 120 mM glucose (phosphotransferase activities, only). (*b*) The rate of hydrolysis of 3 mM glucose-6-P (glucose release) was measured at the indicated pH values, in the absence and presence of 5 mM concentrations of the indicated phosphate compounds. Percentage inhibition = 100 × (activity without inhibitor − activity with inhibitor)/activity without inhibitor. *Source:* (*a*) Nordlie (1971), by permission of Academic Press, Inc., (*b*) Nordlie *et al.* (1971), by permission of American Society of Biological Chemists, Inc.

Table III. Some representative kinetic parameters for synthetic and hydrolytic activities of multifunctional glucose-6-phosphatase-phosphotransferase[a]

Kinetic parameter and dimensions	Maximally activated microsomes[b]				Unactivated microsomes
	Rat	Chicken	Beef	Man	Rat
Glucose-6-P phosphohydrolase					
$K_{m,\ glc\text{-}6\text{-}P}$ (mM)	1.8	1.4	1.8	—	2.2
V_{max}[c]	18.4	34.4	35	—	—
pH Optimum	6.5	6.5	—	6.1	—
Carbamyl-P:glucose phosphotransferase					
$K'_{CP,\ min.}$ (mM)	2.0	1.5	3.0	4.7	15.2
$K'_{glc,\ min.}$ (mM)	50.0	23.0	33.0	28.0	86.0
$K'_{CP,\ max.}$ (mM)	5.0	2.6	8.4	7.8	15.6
$K'_{glc,\ max.}$ (mM)	125.0	42.0	86.0	51.0	88.0
$K_{CP\text{-}glc}$ (M^2)	250×10^{-6}	62×10^{-6}	268×10^{-6}	—	1340×10^{-6}
V_{max}[c]	36.4	47.0	128.0	14.4	—
pH Optimum	5.7	5.2	—	5.6	4.8
Some alternate phosphoryl group donors (pH 5.5)					
$K_{m,\ PP_i}$ (mM)	1.3	—	—	—	2.9
$K_{m,\ ATP}$ (mM)	4.7	—	—	—	—
$K_{m,\ CDP}$ (mM)	3.8	—	—	—	—
$K_{m,\ P\text{-}amidate}$[d] (mM)	2.8	—	—	—	13.0
$K_{m,\ P\text{-}enolpyruvate}$ (mM)	6.9	—	—	—	—
$K_{m,\ CTP}$ (mM)	3.1	—	—	—	—

$K_{m,\ ITP}$ (mM)	6.1	—	—	—	—
$K_{m,\ GTP}$ (mM)	3.6	—	—	—	—
Some metabolite inhibitors					
$K_{i,\ glc}$ (mM)	115.0	63.0	86.0	—	—
$K_{i,\ ATP}$ (mM)	11.1	3.2	—	—	40.0
$K_{i,\ P_i}$ (mM)	10.5	4.0	—	—	18.5[e]
$K_{i,\ HCO_3^-}$ (mM)	20.0	—	—	—	20.0
$K_{i,\ citrate}$ (pH 5.6) (mM)	5.2	—	—	—	—
Some alternate phosphoryl acceptors[f]					
$K_{m,\ mannose}$ (mM)	100.0	—	—	—	—
$K_{m,\ 2\text{-deoxy-D-glucose}}$ (mM)	133.0	—	—	—	—
$K_{m,\ galactose}$ (mM)	571.0	—	—	—	—
$K_{m,\ glycerol}$[g] (mM)	3000.0	—	—	—	—
$K_{m,\ ribitol}$[h] (mM)	5600.0	—	—	—	—

[a] Data are from our own work (Lueck *et al.*, 1972; Nordlie *et al.*, 1968c; Vianna and Nordlie, 1969; Dyson *et al.*, 1969; Nordlie, 1971; Hanson *et al.*, 1970; Herrman and Nordlie, 1973; Gunderson and Nordlie, 1973a, 1973b; Colilla and Nordlie, 1973c; Jorgenson and Nordlie, 1973), at pH 7.0, unless otherwise specifically indicated.

[b] *Maximally activated microsomes* refers to microsomal preparations that had been dispersed with supplemental deoxycholate (to 0.2%, w/v) and partially purified by fractional ammonium sulfate precipitation. Essentially identical kinetic parameters have been obtained with untreated, intact avian hepatic nuclei, where measured (Gunderson and Nordlie, 1973, 1975).

[c] V_{max} values are expressed as micromoles per minute per gram of wet liver.

[d] From Parvin and Smith (1969).

[e] pH 7.5.

[f] Studied with PP_i as phosphoryl donor; studies were at pH 5.2, except those with glycerol.

[g] From Stetten and Rounbehler (1968); pH 5.4.

[h] From Stetten and Kehoe (1971); pH 5.2.

phoanhydrides ATP, ADP, PP_i, to inhibit at alkaline as well as acid pH (in marked contrast with their behavior as substrates Fig. 10*a*) is dramatized. Detailed mechanistic considerations pertaining to the effects of pH on the various reactions are given elsewhere (Nordlie, 1971, 1974).

V. RELEVANCE OF NEWLY CHARACTERIZED FEATURES OF GLUCOSE-6-PHOSPHATASE-PHOSPHOTRANSFERASE IN HEPATIC REGULATION OF BLOOD GLUCOSE LEVELS

Newly elucidated characteristics of glucose-6-phosphatase described above appear eminently suited for integration into the scheme of Cahill *et al.* (1959), described in Section III, for control of blood glucose concentrations at the hepatic level. Such features of the enzyme are not only compatible with previously elaborated theory (Section III) but also appear to offer highly tenable answers to certain shortcomings of extant theory that we pointed out earlier (Section III.B). Specifically: (*a*) newly defined transferase activities of this enzyme appear to be suited as an additional "glucose kinase" the existence of which seems essential where previously defined glucokinase is absent or insufficient (diPietro *et al.*, 1962; Sharma *et al.*, 1963; Viuela *et al.*, 1963; Salas *et al.*, 1963; Walker, 1963; Walker and Parry, 1966); (*b*) a variety of mechanisms for rapidly responsive, highly sensitive, continuously functioning control of glucose-6-P hydrolysis through metabolite interaction at the active site of this enzyme are all inherent in the overall mechanism of action depicted in Fig. 9.

Phosphotransferase activities as a "third glucose kinase" are considered first, and mechanisms for highly sensitive, quickly responsive, continuously functioning control of glucose-6-P hydrolysis are dealt with next. The importance of all these mechanisms of control is emphasized as they may operate under conditions of accelerated gluconeogenesis, particularly as typified by untreated diabetes.

A. Phosphotransferase Activities of Glucose-6-Phosphatase: The "Third Glucose Kinase"

As pointed out earlier (Section III.B.2), glucokinase or the combined action of hepatic glucokinase plus hexokinase appears insufficient to explain observed rates of glucose uptake and utilization in conditions of fasting or diabetes where levels of insulin-dependent glucokinase are very low (diPietro *et al.*, 1962; Sharma *et al.*, 1963; Viñuela *et al.*, 1963; Salas *et al.*, 1963; Walker, 1963).

McCraw (1968), McCraw *et al.* (1967), and Friedmann *et al.* (1967) have suggested that PP_i: glucose phosphotransferase activity of glucose-6-phosphatase may constitute a physiologically important auxilary mechanism for glucose phosphorylation under such circumstances of glucokinase insufficiency. The PP_i liberated in the UDPG synthetase step of hepatic glycogenesis effectively could serve as substrate for phosphorylation of glucose as a prerequisite to incorporation of this hexose into the glycogen polymer (Friedmann *et al.*, 1967). A significant conservation in "high-energy phosphate bond" energy may thus be accomplished correlative with requisite phosphorylation of glucose. The overall concept is outlined in Fig. 11.

Prior to the first report of our discovery of potent phosphotransferase activities of glucose-6-phosphatase, Ballard and Oliver (1964) reported a direct dependence on glucose levels, in the range 20 to 200 mM, of ^{14}C-glucose incorporation into glycogen in slices of sheep liver which they showed to be devoid of glucokinase.

Gordon (1963) has similarly reported a linear relationship between rate of glucose uptake by isolated, perfused livers and perfusate glucose concentrations between 100 and 500 mg %, the extent of the range of concentrations she studied. Livers, which presumably were very low in glucokinase, were from rats fasted for 18 to 20 hours. In ongoing studies in our laboratory, Dr. Fred Alvares (Alvares *et al.*, 1974) has confirmed and extended the work of Gordon (1963). He has directly demonstrated increases in phosphotransferase activity of glucose-6-phosphatase, and a marked decrease in glucokinase (to approximately 10% of normal levels), along with concentration-dependent glucose uptake between 5.6 and 67 mM D-glucose.

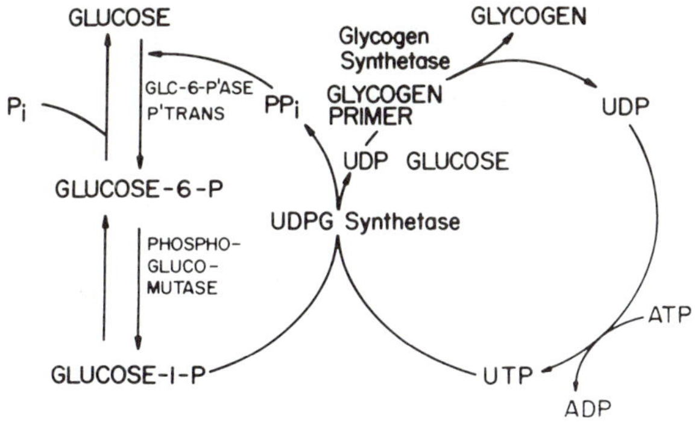

Figure 11. Postulated cyclic process involving phosphorylation of glucose by way of PP_i-glucose phosphotransferase activity of glucose-6-phosphatase. Initially PP_i is released at the UDP glucose synthetase step in glycogenesis, and its further use as indicated may serve as a driving force for the overall glycogenic process. *Source:* Friedmann *et al.* (1967), by permission of The Endocrine Society.

Cahill *et al.* (1958b) have shown a direct, linear dependence on glucose concentrations, between 5 and 50 mM, in the rate of incorporation of glucose into glycogen in rat liver slices. In contrast, the system for formation of CO_2 from glucose in these slices was saturated at relatively low (approximately 15 mM) glucose levels. Perhaps separate systems exist for phosphorylation of glucose preliminary to channeling of resulting glucose-6-P into the two pathways.

Very recently, Lowenstein and his group (Brunengraber *et al.*, 1973) have carried out very careful, detailed studies on the relationship between perfusate glucose concentrations and the rate of fatty acid synthesis in the isolated, perfused liver. Livers from both fed and fasted rats were employed. The investigators observed a direct, functional relationship with both such preparations over the entire range of perfusate glucose concentrations studied (4 to 25 mM) and pointed out that these results are inconsistent with known properties (K_m, glucose, ≅ 10 mM) of hepatic glucokinase. The present author points out further that glucokinase levels in livers of rats fasted for 1 or 2 days, as employed in some of these studies, would be very low, raising additional questions about the adequacy of glucokinase as a generalized hepatic glucose phosphorylating mechanism.

Further quantitative arguments indicating the need for supplemental glucose-phosphorylating capacity in liver have been presented in Section III.B. As illustrated in Fig. 7, plots of rates of glucose phosphorylation by way of hepatic hexokinase plus glucokinase and of glucose-6-phosphohydrolase (unadjusted for glucose inhibition) do not intersect in the range between 0 and 1080 mg % glucose. And even when adjustments are made for glucose inhibition of glucose-6-P hydrolysis (crosses, Fig. 7), such plots either do not cross (chicken, beef, or diabetic rat) or converge at a glucose concentration well in excess of physiological blood glucose values (normal rats).

Further refinements thus appear to be in order if theory is to meet experimental fact.

Additional adjustments have been made in such plots on the basis of newly found properties of glucose-6-phosphatase-phosphotransferase in Figs. 12*a* and 12*b* for normal and diabetic rats, respectively, and for chicken and beef (Figs. 13*a* and 13*b*).

Specifically, the following additional factors have been taken into account in Figs. 12 and 13 as compared with those considered in Fig. 7.

1. Inhibition of glucose-6-P phosphohydrolase by glucose routinely has been incorporated.
2. Carbamyl-P:glucose phosphotransferase activity as a "third glucose kinase" has been added to the complement of enzymes catalyzing glucose phosphorylation, carbamyl-P concentration being assumed physiologically as 0.2 mM (Raijman, 1974; see also Herrman *et al.*, 1971).

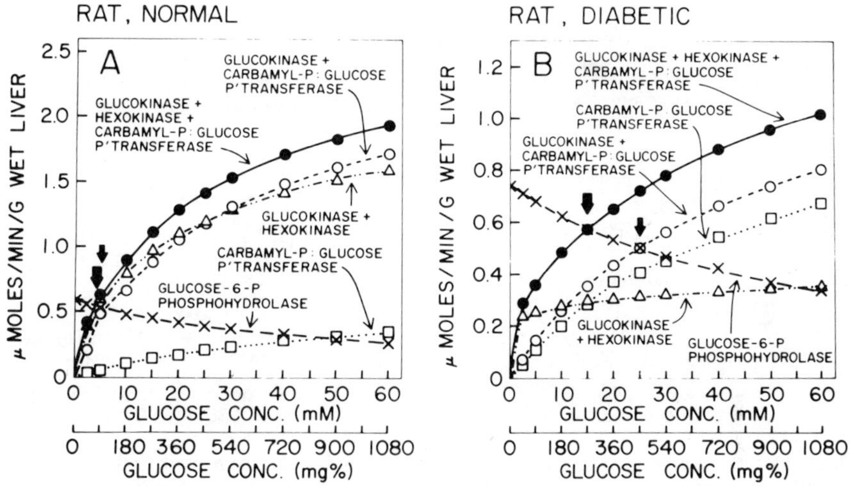

Figure 12. Comparison of rates of various hepatic enzymic activities of glucose metabolism in normal (*a*) and diabetic (*b*) rats. Levels of the indicated activity were calculated as a function of designated, varied glucose concentrations, on the basis of assumed physiological levels of other substrates; activity values have been further adjusted for inhibitions by presumed physiological levels of certain inhibitors as indicated in detail in the text and elsewhere (Nordlie, 1974). The vertical arrows represent concentrations of glucose at which rates of glucose-6-P hydrolysis and glucose phosphorylation by the designated combinations of phosphorylative activities are just equal.

3. Inhibition of both hydrolase and carbamyl-P:glucose phosphotransferase by the indicated, assumed physiological levels of ATP (4.5 mM for normal, 3.2 mM for diabetic), P_i (10 mM for normal, 14.5 mM for diabetic), and HCO_3^- (20 mM for normal, 16 mM for diabetic) is incorporated (see Nordlie, 1974).
4. Inhibition by 0.2 mM carbamyl-P of glucose-6-P phosphohydrolase, and by 0.13 mM (normal) or 0.08 mM (diabetic) glucose-6-P of carbamyl-P:glucose phosphotransferase, is taken into account.
5. Enzymic activity levels are as presumed earlier in Fig. 7 and as indicated in Herrman and Nordlie (1972) for chicken and beef liver (see also Tables II and IV).
6. Kinetic parameters for glucose-6-phosphatase-phosphotransferase are those most directly reflective, in the author's opinion, of the intrinsic catalytic capacity of this enzyme—those noted with maximally activated microsomes, partially purified enzyme preparations, or intact, isolated nuclei (Section IV.E.3; also see Lueck *et al.*, 1972; and Nordlie, 1974).

A minutely detailed consideration of all the assumptions, mechanistic considerations, and other details involved in the calculation of data in Figs.

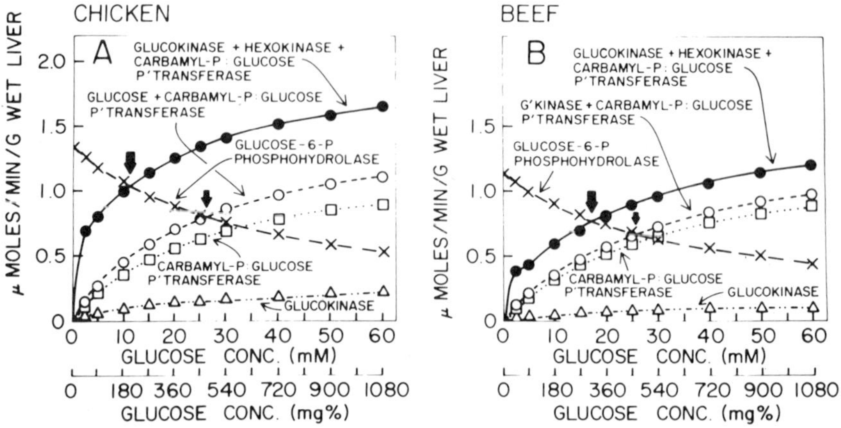

Figure 13. Comparisons of levels of various (*a*) avian and (*b*) bovine hepatic enzymic activities, calculated for indicated, varied glucose levels and presumed physiological substrate concentrations, and adjusted for inhibitions by assumed physiological levels of certain additional metabolites. Other details are as in Fig. 12.

7, 12, and 13 is given by the author elsewhere (Nordlie, 1974). The interested reader is directed to this source; further elaboration is omitted here to avoid redundancy.

As Figs. 12 and 13*a* indicate, intersections of curves for rates of glucose-6-P hydrolysis and for glucose phosphorylation by combined action of glucokinase plus carbamyl-P: glucose phosphotransferase, or the three kinases designated,[10] now intersect at glucose levels rather closely approximating the physiological for normal rat (intersection at 80–90 mg% glucose), diabetic rat (270–450 mg % glucose), and normal domestic chicken (200–460 mg % glucose). These values compare favorably with noted experimental blood glucose values (see Table IV).

These observations are interpreted by this author to validate experimentally the basic theoretical concepts originally postulated by Cahill *et al.* (1959), with the requisite inclusion of additional refinements based on newly defined properties of multifunctional glucose-6-phosphatase-phosphotransferase.

[10] Because hepatic hexokinase is in considerable part associated with cellular particles (mitochondria—see Rose and Warms, 1967, e.g.), and because this activity is extensively inhibited at elevated glucose levels (Grossbard and Schimke, 1966), we question whether this activity should be included in the hypothesized mechanism for regulation of blood glucose concentrations at the hepatic level. We have accordingly routinely presented data in Figs. 12 and 13 both including and excluding hexokinase as a relavant glucose phosphorylating mechanism.

Table IV. Some comparative, contrasting, representative information on the three principal hepatic enzymes of glucose phosphorylation in several species

	Inducible or constitutive	Response in diabetes	Apparent K_m value for glucose	Levels in indicated species[a]				
				Rat				
				Normal	Diabetic	Chicken	Beef	Man
Hexokinase activity[b]	Constitutive	Unchanged	10^{-2} mM[d]	0.24	0.24	0.58	0.25	0.4
Glucokinase activity[b]	Inducible	Decreased > 90%	10–20 mM[e]	1.0	0.1	0.1	0.06	0.4
Carbamyl-P:glucose phosphotransferase activity[b]	Inducible	Increased 2- to 3-fold	23–50 mM[f]	9.7	19.4	28.5	31.4	4.0[c]
Carbamyl-P:glucose phosphotransferase glucokinase ratio	—	—	—	9.7	194.0	285.0	491.0	10.0
Blood glucose levels (mg %)	—	—	—	90.0	250–600	200–300[g]	40.0[h]	100.0

[a] Enzymic activity is expressed as micromoles of glucose-6-P formed per minute per gram of wet liver.

[b] Assay conditions: pH 7 to 7.4; hexokinase, 5 mM ATP-Mg^{2+} and 0.5 mM glucose; glucokinase, 5 mM ATP-Mg^{2+} and 100 mM glucose; carbamyl-P:glucose phosphotransferase, 5 mM carbamyl-P and 100 mM glucose. Detergent-supplemented hepatic homogenates were employed for assay of this last activity. Hexokinase and glucokinase values are from the literature (Borrebaek *et al.*, 1970; Ballard, 1965; Herrman and Nordlie, 1972; Herrman *et al.*, 1971; Molstad and Nordlie, 1973; Salas *et al.*, 1963; Sharma *et al.*, 1963; Viñuela *et al.* 1963; Wallace and Newsholme, 1967; Willms *et al.*, 1970), carbamyl-P:glucose phosphotransferase values are from our laboratory (Herrman and Nordlie, 1972; Herrman *et al.*, 1971).

[c] Based on assay of autopsy samples; probably represents a minimum value.

[d] From Grossbard and Schimke (1966); Katzen and Schimke (1965).

[e] From diPietro *et al.* (1962); Walker (1963); Walker and Parry (1966).

[f] From Herrman and Nordlie (1972); Lueck *et al.* (1972).

[g] From Sturkie (1965).

[h] From Pantelouris (1967); Bergman (1970).

Figure 14 schematizes the situation generally as we propose it to exist. Control of net direction and magnitude of flux of glucose between liver and blood, through the apposed formation and degradation of glucose-6-P, is apparent in all three situations (Figs. 12*a*, 12*b*, and 13*a*), as is also the "retuning" of blood glucose concentrations to meet the three quite contrasting metabolic requirements of the specimens in question. In the diabetic rodent and in birds, blood glucose levels are experimentally much higher than in the normal rat (see Table IV). Concomitantly, crossover of experimental plots depicting rates of glucose phosphorylation and of glucose-6-P hydrolysis are apparent at glucose concentrations considerably higher in birds or diabetic rats than in normal rats. The ratio *carbamyl-P:glucose phosphotransferase levels/glucokinase levels* is much higher in the former two situations than in the normal rat (see Table IV), and a displacement of crossover point to the right in Figs. 12*b* and 13*a* (and also Fig. 13*b*; see below), compared with Fig. 12*a*, arises because of this change in ratio of glucose phosphorylating enzymes and their differences in apparent K_m values for glucose—15 mM for glucokinase and approximately 25 to 40 mM for carbamyl-P:glucose phosphotransferase. Such changes, favoring carbamyl-P:glucose phosphotransferase, in activity ratios for these two

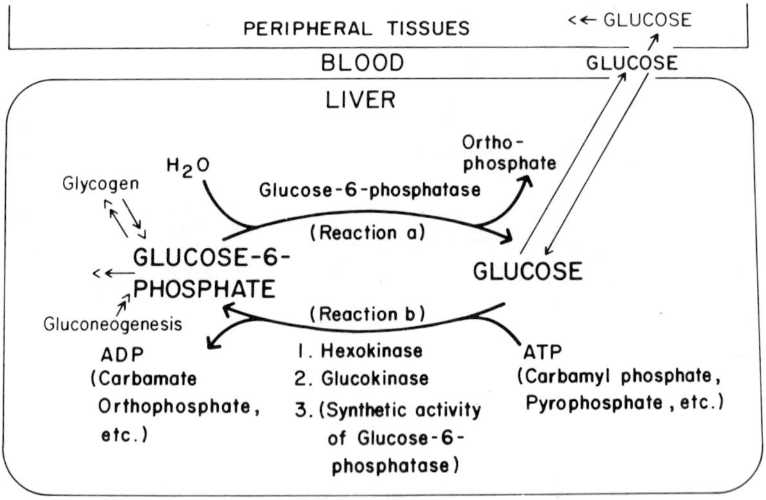

Figure 14. Interrelationships among liver, blood, and peripheral tissues with regard to glucose homeostasis, as modified by Nordlie (1973). Synthetic activities of glucose-6-phosphatase-phosphotransferase are postulated by the author as a requisite adjunct to glucokinase (and possibly hexokinase) for effective hepatic glucose phosphorylation as part of this so-called futile (Clark *et al.*, 1973) or utile (Hue and Hers, 1974) cycle. It is suggested that the relative rate of such combined phosphorylative activities compared with that of glucose-6-P hydrolysis determines the direction and rate of flux of glucose between hepatocyte and blood.

activities with their varying K'_m, glucose values would tend to flatten the experimental hyperbolic curve for glucose phosphorylation, thus effecting the noted displacement toward higher "null point" blood glucose levels.

In the ruminant (beef: Fig. 13*b*), which is included for purpose of contrast with the previously considered three animals, control on a similar basis appears possible, but only at rather high glucose levels (310–460 mg %). In such species propionate rather than glucose is introduced from digested foodstuffs into the blood (Bergman, 1970). Their only source of glucose appears to be that generated gluconeogenically, and normal blood sugar levels in ruminants are thus characteristically low (Bergman, 1970; Pantelouris, 1967). Hence hydrolytic activity only of glucose-6-phosphatase, acting in its traditional capacity as the terminal step in gluconeogenesis, normally seems to be important in cattle, sheep, and the like under physiological conditions. Diabetes may be induced or encountered in ruminants as in other mammals (Bergman, 1970; McCandless *et al.,* 1948). And in such cases, when the animal experiences a high glucose load (e.g., McCandless *et al.,* 1948), control through mechanisms as outlined in Fig. 14 and in the foregoing text appears to be tenable even for the ruminant that lacks glucokinase.

Let us now summarize the newer aspects of control involving phosphotransferase activities of glucose-6-phosphatase. Inclusion of such synthetic activities of glucose-6-phosphatase along with the other indicated kinases (Fig. 14) gives an added element of diversity and flexibility to the complement of enzymes catalyzing hepatic glucose phosphorylation (see Table IV for details) and provides for the continued capacity for significant phosphorylation of glucose, and the control of hepatic glucose release, even in situations (e.g., diabetes) and species (e.g., birds) in which insulin-dependent glucokinase may be low or absent.

A reproportioning of relative levels of these three enzymes, with their diversity of requirements (differing apparent Michaelis constant values; see Table IV) for glucose may result in a "retuning" of blood sugar levels to meet the metabolic demands placed on the organism by the relevant metabolic perturbation, situation, or condition. Because of such differences in Michaelis constant values for glucose for the two enzymes, an increase in the ratio *phosphotransferase activity of glucose-6-phosphatase:glucokinase,* as seen in untreated diabetes, for example (Table IV), would tend to raise that level of glucose at which the rate of hepatic glucose-6-P hydrolysis just equals hepatic glucose phosphorylation. The increase in plateau blood glucose levels thus effected would allow for continued penetration of glucose, by way of mass action, into cells of peripheral tissues (Fig. 14; also see Section V) that become largely inaccessible to this sugar at normal concentration values, in the absence of insulin (see Nordlie, 1968, 1969; West *et al.,* 1966).

Maintenance of blood glucose levels in birds at values normally two to three times those of mammals (Sturkie, 1965) may be similarly explained (see the marked differences in activity ratios for birds as compared with normal mammals in Table IV) and may serve to meet the higher energy demand imposed by the bird's capacity for flight.

We further suggest that such changes in relative enzymic activity levels also may help explain the propensity toward hyperglycemia often observed among the aged. Adelman (1970) has already reported highly significant decreases in hepatic glucokinase levels during aging.

Regulation of blood sugar levels clinically through chemotherapeutically induced changes in relative levels of these various enzymes for hepatic glucose phosphorylation appears to be a distinct possibility (Nordlie, 1973) and is the subject of intensive investigation in the author's laboratory.

B. Additional Mechanisms for Rapidly Responsive, Highly Sensitive, Continuously Functional Control of Net Glucose-6-P Phosphohydrolase Activity

As pointed out in Section III.B.1, a second shortcoming of extant mechanisms for control of blood sugar concentrations at the glucose-6-phosphatase level, in addition to the need for additional mechanisms of glucose phosphorylation, is the sluggishness of response. Rapidly- responsive, highly sensitive, continuously functional mechanisms for control of glucose-6-P hydrolysis involving various metabolite interactions with glucose-6-phosphatase inherent in the multifunctional nature of this catalyst, appear to be distinct physiological possibilities. Indeed, five such mechanisms for control were already introduced as basic assumptions in the treatment of the "third glucose kinase" just considered. Specifically involved were (*a*) apparent inhibition of glucose-6-P phosphohydrolase activity by glucose, (*b*) competitive inhibitions between glucose-6-P and carbamyl-P in both hydrolase and phosphotransferase reactions, and (*c*) competitive inhibitions with respect to glucose-6-P and carbamyl-P by ATP, P_i, and bicarbonate. Such interactions of the enzyme with most of these compounds are obvious, and indeed predictable, from the mechanistic considerations summarized in Fig. 9.

The significance of such metabolite-effected control, in addition to previous considerations, is discussed below from the standpoint of control under conditions of accelerated gluconeogenesis; next, quite briefly, we examine the possible regulatory effects of certain other inhibiting or activating agents.

1. Autoregulation of Glucose-6-P Phosphohydrolase by Glucose. Inhibitions by increasing levels of glucose of net glucose-6-P hydrolysis are readily apparent from data in Figs. 12 and 13. Such inhibitions arise as a result of competition between glucose and water as alternate phosphoryl acceptors (Hass and Byrne, 1960; Segal, 1959; Arion and Nordlie, 1964; Nordlie and Arion, 1964; Lueck *et al.*, 1972; Stetten and Taft, 1964) and are really a manifestation of glucose-6-P: glucose phosphotransferase activity. Such inhibitions (reactions 4 →3 → 4 in Fig. 9) appear to provide a convenient and effective method for the autocatalytic limitation of the excessive, wasteful overproduction of glucose by liver under conditions of stimulated gluconeogenesis. Constraint of accelerated gluconeogenesis at the glucose-6-P phosphohydrolase level may thus be conveniently and rapidly (instantaneously) effected.

2. ***Inhibitions by ATP, ADP, and P_i of Glucose-6-P Phosphohydrolase.*** As indicated in reactions 1 and 4 in Fig. 9, ATP, ADP, and other alternate phosphoryl substrates compete with glucose-6-P for the active enzymic site.

Inhibition by ATP of glucose-6-P phosphohydrolase (Nordlie *et al.*, 1968c) may constitute an effective mechanism for prevention of energetically wasteful nonsense cycling, involving otherwise unconstrained concerted action of glucose-6-phosphatase and hexokinase plus glucokinase (Stadtman, 1966).

Inhibition by nucleotides and P_i also may provide a means of continuously constraining glucose-6-P hydrolysis at levels well below maximal for physiological levels of glucose-6-P (Nordlie *et al.*, 1968c; Vianna and Nordlie, 1969; Nordlie, 1969). For example, as presented in Fig. 13*a*, the rate of hydrolysis of physiological levels (Young, 1966; Nordlie *et al.*, 1968c; Vianna and Nordlie, 1969) of glucose-6-P (0.13 mM) was constrained approximately 56% by combined action of physiological levels of P_i (10 mM; see Sacks, 1963; Vianna and Nordlie, 1969; 47% inhibition by itself) and ATP (4.5 mM; see Nordlie *et al.*, 1968c; 28% inhibition by itself).

Furthermore, increased inhibition by elevated concentrations of the degradation products of ATP (ADP and P_i) as energy charge (Atkinson, 1970) decreases also may constitute a significant constaint mechanism under conditions of accelerated gluconeogenesis. The generation of a single molecule of glucose from two molecules of lactate by way of hepatic (or kidney) gluconeogenic processes occurs at the expense of the hydrolysis of six ATP molecules (Fig. 15). Resynthesis of ATP from ADP and P_i thus generated is accomplished by mitochondrial oxidative phosphorylation. Since ADP and P_i are each approximately as effective in inhibiting glucose-6-P phosphohydrolase activity as ATP (Nordlie *et al.*, 1968c), it may be of particular significance that when use of ATP for gluconeogensis exceeds the

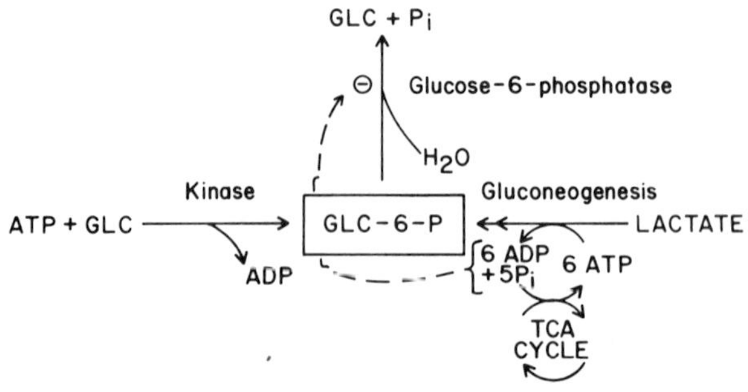

Figure 15. Proposed mechanism for limiting the excessive drainage of energy from the adenylate pool during accelerated gluconeogenesis. If the rate of the latter process exceeds the capacity of mitochondrial oxidative phosphorylation for ATP regeneration, ADP and P_i accumulate. Increased inhibition of the terminal reaction of gluconeogenesis—glucose-6-P hydrolysis—by such accumulated metabolites is then suggested to ameliorate the gluconeogenic process.

capacity of liver for resynthesis of this compound, two molecules of effective inhibitor of the terminal reaction in the gluconeogenic process—ADP and P_i—are present for each original ATP molecule. Increased inhibition by accumulation of ADP and P_i thus would serve as a self-regulating device for limiting gluconeogenesis and would tend to keep the energy-requiring gluconeogenic process in step with the production of energy by way of mitochondrial oxidative phosphorylation.

It is of considerable interest that although ATP and ADP can serve as substrates for phosphotransferase and phosphohydrolase activities of glucose-6-phosphatase only at acid pH (Fig. 10*a*; also see Nordlie and Arion, 1965; Hanson *et al.*, 1970; Nordlie *et al.*, 1968b), they are effective inhibitors (Nordlie *et al.*, 1968c, 1971) at neutral and alkaline as well as acid pH values (Fig. 10*b*). Mechanistic implications of these observations are elaborated on at length by the author elsewhere (Nordlie, 1971, 1974).

3. ***Inhibition by Bicarbonate Ion.*** Observed inhibition by bicarbonate of glucose-6-P phosphohydrolase activity (K_i = approximately 20 mM; Table III and Dyson *et al.*, 1969), together with earlier observations by Anderson and Nordlie (1968; Anderson *et al.*, 1968) and Horne and Nordlie (1971) on the differential effects of bicarbonate on glucose-6-P dehydrogenase (inhibition) and associated glucose dehydrogenase activity (marked activation) offer an attractive rationale for the observations of Hastings and Longmore (1965; Longmore *et al.*, 1969) on the stimulatory role of bicarbonate on the conversion of glucose to glycogen in liver slices. As indicated in Fig. 16, inhibition by bicarbonate would reduce two of the four metabolic pathways (hydrolysis and oxidation) available for glucose-6-P, and an increased rate of flow toward liver glycogen formation might reasonably be expected in the presence of this anion. Concomitant stimulation by HCO_3^- of glucose dehydrogenase activity of glucose-6-P dehydrogenase (Anderson and Nordlie, 1968; Horne and Nordlie, 1971) would provide for continued NADPH generation under such circumstances.

4. Other Factors Limiting Accelerated Gluconeogenesis at the Glucose-6-P Phosphohydrolase Level. It is possible that a variety of compounds or factors that may be perturbed significantly from normal in accelerated gluconeogenesis and have been shown to modify in one way or another glucose-6-phosphatase-phosphohydrolase activity are also of physiological relevance in establishing defined upper limits to the rates of release of blood glucose by way of accelerated gluconeogenic processes. Following are some examples.

(a) Diabetic Acidosis. Acidosis (and possibly dehydration) arising for example, in diabetes, may tend to favor phosphotransferase activity of the enzyme over that for glucose-6-P hydrolysis. The marked dependence of phosphotransferase activities noted with phosphoanhydride substrates (ATP, ADP, PP_i) between pH 6 and 7 (Fig. 10*a*) is especially pronounced and is in sharp contrast with the effects of varied pH on glucose-6-P phosphohydrolase activity as indicated in Fig. 10*b*.

(b) Free Fatty Acids and Long-Chain Fatty Acyl Coenzyme A Esters. Levels of these compounds increase significantly under conditions of accelerated gluconeogenesis (Tubbs and Garland, 1964; Tarrant *et al.*,

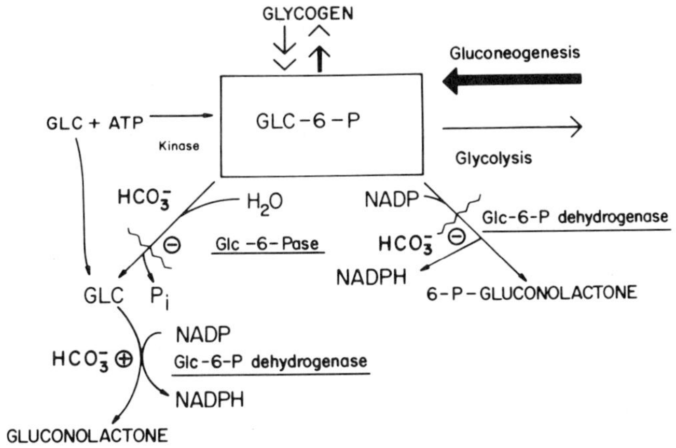

Figure 16. Suggested mechanistic basis for observed stimulatory effects of added bicarbonate on liver glycogen synthesis from gluconeogenic precursors. Inhibitions by bicarbonate are indicated by wavy lines.

1964). Such compounds have been found to exert selective, differential, activity-discriminant effects on glucose-6-phosphatase-phosphotransferase of isolated, otherwise untreated microsomes—certain concentrations, for example, markedly activated phosphotransferase while concomitantly inhibiting significantly (more than 50%, maximally) glucose-6-P phosphohydrolase (Nordlie, *et al.*, 1967). Whether such effects are specific, biologically important phenomena, or simply nonspecific detergent effects, or, when noted with isolated, manufactured fragments of the endoplasmic reticulum, really physiologically relevant at all (Sections IV.E.3), remains to be determined.

(c) Free Amino Acids. Inhibition by free amino acids of glucose-6-P phosphohydrolase appears to involve chelation of enzyme-bound divalent cation (Nordlie and Johns, 1967, 1968). Whether inhibition of glucose-6-phosphatase activity by elevated levels of amino acids as found under conditions of accelerated gluconeogenesis is physiologically important is not clear—the rather high K_i values (>20 mM; Huang, 1969) may contraindicate such a role. Of interest, however, is the independent observation (Cochrane *et al.,* 1956) that the amino acid leucine produces marked hypoglycemic effects.

(d) Citrate. Inhibition by citrate (and isocitrate) of glucose-6-P phosphohydrolase under pronounced acidotic conditions could serve to limit gluconeogenesis at the glucose-6-phosphatase level (Nordlie and Lygre, 1966; Nordlie, 1968).

(e) Variations in Levels of Carbamyl-P and PP_i. A factor of potentially great regulatory significance that still needs much experimental analysis is variation in hepatic carbamyl-P (and PP_i) levels under various metabolic conditions. Accelerated gluconeogenesis is accompanied by a large increase in protein degradation. As the free amino acid pool is thus increased, a marked increase in urea cycle enzyme activity also would be manifest for the removal of ammonia. Carbamyl-P synthesis[11] thus would be dra-

[11] Two types of carbamyl-P synthetase with contrasting properties are established in mammalian tissues. This synthetase of cytosol is believed to produce carbamyl-P for pyrimidine biosynthesis (Tatibaba and Ito, 1967; Hager and Jones, 1967; Nakanishi *et al.*, 1968), whereas the more abundant *N*-acetylglutamate-dependent synthetase of mitochondria generates carbamyl-P for use in the urea cycle (Grisolia and Cohen, 1951; Metzenberg *et al.*, 1957; Cohen, 1962). Natale and Tremblay (1969) have shown that it is physiologically feasible to utilize carbamyl-P synthesized by the latter system for pyrimidine biosynthesis. Lueck and Nordlie (1972) more recently have demonstrated that carbamyl-P generated intramitochondrially may be utilized for extramitochondrial glucose phosphorylation by way of microsomal carbamyl-P:glucose phosphotransferase. Thus both mitochondrial and cytosolic carbamyl-P synthetases must be regarded as potential generators of carbamyl-P for glucose phosphorylation.

matically accelerated, and increased levels of this compound ensuing could influence blood sugar production both through increased competitive inhibition of glucose-6-P hydrolysis and through acceleration in the rate of hepatic glucose rephosphorylation by way of carbamyl-P:glucose phosphotransferase.

(f) Various Anions and Divalent Cations. More than 15 inorganic and organic anions have been observed to inhibit the various activities of this enzyme competitively with respect to phosphate substrates. Presumably, ion pairing with enzyme-bound divalent cation is involved. Although K_i values are somewhat large, several such anions (e.g., Cl^-) may exert physiologically significant effects (Colilla *et al.*, 1974).

Also of potential physiological regulatory importance, Mg^{2+} has been found to exert activity-specific effects (more specifically, phosphate substrate) on activities of this enzyme. Activities with glucose-6-P are unaffected, those with carbamyl-P are modestly reduced, and activities with PP_i or nucleotides are extensively inhibited. Chelation of Mg^{2+} with certain such phosphate substrates appears to be the mechanism involved, for noted extents of inhibition correlate directly with the affinity of the various phosphate compounds for Mg^{2+} (Johnson, *et al.*, 1974).

In ongoing studies, Cu^{2+} has been noted to exert differential, discriminating effects on activities of the enzyme. For example, 1 to 10 μM Cu^{2+} activates carbamyl-P:glucose phosphotransferase while inhibiting glucose-6-P phosphohydrolase (Johnson and Nordlie, 1975).

VI. GLUCONEOGENIC IMPLICATIONS OF SYNTHETIC AND HYDROLYTIC ACTIVITIES OF GLUCOSE-6-PHOSPHATASE OF LIVER, KIDNEY, SMALL INTESTINE, PANCREAS, BRAIN, ADRENALS, AND TESTES: AN ATTEMPTED SYNTHESIS

Studies over the past decade indicate the presence of both hydrolytic and synthetic activities of glucose-6-phosphatase in a variety of vertebrate and invertebrate species [all of 21 species thus far studied with respect to both activities (Herrman and Nordlie, 1972)], in quite a number of organs [liver, kidney, small intestinal mucosal cells, pancreas, adrenals, testes, and brain (see Nordlie, 1971; Colilla and Nordlie, 1973b, 1975)], and in membranes of several organelles within the cell [smooth and rough endoplasmic reticulum, nuclei, probably plasma membrane (Emmelot and Bos, 1969, 1970; Nordlie and Jorgenson, 1975)], and possibly mitochondria (Nordlie and Gunderson, 1973b, 1973c).

This wide distribution, as well as the newly elucidated multifunctionality of the enzyme, strongly support the concept that the physiological roles for this complex biological catalyst extend far beyond the simple hydrolysis of glucose-6-P as the terminal step in gluconeogenesis and glycogenolysis in liver and kidney.

Further considerations in this area are largely speculative; and we have speculated extensively on this topic elsewhere (Nordlie, 1974). Here we summarize briefly some of our current ideas of the integrated function of glucose-6-phosphatase-phosphotransferase activities variously located in the body, as they may contribute to the control of blood glucose levels in health and disease. Figure 17 depicts schematically some interrelationships among relevant tissues and organs in the body. Most of the tissues in which the multifunctional enzyme has been found either contribute directly to blood glucose levels or modify the production or utilization of such blood glucose within the body.

A. Suggested Roles of Multifunctional Glucose-6-Phosphatase-Phosphotransferase in Various Organs

1. Liver. The function of the liver in regulating blood glucose levels through gluconeogenesis, glycogenolysis, and glycogenesis, and postulated roles for hydrolytic and synthetic activities of glucose-6-phosphatase in this overall process, have been considered at length in this chapter and elsewhere (Nordlie, 1974).

2. Kidney. Kidney contributes to blood glucose both through glycogenolytic and gluconeogenic processes and also through tubular reabsorption of this hexose.

On an "activity per gram of tissue" basis, kidney exceeds liver in gluconeogenic capacity. Because of its relatively smaller mass, the kidney's contribution to total glucose formation is much the smaller, however. The same arguments relating to roles and regulation of synthetic and hydrolytic activities of hepatic glucose-6-phosphatase in liver may obtain also for kidney, except that phosphoryl donors other than carbamyl-P, which is not produced in large amounts in renal tissue (Jones *et al.*, 1961), appear to predominate.

The fact that glucose-6-phosphatase is a membrane-bound enzyme that is confined to tissues involved in glucose uptake from or release into extracellular fluid has suggested to a number of authors (Ernster *et al.*, 1962; deDuve, 1962; Stetten and Burnett, 1967; Nordlie and Soodsma, 1966; Nordlie, 1971) that the enzyme may be involved in phosphorylating-dephosphorylating mechanisms for transport of the hexose across cell membranes. Phlorizin has been observed to inhibit activities of this enzyme *in vitro* and *in vivo* by Soodsma *et al.*, (1967) and by Lygre and Nordlie (1969a), sup-

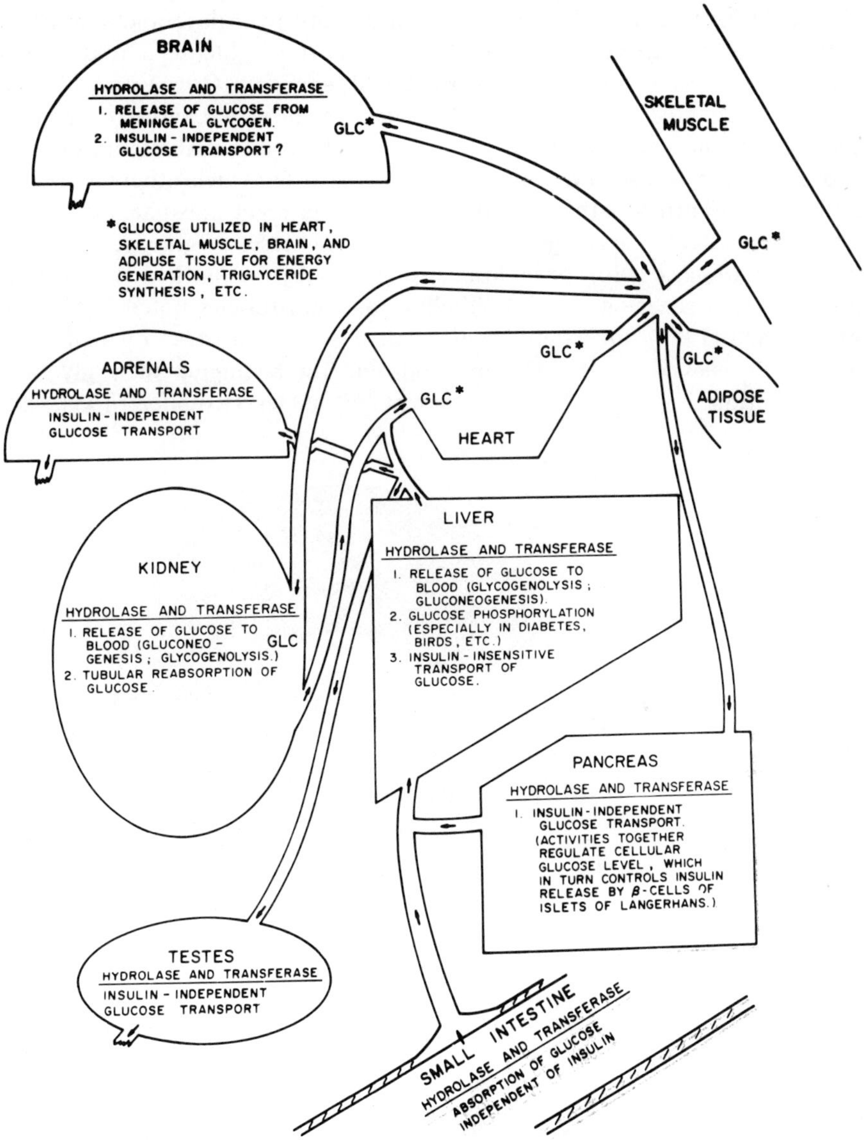

Figure 17. Highly schematic anatomical representation of various sites for the occurrence of multifunctional glucose-6-phosphatase-phosphotransferase in the body, some interrelationships among these sites, and hypothesized (and established) physiological functions for such activities in these tissues and organs.

porting this concept. Also, sulfhydryl groups are involved (Colilla and Nordlie, 1973a). A scheme for participation of both hydrolase and synthetase activities of the enzyme in facilitated transport of glucose across kidney tubular cells, as suggested by Nordlie and Soodsma (1966) and Nordlie (1971), is depicted in Fig. 18. Such a mechanism would serve to promote the maintenance of elevated levels of blood glucose in diabetic animals by promoting the reabsorption of some of this hexose, which otherwise would be lost in the urine of the diabetic animal. Significant phosphorylation, by phosphotransferase activity of glucose-6-phosphatase, of glucose in organs where glucokinase is low or lacking is suggested by the marked pathological glycogen vacuolization of the epithelium of renal tubules noted by Ritchie and Waugh (1957), for example, in severe cases of diabetes in which blood glucose levels exceeded 1000 mg/100 ml. Interestingly, such glycogen vacuolization appears to be prominent in nuclei (Rouiller, 1963, and

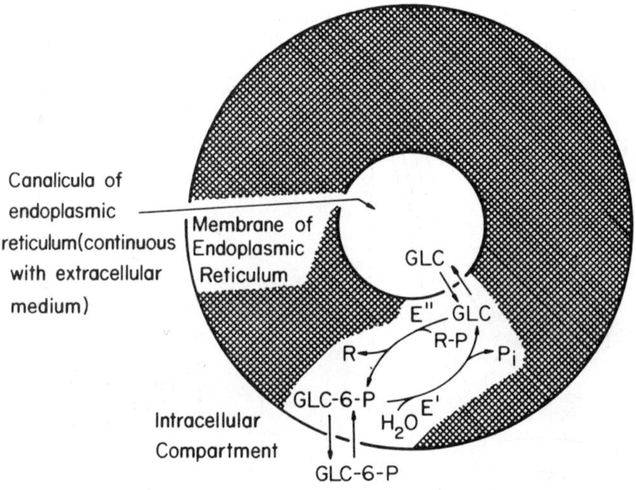

Figure 18. Proposed roles for synthetic and hydrolytic activities of multifunctional glucose-6-phosphatase-phosphotransferase of endoplasmic reticulum in bidirectional transport of glucose between the cell and the blood: E′ and E″, respectively, represent forms of the enzyme functioning predominantly as glucose-6-P phosphohydrolase and as phosphotransferase. Under relatively hypoglycemic conditions, the former is postulated to predominate and to serve to expedite the flow of glucose from the liver (or kidney) cell to the blood by way of the canaliculus of the endoplasmic reticulum. In contrast, under hyperglycemic conditions (e.g., following ingestion of a meal rich in carbohydrate), uptake of glucose from the blood by the liver is hypothesized to be facilitated by initial phosphorylation, by way of E″ action, of glucose within the canaliculus of the endoplasmic reticulum. *Source:* Modified from Nordlie (1971), by permission of Academic Press, Inc. Further details are given in the text and elsewhere (Nordlie, 1971).

references given there) that have been shown in recent studies to possess multifunctional glucose-6-phosphatase in their outer membranes (Gunderson and Nordlie, 1973a, 1975).

*3. **Mucosal Cells of Small Intestine.*** The possible function of glucose-6-phosphatase-phosphotransferase in glucose absorption in the small intestine has been suggested (Lygre and Nordlie, 1968, 1969a). Essential enzymes of gluconeogenesis such as phosphoenolpyruvate carboxykinase and pyruvate carboxylase are absent from this tissue (Utter, 1959, 1961; Utter and Keech, 1963), and alternative functions appear logical (Lygre and Nordlie, 1968). Because of the properties of this multifunctional enzyme and its strategic location, a role in glucose transport, as outlined previously also for the renal tubular enzyme (Section VI.A.1; also Nordlie and Soodsma, 1966; Soodsma *et al.*, 1967; Nordlie, 1968, 1971), appears likely to this author.

*4. **Pancreas.*** The rate of release of insulin from the islets of Langerhans of the pancreas appears to be triggered by levels of free blood glucose (Taylor and Montague, 1970). The idea that pancreatic glucose-6-P phosphohydrolase may be involved in the metabolic regulation of insulin release through its action in liberating free glucose has been explored (Täljedal, 1969; Ashcroft and Randle, 1964); Ashcroft and Randle, but not Täljedal have concluded that the enzyme may play such a role. Involvement of glucose phosphorylation in such processes also has been considered (e.g., Coore and Randle, 1964). Scott and Jones (1970) have demonstrated that bovine pancreas islet cell glucose-6-phosphatase catalyzes PP_i-glucose phosphotransferase as well as phosphohydrolase activity, and they suggest that under various conditions, both types of activity of the enzyme may contribute to net glucose concentrations in the pancreas, hence may be instrumental in regulating insulin release from β-cells.

Very recently Colilla and Nordlie (1973b, 1975) have identified carbamyl-P:glucose phosphotransferase in beef and rat pancreas, thus stengthening the argument for possible involvement of synthetic and hydrolytic activities of glucose-6-phosphatase in a mechanism for ensuring that intracellular concentrations of glucose in pancreatic β-cells reflect extant blood glucose levels, hence convey to sensors within these cells the immediate needs for insulin in the body. This concept is supported by the observation that glucose tolerance curves observed with von Gierke patients resemble those of prediabetics or diabetics (Cornblath and Schwartz, 1966; Nelson *et al.*, 1969). A high initial elevation following ingestion of glucose is noted, with an ensuing slow decrease over several hours. Quite probably, this slow decrease reflects a delayed production and/or release of insulin, which in turn may indicate diminished or absent glucose-6-phosphatase-phosphotransferase in β-cells of the pancreas of the von Gierke patient.

5. ***Brain.*** In an interesting article, Rovainen (1970) has described the presence of glucose-6-phosphatase in meningeal tissue of lamprey larvae and has suggested that the enzyme may play a role in releasing glucose for use by the brain during periods when the lamprey larvae undergo extreme physical stress. Colilla and Nordlie (1973b, 1975) recently have described carbamyl-P: glucose phosphotransferase activity, which is due to glucose-6-phosphatase-phosphotransferase, in rat brain preparations.

6. ***Adrenals and Testes.*** Studies of Colilla and Nordlie (1973b, 1975) have established the presence of both carbamyl-P: glucose phosphotransferase and glucose-6-P phosphohydrolase in rat testes and rat and beef adrenals, both organs in which glucose metabolism is highly important. The possible significance of these observations is considered in Section VI.B.

B. Some Final Speculative Generalizations

Generalizing, we suggest as a working hypothesis that multifunctional glucose-6-phosphatase-phosphotransferase may function as a mechanism for facilitated glucose transport into cells (and perhaps intracellular organelles such as nuclei) of those tissues in which glucose transport independent of insulin (or indeed at an accelerated rate in the absence of insulin) is a physiological necessity.

Certainly, if glucose concentration in pancreatic β-cells is indeed the triggering mechanism for insulin release, these cells must be freely and immediately accessible to blood glucose in a manner totally independent of circulating insulin (Mayhew *et al.*, 1969).

The free permeability of hepatic cells with respect to glucose, independent of insulin requirement, appears to be well established (Cahill *et al.*, 1958a) and is entirely consistent with this organ's role in accelerated glucose production in diabetes or fasting.

Utilization of glucose for energy production in the brain, regardless of the presence or absence of insulin, appears to occur.

In both the adrenal glands and testes, glucose-6-P oxidation for NADPH production and for genesis of acetyl coenzyme A are prerequisites for the organs' important steroidogenic function. Glucocorticoids formed in the adrenals, in turn, possess the ability to modify dramatically hepatic carbohydrate metabolism; stimulation of gluconeogenesis by cortisone, hydrocortisone, and the like is well documented. Indeed, increased synthesis and release of glucocorticoids seems to be dictated in diabetes; under such circumstances continued accessibility by such cells to circulating glucose in the absence of insulin appears to be essential if accelerated steroidogenesis is to occur.

Continued intestinal and kidney renal tubular reabsorption of glucose is necessary if elevated blood glucose levels are to be maintained in untreated diabetes—and they must be maintained if life is to continue under such circumstances (Cahill *et al.*, 1959; Nordlie, 1968, 1969; West *et al.*, 1966).

An enhanced sex drive in the early stages of starvation, where circulating insulin levels are low, would also contraindicate a diminished steroidogenesis in the gonads.

Glucose transport into nuclei within the hepatic or kidney cell appears to be independent of insulin requirement, since this peptide hormone apparently does not enter these cells.

It appears to the author that the exciting glucose-6-phosphatase story is just beginning, with our discovery of its multiplicity of functions and our demonstration that the distribution of this multifunctional, membrane-bound, metabolically important biological catalyst is considerably wider than previously suspected.

VII. EPILOGUE

We hope that the reader has shared with us some of the pleasure and excitement we have experienced over the past 14 years as this continuing saga has begun to unfold.

The story began with our introduction to inorganic pyrophosphate metabolism and the study of gluconeogenesis in the laboratory of Dr. Henry Lardy at the Institute for Enzyme Research in 1960. Any small successes we have experienced over the intervening years are in large part attributable to lessons in philosphy of research (and life) that it was our privilege to experience during two most pleasant and intellectually stimulating years—those spent as an associate of an outstanding gentleman and scientist, Professor Lardy.

We hope that our modest accomplishments as outlined in this chapter succeed in a small way, in saying "Thanks, Hank, for the confidence you always placed in all of us who have had the privilege of being associated with you."

ACKNOWLEDGMENTS

Studies from the author's laboratory described here were supported in part by grants from the Hill Family Foundation; the National Institute of Arthritis, Metabolism, and Digestive Diseases (Research Grant AM 07141); the National Science Foundation; the American Diabetes Association; and the North Dakota Heart Association. The author is deeply indebted to Dr. John Exton for generously making available highly pertinent manuscripts prior to their publication.

ADDENDUM

Very recent work by Katz and his group with isolated hepatic parenchymal cells (Clark *et al.*, 1973; Rognstad *et al.*, 1973) and by Hue and Hers, who studied metabolic flow in intact animals (Hue and Hers, 1974a, 1974b), strongly supports a controlled isotopic flow between hepatic glucose and glucose-6-P (site III, Fig. 2)—so-called futile cycling. The utility of this "futile" process in providing for control of blood glucose through metabolite-sensitive bidirectional hepatic mechanisms has been pointed out by Hue and Hers (1974b), who also have raised serious questions, from the experimental point of view, regarding the existence of comparable recyling at the fructose-6-P/fructose-1,6-P_2 stage (site II, Fig. 2) (Hue and Hers, 1974a). The present author (Nordlie, 1974) has recently considered at length mechanisms that may control blood glucose. Such proposed mechanisms, operative at the hepatic level, involve the regulated, apposed, enzymic synthesis and degradation of glucose-6-P (i.e., finely controlled "futile cycling").

REFERENCES

Adelman, R. C. (1970). *J. Biol. Chem.* **245**, 1032–1035.

Alvares, F. L., Gunderson, H. M., and Nordlie, R. C. (1974). *Fed. Proc. Fed. Am. Soc. Exp. Biol.* **33**, 1559.

Anchors, J. M. and Karnovsky, M. L. (1974). *Fed. Proc. Fed. Am. Soc. Exp. Biol.* **33**, 1410.

Anderson, W. B. and Nordlie, R. C. (1967). *J. Biol. Chem.* **242**, 114–119.

Anderson, W. B. and Nordlie, R. C. (1968). *Biochemistry* **7**, 1479–1485.

Anderson, W. B., Horne, R. N., and Nordlie, R. C. (1968). *Biochemistry* **7**, 3997–4004.

Arion, W. J. and Nordlie, R. C. (1964). *J. Biol. Chem.* **239**, 2752–2757.

Arion, W. J. and Nordlie, R. C. (1965). *Biochem. Biophys. Res. Commun.* **48**, 694–699.

Arion, W. J. and Nordlie, R. C. (1967). *J. Biol. Chem.* **242**, 2207–2210.

Arion, W. J., Wallin, B. K., Carlson, P. W., and Lange, A. J. (1972). *J. Biol. Chem.* **247**, 2558–2565.

Ashcroft, S. J. H. and Randle, P. J. (1964). *Nature* (*Lond.*) **219**, 857–858.

Ashmore, J. and Nesbett, F. B. (1955). *Proc. Soc. Exp. Biol. Med.* **89**, 78–81.

Ashmore, J. and Weber, G. (1959). *Vitam. Horm.* **17**, 91–132.

Ashmore, J., Hastings, A. B., and Nesbett, F. B. (1954). *Proc. Nat. Acad. Sci. U.S.* **40**, 673–678.

Ashmore, J., Hastings, A. B., Nesbett, F. B., and Renold, A. E. (1956). *J. Biol. Chem.* **218,** 77–88.

Atkinson, D. E. (1970). In *The Enzymes,* Vol. 1, 3rd ed. (P. D. Boyer, ed.), Academic Press, New York, pp. 461–489.

Ballard, F. J. (1965). *Comp. Biochem. Physiol.* **14,** 437–443.

Ballard, F. J. and Oliver, I. T. (1964). *Biochem. J.* **92,** 131–136.

Beaufay, H. and deDuve, C. (1954a). *Bull. Soc. Chim. Biol.* **36,** 1525–1537.

Beaufay, H. and deDuve, C. (1954b). *Bull. Soc. Chim. Biol.* **36,** 1551–1568.

Beaufay, H., Hers, H. G., Berthet, J., and deDuve, C. (1954). *Bull. Soc. Chim. Biol.* **36,** 1539–1550.

Berezney, R., Macaulay, L. K., and Crane, F. L. (1972). *J. Biol. Chem.* **247,** 5549–5561.

Bergman, E. N. (1970). In *Dukes' Physiology of Domestic Animals* (M. J. Swenson, ed.), Cornell University Press, Ithaca, N.Y., pp. 595–618.

Blobel, G. and Potter, V. R. (1966). *Science* **154,** 1662–1665.

Borrebaek, B., Holtman, E., Nilsson, L. H., Roch-Nordlund, A. E., and Spdevold, Ø. (1970). *Biochem. Med.* **4,** 469–475.

Bot, G. and Vereb, G. (1966). *Acta Biochim. Biophys. Acad. Sci. Hung.* **1,** 169–177.

Bot, G., Andrassay, K. O., and Kovacs, E. F. (1960). *Acta Physiol. Acad. Sci. Hung.* **17,** 377–381.

Broh-Kahn, R. H. and Mirsky, I. A. (1948). *Arch. Biochem. Biophys.* **16,** 87–107.

Brunengraber, H., Boutry, M., and Lowenstein, J. M. (1973). *J. Biol. Chem.* **248,** 2656–2669.

Byrne, W. L. (1961). In *The Enzymes,* Vol. 5, 2nd ed. (P. D. Boyer, H. Lardy, and K. Myrbäck, eds.), Academic Press, New York, pp. 73–78.

Cahill, G. F., Jr., Ashmore, J., Earle, A. S., and Zottu, S. (1958a). *Am. J. Physiol.* **192,** 491–496.

Cahill, G. F., Jr., Hastings, A. B., Ashmore, J., and Zottu, S. (1958b). *J. Biol. Chem.* **230,** 125–135.

Cahill, G. F., Jr., Ashmore, J., Renold, A. E., and Hastings, A. B. (1959). *Am. J. Med.* **26,** 264–282.

Cater, B. R., Poulter, J., and Hallinan, T. (1970). *FEBS Lett.* **10,** 346–348.

Clark, D. G., Rognstad, R., and Katz, J. (1973). *Biochem. Biophys. Res. Commun.* **54,** 1141–1148.

Cochrane, W. A., Payne, W. W., Simpkiss, M. J., and Woolf, L. I. (1956). *J. Clin. Invest.* **35,** 411–422.

Cohen, P. P. (1962). In *The Enzymes,* Vol. 6, 2nd ed. (P. D. Boyer, H. Lardy, and K. Myrbäck, eds.), Academic Press, New York, pp. 477–494.

Cohn, R. M., Herman, R. H., and Zakim, D. (1969). *Am. J. Clin. Nutr.* **22,** 1204–1210.

Colilla, W. and Nordlie, R. C. (1973a). *Biochim. Biophys. Acta* **309,** 328–338.

Colilla, W. and Nordlie, R. C. (1973b). American Chemical Society, 166th National Meeting, Chicago, September 1973 (*Abstr.* BIOL 52).

Colilla, W. and Nordlie, R. C. (1973c). Unpublished observations.

Colilla, W., Johnson, W. T., and Nordlie, R. C. (1974). *Biochim. Biophys. Acta* **364,** 78–87.

Colilla, W. and Nordlie, R. C. (1975). *Biochim. Biophys. Acta* **377,** 117–125.

Coore, H. G. and Randle, P. J. (1964). *Biochem. J.* **93,** 66–78.

Cori, G. T. (1952–1953). *Harvey Lect.* **48,** 145–171.

Cori, G. T. and Cori, C. F. (1938). *Proc. Soc. Exp. Biol. Med.* **39,** 337–338.

Cori, G. T. and Cori, C. F. (1952). *J. Biol. Chem.* **199,** 661–667.

Cornblath, M. and Schwartz, R. (1966). *Disorders of Carbohydrate Metabolism in Infancy,* Saunders, Philadelphia, p. 124.

Crane, R. (1955). *Biochim. Biophys. Acta* **17,** 443–444.

deDuve, C. (1962). *Ciba Foundation Symposium, Enzymes and Drug Action* p. 505.

deDuve, C. and Beaufay, H. (1951). *Bull. Soc. Chim. Biol.* **33,** 421.

deDuve, C., Berthet, J., Hers, H. G., and Dupret, L. (1949a). *Proceedings of the First International Congress on Biochemistry,* Cambridge. Abstr. Commun., p. 403.

deDuve, C., Berthet, J., Hers, H. G., and Dupret, L. (1949b). *Bull. Soc. Chim. Biol.* **31,** 1242–1253.

diPietro, D. L., Sharma, C., and Weinhouse, S. (1962). *Biochemistry* **1,** 455–461.

Drabkin, D. L. (1948). *Proc. Am. Diabetes Ass.* **8,** 173–212.

Dyson, J. E. D., Anderson, W. B., and Nordlie, R. C. (1969). *J. Biol. Chem.* **244,** 560–566.

Emmelot, P. and Bos, C. J. (1969). *Int. J. Cancer* **4,** 705–722.

Emmelot, P. and Bos, C. J. (1970). *Biochim. Biophys. Acta* **211,** 169–183.

Erickson, R. P., Gluecksohn-Waelsch, S., and Cori, C. F. (1968). *Proc. Nat. Acad. Sci. U.S.* **59,** 437–444.

Ernster, L., Siekevitz, P., and Palade, G. E. (1962). *J. Cell Biol.* **15,** 541–578.

Exton, J. H. (1973). Personal communication.

Exton, J. H., Mallette, L. E., Jefferson, L. S., Wong, E. H. A., Friedmann, N., Miller, T. B., Jr., and Park, C. R. (1970). *Recent Progr. Horm. Res.* **26,** 411–461.

Exton, J. H., Harper, S. C., Tucker, A. L., and Ho, R.-J. (1973a). *Biochim. Biophys. Acta* **329,** 23–40.

Exton, J. H., Harper, S. C., Tucker, A. L., Flagg, J. L., and Park, C. R. (1973b). *Biochim. Biophys. Acta* **329,** 41–57.

Fantl, P. and Rome, M. N. (1945). *Aust. J. Exp. Biol. Med. Sci.* **23,** 20–27.

Fisher, C. J. and Stetten, M. R. (1966). *Biochim. Biophys. Acta* **121,** 102–109.

Freedland, R. A., and Harper, A. E. (1958). *J. Biol. Chem.* **230,** 3338–3342.

Friedmann, B., Goodman, E. H., Jr., and Weinhouse, S. (1967). *Endocrinology,* **81,** 486–496.

Garland, R. C. and Cori, C. F. (1972). *Biochemistry* **11,** 4712–4718.

Gordon, E. R. (1963). *Can. J. Biochem. Physiol.* **41,** 1611–1620.

Greengard, O. (1969a). *Biochem. J.* **115,** 19–24.

Greengard, O. (1969b). *Science* **163,** 891–895.

Grisolia, S. and Cohen, P. P. (1951). *J. Biol. Chem.* **191,** 189–202.

Grossbard, L. and Schimke, R. T. (1966). *J. Biol. Chem.* **241,** 3546–3560.

Gunderson, H. M. and Nordlie, R. C. (1973). *Biochem. Biophys. Res. Commun.* **52,** 601–607.

Gunderson, H. M. and Nordlie, R. C. (1975). *J. Biol. Chem.* **250,** 3552–3559.

Hager, S. E. and Jones, M. E. (1967). *J. Biol. Chem.* **242,** 5674–5680.

Hanson, T. L., Lueck, J. D., Horne, R. N., and Nordlie, R. C. (1970). *J. Biol. Chem.* **245,** 6078–6084.

Hass, L. F. and Byrne, W. L. (1960). *J. Am. Chem. Soc.* **82,** 947–954.

Hastings, A. B. and Longmore, W. J. (1965). *Adv. Enzyme Regul.* **3,** 147–159.

Herrman, J. L. and Nordlie, R. C. (1972). *Arch. Biochem. Biophys.* **152,** 180–186.

Herrman, J. L. and Nordlie, R. C. (1973). Unpublished observations.

Herrman, J. L., Nordlie, P. E., and Nordlie, R. C. (1971). *FEBS Lett.* **18,** 241–244.

Hers, H. G. and deDuve, C. (1950). *Bull. Soc. Chim. Biol.* **32,** 20–29.

Hers, H. G., Berthet, J., Berthet, L., and deDuve, C. (1951). *Bull. Soc. Chim. Biol.* **33,** 21–41.

Horne, R. N. and Nordlie, R. C. (1971). *Biochim. Biophys. Acta* **242,** 1–13.

Huang, S.-M. (1969). Master's thesis, University of North Dakota.

Hue, L. and Hers, H. G. (1974a). *Biochem. Biophys. Res. Commun.* **58,** 532–539.

Hue, L. and Hers, H. G. (1974b). *Biochem. Biophys. Res. Commun.* **58,** 540–548.

Illingworth, B. and Cori, C. F. (1965). *Biochem. Biophys. Res. Commun.* **19,** 10–14.

Johnson, W. T. and Nordlie, R. C. (1973). Unpublished observations.

Johnson, W. T., Colilla, W., and Nordlie, R. C. (1974). *Arch. Biochem. Biophys.* **163,** 297–305.

Johnson, W. T. and Nordlie, R. C. (1975). *Fed. Proc. Fed. Am. Soc. Exp. Biol.* **34,** 575.

Jones, M. E., Anderson, A. D., Anderson, C., and Hudes, S. (1961). *Arch. Biochem. Biophys.* **95,** 499–507.

Jorgenson, R. A. and Nordlie, R. C. (1973). Unpublished observations.

Kalckar, H. M. (1937). *Enzymologia* **2,** 47–52.

Kanamura, S. (1971). *J. Histochem. Cytochem.* **19,** 320–321.

Kanamura, S. (1973). *J. Histochem. Cytochem.* **21,** 1086–1089.

Kashnig, D. M. and Kasper, C. B. (1969). *J. Biol. Chem.* **244,** 3786–3792.

Katzen, H. M. and Schimke, R. T. (1965). *Proc. Nat. Acad. Sci. U.S.* **54,** 1218–1225.

Kilsheimer, G. S., Weber, D. R., and Ashmore, J. (1960). *Proc. Soc. Exp. Biol. Med.* **104,** 515–518.

Krebs, H. A. (1954). *Bull. Johns Hopkins Hosp.* **95,** 19–33.

Lardy, H. A. (1965). *Harvey Lect.* **60,** 261–278.

Longmore, W. J., Niethe, C. M., and McDaniel, M. J. (1969). *J. Biol. Chem.* **244,** 6451–6457.

Lueck, J. D. and Nordlie, R. C. (1970). *Biochem. Biophys. Res. Commun.* **39,** 190–196.

Lueck, J. D. and Nordlie, R. C. (1972). *FEBS Lett.* **20,** 195–198.

Lueck, J. D., Herrman, J. L., and Nordlie, R. C. (1972). *Biochemistry* **11,** 2792–2799.

Lundsgaard, E. (1933). *Biochem. Z.* **264,** 209–220.

Leskes, A., Siekevitz, P., and Palade, C. E. (1971). *J. Cell Biol.* **49,** 264–287.

Lygre, D. G. and Nordlie, R. C. (1968). *Biochemistry* **7,** 3219–3226.

Lygre, D. G. and Nordlie, R. C. (1969a). *Biochim. Biophys. Acta* **185,** 360–366.

Lygre, D. G. and Nordlie, R. C. (1969b). *Biochim. Biophys. Acta* **178,** 389–391.

Maley, F. and Lardy, H. A. (1956). *J. Am. Chem. Soc.* **78,** 1393–1397.

Mayhew, D. A., Wright, P. H., and Ashmore, J. (1969). *Pharm. Rev.* **21,** No. 3, 183–211.

McCandless, E. L., Woodward, B. A., and Dye, J. A. (1948). *Am. J. Physiol.* **154,** 94–106.

McCraw, E. F. (1968). *Metab. Clin. Exp.* **17,** 833–837.

McCraw, E. F., Peterson, M. J., and Ashmore, J. (1967). *Proc. Soc. Exp. Biol. Med.* **126,** 232–236.

Metzenberg, R. L., Hall, L. M., Marshall, M., and Cohen, P. P. (1957). *J. Biol. Chem.* **229,** 1019–1029.

Molstad, C. and Nordlie, R. C. (1973). Unpublished observations.

Morton, R. K. (1955). *Disc. Faraday Soc.* **20,** 149–156.

Nakanishi, S., Ito, K., and Tatibana, M. (1968). *Biochem. Biophys. Res. Commun.* **33,** 774–781.

Natale, P. J. and Tremblay, G. C. (1969). *Biochem. Biophys. Res. Commun.* **37,** 512–516.

Nelson, W. E., Vaughan, V. C., and McKay, R. J. (1969). *Textbook of Pediatrics,* 9th ed., Saunders Philadelphia, pp. 443–444.

Nelson, P., Wagle, S. R., and Ashmore, J. (1970). *Metab, Clin. Exp.* **19,** 349–353.

Newsholme, E. A., and Gevers, W. (1967). *Vitam. Horm.* **25,** 1–87.

Nordlie, R. C. (1968). In *Control of Glycogen Metabolism* (W. J. Whelan, ed.), Academic Press, New York, pp. 153–168.

Nordlie, R. C. (1969). *Ann. N.Y. Acad. Sci.* **166,** 699–718.

Nordlie, R. C. (1971). In Vol. 4, 3rd ed. *The Enzymes,* (P. D. Boyer, ed.), Academic Press, New York, pp. 543–609.

Nordlie, R. C. (1973). *Guidelines to Metabolic Therapy* (Upjohn Co., Kalamazoo, Mich.). Vol. 2, No. 2, pp. 1–3.

Nordlie, R. C. (1974). In *Current Topics in Cellular Regulation,* Vol. 8 (B. L. Horecker and E. R. Stadtman, eds.), Academic Press, New York, pp. 33–117.

Nordlie, R. C. and Arion, W. J. (1964a). *J. Biol. Chem.* **239,** 1680–1685.

Nordlie, R. C. and Arion, W. J. (1964b). *Fed. Proc. Fed. Am. Soc. Exp. Biol.* **23,** 534.

Nordlie, R. C. and Arion, W. J. (1965). *J. Biol. Chem.* **240,** 2155–2164.

Nordlie, R. C. and Arion, W. J. (1966). *Methods Enzymol.* **9,** 619–625.

Nordlie, R. C. and Gehring, A. W. (1963). *Proc. N. Dak. Acad. Sci.* **17,** 13.

Nordlie, R. C. and Gunderson, H. M. (1973a). Unpublished observations.

Nordlie, R. C. and Gunderson, H. M. (1973b). *Abstracts of the Ninth International Congress on Biochemistry,* Stockholm, July 1–7, p. 377.

Nordlie, R. C. and Gunderson, H. M. (1973c). *Fed. Proc. Fed. Am. Soc. Exp. Biol.* **32,** 514 Abs.

Nordlie, R. C. and Johns, P. T. (1967). *Biochem. J.* **104,** 37*P*–38*P*.

Nordlie, R. C. and Johns, P. T. (1968). *Biochemistry* **7,** 1473–1479.

Nordlie, R. C. and Jorgenson, R. A. (1975). In *The Enzymes of Biological Membranes* (A. Martonosi, ed.), Plenum Press, New York. In press.

Nordlie, R. C. and Lardy, H. A. (1961). *Biochem. Biophys. Acta* **53,** 309–323.

Nordlie, R. C. and Lygre, D. G. (1966). *J. Biol. Chem.* **241,** 3136–3141.

Nordlie, R. C. and Snoke, R. E. (1967). *Biochim. Biophys. Acta* **148,** 222–232.

Nordlie, R. C. and Soodsma, J. F. (1966). *J. Biol. Chem.* **241,** 1719–1724.

Nordlie, R. C., Arion, W. J., and Glende, E. A., Jr. (1965). *J. Biol. Chem.* **240,** 3479–3484.

Nordlie, R. C., Hanson, T. L., and Johns, P. T. (1967). *J. Biol. Chem.* **242,** 4144–4148.

Nordlie, R. C., Arion, W. J., Hanson, T. L., Gilsdorf, J. R., and Horne, R. N. (1968a). *J. Biol. Chem.* **243,** 1140–1146.

Nordlie, R. C., Gilsdorf, J. R., Horne, R. N., and Paur, R. J. (1968b). *Biochim. Biophys. Acta* **158,** 157–159.

Nordlie, R. C., Hanson, T. L., Johns, P. T., and Lygre, D. G. (1968c). *Proc. Nat. Acad. Sci. U.S.* **60,** 590–597.

Nordlie, R. C., Lueck, J. D., Hanson, T. L., and Johns, P. T. (1971). *J. Biol. Chem.* **246,** 4807–4812.

Ostern, P., Herbert, D., and Holmes, E. (1939). *Biochem. J.* **33,** 1858–1878.

Pandhi, P. N. and Baum, H. (1967). *Nature* (*Lond.*) **216,** 1324–1325.

Pandhi, P. N. and Baum, H. (1970). *Life Sci.* **9,** Part II, 87–92.

Pantelouris, E. M. (1967). *Introduction to Animal Physiology and Physiological Genetics*, Pergamon Press, New York, pp. 316–321.

Parvin, R. and Smith, R. A. (1969). *Biochemistry*, **8,** 1748–1755.

Pollak, J. K., Malor, R., Morton, M., and Ward, K. A. (1971). In *Autonomy and Biogenesis of Mitochondria and Chloroplasts* (N. K. Boardman, A. W. Linnane, and R. M. Smillie, eds.), North-Holland Press, Amsterdam, pp. 27–41.

Pollak, J. K. and Shorey, C. D. (1968). *Dev. Biol.* **17,** 536–543.

Proctor, E. B. and Nordlie, R. C. (1972). *Arch. Int. Physiol. Biochim.* **80,** 461–467.

Rafter, G. W. (1958). *J. Biol. Chem.* **230,** 643–648.

Rafter, G. W. (1960). *J. Biol. Chem.* **235,** 2475–2477.

Raijman, L. (1974). *Biochem. J.* **138,** 225–232.

Renold, A. E., Hastings, A. B., Nesbett, F. B., and Ashmore, J. (1955). *J. Biol. Chem.* **213,** 135–146.

Ritchie, S. and Waugh, D. (1957). *Am. J. Pathol.* **33,** 1035–1057.

Rognstad, R., Clark, D. G., and Katz, J. (1973). *Biochem. Biophys. Res. Commun.*, **54,** 1149–1156.

Rose, I. A. and Warms, J. V. B. (1967). *J. Biol. Chem.* **242,** 1635–1645.

Roseman, S. (1969). *J. Gen. Physiol.* **54,** 138s–180s.

Rosen, S. I. (1970a). *Acta Histochem.* **36,** 44–53.

Rosen, S. I. (1970b). *Acta Histochem.* **38,** 199–217.

Rouiller, C. (1963). *The Liver. Morphology, Biochemistry, Physiology*, Vol. 1, Academic Press, New York, p. 146.

Rovainen, C. M. (1970). *Science* **167,** 889–890.

Sacks, J. (1953). *Am. J. Physiol.* **172,** 93–99.

Salas, M., Viñuela, E., and Sols, A. (1963). *J. Biol. Chem.* **238,** 3535–3538.

Scott, D. B. M. and Jones, G. (1970) In *Enzymes and Isozymes* (D. Shugar, ed.), Academic Press, New York, Abstr. 364.

Segal, H. L. (1959). *J. Am. Chem. Soc.* **81,** 4047–4050.

Segal, H. L. and Washko, M. E. (1959). *J. Biol. Chem.* **234,** 1937–1941.

Sharma, C., Manjeshwar, R., and Weinhouse, S. (1963). *J. Biol. Chem.* **238,** 3840–3845.

Snoke, R. E. and Nordlie, R. C. (1967). *Biochim. Biophys. Acta* **139,** 190–192.

Snoke, R. E. and Nordlie, R. C. (1972). *Biochim. Biophys. Acta* **258,** 188–205.

Soodsma, J. F. and Nordlie, R. C. (1969). *Biochim. Biophys. Acta* **191,** 636–643.

Soodsma, J. F. Legler, B., and Nordlie, R. C. (1967). *J. Biol. Chem.* **242,** 1955–1960.

Soskin, S., Essex, H. E., Herrick, J. P., and Munn, F. C. (1938). *Am. J. Physiol.* **124,** 558–567.

Stadtman, E. R. (1966). *Adv. Enzymol.* **28,** 41–154.

Steele, R. (1966). *Ergeb. Physiol. Biol. Chem. Exp. Pharmakol.* **57,** 91–189.

Stetten, M. R. (1964a). *Fed. Proc. Fed. Am. Soc. Exp. Biol.* **23,** 535.

Stetten, M. R. (1964b). *J. Biol. Chem.* **239,** 3576–3583.

Stetten, M. R. (1965). *J. Biol. Chem.* **240,** 2248–2253.

Stetten, M. R. and Burnett, F. F. (1967). *Biochim. Biophys. Acta* **139,** 138–147.

Stetten, M. R. and Goldsmith, P. K. (1974). *Biochim. Biophys. Acta* **327,** 82–91.

Stetten, M. R. and Kehoe, J. J. (1971). *Biochim. Biophys. Acta* **250,** 501–513.

Stetten, M. R. and Rounbehler, D. (1968). *J. Biol. Chem.* **243,** 1823–1832.

Stetten, M. R. and Taft, H. L. (1964). *J. Biol. Chem.* **239,** 4041–4046.

Sturkie, P. D. (1965). *Avian Physiology,* 2nd ed., Cornell University Press, Ithaca, N.Y., p. 43.

Swanson, M. A. (1949). *Fed. Proc. Fed. Am. Soc. Exp. Biol.* **8,** 258.

Swanson, M. A. (1950). *J. Biol. Chem.* **184,** 647–659.

Swanson, M. A. (1955). *Methods Enzymol.* **2,** 541–543.

Täljedal, I.-B. (1969). *Biochem. J.* **114,** 387–394.

Tarrant, M. E. Mahler, R., and Ashmore, J. (1964). *J. Biol. Chem.* **239,** 1714–1719.

Tatibaba, M. and Ito, K. (1967). *Biochem. Biophys. Res. Commun.* **26,** 221–227.

Taylor, K. W. and Montague, W. (1970). In *Pathogenesis of Diabetes Mellitus: Nobel Symposium 13* (E. Cerasi and R. Luft, eds.), Wiley, New York, pp. 123–135.

Tubbs, P. K. and Garland, P. B. (1964). *Biochem. J.* **93,** 550–557.

Utter, M. F. (1959). *Ann. N.Y. Acad. Sci.* **72,** 451–461.

Utter, M. F. (1961) In *The Enzymes,* Vol. 5, 2nd ed. (P. D. Boyer, H. Lardy, and K. Myrbäck, eds.), Academic Press, New York, pp. 319–340.

Utter, M. F. and Keech, D. B. (1963). *J. Biol. Chem.* **238,** 2603–2608.

Vianna, A. L. and Nordlie, R. C. (1969). *J. Biol. Chem.* **244,** 4027–4032.

Viñuela, E., Salas, M., and Sols, A. (1963). *J. Biol. Chem.* **238,** PC1175–PC1177.

Walker, D. G. (1963). *Adv. Enzyme Regul.* **3,** 163–184.

Walker, D. G. and Parry, M. J. (1966). *Methods Enzymol.* **9,** 381–392.

Wallace, J. C. and Newsholme, E. A. (1967). *Biochem. J.* **104,** 378–384.

Weber, G. and Cantero, A. (1955). *Cancer Res.* **15,** 105–108.

Weber, G. and Cantero, A. (1957). *Endocrinology* **61,** 701–712.

West, E. S., Todd, W. R., Mason, H. S., and Van Bruggen, J. T. (1966). *Textbook of Biochemistry,* 4th ed., Macmillan, New York, pp. 1114–1121.

Williamson, J. R. (1967). *Adv. Enzyme Regul.* **5,** 229–255.

Willms, B., Ben-Ami, P., and Söling, H. D. (1970). *Horm. Metab. Res.* **2,** 135–141.

Young, D. A. (1966). *Arch. Biochem. Biophys.* **114,** 309–313.

Zakim, D. (1970). *J. Biol. Chem.* **245,** 4953–4961.

Zerr, C. and Novoa, W. B. (1968). *Biochem. Biophys. Res. Commun.* **32,** 129–133.

CHAPTER FOUR

Apparent K_ms and Apparent Concentrations: An Apparent Conundrum

PAUL A. SRERE

I. INTRODUCTION

Lardy (1965) has shown that gluconeogenesis proceeds in certain species by a pathway in which mitochondrial pyruvate is converted to oxalacetate (by way of an acetyl CoA activated enzyme, pyruvate carboxylase), which is then transported to the cytosol either as malate or aspartate. It is clear that the free concentrations of mitochondrial oxalacetate and acetyl CoA are potentially important in controlling the rate of gluconeogenesis. Since they are also substrates for citrate synthase, their concentrations are important for control of the Krebs cycle as well. It is a goal of enzymologic research to be able to explain and predict the metabolic behavior of cells and tissues using data gathered from studies on pure enzymes. In the case of citrate synthase of rat liver we now have enough metabolic and enzymologic data to determine whether there is agreement between the two types of information.

I have calculated the rate of citrate synthase in rat liver using the rate and binding data obtained on the pure enzyme (Moriyama and Srere, 1971; Matsuoka and Srere, 1973) combined with data on its substrate concentrations in rat liver calculated from data by Krebs and Veech (1970) and Veech and Guynn (1972) based on the "near-equilibrium" assumption of Hohorst *et al.* (1959) and D. H. Williamson *et al.* (1967). This calculated rate is much lower than the rate of the citrate synthase reaction *in situ,* calculated from the Q_{O_2} of the tissue. This chapter outlines the problem and discusses possible reasons for this discrepancy.

II. RAT LIVER CITRATE SYNTHASE (APPARENT K_ms)

We have purified four different rat citrate synthases: heart, liver, kidney, and brain (Moriyama and Srere, 1971; Matsuoka and Srere, 1973). The four enzymes have the same molecular weight (10^5) and are composed of two identical subunits. Their behavior in polyacrylamide gel electrophoresis is identical, and the kinetic behavior that has been compared among the four enzymes also has been identical. Antibodies were prepared to the rat heart enzyme and neutralization curves for the other enzymes were identical to that for rat heart enzyme. Precipitin lines on enzyme antibody interaction in Ouchterlony plates showed that the four enzymes behaved identically in that continuous lines appeared without spur formation. We have concluded on these bases that the four enzymes were identical (Moriyama and Srere, 1971; Matsuoka and Srere, 1973).

Kinetics for limited substrate ranges in the forward direction have been determined for all four enzymes. The reverse reaction has been studied with the brain and kidney enzyme, and a complete kinetic study has been performed with the kidney enzyme. The results of such kinetic studies, along with kinetic studies with a variety of inhibitors, indicated that a rapid equilibrium random mechanism with two dead-end inhibitors could account for the observed kinetics (Moriyama and Srere, 1971; Matsuoka and Srere, 1973) (Fig. 1).

This mechanism is different from the one proposed by Weidman *et al.* (1973) for pig heart citrate synthase. Those workers proposed an ordered mechanism with OAA adding to the enzyme first. Their mechanism was based on electron spin resonance (ESR) changes of a bound spin labeled AcCoA analog observed in the presence of various substrate analogs. We assume that the mechanism derived from initial-velocity–steady-state assumptions (rapid equilibrium, random) is the correct one.

Table I gives the kinetic constants for the substrates of rat kidney citrate synthase. When used in the Haldane equation, these figures yield a value for

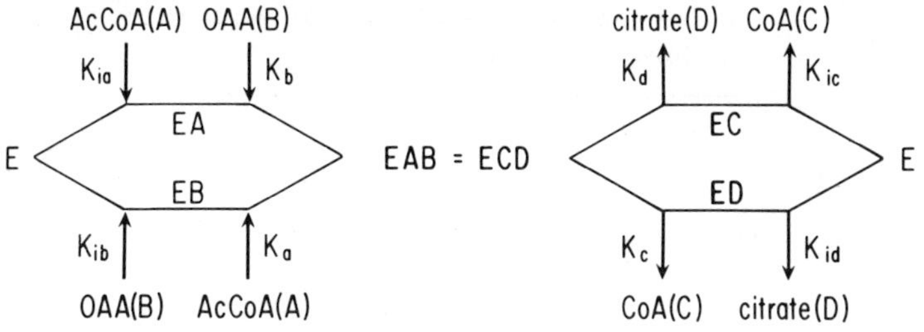

$$EAcCoA + cit = EAcCoA \cdot cit$$

$$EOAA + CoA = EOAA \cdot CoA$$

Figure 1

the equilibrium constant of the reaction of 1.8×10^7 that is close to 1.1×10^7 (pH 8), obtained by Guynn *et al.* (1973).

A peculiar kinetic behavior is observed for acetyl CoA with rat citrate synthase; the K_m for AcCoA changes with the concentration range of AcCoA studied. The K_ms for acetyl CoA at various acetyl CoA concentrations are shown in Table II. When a Lineweaver-Burk plot of these data is made, the changes in K_m appear to occur at discrete substrate concentrations; as a result, these plots are not straight lines but contain at least one abrupt change of slope. Kinetic behavior of this kind (i.e., abrupt breaks in the Lineweaver-Burk plot) has been reported for a number of different enzymes (Dalziel and Engel, 1968). Engel and Ferdinand (1973) present a model of enzyme action in which both negative and positive cooperativity exist; such a model exhibits similar kinetic behavior. Down to the 1 μM range no similar abrupt breaks in Lineweaver-Burk plots are obtained when OAA is the variable substrate. It is clear that if V_{max} and K_m for AcCoA both change, the Haldane relationship will still be valid.

The K_ms for the substrates of rat citrate synthase are very closely equal to the K_{DISS} for those substrates. This unusual situation is related to the fact that intersection points in reciprocal plots are on the X-axis. When this occurs and the mechanism is a random one; it has been demonstrated by Lueck and Fromm (1973) that $K_m = K_{DISS}$. This has been verified for OAA in the citrate synthase reaction using an independent method to measure K_{DISS} for OAA that was found to be near in value to its kinetically determined K_m.

A large number of naturally occurring intermediates have been shown to affect the activity of animal citrate synthases. These include ATP (Hathaway and Atkinson, 1965), succinyl CoA (LaNoue *et al.,* 1972), propionyl CoA (Srere, 1966), acetoacetyl CoA (Srere and Matsuoka, 1972)

Table I. Kinetic constants[a] for rat citrate synthase

	μM
K_a	5.0 ±0.4
K_{ia}	5 ±1
K_b	4.5 ±0.5
K_{ib}	4.5 ±0.7
K_c	39 ±4
K_{ic}	56 ±9
K_d	3000 ±360
K_{id}	4300 ±600

$V_f = 3 \times 10^4$ moles/min/mole enzyme

$V_r = 12.5$ moles/min/mole enzyme

$$K_{eq} = \frac{V_f K_c K_{id}}{V_r K_{ia} K_b} = 1.8 \times 10^7$$

[a] Subscripts as follows: a = acetyl CoA, b = oxalacetate, c = CoA, d = citrate.

NADH (Srere *et al.*, 1973a), NADPH (Srere *et al.*, 1973a), which are all competitive for acetyl CoA and malate, and α-ketoglutarate and isocitrate (Srere *et al.*, 1973a), which are competitive for oxalacetate (see Table III for some K_i values). It is apparent that these are inhibitory because of their structural similarities to the substrates for citrate synthase, not because of special "regulatory" interactions with it. To date no convincing evidence has been presented to indicate which of these compounds, if any, par-

Table II. K_ms for acetyl CoA and OAA at several concentration ranges of acetyl CoA

AcCoA concentration range (μM)	V_{max} [a]	K_a (μM)	K_b (μM)
> 50	3.2	11.8 ±0.9	4.8 ±0.3
5–50	2.9	5.0 ±0.4	4.5 ±0.5
< 5	1.1	1.3 ±0.1	3.0 ±0.2

[a] V_{max}, nmole/min/10^{-8} g.

Table III. Inhibitors of citrate synthase

	K_i
	(Competitive against AcCoA)
ATP	0.1–0.7 mM (concentration dependent)
Propionyl CoA	6.5 μM
NADH	1.5 mM
NADPH	0.8 mM
	(Competitive against OAA)
Malate	16 mM
α-KG	10 mM

ticipates in citrate synthase regulation; thus we ignore their possible roles in the control of this activity.

The pH dependence and temperature dependence of the synthase activity has been studied thoroughly only with the pig heart enzyme. The results indicate a pH optimum of about 8.1 (rather broad) and a temperature dependence between 28 and 37°C of 5.7%/deg or about 51% increase for that change (Kosicki, 1961; Kosicki and Srere, 1961). In the absence of specific data on the rat enzyme, these values for the pig heart enzyme are used.

III. CONCENTRATION OF OXALACETATE IN RAT LIVER (APPARENT CONCENTRATION)

With the pertinent kinetic information available for rat liver citrate synthase, if the free concentrations of its substrates were known it would be possible to calculate the rate of citrate synthesis and compare this to the rate of Krebs cycle activity of rat liver as estimated from its Q_{O_2}.

The value for free concentration of oxalacetate in rat liver mitochondria has been estimated by a number of laboratories, and most of the values are in agreement at about 10^{-7} to 10^{-8}M (J. R. Williamson, 1965; Gumaa *et al.*, 1971; Krebs and Veech, 1970). This calculation uses the equilibrium constant of the malate dehydrogenase reaction as 2×10^{-5} and the values of malate concentration of 0.3 mM and NAD/NADH ratios of 7 (Krebs and Veech, 1970), determined from experiments with freeze-clamped normal rat liver, after making near-equilibrium assumptions. Then $[OAA] = K_{eq}$ [malate] [NAD]/[NADH] $= 2 \times 10^{-5} [3 \times 10^{-4}] (7) = 4.2 \times 10^{-8}$ M, or about

$10^{-2}K_m$ (citrate synthase). Even if one used the total oxalacetate concentration of mitochondria as about 1 μM, the *free* concentration would be much lower because of the high concentrations (> μM) of enzymes that bind oxalacetate in the mitochondria (Srere, 1969b). No accurate estimate for free acetyl CoA concentration is available.

IV. CALCULATED AND ACTUAL RATE OF KREBS CYCLE IN LIVER (APPARENT CONUNDRUM)

Using the free concentration of OAA and assuming acetyl CoA to be saturating, it is possible to estimate the fraction of V_{max} activity of citrate synthase just due to one lowered substrate concentration. If we assume that the relation between the velocity and substrate concentration is approximately linear when S $\ll K_m$, so that $V = V_{max}/2$ ([S]/K_m), then at [oxalacetate] = 10^{-2} K_m, then $V = V_{max}$ (0.5×10^{-2}). The Q_{O_2} of rat liver tissue expressed as micromoles of C_2 utilized per minute per gram of tissue at 25°C is 0.92 (Srere, 1969a). The V_{max} of citrate synthase (pH 8.1) is 14 μmole/min/g tissue at 25°C. Therefore the actual rate of citrate synthase in liver = V_{max} (0.65×10^{-1}), or about one order of magnitude higher than our calculated rate. Corrections due to other factors such as pH, ionic strength, viscosity, effectors (all known ones are negative), and corrections necessitated because [E] = [K_m] (Cha, 1970) would all tend to lower the calculated velocity and to make the discrepancy greater. This all also assumes that acetyl CoA is in saturating concentrations, a matter that is not yet certain (Veech and Guynn, 1972).

V. POSSIBLE RESOLUTIONS OF THE PROBLEM

We can look for the reasons for this discrepancy in two places: either V_{max} and K_m are different for the isolated enzyme and for the enzyme *in situ,* or the calculated apparent free concentration of oxalacetate is not the concentration *seen by the enzyme.* The kinetic constants observed for the rat citrate synthase satisfy the Haldane relationship. This is an expression of a thermodynamic property of the reaction and cannot be different for the *in situ* and free pure enzyme. This means if we propose that the K_m is lower *in situ* either the V_{max} would have to be lower or the K_m for acetyl CoA would be higher or those of the products lower. These conditions would all tend to limit the citrate synthase reaction for other reasons, and it is unlikely therefore that our dilemma can be resolved on this basis.

We are left with the proposition of abandoning the results obtained from near-equilibrium assumptions. This is not a happy prospect, since apparent consistency is obtained when several cytosolic and mitochondrial redox reactions are used as markers. The likelihood of obtaining agreement by chance with three separate enzyme redox-couples is small. If we assume that these calculations give an average redox for many different microenvironments, we are able to construct a model that gives a Krebs cycle working at the rate observed. This model includes the assumption that the Krebs cycle enzymes are arranged as a multienzyme system next to or on the inner surface of the inner mitochondrial membrane (Srere, 1972), permitting locally high concentrations of intermediates to exist in the vicinity of each active site, as they are thought to do in conventional multienzyme systems.

We cannot just say that the oxalacetate concentration in the environment of citrate synthase is near its K_m for that enzyme (10^{-6} M) because then the environment for malate dehydrogenase must contain that same concentration of OAA. It is clear that the MDH reaction ($K_{eq} = 2 \times 10^{-5}$) cannot operate in the forward direction if [OAA] = 10^{-6} M and NAD/NADH = 7. Thus if we postulate 10^{-6} M OAA, we are forced to assume either a very high malate concentration (30 mM) or that a free redox potential in the microenvironment of MDH is much different from a value of 7. A different redox value can be attained if the formed NADH immediately transfers electrons to the inner membrane and thus maintains a favorable redox potential that allows MDH to operate.

It is not particularly satisfying (given our penchant to follow Occam's dictum) to proliferate special intracellular conditions such as the specification of the microenvironment of each enzyme. However, compartmentation has been demonstrated on tissue, cell, organelle, and multienzyme levels and is believed to be a necessity for the orderly and controlled operation of cellular reactions. It is not too great an extension of these ideas to believe that compartmentation exists even for enzymes whose interactions are too weak to demonstrate.

It is difficult at present to design experiments to test for the existence of these postulated microenvironments. Model systems of immobilized enzymes, however, provide a means of constructing microenvironments with multienzyme systems. This makes it possible to determine the kinetic differences between the immobilized multienzyme system and a corresponding free enzyme system in a single environment. Thus malate dehydrogenase and citrate synthase immobilized on Sepharose (or trapped in acrylamide) were compared with the corresponding free enzymes (Srere *et al.*, 1973b). It was shown that at limiting malate concentrations the immobilized enzyme system, where the average distance between enzymes was smaller than in a free system, operated at about twice the rate as the comparable system of free enzymes. Furthermore, when lactate dehydrogenase was immobilized in the same system and pyruvate added as an electron acceptor, rate enhancements of fourfold over the free system were measured (Srere *et al.*, 1973b).

Utter (1969) has indicated the central role that oxalacetate seems to occupy in metabolism. Its importance appears to be even greater when we see that its total concentration is so much lower than that of each of the other intermediates in glycolysis, gluconeogenesis, Krebs cycle, and the pentose shunt pathways. It was not possible before the advent of the "near-equilibrium" experiments and calculations to even guess at the free concentration of the various intermediates in the various cellular compartments. These experiments and assumptions have indicated, not surprisingly, that the *free* concentration of oxalacetate is *much* lower than its total concentration; thus it appears to be rate limiting for pathways such as gluconeogenesis and citrate oxidation. There is at present no alternate method for obtaining information on the free concentration of intermediates, and it is surprising that the test of its validity by the calculations I have presented here has failed to find the apparent free concentrations in a permissive range to explain known metabolic rates. It is important that other systems be examined in a similar manner.

It is still not easy to dismiss all the "near-equilibrium" data. Perhaps they are still consistent in that they indicate that in spite of varying microenvironments in the same structural compartment of the cell, the overall balance is maintained to keep the *net* thermodynamic state near equilibrium.

ACKNOWLEDGMENTS

This work was supported by a grant from the U.S. Public Health Service. I am indebted to Dr. Richard Veech for his constant articulate opposition to these ideas.

REFERENCES

Cha, S. (1970). *J. Biol. Chem.* **245,** 4814.

Dalziel, K. and Engel, P. C. (1968). *FEBS Lett.* **1,** 349.

Engel, P. C. and Ferdinand, W. (1973). *Biochem. J.* **131,** 97.

Gumaa, K. A., McLean, P., and Greenbaum, A. L. (1971). In *Essays in Biochemistry*, Vol. 7 (P. N. Campbell and F. Dickens, eds.), Academic Press, London, pp. 39–86.

Guynn, R. W., Gelberg, H. J., and Veech, R. L. (1973). *J. Biol. Chem.* **248,** 6957.

Hathaway, J. A. and Atkinson, D. E. (1965). *Biochem. Biophys. Res. Commun.* **20,** 661.

Hohorst, H., Kreutz, F. H., and Bucher, T. (1959). *Biochem. Z.* **332,** 18.

Kosicki, G. W. (1961). Ph.D. thesis, University of Michigan, Ann Arbor.

Kosicki, G. W. and Srere, P. A. (1961). *J. Biol. Chem.* **236,** 256.

Krebs, H. A. and Veech, R. L. (1970). In *Pyridine Nucleotide-Dependent Dehydrogenases* (H. Sund, ed.), Springer-Verlag, Berlin, pp. 413–434.

LaNoue, K. F., Bryla, J., and Williamson, J. R. (1972). *J. Biol. Chem.* **247,** 667.

Lardy, H. A., Paetkau, V., and Walter, P. (1965). *Proc. Nat. Acad. Sci. U.S.* **53,** 1410.

Lueck, J. O. and Fromm, H. J. (1973). *FEBS Lett.* **32,** 184.

Matsuoka, Y. and Srere, P. A. (1973). *J. Biol. Chem.* **248,** 8022.

Moriyama, T. and Srere, P. A. (1971). *J. Biol. Chem.* **246,** 3217.

Srere, P. A. (1966). *J. Biol. Chem.* **241,** 2157.

Srere, P. A. (1969a). *Biochem. Med.* **3,** 61.

Srere, P. A. (1969b). *Science* **158,** 936.

Srere, P. A. (1972). In *Energy Metabolism and the Regulation of Metabolic Processes in Mitochondria* (M. A. Mehlman and R. W. Hanson, eds.), Academic Press, New York, pp. 79–91.

Srere, P. A. and Matsuoka, Y. (1972). *Biochem. Med.* **6,** 262.

Srere, P. A., and Matsuoka, Y., and Mukherjee, A. (1973a). *J. Biol. Chem.* **248,** 8031.

Srere, P. A., Mattiasson, B., and Mosbach, K. (1973b). *Proc. Nat. Acad. Sci. U.S.* **70,** 2534.

Utter, M. F. (1969). In *Citric Acid Cycle* (J. M. Lowenstein, ed.), Dekker, New York, pp. 249–296.

Veech, R. L. and Guynn, R. A. (1972). PAABS Winter School, Miami, Fla.

Weidman, S. W., Drysdale, G. R., and Mildvan, A. S. (1973). *Biochemistry* **12,** 1874.

Williamson, D. H., Lund, P., and Krebs, H. A. (1967). *Biochem. J.* **103,** 514.

Williamson, J. R. (1965). In *Control of Energy Metabolism* (B. Chance, R. W. Estabrook, and J. R. Williamson, eds.), Academic Press, New York, p. 333.

PART II

Mitochondrial Interactions in Gluconeogenesis

CHAPTER FIVE

Role of Anion Transport in the Regulation of Metabolism

JOHN R. WILLIAMSON

1. INTRODUCTION

Kinetic studies of oxidation-reduction reactions in the respiratory chain have shown that the interaction of substrates with mitochondrial dehydrogenases is the slowest step of electron transport (Chance *et al.*, 1967). Further studies have established that the inner mitochondrial membrane provides a permeability barrier to free diffusion of ions between the cytosol and the mitochondrial matrix (Chappell, 1968, 1969). Anion transport is thus a consequence of compartmentation of functions between the mitochondria and cytosol (Bücher and Klingenberg, 1958). Hence the possibility exists that under conditions of *in vivo* metabolism, the transport of

Supported by grants from the United States Public Health Service AM-15120, HL-14461, and AA-00292.

metabolites across the mitochondrial membrane participates in the overall regulation of flux through metabolic pathways. The properties of anion translocation mechanisms as revealed by studies with isolated respiring and nonrespiring mitochondria are reviewed as a basis for understanding the more complex interactions observed in the intact cell. Attention is focused on mechanisms responsible for the maintenance of anion gradients and of redox potential differences between the NAD and NADP systems in the separate cytosolic and mitochondrial compartments of the cell. Finally, evidence is evaluated concerning the role of anion translocation across the mitochondrial membrane in the regulation of hepatic biosynthetic pathways and transport of reducing equivalents across the mitochondrial membrane.

II. COMPARTMENTATION AND ANION TRANSPORT: OVERVIEW

Compartmentation of metabolic processes between the cytosol and mitochondria by virtue of the cellular location of specific enzymes and the impermeability of the mitochondrial membrane to certain cellular constituents necessitates the transport of various anions of intermediary metabolism across the inner mitochondrial membrane. The requirements for transport of specific anions vary somewhat with individual organs and even with the same organ in different species, as described further below. Certain pathways involving anion translocations across the mitochondrial membrane are fundamental to mammalian metabolism. Of these, the most ubiquitous is transport of reducing equivalents from the cytosol to mitochondria. This function is a direct consequence of the fact that glycolysis produces NADH in the cytosol, whereas the energy conservation pathway for NADH reoxidation is located in the inner mitochondrial membrane and is not directly accessible to external NADH. Two major mechanisms have been elucidated that appear to account satisfactorily for the indirect transfer of reducing equivalents in the direction from cytosol to mitochondria. These are the malate-aspartate cycle (Borst, 1963) and the α-glycerophosphate cycle (Bücher and Klingenberg, 1958), depicted schematically in Fig. 1. The essential features of the malate-aspartate cycle are: oxidation of cytosolic NADH by way of malate dehydrogenase; entry of malate into mitochondria, where it produces reduction of NAD by way of mitochondrial malate dehydrogenase; transamination of intramitochondrial oxalacetate by aspartate aminotransferase, accompanied by efflux of aspartate and α-ketoglutarate and influx of glutamate across the mitochondrial membrane. Transamination of aspartate and α-ketoglutarate in the cytosol with the regeneration of glutamate and oxalacetate completes the cycle. The α-glycerophosphate cycle, on the other hand, is not associated with transport of anions across the mitochondrial membrane, since the mitochondrial flavin-linked α-glycerophosphate dehydrogenase apparently has its substrate-binding site on the outer face of the inner mitochondrial membrane (Donnellan *et al.*, 1970; Klingenberg and Buchholz, 1970). The regulation and energetics of these hydrogen "shuttles" are treated in a later section.

Transport of various anions across the mitochondrial membrane is an obligatory step in nearly all the biosynthetic pathways of liver metabolism. Pyruvate carboxylase, the first enzyme of the gluconeogenic pathway in liver and kidney, is a mitochondrial enzyme in all species investigated. Oxalacetate, the product of this reaction, is transported very slowly across the mitochondrial membrane at physiological oxalacetate concentrations. Hence efflux of oxalacetate from the mitochondria is unable to compete kinetically with the alternative processes of malate and aspartate formation accompanied by efflux of these anions from the mitochondria (Lardy *et al.*, 1965). In the intact liver cell, oxalacetate is reformed from malate or aspartate in the cytosol by malate dehydrogenase and aspartate aminotransferase, respectively, and provides the substrate for P-enolpyruvate forma-

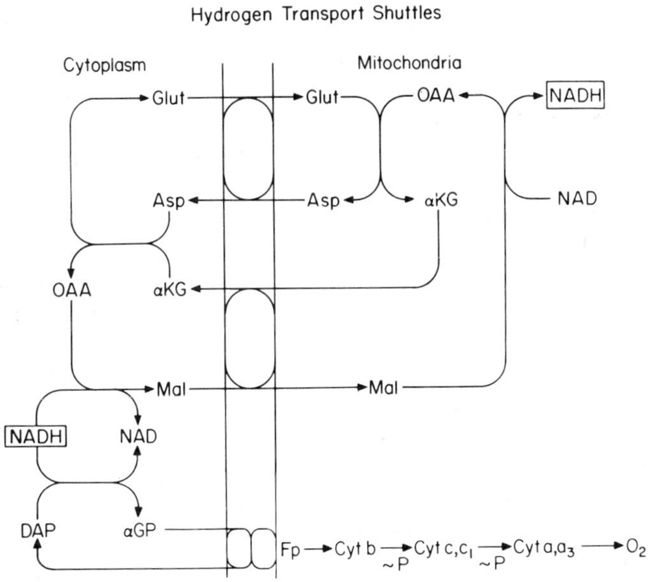

Figure 1. Schematization of transport of reducing equivalents from cytosol to mitochondria by way of the malate-aspartate and α-glycerophosphate cycles. Abbreviations: Glut, glutamate; Asp, aspartate; OAA, oxalacetate; αKg, α-ketoglutarate; Mal, malate; DAP, dihydroxyacetone-P; αGP, α-glycerophosphate.

tion and finally glucose. Unlike liver from most other species, rat liver contains very little P-enolpyruvate carboxykinase in the mitochondria; thus transport of malate and aspartate into the cytosol is obligatory for gluconeogenesis from 3-carbon precursors such as lactate, pyruvate, and alanine. In species such as guinea pig (Nordlie and Lardy, 1963; Nordlie *et al.*, 1965) and man (Brech *et al.*, 1970), which contain both cytosolic and mitochondrial P-enolpyruvate carboxykinase, high rates of P-enolpyruvate formation can readily be demonstrated in isolated mitochondria. Under these conditions, relative rates of P-enolpyruvate, malate, and aspartate production are determined by substrate availability and a variety of factors affecting competition between the mitochondrial enzymes for oxalacetate. In rabbit and pigeon liver, P-enolpyruvate carboxykinase is mainly located in the mitochondria (Lardy, 1975), which means that efflux of P-enolpyruvate from the mitochondria is obligatory for gluconeogenesis.

Theoretically, the stoichiometry between malate and aspartate efflux from the mitochondria in rat liver depends on the oxidation state of the gluconeogenic precursor relative to glucose and whether sufficient NADH is generated in the cytosol to meet the requirements of carbon flux through glyceraldehyde-3-P dehydrogenase (Krebs *et al.*, 1967, and see Williamson *et al.*, 1968 for diagrams). Thus with lactate as substrate, NADH is formed directly in the cytosol during oxidation of lactate to pyruvate, and transport of aspartate from the mitochondria accounts for a stoichiometric conversion of lactate to glucose. Likewise, with pyruvate as substrate, transport of malate from the mitochondria provides both carbon and reducing equivalents for further metabolism to glucose in the cytosol. During conversion to glucose of amino acid substrates such as alanine, no reducing equivalents are generated in the cytosol directly by dehydrogenases, but an indirect generation is achieved due to conversion of aspartate to malate by reactions of the urea cycle (Chamalaun, 1971; J. R. Williamson *et al.*, 1974a). This is brought about by conversion of aspartate and citrulline to urea, ornithine, and malate by way of argininosuccinate, arginine, and fumarate (see Fig. 3). Hence in the absence of any net accumulation of other amino acids, alanine conversion to glucose and urea requires equal rates of transport of both malate and aspartate from the mitochondria. Production of P-enolpyruvate in the mitochondria necessitates an independent supply of NADH in the cytosol for the reductive step of gluconeogenesis. With lactate as substrate, this is supplied directly by dismutation of lactate and glyceraldehyde-3-P dehydrogenases. However, with oxidized substrates such as pyruvate, cytosolic NADH must be supplied by transport of reducing equivalents from the mitochondria. We find later that transfer of reducing equivalents out of mitochondria cannot occur by way of the malate-aspartate cycle because of the slow rate of aspartate entry into mitochondria.

An alternative mechanism that has been elucidated in rat liver involves carboxylation of pyruvate and reduction of oxalacetate to malate in the mitochondria, efflux of malate to the cytosol, oxidation to oxalacetate, and regeneration of pyruvate by way of P-enolpyruvate carboxykinase and pyruvate kinase (Meijer and Williamson, 1974; Rognstad and Katz, 1972). This cyclic pathway (Fig. 2) requires the utilization of one ATP equivalent per NADH equivalent transported across the mitochondrial membrane. It is of interest that pyruvate, unlike lactate, is a very poor gluconeogenic precursor in pigeon liver (Söling *et al.*, 1973), possibly because the aforementioned hydrogen transport system is lacking as a result of the virtual absence of cytosolic P-enolpyuvate carboxykinase.

Evidence supporting these conclusions concerning anion transport across the mitochondrial membrane in the intact tissue has been obtained by use of known inhibitors of the mitochondrial transport systems (Meijer and Van Dam, 1974). Thus butylmalonate, an inhibitor of malate:phosphate exchange (see later section), has been shown to inhibit gluconeogenesis from pyruvate and alanine in rat liver (Chamalaun, 1971; Williamson *et al.*, 1970). Urea formation from ammonia in the absence of carbon substrates requires entry of malate into mitochondria in order to regenerate aspartate (Fig. 3) and has also been shown to be inhibited by butylmalonate (Chamalaun, 1971; Williamson *et al.*, 1974a). Likewise, enzyme inhibitors have

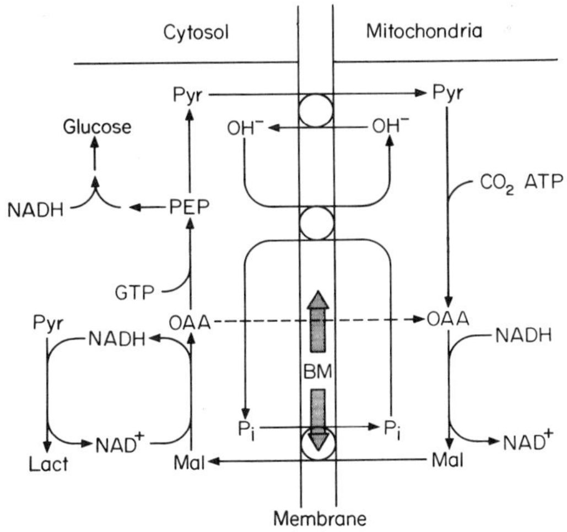

Figure 2. Schematized mechanism for transport of reducing equivalents from mitochondria to cytosol in liver. Abbreviations: Pyr, pyruvate; PEP, P-enolpyruvate; OAA, oxalacetate; Lact, lactate; Mal, malate. Inhibition of the oxalacetate and malate transport across the mitochondrial membrane by butylmalonate (BM) is depicted by the solid arrows.

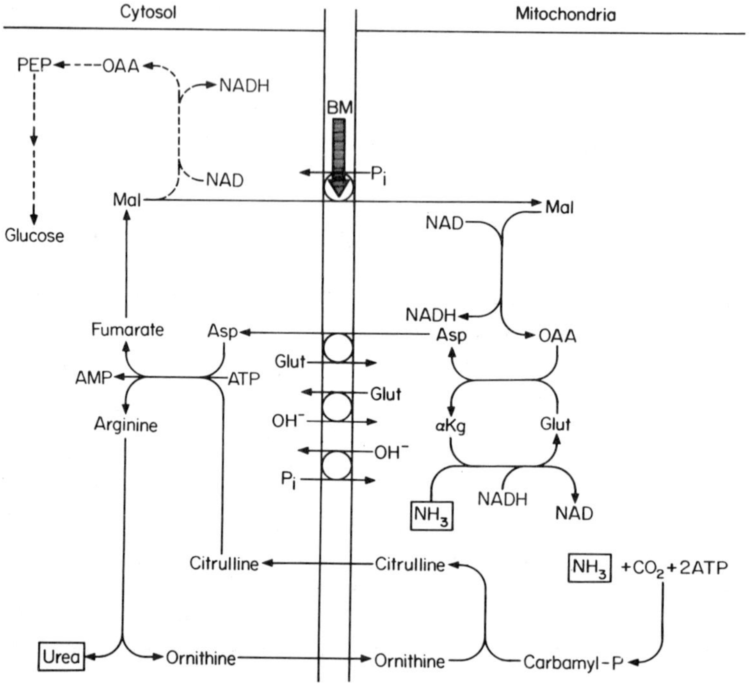

Figure 3. Metabolic scheme for urea synthesis from ammonia in liver by way of intramitochondrial aspartate formation. This pathway is operative in livers from starved rats in the absence of gluconeogenic precursors. Inhibition of malate transport by butylmalonate (BM) is depicted by the solid arrow. Abbreviations as in Figs. 1 and 2.

been useful in establishing pathways of anion transport across the mitochondrial membrane for biosynthetic reactions. Addition of the transaminase inhibitor aminooxyacetate to rat kidney cortex slices (Longshaw *et al.*, 1972; Rognstad and Katz, 1970) or perfused rat liver (Anderson *et al.*, 1971) causes almost complete inhibition of gluconeogenesis from lactate but very little inhibition of gluconeogenesis from pyruvate, thereby confirming that aspartate and malate are the primary precursors for cytosolic oxalacetate with lactate and pyruvate as substrates, respectively. On the other hand, with guinea pig liver, a less complete inhibition of gluconeogenesis from lactate was obtained by addition of aminooxyacetate, but some inhibition was obtained with pyruvate as substrate (Arinze *et al.*, 1973; J. R. Williamson and C. Refino, unpublished data). These data indicate a contribution of the mitochondrial as well as the cytosolic P-enolpyruvate carboxykinase to the overall rate of gluconeogenesis. The progression is continued with rabbit liver (Arinze *et al.*, 1973) and pigeon liver (Söling *et al.*, 1973), where gluconeogenesis from lactate is insensitive to inhibition by aminooxyacetate. A similar conclusion has been reached using tryptophan or quinolinate to inhibit P-enolpyruvate carboxykinase, the cytosolic enzyme being more readily inhibited than the mitochondrial enzyme. Thus gluconeogenesis in guinea pig liver was partially sensitive to inhibition by quinolinate, but gluconeogenesis in the pigeon liver was insensitive at similar concentrations of the inhibitor (Söling *et al.*, 1970). In relation to other pathways, an obligatory role for citrate and malate transport across the mitochondrial membrane during lipogenesis in rat liver has been ascertained by use of butylmalonate and hydroxycitrate, the latter compound being an inhibitor of citrate cleavage enzyme (Lowenstein, 1971; Watson and Lowenstein, 1970).

III. ANION TRANSPORT IN ISOLATED MITOCHONDRIA

A. Types of Translocators

Considerable evidence has accumulated over the past few years to substantiate that most anionic substrates enter the mitochondria in exchange for intramitochondrial anions by way of a number of specific translocators (for reviews see Chappell, 1969; Klingenberg, 1970a, 1970b; Meijer and Van Dam, 1974; Papa *et al.*, 1969; Van Dam and Meijer, 1971). All these, except the ADP:ATP and glutamate:aspartate exchanges, proceed by a 1:1 exchange diffusion process and are not directly dependent on the energy state of the mitochondria (Papa *et al.*, 1970). The well-recognized translocators that have been studied in some detail are depicted in Fig. 4. In addition to these, transport systems have been described for ornithine (Gamble and Lehninger, 1973) and glutamine (Crompton and Chappell, 1973). The relatively high specificity of the different translocators for substrates and inhibitors (see later sections), together with their kinetic behavior, strongly suggests that transport is carrier mediated, possibly involving low molecular weight mobile proteins in the membrane (Julliard and Gautheron, 1973). Attempts are being made to achieve the isolation and purification of the postulated carrier proteins (Azzone *et al.*, 1974). As illustrated in Fig. 4, the translocators may be placed into three categories: (*a*) electrogenic, (*b*) electroneutral but proton compensated, and (*c*) electroneutral. Electrogenic transport implies that the substrate binding sites on the carrier have a specificity for anions of different charge on either side of the membrane. Translocation (with exchange) of the anions, therefore, results in alteration of the charge distribution across the membrane. Transport ceases unless the equilibrium is disturbed by altering

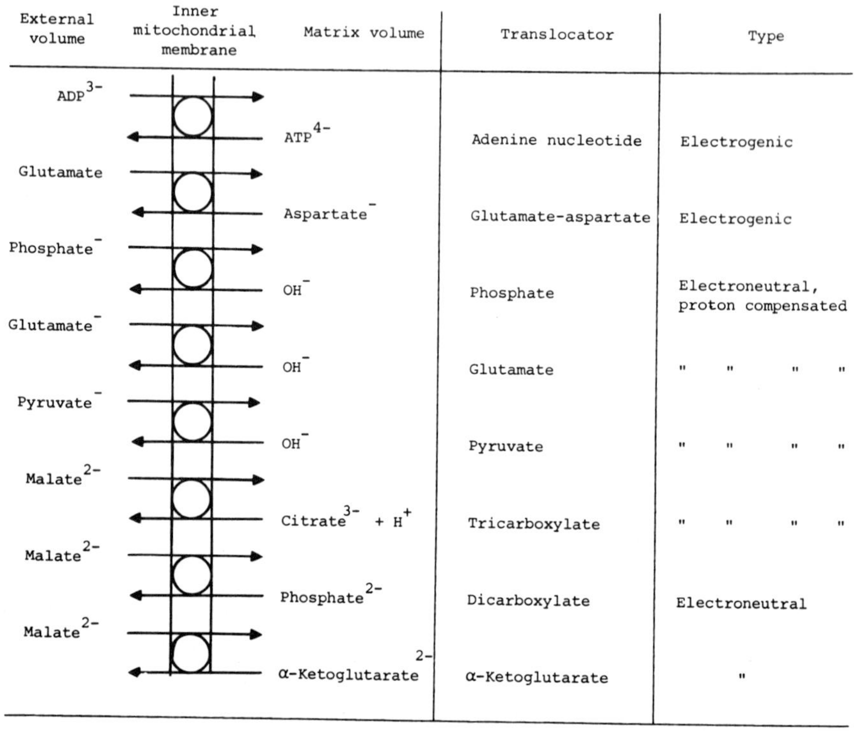

Figure 4. Anion translocators across the mitochondrial membrane.

the energy state of the system. As discussed subsequently, in the energized state the mitochondrial membrane is associated with maintenance of an electrial potential (negative inside) and a ΔpH (more alkaline inside). Therefore, exchanges of adenine nucleotides and glutamate-aspartate as depicted in Fig. 4 are energy consuming and vectorial, and transport responds to the electrical potential difference across the mitochondrial membrane. Three of the four electroneutral but charge-compensated translocators appear in Fig. 4 as antiports with hydroxyl ions. However, this is formally equivalent to co-transport with protons. These translocators together with the tricarboxylate carrier are reversible, and since transport is associated with simultaneous proton movements, exchange is dependent on the ΔpH across the mitochondrial membrane (McGivan and Klingenberg, 1971). In contrast, exchange by way of the dicarboxylate and α-ketoglutarate carriers should in principle be pH independent because the exchange is electroneutral (McGivan and Klingenberg, 1971; Papa *et al.*, 1970). However, binding of substrates to these carriers may be altered by both H^+

and cations, with an increase of pH increasing the apparent K_m values without affecting V_{max} (Meisner *et al.*, 1972; Palmieri *et al.*, 1971b; Palmieri *et al.*, 1972a). With reversible, electroneutral carriers, the direction of net flux is a function of the relative gradients of the exchanging species across the mitochondrial membrane.

A factor of particular importance for operation of the anion exchange reactions as systems providing net anion transport is their coupling through common anions to P_i transport and eventually to proton movements generated by the electron transport chain. Thus efflux of α-ketoglutarate in exchange with malate can be coupled with the malate:phosphate and phosphate:hydroxyl exchanges to give a net transport of α-ketoglutarate in exchange with either phosphate or hydroxyl ions. In the latter case, net efflux of α-ketoglutarate will be pH sensitive because of the involvement of the phosphate:hydroxyl exchange. Similar considerations apply to net transport of citrate and other anions. Values close to theoretical have been obtained for the stoichiometries of H^+/malate and H^+/P_i for malate and phosphate movements in nonrespiring mitochondria (McGivan and Klingenberg, 1971), but clearly under conditions of active respiration, net H^+ (or OH^-) movements will be counterbalanced by transport of the metabolite precursors and bicarbonate, as well as by P_i plus ADP conversion to ATP by oxidative phosphorylation. Furthermore, the degree of coupling between the different anion exchange reactions in mitochondria from different organs will depend on the kinetic characteristics of the translocators, their relative V_{max} values, and the tissue concentrations of the various anions. Coupling between the anion translocators is also important in considerations of the energetics of anion transport. Thus the free energy change associated with the concentration gradient of a particular anion across the mitochondrial membrane can be used to drive the transport of other anions. Ultimately, the driving force for anion movement is provided by oxidative phosphorylation.

The transport across the mitochondrial membrane of those monocarboxylic acids with high pK_a's (i.e., acetate) may be expected to be determined by the concentration of the undissociated acid in the aqueous phanse, its solubility, and its diffusion coefficient in the lipid phase of the membrane. The concentration of the undissociated acid will be the same on either side of the membrane, whereas the distribution of the anionic form will be equal to the ratio of the H^+ concentrations outside and inside the mitochondria. On the other hand, pyruvate, malate, and succinate, as well as other carboxylic acids with low pK_a's have a very low concentration of undissociated acid in the aqueous phase. Although it has not been resolved whether these intermediates are transported across the mitochondrial membrane in the nondissociated form or as anions (Klingenberg, 1970b), it is clear that net flux during metabolism must be associated with osmotic and electrical equilibria.

B. Energetics of Ion Transport

When freshly isolated from isotonic sucrose medium in the cold, mitochondria retain K^+ and metabolic anions against large concentration gradients. The mitochondrial membrane has a low permeability to passive movements of K^+, but energy-linked transport of K^+ can be induced by valinomycin and other ionophores (Pressman, 1965). Calcium and other divalent cations are also taken up rapidly by mitochondria when provided with an energy source, and a high matrix Ca^{2+} content can be achieved, together with water uptake, if adequate concentrations of permeant anions such as phosphate or acetate are present. In the absence of suitable anions, Ca^{2+} uptake is associated with H^+ efflux, resulting in alkalinization of the mitochondrial matrix. The apparent demonstration of an active primary movement of cations prompted the suggestion that an electrical potential (positive inside) was developed across the mitochondrial membrane, with the anion distribution passively following the electrochemical gradient according to the Nernst equation (Harris and Pressman, 1969). This hypothesis now appears unlikely for a number of reasons. It has been found that Ca^{2+} (Rottenberg and Scarpa, 1974) and K^+ movements (either with or without valinomycin) in fact represent passive transport towards electrochemical equilibrium and that anions transported together with K^+ or in electroneutral exchange with other anions become distributed according to the ΔpH across the mitochondrial membrane, with the distribution bearing no relation to the potassium concentration ratio (Quagliariello and Palmieri, 1970; Rottenberg, 1973; Skulachev, 1971). A widely accepted alternative model proposed by Mitchell (Greville, 1969; Mitchell, 1969) assumes that the oxidation-reduction reactions at the three energy conservation sites of the electron transport chain generate across the mitochondrial membrane a protonmotive force that is due to an outward-directed proton pump. Thus a pH difference is generated across the membrane together with an electrical potential difference resulting from the inward movement of electrons. The overall force generated, termed the protonmotive force Δp, is defined by the expression

$$\Delta p = \Delta\Psi - 2.3\frac{RT}{F}\Delta\text{pH millivolts} \qquad (1)$$

where $\Delta\psi$ is the membrane potential in millivolts. According to this model, cations move down the electrochemical gradient. However, the membrane potential component of the protonmotive force provides a vectorial driving force for electrogenic anion exchange (see section on kinetics of anion translocators).

As previously mentioned, the distribution of nonelectrogenically driven anions is determined by the ΔpH across the mitochondrial membrane. In isolated mitochondria in which respiration and phosphorylation are inhibited by rotenone and oligomycin, the uptake of acetate, phosphate, pyruvate, and citric acid cycle intermediates at equilibrium increased upon lowering the external pH over the range from 8 to 6, and the ΔpH across the mitochondrial membrane was shown to be proportional to the log of the anion-distribution ratio inside and outside the mitochondria (Palmieri *et al.*, 1970; Palmieri *et al.*, 1971a; Quagliariello and Palmieri, 1970; Papa and Paradies, 1974). At low (0.1–0.2 mM) external anion concentrations, the slope of ΔpH dependence was approximately equal to the charge of the anions. However, at higher substrate concentrations (1 mM) relatively small anion gradients were observed at low as well as at high pH values. These discrepancies were interpreted in terms of a limited total capacity of the mitochondrial matrix for anions (i.e., because of a fixed K^+ concentration) and activity coefficients of less than unity for mitochondrial anions (Palmieri *et al.*, 1970). Furthermore, anion gradients of unity were not obtained when the ΔpH was zero. Nevertheless, these and other studies performed with respiring mitochondria (Quagliariello *et al.*, 1971) have provided impressive evidence supporting the conclusions (*a*) that with limited external anion concentration, the distribution of anions across the mitochondrial membrane is determined by the ΔpH, and (*b*) that net transport of anions can utilize the portion of energy generated from electron transport that is manifested in maintenance of a Δψ or a ΔpH across the membrane.

C. Regulation of Anion Transport by ΔpH and Δψ

The experimental findings just outlined are of relevance primarily to the mechanism of cation and anion transport. These basic considerations describe the energy linkage with the oxidation-reduction reactions of the respiratory chain associated with maintenance of an electrochemical potential across the mitochondrial membrane. Extrapolation of these findings to normal respiring mitochondria, and particularly to mitochondrial function in the intact cell, requires a knowledge of the magnitude of the ΔpH and potential gradients. If these parameters change under different metabolic conditions, they may provide regulation of anion transport flux. Relatively small values for the ΔpH have been reported during substrate-supported respiration with isolated mitochondria. Thus ΔpH values of 0.2 were found with beef heart mitochondria in state 3, with pyruvate and malate as substrates in the presence of 20 mM Mg^{2+}, compared with values of about 0.4 for state 4 conditions (Addanki *et al.*, 1968). Similar small values have been reported by Rottenberg (Paden and Rottenberg, 1973; Rottenberg, 1970). In the study by Mitchell and Moyle (1969a), ΔpH values were about 1 under conditions of both state 4 and state 3 respiration. Relatively high ΔpH values (0.7–0.8) were also found with rotenone and antimycin A inhibited rat liver mitochondria supported by ascorbate and TMPD respiration (Quagliariello *et al.*, 1971). However, it is well documented that the ΔpH decreases with increasing external anion concentrations at a fixed external pH (Palmieri *et al.*, 1970; Rottenberg, 1973). Furthermore, the ΔpH decreases with increasing external pH, and it is estimated to be zero with an external pH of 7.4 (Palmieri *et al.*, 1970). Consequently, since many anions including phosphate and bicarbonate are present in the cytosol at millimolar concentrations, it may be anticipated that pH gradients across the mitochondrial membrane will be small (e.g., ΔpH 0.2–0.5) and constant at a constant extracellular pH. Nevertheless, the size of the anion gradient should increase with its charge according to the relationship $\log(A_i^{n-}/A_e^{n-}) = n\Delta pH$, where A_i and A_e are the internal and external concentrations of anion A with charge n. It may be noted that a fourfold gradient of citrate^{3-} is predicted for a ΔpH as low as 0.2. Studies with isolated, actively metabolizing mitochondria have in fact shown that the gradient of citric acid cycle intermediates across the mitochondrial membrane at equilibrium lies between 1 and 5. The concentration of citric acid cycle anions in the mitochondrial matrix is normally relatively low, varying over the range from 0.2 to 3 mM (Bryla *et al.*, 1973; LaNoue *et al.*, 1972; LaNoue *et al.*, 1973).

Although there are considerable technical difficulties in obtaining accurate measured values for ΔpH and $\Delta\psi$, it is generally recognized that the contribution of the ΔpH term to the overall protonmotive force Δp (Eq. 1) is small (Greville, 1969). Estimates of the size of the electrical potential across the membrane of liver mitochondria incubated under different respiratory states have been made from the K^+ distribution, using valinomycin to increase the permeability of the membrane to K^+. Padan and Rottenberg (1973) report values for $\Delta\psi$ of 125 and 124 mV with rat liver mitochondria incubated with succinate at 0.25 mM K^+ in state 4 and state 3, respectively, and Δp values of 148 and 143 mV. Values given by Mitchell and Moyle (1969a) for β-hydroxybutyrate-supported respiration of rat liver mitochondria are somewhat higher, being 169 mV for $\Delta\psi$ in state 4 (Δp = 225 mV) and 145 mV for $\Delta\psi$ in state 3, where Δp remains constant.

Values of ΔpH and $\Delta\psi$ across the mitochondrial membrane can be varied separately by means of suitable lipid soluble ionophores (see Greville, 1969 for references). Nigericin contains an ionizable carboxyl group and transports K^+ across the mitochondrial membrane by forming an uncharged complex. Consequently, a coupled electroneutral exchange of K^+ with H^+ is promoted, which collapses the ΔpH but leaves the membrane potential unaffected. Because of the decrease of ΔpH, the high gradients of glutamate, malate, phosphate, and other anions observed in respiration-inhibited mitochondria incubated at low pH are abolished (Palmieri *et al.*, 1971a; Quagliariello and Palmieri, 1970). With respiring mitochondria under conditions of low substrate availability, nigericin can inhibit respiration by causing depletion of the intramitochondrial substrates. On the other hand, antibiotics like valinomycin, which are uncharged, form a charged complex with K^+, and at high external K^+ concentrations they produce collapse of the membrane potential but not the ΔpH. Under suitable conditions in the presence of valinomycin, anions may be co-transported into mitochondria with K^+, causing an increase of the anion concentration gradient across the mitochondrial membrane and a stimulation of respiration. Uncoupling agents increase the permeability of the membrane to protons and collapse both the ΔpH and the membrane potential. Inhibition of anion accumulation results, and limited availability of intramitochondrial anions may lead to inhibition of respiration. Agents that demonstrably collapse the membrane potential also inhibit the electrogenic transport of the ADP:ATP and glutamate:aspartate exchange.

Measurements of the membrane potential made in the presence of valinomycin may underestimate the true Δp of normally respiring mitochondria because of energy leaks (Slater, *et al.*, 1973). Nevertheless, it is apparent that the membrane potential, like the ΔpH, is unlikely to change markedly with different conditions of substrate supply or respiratory activity encountered by mitochondria *in vivo*. It appears, therefore, that regulation of anion transport flux is more likely to be achieved by other mechanisms that modify the kinetic properties of the transport carriers in much the same way as enzyme reactions are regulated. We propose in this chapter that the rate of transmembrane anion flux, hence flux through metabolic pathways, is regulated mainly by alterations of the concentration of anions in the mitochondria or cytosol. The anions that are transported across the mitochondrial membrane are also directly involved in dehydrogenase reactions on one or both sides of the membrane, and there is a link between the phosphate potential and the ratio of free NAD/NADH by way of near-equilibrium reactions (Krebs and Veech, 1969; Stubbs *et al.*, 1972; Veech *et al.*, 1970). Therefore, concentrations of certain anions, hence their transport rates, may be expected to be regulated by changes of the pyridine nucleotide redox potential and the adenine nucleotide phosphate potential.

D. Kinetics of Anion Translocators

The kinetic properties of a number of translocators for anionic substrates have been studied in some detail, and values for the kinetic constants are summarized in Table I. Additional details of the properties of the individual translocators are given below.

Table I. Kinetic constants for anion translocators[a]

Exchange reaction (external:internal)	K_m for external substrate	V_{max} (nmole/min/mg)	Temperature (°C)	Activation energy (kcal/mole)	V_{max}, 37°C	References
ADP:ATP	1–12 μM	200	20	21	1440	Klingenberg (1970b), Pfaff and Klingenberg (1968), Souverijn *et al.* (1973), Vignais *et al.* (1973b), Davis and Lumeng (1975)
Glutamate:aspartate (heart)	6–8 mM	335	28			La Noue and Tischler, unpublished observations, J. R. Williamson *et al.* (1973)
Glutamate:aspartate	0.5–1 mM	>100	28	15.3		La Noue and Tischler, unpublished observations
Glutamate:hydroxyl	5 mM	23	25	17.5	72	Meyer and Vignais (1973)
Glutamate:hydroxyl	4 mM	9.1	20	8.8	21	Bradford and McGivan (1973)
Phosphate	1.6 mM	205	0			Coty and Pedersen (1974)
Dicarboxylate	0.23 mM for malate	70	9	22	2400	Quagliariello *et al.* (1969), Palmieri *et al.* (1971b)
Tricarboxylate	0.12 mM for citrate	22.5	9	20.1	570	Palmieri *et al.* (1972b)
α-Ketoglutarate:malate	See text	43	9	20.5	1200	Palmieri *et al.* (1972a)

[a] Except where noted, values were measured with liver mitochondria.

1. Adenine Nucleotides. Transport of adenine nucleotides has been intensively investigated and reviewed (Klingenberg, 1970b; Klingenberg, 1972; Klingenberg *et al.*, 1973; Pfaff *et al.*, 1969; Pfaff and Klingenberg, 1968; Vignais *et al.*, 1973b) and appears to be the rate-limiting step for oxidative phosphorylation with intact mitochondria (Davis and Lumeng, 1975). The adenine nucleotide translocator is specific for ATP and ADP, whereas AMP is not transported. It is of special interest that the energized state of the mitochondria imposes vectorial properties on the ATP:ADP exchange because of maintenance of a potential difference across the mitochondrial membrane. Thus exchange of endogenous ATP with exogenous ADP in the coupled state is up to 10 times faster than exchange in the reverse direction. This difference of translocation activities is abolished by addition of uncoupling agents or K^+ ionophores in the presence of high K^+ concentrations that collapse the membrane potential. The asymmetry of the carrier in the energized state is expressed by the ratio of the K_m's for ATP and ADP being greater than unity for inward translocation, whereas for outward translocation, ATP and ADP are equally active and independent of the energy state. External ATP inhibits ADP uptake by mitochondria, and the apparent K_i for ATP has been reported to decrease from 200 to 1.5 μM under high energy and uncoupled conditions, respectively, while the K_m for ADP did not change (Souverijn *et al.*, 1973; see also Davis and Lumeng, 1975). Various models for adenine nucleotide transport, involving either a mobile carrier or an energy-linked conformational change that alters nucleotide binding, have been discussed by Vignais *et al.*, (1973b).

The energetics of adenine nucleotide exchange are not fully explained, but it has been concluded that the exchange is partially H^+ compensated and electroneutral and partially electrogenic (Klingenberg, 1970b, 1972). This implies that the carrier should be able to accept either ATP^{4-} or ATP^{3-} for transport from the mitochondrial matrix to the medium. Inward movement of ADP^{3-} in exchange with efflux of ATP^{4-} produces electrogenic transport, which will tend to collapse a preestablished membrane potential. Exchange of ADP^{3-} with ATP^{3-} is electroneutral, and net charge compensation is achieved by H^+ efflux or by entry of P_i^- on the phosphate carrier (Klingenberg, 1972). The electrogenic proportion of the overall ADP:ATP exchange varies with the experimental conditions, including the presence of P_i, between 50 and 70% (Klingenberg, 1972). It is apparent, therefore, that both a membrane potential and a pH gradient across the mitochondrial membrane will influence transport of the nucleotides. As pointed out by Klingenberg *et al.* (Heldt *et al.*, 1972; Klingenberg, 1970b, 1972), an electrogenic exchange of ADP^{3-} with ATP^{4-} results, at equilibrium, in the external ADP/ADP ratio being greater than the internal ATP/ADP ratio. A gradient of $[H^+]$ from outside to inside would have the reverse effect for a proton-compensated $ADP^{3-}:ATP^{3-}$ exchange. The actual nucleotide distribution will represent the sum of both effects. Control of the exchange by the membrane potential predominates, since a considerably higher external than internal ATP/ADP ratio has been observed with isolated rat liver mitochondria (Davis *et al.*, 1974, 1975; Heldt *et al.*, 1972; Slater *et al.*, 1973) and intact rat liver (Elbers *et al.*, 1974; Zuurendonk and Tager, 1974; also see later section). The P_i distribution is controlled by the ΔpH (Palmieri *et al.*, 1970), so that the intramitochondrial P_i concentration is greater than that outside. Consequently, for net phosphorylation of ADP plus P_i to ATP in isolated liver mitochondria, the external phosphate potential (ATP/ADP $\times$ P_i) may be 50 times greater than that inside (Heldt *et al.*, 1972). Furthermore, the internal ATP/ADP ratio is relatively independent of the external ATP/ADP ratio and the respiratory rate (Davis *et al.*, 1974, 1975; Klingenberg, 1972). Transport of ATP against a phosphate potential gradient, therefore, requires energy under conditions of steady state electron transport flux. This energy is derived from the same source as that used for the phosphorylation of ADP, and it produces a fall of the P/O ratio from the theoretical value of 1 per phosphorylation site (Davis *et al.*, 1974; Klingenberg, 1972). Calculations of the sum of the energy requirements for transport of adenine nucleotides and phosphate for *in vitro* conditions have provided estimates of 2 to 4 kcal, or about 15% of the free energy of formation of ATP (Davis *et al.*, 1974; Heldt *et al.*, 1972; Klingenberg, 1970b; Klingenberg, 1972; Slater *et al.*, 1973).

There is evidence that the rate of respiration of isolated mitochondria is controlled by the external phosphate potential (ATP/ADP $\times$ P_i) (Davis *et al.*, 1974, 1975; Owen and Wilson, 1974; Slater *et al*, 1973), but the difference between the internal and external phosphate potentials of actively respiring mitochondria, hence the net energy expenditure for translocation of ADP plus P_i in exchange for ATP, has not been adequately elucidated. The phosphorylation potential

$$\left(\Delta Gp = \Delta G_0' + \frac{RT}{nF} \ln \frac{[\mathrm{ATP}]}{[\mathrm{ADP}][\mathrm{P}_i]}\right)$$

where $\Delta G_0'$ is the standard free energy charge for ATP hydrolysis, was determined by Slater *et al.* (1973) to be 16.1 kcal/mole for liver mitochondria under equilibrium conditions of state 4 respiration. For a stoichiometry of 2 H^+/ATP, the above mentioned value for the phosphorylation potential corresponds to a calculated proton electrochemical gradient of 350 mV ($\Delta G = 2F\Delta E$). The large discrepancy between the calculated and measured proton electrochemical gradients can be accounted for by the presence of energy leaks in the valinomycin system used to measure membrane potential, inaccuracies of measurement, or more probably by a H^+/ATP stoichiometry greater than 2 (Mitchell, 1969).

2. ***Glutamate: Aspartate.*** The glutamate: aspartate carrier has been intensively investigated by LaNoue and colleagues (LaNoue and Tischler, 1974; LaNoue and Williamson, 1971; LaNoue *et al.*, 1973; LaNoue *et al.*, 1974a; LaNoue *et al.*, 1974b; Williamson *et al.*, 1973). Earlier studies (DeHaan *et al.*, 1967; Klingenberg, 1962; Klingenberg *et al.*, 1965) indicated an energy-dependent step in the formation of aspartate from glutamate transamination in rat liver mitochondria. This step was later identified as an energy requirement for aspartate efflux, as shown by accumulation of intramitochondrial aspartate with inhibited flux after addition of uncoupling agents (LaNoue and Williamson, 1971). Further studies revealed that aspartate transport can be energized by electron transport (insensitive to inhibition by oligomycin) or by externally added ATP in an oligomycin sensitive manner (LaNoue *et al.*, 1973; LaNoue *et al.*, 1974a). An unusual feature of glutamate: aspartate exchange, besides its energy dependence, is that the transport is asymmetric. With mitochondria in an energized state, efflux of aspartate is very much faster than aspartate entry (LaNoue and Williamson, 1971). When the mitochondria are deenergized by addition of uncoupling agents or by high K^+ in the presence of valinomycin, aspartate efflux is prevented and entry is promoted (LaNoue and Williamson, 1971; La Noue *et al.*, 1974a). In contrast to the adenine nucleotide translocator, therefore, aspartate transport is asymmetric even in the deenergized state. Efflux of aspartate is obligatorily linked to entry of glutamate with a stoichiometry of unity as shown with aspartate-loaded rat liver mitochondria (La Noue and Tischler, 1974; La Noue *et al.*, 1974b) and also with glutamate-loaded submitochondrial particles (Brouwer *et al.*, 1973).

The studies mentioned suggested that the glutamate: aspartate exchange is electrogenic because it responds to agents that lower the membrane potential, even though glutamate and aspartate carry similar charges. Electrogenic efflux of aspartate energized by the membrane potential (negative inside) implies co-transport of glutamate in an ionic form that carries one less negative charge than aspartate. Further studies showed that for limited transport, the glutamate: aspartate exchange was accompanied by an equivalent uptake of protons and release of K^+ (La Noue and Tischler, 1974). For continuous transport under energized conditions, no net movements of H^+ or K^+ occur, and transport is driven by the energy expenditure needed to maintain the proton electrochemical gradient across the mitochondrial membrane, as with adenine nucleotide transport. The contribution of the ΔpH to the driving force is small, since addition of the ionophore nigericin in the presence of K^+ to collapse the membrane ΔpH had little inhibitory effect on aspartate transport (LaNoue *et al.*, 1974a). It is evident, therefore, that aspartate transport should use energy otherwise employed to generate ATP, thus resulting in a lowered P/O ratio. This has been observed experimentally, and although the exact energy stoichiometry has not yet been ascertained, it is expected to amount to no more than a 15% decrease from the theoretical P/O ratio of 3 (LaNoue *et al.*, 1974b). Although aspartate transport is not affected by the ΔpH across the mitochondrial membrane, it is pH sensitive, being inhibited at acid pH (Williamson *et al.*, 1973), with the pH sensitive site for aspartate efflux being on the matrix side of the mitochondrial membrane (Tischler *et al.*, 1975). In heart mitochondria, which are deficient in the glutamate: hydroxyl carrier, glutamate metabolism is regulated by the aspartate carrier and glutamate influx is equal to aspartate efflux (Safer and Williamson, 1973; Williamson *et al.*, 1973). Aspartate is a noncompetitive inhibitor of glutamate entry having a K_i of 3 to 5 mM in rat heart mitochondria (Williamson *et al.*, 1973). The K_m for glutamate exchange with aspartate is lower in rat liver (about 1 mM) than rat heart mitochondria (6–8 mM), and glutamate entry is also inhibited by aspartate (Tischler *et al.*, 1975).

3. ***Glutamate: Hydroxyl.*** The kinetics of the glutamate: hydroxyl carrier have been studied by two groups of workers (Table I) who obtained similar values for the K_m for glutamate but different V_{max} and energies of activation. The difference may be related to different amounts of glutamate transport on the glutamate: aspartate carrier, although both groups used rotenone-inhibited mitochondria. The V_{max} of the glutamate carrier is more than an order of magnitude lower than those of the other translocators (Bradford and McGivan, 1973). A lipoprotein has been isolated from pig heart mitochondria, which stimulates the transport of glutamate across an artificial phospholipid membrane (Julliard and Gautheron, 1973) and may be related to the mitochondrial glutamate carrier. An unexplained energy dependence for glutamate uptake apparently by the glutamate: hydroxyl carrier has been reported (Brouwer *et al.*, 1973).

4. ***Phosphate: Hydroxyl.*** The kinetics of phosphate transport have been measured at 0°C using $^{32}P_i$ isotope exchange combined with the inhibitor stop techniques (Coty and Pedersen, 1974). The overall activity of the phosphate: hydroxyl exchange appears to be higher than that of the other translocators, including phosphate transport on the dicarboxylate carrier (Table I).

5. ***Pyruvate: Hydroxyl.*** A specific pyruvate: hydroxyl exchange carrier has been proposed (Brouwer *et al.*, 1973; Halestrap, 1975; Halestrap and Denton, 1974; Mowbray, 1975; Papa and Paradies, 1974; Papa *et al.*, 1971). Recent studies by Halestrap (1975), using the inhibitor-stop technique at 6°C, yielded values of 0.15 mM and 0.54 nmole/min/mg for the K_m and V_{max}, respectively and an activation energy of 27 kcal/mole. Papa and Paradies (1974), however, obtained data at 0°C of 0.64 mM and 32 nmole/min/mg with a multilayer centrifugation-filtration technique. Additional experiments have shown that pyruvate uptake appears to follow first-order kinetics (Halestrap, 1975; Mowbray, 1975) and that acetoacetate may also be transported by the same carrier (Papa and Paradies, 1974).

*6. **Dicarboxylate : Dicarboxylate.*** The uptake of dicarboxylate anions on the dicarboxylate carrier in rat liver mitochondria has been studied with inhibitor-stop techniques by Palmieri *et al.* (1971b). The apparent K_m for malate was 0.23 mM compared with values of 0.37 and 1.17 mM for malonate and succinate, respectively. The maximum rate for phosphate transport on the dicarboxylate carrier in the presence of *N*-ethylmaleimide to inhibit phosphate:hydroxyl exchange was about the same as that for malate and succinate, but the K_m for P_i was higher (1.5 to 2 mM). Competitive inhibition for uptake was observed between malate, malonate, and succinate with K_i values similar to their respective K_m's, as well as with substrate analogs such as 2-phenylsuccinate (K_i of 0.7 mM). Phosphate inhibited malate uptake on the dicarboxylate carrier with mixed kinetics, and it was concluded that the carrier has two separate binding sites, one specific for P_i and the other specific for dicarboxylate anions (Palmieri *et al.*, 1971b). Oxalacetate is transported on the dicarboxylate carrier in exchange with malate or phosphate, the transport being inhibited by butylmalonate (Gimpel *et al.*, 1973). The V_{max} was reported in the range of 80 to 270 nmole/min/mg protein at 30°C, with a K_m of 80 to 130 μM (Gimpel *et al.*, 1973). The high K_m probably severely limits oxalacetate transport under physiological conditions.

*7. **Tricarboxylate Exchange.*** Measurements of the kinetics of citrate uptake by malate-loaded rat liver mitochondria provided values of 0.12 mM for the apparent K_m of citrate. The citrate exchange is inhibited competitively by citrate analogs such as 1, 2, 3-benzene tricarboxylate and also by dicarboxylic acids of the citric acid cycle and their analogs (Palmieri *et al.*, 1972b; Robinson *et al.*, 1971a, 1971b). The apparent K_i of citrate uptake for malate was 0.7 mM, and that for P-enolpyruvate was 0.11 mM (Palmieri *et al.*, 1972b). A single binding site for tricarboxylic acids, P-enolpyruvate, and dicarboxylic acids is postulated (Palmieri *et al.*, 1972b).

8. α-Ketoglutarate: Malate. The kinetics of the α-ketoglutarate carrier has been studied in rat liver and heart mitochondria (Palmieri, *et al.*, 1972a; Sluse *et al.*, 1972; Sluse *et al.*, 1973). For the exchange of extramitochondrial α-ketoglutarate with malate in liver mitochondria, values for the apparent K_m for α-ketoglutarate of 46 μM were obtained (Palmieri *et al.*, 1972a). Competitive inhibiton by a variety of dicarboxylic acids was observed; malate had the highest affinity (K_i of 0.12 mM). Tricarboxylic acids also inhibited but with a relatively lower affinity (K_i for citrate = 3.6 mM), and the data were interpreted in terms of competition with α-ketoglutarate for binding only, whereas malate competed for both binding and transport. An extensive study of the kinetics of the α-ketoglutarate carrier in rat heart mitochondria has been reported in which exchange reactions were treated as analogous to a two-substrate, two-product, enzyme-catalyzed reaction (Sluse *et al.*, 1972; Sluse *et al.*, 1973). The rapid equilibrium random bi-bi mechanism proposed by Sluse *et al.* (1972) implies that the binding of the first ligand to the carrier does not alter the affinity for the second ligand, and that the simultaneous coupled movement of malate and α-ketoglutarate in opposite directions across the membrane is the limiting step. However, the second ligand acts as a competitive inhibitor for the first ligand. Thus for malate entry in exchange with α-ketoglutarate, extramitochondrial α-ketoglutarate increases the apparent K_m for malate transport. The data showed that the affinity of the carrier for external substrates was much higher than that for internal substrates. For the α-ketoglutarate: malate exchange, dissociation constants of 2.3 μM and 1.5 mM were calculated for external and internal α-ketoglutarate; corresponding values for malate were 11.6 μM and 3.6 mM (Sluse *et al.*, 1972). However, when the exchange of internal malate was measured over a wider range of external α-ketoglutarate concentrations, breaks were obtained in the double reciprocal plots indicating higher apparent K_ms for the external anion (Sluse *et al.*, 1975). This is more in accordance with the K_m value of 0.32 mM measured for malate activation of α-ketoglutarate efflux in respiring rat heart mitochondria (LaNoue *et al.*, 1973).

E. Inhibitors of Anion Transport

A variety of inhibitors have been very useful in discriminating between the different transport systems, particularly in studies of the kinetics of the exchange diffusion process. Thus exchanges of dicarboxylates with phosphate are inhibited by 2-*n*-butylmalonate, 2-*n*-pentylmalonate, *p*-iodobenzylmalonate or 2-phenylsuccinate (Palmieri *et al.*, 1971b; Robinson and Chappell, 1967; Robinson and Williams, 1970; Robinson *et al.*, 1971a), with 2-phenylsuccinate being the most effective in also inhibiting the α-ketoglutarate translocator (Palmieri *et al.*, 1972a). The tricarboxylate carrier is specifically inhibited by 2-ethylcitrate, benzene-1,2,3-tricarboxylate, and other analogs as well as by *p*-iodobenzylmalonate (Kleineke *et al.*, 1973; Palmieri *et al.*, 1972b; Robinson, 1971; Robinson *et al.*, 1970; 1971a; 1971b). Sulfhydryl binding agents such as mersalyl or mercuribenzoates inhibit phosphate exchange with hydroxyl ions at low concentrations and phosphate exchange with dicarboxylates at higher concentrations (Fonyó, 1968; McGivan and Klingenberg, 1971; Meyer and Tager, 1969; Tyler, 1969), while *N*-ethylmaleimide only inhibits the exchange of phosphate with hydroxyl ions (Johnson and Chappell, 1973; Meijer *et al.*, 1970). Thus all sulfhydryl blocking agents inhibit the phosphate:hydroxyl translocator, while butylmalonate has no effect on this exchange but inhibits the phosphate:dicarboxylate translocator. Exposed sulfhydryl groups of the mitochondria form covalent bonds with the –SH reagents and the inhibition is irreversible. Glutamate-hydroxyl ion exchange is not affected by mersalyl or mercurials but is strongly inhibited by *N*-ethylmaleimide and fuscin (Meijer *et al.*, 1970; Meyer and Vignais, 1973; Vignais and Vignais, 1972). The antibiotic avenaciolide, previously thought to inhibit glutamate transport because of its steric analogy with glutamate (McGivan and Chappell, 1970), functions as a –SH reagent, since it reacts with mitochondrial glutathione like *N*-ethylmaleimide and fuscin (Vignais and Vignais, 1973). Avenaciolide, fuscin, and *N*-ethylmaleimide are uncharged and penetrate the mitochondrial membrane more readily than mersalyl, which contains a free carboxylic acid group (Meyer and Vignais, 1973). It has been concluded, therefore, that the thiol groups of the phosphate:hydroxyl carrier are exposed on the outer surface of the inner mitochondrial membrane, whereas those of the glutamate:hydroxyl carrier are located in the hydrophobic part of the membrane (Meyer and Vignais, 1973; see also Fonyó *et al.*, 1975). Relatively high concentrations of mercurials have been shown to inhibit the α-ketoglutarate and citrate translocators (Quagliariello and Palmieri, 1972).

The adenine nucleotide translocator is also sensitive to inhibition by –SH reagents in the presence of micromolar concentrations of ADP, which is required to expose the sulfhydryl groups by inducing a conformational change in the mitochondrial membrane (LeBlanc and Clauser, 1972; Vignais and Vignais, 1972). Atractyloside, carboxyatractyloside, and bongkrekic acid are particularly effective inhibitors of the adenine nucleotide exchange and bind tightly to specific sites on the inner mitochondrial membrane (Bruni *et al.*, 1962; Erdelt *et al.*, 1972; Henderson and Lardy, 1970; Klingenberg and Buchholz, 1973; Klingenberg *et al.*, 1972; Klingenberg *et al.*, 1973; Söling and Seck, 1975; Vignais *et al.*, 1973a, 1973b). Atractyloside inhibition is competitive with ADP, whereas carboxyatractyloside and bongkrekic acid are noncompetitive inhibitors. Unlike atractyloside and carboxyatractyloside, bongkrekic acid penetrates the mitochondrial membrane and inhibits ADP binding and transport from the matrix side. The reverse is found with submitochondrial particles, which exhibited a 1:1 exchange of ATP for ADP (Shertzer and Racker, 1974). Long-chain acyl-CoA esters inhibit adenine nucleotide translocation with high sensitivity (Lerner *et al.*, 1972; Pande and Blanchaer, 1971; Vaartjes *et al.*, 1972), and bongkrekic acid inhibition may depend partly on its conversion to the CoA derivative (Harris *et al.*, 1972; Vaartjes *et al.*, 1972). Of interest is the finding that P-enolpyruvate can be exchanged with ATP on the adenine nucleotide carrier, although it is preferentially transported on the tricarboxylate carrier (Chudapongse and Haugaard, 1973; Shug and Shrago, 1973). P-enolpyruvate transport on both the adenine nucleotide and tricarboxylate translocators is inhibited by atractyloside and long-chain fatty acyl-CoA esters (Shug and Shrago, 1973), the latter compounds also inhibiting citrate transport on the tricarboxylate carrier (Halperin *et al.*, 1972). Benzene-1,2,3-tricarboxylate, an inhibitor of P-enolpyruvate exchange on the tricarboxylate carrier (Kleineke *et al.*, 1973; Robinson, 1971) did not inhibit P-enolpyruvate transport on the adenine nucleotide carrier (Shug and Shrago, 1973). In a kinetic evaluation of the effect of acyl-CoAs with rat liver mitochondria, Morel *et al.* (1974) found maximal inhibitory effects on the different translocators for chain lengths between C_{12} and C_{18} with little effect of the degree of unsaturation of the fatty acid. Inhibitory constants were 0.1 to 0.25 μM for the adenine nucleotide carrier, 3 to 9 μM for the dicarboxylate and tricarboxylate carriers, and 25 μM for the phosphate carrier (Morel *et al.*, 1974; see also Davis and Lumeng, 1974). Carboxyatractyloside inhibited citrate transport weakly (K_i of 17 μM) compared with a K_i of 10 nM for inhibition of ADP transport, and had no effect on malate transport (Morel *et al.*, 1974). Inhibition of anion transport by long-chain acyl-CoA derivation is probably due to interaction between the acyl moiety of the acyl-CoAs with the carriers. The physiological significance of the inhibitory effect of fatty acyl-CoAs on transport processes is a subject of continuing debate because of the wide range of the inhibition constants (Morel *et al.*, 1974) and because the inhibitory effects are largely abolished by binding of the CoA esters to protein.

A sulfonyl urea derivative (glisoxepide) has recently been reported to inhibit the glutamate:aspartate translocator in rat liver mitochondria and gluconeogenesis from alanine and lactate in isolated rat liver cells (Söling and Seck, 1975). Sulfate and sulfite are transported into mitochondria on the dicarboxylate carrier, as shown by their sensitivity to inhibition by butylmalonate, phenylsuccinate, and mersalyl (Crompton *et al.*, 1974). The heavy metal chelating agent bathophenanthroline sulfonate is a competitive inhibitor with substrates for transport on the dicarboxylate, tricarboxylate, and α-ketoglutarate carriers (but not the phosphate carrier) with inhibitory constants of 25 to 30 μM (Passerella *et al.*, 1973). The sensitive carriers may thus contain a metal ion that chelates with carboxylic acid groups of the substrates (Tyler and Newton, 1970). The pyruvate-hydroxyl is inhibited by analogs of α-cyano-4-hydroxycinnamate (Halestrap, 1975; Halestrap and Denton, 1974). Unfortunately, however, many inhibitors of anion transport with isolated mitochondria are not suitable for studies with whole tissues because of their poor penetration through the plasma membrane or their nonspecific inhibitory effects on enzymes.

All of the carriers listed have been described in rat liver mitochondria, but there is a certain degree of organ and possibly species variation. Thus rat or beef heart mitochondria have only low activities of the citrate: malate, malate: phosphate, and glutamate: hydroxyl exchange reactions (Brouwer, *et al.*, 1973; LaNoue *et al.*, 1973; Scott *et al.*, 1972; Sluse *et al.*, 1971). The virtual absence of the di- and tricarboxylate carriers in cardiac muscle is in accordance with the lack of gluconeogenic and *de novo* fatty acid synthetic functions in this tissue. Consequently, net transport of malate and glutamate across the mitochondrial membrane is approximately equal to that of α-ketoglutarate and aspartate, respectively, as required for the malate-aspartate cycle, although flux through both translocators can be different in transition states (Safer and Williamson, 1973). The tricarboxylate carrier also appears to be absent in rat skeletal muscle (Bouwer *et al.*, 1973) and beef adrenal cortex mitochondria (Michejda *et al.*, 1973). Blowfly and housefly flight muscle mitochondria represent an extreme example of deficient transport functions, since they lack the dicarboxylate, tricarboxylate, and α-ketoglutarate translocators (Tulp and Van Dam, 1969; Van Den Bergh and Slater, 1962). Possible hormonal regulation and developmental aspects are relatively unexplored areas (Meijer and Van Dam, 1974). Increased activity of the adenine nucleotide translocator in rat liver mitochondria after thyroid administration *in vivo* has been reported (Babior *et al.*, 1973), as has higher activity in the liver mitochondria of postnatal compared with prenatal rats (Nakazawa *et al.*, 1973). The activity of glutamate-hydroxyl exchange in rat liver mitochondria is increased two- to threefold by feeding rats a high protein diet (McGivan *et al.*, 1974), which may have significance for optimum rates of urea synthesis under some conditions.

IV. CONCENTRATION GRADIENTS OF ANIONS ACROSS THE MITOCHONDRIAL MEMBRANE IN THE INTACT TISSUE

The total volume of the mitochondrial matrix water space is about 10% of the intracellular volume in liver. Thus if an anion has approximately equal concentrations in the mitochondria and the cytosol, by far the greater part of the total tissue content will be present in the cytosolic space. Conversely, an approximately equal distribution between mitochondria and cytosol will be associated with about a tenfold concentration gradient across the mitochondrial membrane, assuming that the activity coefficients of solutes in the two spaces are equal. In this section, values for the cellular distribution of the limited number of metabolic anions that have been directly determined in liver are compared with values calculated on the basis of assumptions of equilibrium of enzyme reactions in the separate cytosolic and mitochondrial spaces. Large discrepancies occur between anion gradients either directly measured or estimated on the basis of a small pH gradient across the mitochondrial membrane and those calculated from assumptions of near-equilibrium networks. Further discussion is concerned with identifying the assumptions made in the calculation methods that may be erroneous and how this information relates to current concepts of redox potential differences between the pyridine nucleotide systems in the cytosol and mitochondria.

A. Direct Measurements of Anion Distributions in Liver

Limited but highly important progress has been made toward the separation of mitochondria from intact liver without disturbing the content of metabolites in the different cellular compartments. In a study by Elbers *et al.* (1974), blood-free, perfused livers were rapidly frozen, and the pulverized tissue was lyophylized and homogenized in a mixture of carbon tetrachloride and heptane. Enriched mitochondrial and cytosolic fractions, as determined from an increased ratio of mitochondrial to cytosolic marker enzymes, were obtained by centrifugation using an exponential gradient of carbon tetrachloride: heptane mixtures with densities between 1.28 and 1.33 g/ml. From measurements of metabolite contents and marker enzyme activities in the different mixed fractions, extrapolations were made to obtain contents in the mitochondria and cytosol without the need for complete purification. Several important results obtained in this study are summarized in Table II. The first is that whereas the ATP concentration is approximately the same in the mitochondrial and cytosolic compartments, the concentration of ADP is much lower in the cytosol. Consequently, the ATP/ADP ratio in the cytosol is about 30 times higher than that in the

Table II. Intracellular distribution of metabolites in perfused rat liver by direct measurements[a]

Conditions	Contents (μmole/g protein)	ATP	ADP	ATP/ADP	Malate	Citrate	Glutamate	Aspartate
Fed rat	Mitochondrial	0.96	6.3	0.15	0.61	1.5	0.1	0.1
	Extramitochondrial	18.9	4.1	4.6	1.58	1.1	24.0	4.5
	Gradient (m/c)	0.3	8		2	7	<0.02	0.1
Fasted rat	Mitochondrial	3.9	7.4	0.53	<0.01	n.m.	0.12	<0.1
	Extramitochondrial	13.4	4.6	2.9	0.84	n.m.	9.1	5.0
	Gradient (m/c)	1.5	8	—	<0.1	—	0.1	<0.1

[a] Mean values from three different livers in each group, perfused with noncirculating Krebs-Henseleit-bicarbonate buffer for 1 hour. Concentration gradients were calculated based on the assumption that the mitochondrial protein concentration is five times higher than the extramitochondrial protein concentration. The livers from fed rats were perfused with 5 mM glucose, 2 mM lactate, and 0.2 mM pyruvate. The livers from rats fasted for 24 hours were perfused with a substrate-free medium.

Source: Data from Soboll, Heldt, Elbers, and Scholz, unpublished observations. See also: *Heldt et al in Use of Isolated Liver Cells and Kidney Tubules in Metabolic Studies,* S. M. Tager, H-D. Söling and J. R. Williamson (Eds.) North Holland Publishing Co., in press.

mitochondria. The implications of this finding have already been discussed. Of the other metabolite distributions that were measured, Table II shows that the concentration gradients of malate and citrate across the mitochondrial membrane were 2 and 7, respectively, for the fed rat. Glutamate and aspartate appeared to be concentrated in the cytosol. In addition, the metabolite gradients appeared to be dependent on the metabolic state of the rat, with the mitochondrial contents becoming relatively more depleted after starvation.

In another method described by Zuurendonk and Tager (1974), the plasma membrane of isolated liver cells was disrupted by brief treatment with low concentrations of digitonin followed by centrifugation to obtain a supernatant cytosolic fraction. By subtracting cytosolic metabolite levels from total cell levels, mitochondrial contents can be measured. This technique provided similar data to those reported by Elbers *et al.* (1974) for adenine nucleotide distributions. In addition, gluconeogenesis from pyruvate was associated with a concentration gradient of malate from mitochondria to cytosol, which was increased after addition of butylmalonate (Zuurendonk, 1974), as previously suggested by calculations of malate distribution in perfused rat livers (J. R. Williamson *et al.*, 1970).

B. Indirect Estimates of Anion Distributions in Liver

There have been several indirect approaches to the problem of metabolite compartmentation in the intact tissue (Anderson and Garfinkel, 1971; Greenbaum *et al.*, 1971; Gumaa *et al.*, 1971; Krebs and Veech, 1969; J. R. Williamson, 1969). These involve measurements of total metabolite contents in quick-frozen tissue combined with calculations of their contents in the separate mitochondrial and cytosolic spaces by assuming that certain reactions in which they participate as substrates or products are maintained close to equilibrium. All the indirect methods are based on the calculation of the cytosolic ratio of free NAD/NADH from the lactate dehydrogenase equilibrium and the calculation of the mitochondrial ratio of free NAD/NADH from the β-hydroxybutyrate dehydrogenase equilibrium (D. H. Williamson *et al.*, 1967).

For normal rat liver *in vivo,* average values for the cytosolic free NAD/NADH ratio are 700 to 1000 and for the mitochondrial free NAD/NADH ratio are 6 to 8 (Greenbaum *et al.*, 1971; Krebs and Veech, 1969; D. H. Williamson *et al.*, 1967), as calculated from measured values of the lactate/pyruvate ratio and the β-hydroxybutyrate/acetoacetate ratio. These values correspond to an NAD redox potential of about -250 mV in the cytosol and about -310 mV in the mitochondria (using a value of -337 mV for the midpotential of the NAD:NADH couple at 37°C and pH 7). Rat livers perfused with lactate or alanine as substrates give similar values to those just cited for the cytosolic free NAD/NADH ratio, but two- to threefold higher values are observed for the mitochondrial free NAD/NADH ratio unless fatty acids are also added to the perfusate (J. R. Williamson *et al.*, 1969a). Recent estimates of free NAD/NADH ratios in the cytosolic and mitochondrial spaces of perfused rat heart using a different approach agree closely with values reported for rat livers perfused in the absence of oleate (Mowbray and Ottaway, 1973). Uncertainties arise in the precise calculation of free NAD/NADH ratios from substrate redox couple indicators because of possible permeability limitations for rapid transport of the substrates across the plasma membrane or inner mitochondrial membrane and because of the assumption of a constant pH between the cytosol and the mitochondrial matrix; however, a large difference between the redox potentials of the NAD systems in the cytosolic and mitochondrial spaces appears to be indisputable.

1. ***Oxalacetate and Malate.*** Malate dehydrogenase is present at high activities in both the cytosolic and mitochondrial spaces; assuming near-equilibrium conditions, a uniform malate concentration, and a constant pH of 7 throughout the cell, it has been suggested that in livers from well-fed rats *in vivo*, the mitochondrial oxalacetate concentration is about 100 times lower than that of the cytosol (Krebs and Veech, 1969; D. H. Williamson *et al.*, 1967). However, according to the equation

$$\frac{[\mathrm{NAD}]_c}{[\mathrm{NADH}]_c} \cdot \frac{[\mathrm{NADH}]_m}{[\mathrm{NAD}]_m} = \frac{[\mathrm{Mal}]_m}{[\mathrm{Mal}]_c} \cdot \frac{[\mathrm{OAA}]_c}{[\mathrm{OAA}]_m} \cdot \frac{[\mathrm{H}^+]_c}{[\mathrm{H}^+]_m} \approx 100 \qquad (2)$$

where the subscripts c and m refer to the cytosol and mitochondrial spaces, respectively, it is clear that the asymmetry of the oxalacetate distribution may be offset by a gradient of malate in favor of the mitochondria or by a ΔpH across the mitochondrial membrane. As previously discussed, the latter two variables are linked by the general relationship between the distribution of nonelectrogenically transported anions and the proton gradient by the relationships (Palmieri *et al.*, 1970):

$$\frac{\mathrm{A}_i^{n-}}{\mathrm{A}_e^{n-}} = F\left(\frac{[\mathrm{H}^+]_e}{[\mathrm{H}^+]_i}\right)^n \qquad (3)$$

or

$$\log \frac{\mathrm{A}_i^{n-}}{\mathrm{A}_e^{n-}} = \log F + n\Delta\mathrm{pH} \qquad (4)$$

where A_i and A_e are the internal and external concentrations of anion A with a charge of n, and F is the ratio of the extra- and intramitochondrial activity coefficients of the anion. If F is considered to be unity and Eq. 3 is applied to the divalently charged malate ion, Eq. 2 reduces to

$$\frac{[\mathrm{NAD}]_c}{[\mathrm{NADH}]_c} \cdot \frac{[\mathrm{NADH}]_m}{[\mathrm{NAD}]_m} = \frac{[\mathrm{OAA}]_c}{[\mathrm{OAA}]_m} \cdot \left(\frac{[\mathrm{H}^+]_c}{[\mathrm{H}^+]_m}\right)^3 \approx 100 \qquad (5)$$

Thus the estimated oxalacetate gradient will be greatly diminished by even a small difference of H^+ ion concentration across the mitochondrial membrane, and a value of F greater than unity will have a similar effect. For a ΔpH of 0.5 unit, the calculated gradient of oxalacetate from Eq. 5 decreases from the value of 100 calculated by Krebs and Veech (1969) to about 10. Clearly oxalacetate cannot be treated as a freely penetrating anion, since Eq. 2 would then require a ΔpH of 2 units to account for the difference between the NAD/NADH ratios in the two compartments.

Alternative approaches to calculating oxalacetate and malate gradients have been suggested by J. R. Williamson (1969), Greenbaum *et al.* (1971) and Gumaa *et al.*, 1971. The method of Williamson is based on knowledge of the ratios of free NAD/NADH in the two compartments and the assumption that malate dehydrogenase in the two spaces is in equilibrium. The ratio of malate/oxalacetate in each compartment can then be calculated, hence the separate distributions of malate and oxalacetate are ascertainable from knowledge of their total tissue contents. The method used by Greenbaum is more indirect, but the user does not need to know the total tissue oxalacetate content. The cytosolic malate concentration is first calculated by assuming that pyruvate is evenly distributed and that malic enzyme and 6-P-gluconate dehydrogenase are in equilibrium with the same pool of NADP in the cytosol. The cytosolic NADP/NADPH ratio is obtained from knowledge of the equilibrium constant of 6-P-gluconate dehydrogenase and the cell contents of 6-P-gluconate and ribulose-5-P (assumed cytosolic); hence cytosolic malate is calculated from the equilibrium equation for malic enzyme. Using the total cell malate content and assuming equilibrium of malate dehydrogenase in the cytosol and mitochondria, the oxalacetate distribution is calculated. The two methods give substantially different results only when malic enzyme activity in the tissue is low (e.g., the starved state). Gradients calculated for a number of different metabolic states for rat liver *in vivo* (Gumaa *et al.*, 1971) are summarized in Table III. Under most conditions the malate gradient from mitochondria to cytosol was about 6; the oxalacetate gradient was about the same size but in the reverse direction and much smaller than that calculated by Krebs and Veech (1969). Malate gradients of 1 to 2 were calculated by Williamson for rat livers perfused with lactate or alanine in the presence of fatty acids (J. R. Williamson *et al.*, 1968, 1969b).

2. ***Glutamate, Aspartate, and α-Ketoglutarate.*** As with oxalacetate and malate, different approaches have been reported for making estimations involving glutamate, aspartate, and α-ketoglutarate. Anderson and Garfinkel (1971) used the five equations representing equilibria of cytosolic and mitochondrial malate dehydrogenase and aspartate aminotransferase plus cytosolic alanine aminotransferase, together with total tissue metabolite contents, to calculate the distributions of the five intermediates, malate, oxalacetate, glutamate, aspartate, and α-ketoglutarate. On applying this method to data from Krebs' laboratory for *in vivo* livers from fed and starved rats, concentration gradients from mitochondria to cytosol of about 20 were obtained for aspartate, 17 to 38 for glutamate, and 0.07 to 0.25 for α-ketoglutarate. In the method developed by Greenbaum *et al.* (1971), glutamate dehydrogenase is assumed to be in equilibrium with the same pool of NAD as β-hydroxybutyrate; knowing the total tissue contents of glutamate, aspartate, and α-ketoglutarate and assuming that aspartate aminotransferase is in equilibrium in the separate cytosolic and mitochondrial spaces, it is possible to calculate the distributions of aspartate, glutamate, and α-ketoglutarate. According to Greenbaum's method, in the fed rat, gradients of 0.04, 0.36, and 0.6 were obtained for aspartate, glutamate, and α-ketoglutarate, respectively (Table III). J. R. Williamson (1969) assumed no concentration gradients for aspartate and glutamate and

Table III. Calculated concentration gradients of anions between mitochondria and cytosol in rat livers *in Vivo*

Metabolite	Fed	Starved	Starved, prefed carbohydrate	Diabetic	Diabetic plus insulin
Malate[b]	6.3	63	7.4	7.5	6.1
Malate[a]	6.8	5.1	12.1	8.3	6.4
Oxalacetate[b]	0.062	0.42	0.12	0.29	0.15
Oxalacetate[a]	0.067	0.034	0.20	0.33	0.16
Glutamate[a]	0.36	—	1.1	0.22	0.09
Aspartate[a]	0.04	0.07	0.05	0.06	0.03
α-Ketoglutarate[a]	0.60	—	4.38	1.16	0.42
Citrate[a]	5.1	—	16.8	12.1	4.0

[a] Values were obtained from the study by Greenbaum *et al.* (1971), where the methods of calculation are fully described.
[b] Calculated according to the method described by J. R. Williamson (1964).

calculated an α-ketoglutarate gradient of 12 to 35 for rat livers perfused with lactate with and without oleate. Since Krebs and Veech (1969) assume that α-ketoglutarate and glutamate are present at equal concentrations on either side of the mitochondrial membrane, equilibria of aspartate aminotransferase on either side of the membrane can be achieved only if the aspartate gradient is the same as the postulated oxalacetate gradient, namely, 0.01.

3. ***Citrate and Isocitrate.*** Both J. R. Williamson (1969) and Greenbaum *et al.* (1971) assumed equilibrium of cytosolic NADP-linked isocitrate dehydrogenase, and these workers arrived at cytosolic isocitrate values from ratios of cytosolic NADP/NADPH calculated from malic enzyme or 6-P-gluconate dehydrogenase equilibria and previously calculated values for cytosolic α-ketoglutarate. Mitochondrial values were obtained from the difference between total values and cytosolic values. Citrate is assumed to be in equilibrium with isocitrate in each compartment by way of aconitase. For fed rat livers, the citrate gradient from mitochondria to cytosol is calculated to be about 5 (Table III). Krebs and Veech (1969), on the other hand, are forced to postulate a gradient for isocitrate (and citrate) of 1000 in favor of the cytosol if the NADP-linked mitochondrial isocitrate dehydrogenase is in equilibrium, since they consider that the NADP/NADPH ratio in the cytosol is 1000 times lower than that in the mitochondria.

V. CRITICISM OF METHODS FOR ESTIMATING METABOLITE CONCENTRATION GRADIENTS ACROSS THE MITOCHONDRIAL MEMBRANE

A. Direct Methods

Very few data have been published using the direct assay methods for obtaining tissue distributions of metabolites, and it is hoped that this approach will eventually resolve the present ambiguities. However, inherent assumptions may cause both published methods to provide artifactual results for some metabolites. As pointed out by Elbers *et al.* (1974), the nonpolar solvent extraction method assumes that during freeze-drying the metabolites are deposited in close contact with the soluble proteins of the marker enzymes, that they do not become separated from each other during purification, and that they are not extracted by the nonpolar solvents. At present no information is available to resolve these possible ambiguities. The digitonin-centrifugation method (Zuurendonk and Tager, 1974) requires up to 30 seconds for cell breakage and centrifugation of the mitochondria and cell debris, which may result in alterations of the concentrations of some metabolites. Further data are needed before either method can be adequately assessed.

B. Disequilibrium of Mitochondrial Reactions

The calculation methods of necessity assume equilibria of a large number of enzymes and sometimes rely on enzymes that are present in rather low amounts. Since equilibrium conditions by definition require flux to be zero, the basic problem in evaluating the validity of the calculations relates to determining whether any of the assumed near-equilibrium reactions are sufficiently displaced from equilibrium to cause a significant error term. Available evidence (e.g., Greenbaum *et al.*, 1971; Krebs and Veech, 1969) suggests in fact that cytosolic NAD- and NADP-linked dehydrogenases and aspartate aminotransferase maintain an approximate equilibrium between the various reactants of these enzymes in the cytosol. Near-equilibrium for the mitochondrial reactions, however, is less likely. Thus [^{14}C]-isotope labeling patterns of metabolic intermediates in rat liver *in vivo* have suggested that mitochondrial malate dehydrogenase is not in equilibrium (Heath, 1968). Second, other data based on isotope distributions between intermediates in perfused rat liver have indicated that the same pool of oxalacetate cannot be in equilibrium with malate, aspartate, and citrate (Anderson *et al.*, 1971). Third, direct measurements of the oxalacetate content of respiring, isolated rat heart mitochondria (J. R. Williamson *et al.*, 1972) show that whereas intramitochondrial aspartate aminotransferase appears to be close to equilibrium, malate dehydrogenase and the malate dehydrogenase–citrate synthase couple are about 2 orders of magnitude displaced from equilibrium. The directly measured intramitochondrial oxalacetate content corresponded to a concentration of 36 μM, which is about 500 times greater than the concentration calculated by Krebs and Veech (1969). However, the measured oxalacetate content considerably overestimates the free oxalacetate concentration because of its binding to intramitochondrial enzymes such as succinate dehydrogenase, aspartate aminotransferase, malate dehydrogenase, and citrate synthase, all of which are present at concentrations between 10^{-4} and 10^{-6} M (Sols & Marco, 1970). It appears probable, therefore, that a variety of factors including the possibility of direct enzyme-enzyme interactions involving bound oxalacetate (Srere, 1972) as well as kinetic restrictions, provide serious obstacles to the treatment of reactions occurring in the mitochondria as a near-equilibrium network that links common intermediates between various enzymes in accordance with their equilibrium constants (Krebs and Veech, 1969).

C. Conclusions Regarding Metabolite Distributions in the Intact Liver in Relation to ΔpH and Membrane Potential

Because data obtained in the direct measurements of tissue metabolite distributions are only preliminary, and because of the uncertainties inherent in the calculation methods, our conclusions must be tentative. Since the underlying assumptions made in the approach used by Greenbaum *et al.* (1971) appear to be the most reasonable, this calculation method serves as a basis for discussion. From considerations of the general principles controlling anion distributions derived from studies with isolated mitochondria, as treated in previous sections, we can conclude that to a first approximation, permeable anions other than the adenine nucleotides and aspartate will be distributed between the mitochondrial matrix space and the external medium in relation to their charge and the difference of pH across the mitochondrial membrane. In the presence of physiological concentrations of phosphate and particularly bicarbonate and with cytosolic anion concentrations in the millimolar range, the mitochondrial pH gradient is expected to be small; hence anion concentration gradients of unity or slightly greater may be anticipated. Exceptions to this rule will occur if the anion translocator has a low activity or is kinetically controlled by the anion concentrations.

The measured malate gradient in fed rat livers is in fact 2 (Table II), in contrast to calculated values of 6 to 7 (Table III), which correspond to a ΔpH of 0.4. On the other hand, the calculated α-ketoglutarate gradient under most conditions is about unity or slightly less, in agreement with a negligible ΔpH. The dicarboxylate and α-ketoglutarate carriers are very active in liver; hence malate and α-ketoglutarate (together with phosphate) would be expected to follow the ΔpH relationship. However, under some conditions—for example, in starved rat livers (Table III), after oleate addition to rat livers perfused with pyruvate (J. R. Williamson *et al.*, 1969b), and after addition of butylmalonate to liver cells (Zuurendonk, 1974) or perfused rat liver (J. R. Williamson *et al.*, 1970)—the calculated or measured malate gradient is much larger. Although the calculated malate gradients may be in error because of deviation of mitochondrial malate dehydrogenase from equilibrium, it is also possible that the higher gradients reflect a kinetic limitation on malate transport that is attributable to unfavorable values of the cytosolic substrate concentrations relative to the apparent K_m values of the translocators. In support of this suggestion, we note that addition of butylmalonate to inhibit the malate: phosphate exchange apparently increases the malate gradient. The calculated and measured citrate gradients agreed closely and corresponded to a small ΔpH. Possible kinetic control of citrate transport is indicated by the changes of the calculated citrate gradient with metabolic state of the rat (Table III). Distribution of glutamate according to the ΔpH depends on the activity of the glutamate: hydroxyl translocator. This has a high K_m for glutamate and a relatively low activity (Table I), which may explain why under some conditions both the measured and calculated glutamate gradients are considerably less than unity. The magnitude of the glutamate gradient across the mitochondrial membrane in the intact cell, therefore, may be regulated less by the pH gradient than by the flux through the glutamate: aspartate translocator and by whether glutamate is formed primarily in the cytosol or mitochondria. Likewise, as previously discussed, oxalacetate distribution is expected to be unrelated to the ΔpH because of the poor permeability of this substance. Since aspartate transport is electrogenic, its distribution at equilibrium should be determined by the membrane potential, which would correspond to a very low gradient for a $\Delta\psi$ of several hundred millivolts. Because of the presence of enzymes reacting with aspartate on either side of the mitochondrial membrane, equilibrium of the aspartate gradient with the membrane potential is not to be expected. Measured and calculated aspartate gradients for fed rats ranged from 0.04 to 0.1, in accordance with a primarily cytosolic distribution (cf. Tables II and III). The aspartate gradient in the intact liver may vary over a wide range below and possibly above unity, depending on the type of substrates being metabolized. Thus active urea synthesis due to increased carbamyl phosphate availability may increase the gradient by removal of cytosolic aspartate to form argininosuccinate, while a low cytosolic NADH/NAD ratio accompanied by a high cytosolic oxalacetate concentration and inactive aspartate transamination may decrease the gradient.

In summary, several of the major metabolic anions appear to be distributed with a gradient across the mitochondrial membrane slightly greater than unity in accordance with a small ΔpH, although oxalacetate and aspartate are notable exceptions. However, the gradient of particular anions probably changes with the metabolic state of the rat and the nature of the substrates being metabolized by the liver because of kinetic restrictions on the anion translocators.

VI. REDOX POTENTIALS OF CYTOSOLIC AND MITOCHONDRIAL NADP SYSTEMS

By far the greatest discrepancy among the different calculation methods covered previously relates to the gradients of citrate and isocitrate across the mitochondrial membrane. Krebs and Veech (1969) require a gradient of approximately 1000 in the direction of cytosol to mitochondria (or a 10^3- to 10^4-fold deviation of mitochondrial NADP-linked isocitrate dehydrogenase from equilibrium). Other estimates, including the directly measured citrate distribution, indicate a gradient of about one order of magnitude in the direction from mitochondria to cytosol. This latter value is also anticipated if a small pH gradient exists across the mitochondrial membrane *in vivo*. The basis for the reported differences between isocitrate distribution relate directly to estimates of the redox potentials of the mitochondrial and cytosolic NADP systems, which therefore are discussed in some detail.

An extension of the treatment of postulated near-equilibrium of dehydrogenases to cytosolic NADP-linked enzymes such as 6-phosphogluconate dehydrogenase, malic enzyme, and isocitrate dehydrogenase has permitted calculation of the cytosolic free NADP/NADPH ratio (Greenbaum *et al.*, 1971; Krebs and Veech, 1969; J. R. Williamson, 1969). Values reported for normal fed rat liver are in the order of 0.01, which corresponds to a redox potential of about −400 mV for the cytosolic NADP system (i.e., about 150 mV more negative than the cytosolic NAD system). Comparable values are obtained whether total tissue metabolite contents (Krebs and Veech, 1969) or values calculated for the cytosol by various methods (Greenbaum *et al.*, 1971; J. R. Williamson, 1969) are used. On the other hand, there is less certainty regarding the redox potential of the mitochondrial NADP system. The ratio of free NADP/NADPH in the mitochondrial compartment cannot be calculated directly from substrate redox couple indicators, since there is no mitochondrial dehydrogenase (corresponding to β-hydroxybutyrate dehydrogenase) that reacts solely with the mitochondrial NADP pool. Krebs and Veech (1969) argue that since isolated glutamate dehydrogenase reacts equally well with NAD and NADP *in vitro*, it must react likewise in the mitochondria and will therefore function as a transhydrogenase to establish near-equilibrium of the overall reaction:

$$\text{NADH} + \text{NADP} \rightleftarrows \text{NAD} + \text{NADPH} \tag{6}$$

The equilibrium constant for this reaction as determined by Engel and Dalziel (1967) is 1.56, hence Krebs and Veech calculate the mitochondrial free NADP/NADPH ratio by multiplying the mitochondrial NAD/NADH ratio by this factor. For the fed rat liver *in vivo* (Krebs and Veech, 1969) reported that mitochondrial NAD/NADH and NADP/NADPH ratios correspond to redox potentials of −310 and −304 mV for the NAD and NADP systems, respectively. Krebs and Veech thus postulate a potential difference of about 100 mV across the mitochondrial membrane for the NADP system, with the cytosolic system more reduced than that of the mitochondria. Since NADP-linked isocitrate dehydrogenase is present in high activity in both cytosolic and mitochondrial spaces, a large difference of redox potential between the cytosolic and mitochondrial NADP systems would necessitate restricted permeability of the mitochondrial membrane to either α-ketoglutarate or isocitrate, or energy-linked transport of either anion. This follows from Eq. 7, for the isocitrate dehydrogenase reaction in the cytosol and mitochondria:

$$\frac{[\text{NADP}]_c}{[\text{NADPH}]_c} \cdot \frac{[\text{NADPH}]_m}{[\text{NADP}]_m} = \frac{[\alpha\text{-Kg}]_c}{[\alpha\text{-Kg}]_m} \frac{[\text{Isocit}]_m}{[\text{Isocit}]_c} \frac{[\text{CO}_2]_c}{[\text{CO}_2]_m} \tag{7}$$

where the subscripts c and m refer to the cytosolic and mitochondrial spaces, respectively. According to Krebs and Veech (1969), the value for the left-hand side of the equation is about 10^{-3}. If the two enzymes are at near-equilibrium and the CO_2 concentration is uniform throughout the cell, it is apparent from Eq. 7 that α-ketoglutarate should have a mainly mitochondrial distribution or isocitrate a mainly cytosolic distribution. This conclusion contradicts the observed rapid transport of α-ketoglutarate and isocitrate in isolated liver mitochondria by the α-ketoglutarate and tricarboxylate carriers, respectively, which should allow equilibration of the anions and the NADP redox potential difference. Furthermore, extensive studies with mitochondria furnish no evidence that the transport of either anion is energy linked. As previously discussed, the gradients of α-ketoglutarate and isocitrate are both expected to be greater than unity if they are distributed according to the ΔpH (Eq. 4). A successful reconstitution of an α-ketoglutarate–isocitrate cycle depicted by Eq. 7 has in fact been described recently: isolated rat liver mitochondria were used, supplemented with bicarbonate, α-ketoglutarate, NADP, and isocitrate dehydrogenase, which indicated that at equilibrium the extramitochondrial NADP/NADPH ratio was similar to the ratio of the intramitochondrial NADP and NADPH contents (Hoek and Ernster, 1974). The data clearly indicate that the redox potential of the intramitochondrial NADP system was very responsive to that of the extramitochondrial NADP system, thereby suggesting relatively unrestricted transport of α-ketoglutarate and isocitrate across the mitochondrial membrane. Likewise, intramitochondrial NADP is rapidly reduced by addition of isocitrate to isolated rat liver mitochondria (Chappell, 1969). It appears very unlikely, therefore, that a large NADP redox potential could be maintained across the mitochondrial membrane in the intact liver.

To resolve this contradiction, the assumptions underlying the calculation of the NADP redox potentials in the mitochondria and cytosol should be reexamined. Krebs and Veech (1969) have pointed out that when two dehydrogenases share a common reactant, their redox potentials are not independent but are fixed by the equilibrium constants of the two systems and the concentrations of the substrates. The cytosolic NAD and NADP systems are linked by a number of enzymes and common substrates, and relatively good agreement has been obtained for the ratio:

$$\frac{[\mathrm{NAD}]_c}{[\mathrm{NADH}]_c}\frac{[\mathrm{NADPH}]_c}{[\mathrm{NADP}]_c} \approx 10^5$$

when calculated by different procedures for a given metabolic state of the rat liver. Thus cytosolic NADP-isocitrate dehydrogenase is linked to cytosolic NAD malate dehydrogenase through the common intermediates α-ketoglutarate and oxalacetate by way of aspartate aminotransferase, and at equilibrium the following relationship applies for the cytosol (Krebs and Veech, 1969):

$$\frac{[\mathrm{NAD}]_c[\mathrm{NADPH}]_c}{[\mathrm{NADH}]_c[\mathrm{NADP}]_c} = \frac{[\mathrm{Isocit}][\mathrm{Asp}]}{[\mathrm{Glut}][\mathrm{Mal}][\mathrm{CO_2}]} \times \frac{K_{\mathrm{ICDH}}}{K_{\mathrm{AAT}} \cdot K_{\mathrm{MDH}}} \tag{8}$$

where K_{ICDH}, K_{AAT}, and K_{MDH} are the equilibrium constants for isocitrate dehydrogenase, aspartate aminotransferase, and malate dehydrogenase, respectively. From the values for the equilibrium constants under standard conditions given by Krebs and Veech (1969), the equilibrium constant term in Eq. 8 is about 5×10^6; if total tissue metabolite contents are taken as representative of their concentrations in the cytosol, the substrate term is about 2×10^{-2}. Consequently, the cytosolic NAD/NADH ratio is calculated to be about 10^5 times greater than the cytosolic NADP/NADPH ratio (i.e., equivalent to a potential difference of about 150 mV). The basic assumptions made for the calculations of the cytosolic NAD and NADP redox potentials all appear to be reasonable, and confidence in their validity is attributable to the consistency obtained by use of different enzyme systems involving different metabolite measurements and equilibrium constants. Furthermore, the lower redox potential for the NADP system is rational physiologically, since it provides greater reducing power for fatty acid synthesis and allows separation of control features for the two major biosynthetic pathways of gluconeogenesis and lipogenesis. Assumptions of near-equilibrium for dehydrogenase and transaminase reactions occurring in the cytosol therefore seem to be appropriate except under particular nutritional states such as starvation, when individual enzymes (e.g., malic enzyme) may have a total activity too low for near-equilibrium to be achieved (Greenbaum *et al.*, 1971).

The remaining possibility is that the calculated mitochondrial NADP redox potential is more positive than the real value in equilibrium with mitochondrial NADP-linked isocitrate dehydrogenase. Krebs and Veech (1969) found that under a number of different dietary conditions the ratio of NAD/NADH in the intact liver calculated from the β-hydroxybutyrate/acetoacetate ratio was similar to that calculated from the glutamate/α-ketoglutarate $\times$ NH_4^+ ratio, using values of 4.93×10^{-2} and

3.87×10^{-6} M for the equilibrium constants of β-hydroxybutyrate dehydrogenase and glutamate dehydrogenase, respectively, at pH 7.0, 38°C, and ionic strength of 0.25. On the basis of these observations, the conclusion was reached that glutamate dehydrogenase *in vivo* is in equilibrium with the mitochondrial NAD system. Although other workers have obtained greater discrepancies between values for the mitochondrial NAD/NADH ratio calculated from β-hydroxybutyrate and glutamate dehydrogenases (Chamalaun and Tager, 1970; Henley and Laughrey, 1970; Hoek and Ernster, 1974), these are too small to account for a 100 mV difference between the redox potentials of the NAD and NADP systems on the assumption that glutamate dehydrogenase reacts preferentially with the NADP system as previously postulated (Papa *et al.,* 1969; Tager, 1969). Furthermore, Hoek and Tager (1973) found in studies with isolated rat liver mitochondria that the substrates of glutamate dehydrogenase were in near equilibrium with the free NAD/NADH ratio calculated from β-hydroxybutyrate dehydrogenase and similar to the analytically determined total NAD/NADH ratio. On the other hand, the total NADP/NADPH ratio was 10 to 50 times lower, corresponding to a lower redox potential, when the mitochondria were in a high energy state. Equilibration of the NAD and NADP redox potentials occurred in the uncoupled state. It is generally recognized that the redox potential of the mitochondrial total NADP system is maintained more reduced than the total NAD system by the energy-linked transhydrogenase (Danielson and Ernster, 1963). This enzyme is tightly bound to the inner mitochondrial membrane and utilizes energy from oxidation reduction reactions of the electron transport chain at the energy conservation sites for the reduction of NADP by NADH. In contrast to the non-energy-linked transhydrogenase reaction, which has an equilibrium constant of 0.79 (Kaplan, *et al.,* 1953), the apparent equilibrium constant of the energy-linked reaction ($K' = [NAD][NADPH]/[NADH][NADP]$) has been reported to be in the order of 500 (Lee and Ernster, 1964). This is not a true equilibrium constant but represents a steady state for the sum of energy-linked and non-energy-linked transhydrogenation under given conditions (Rydström *et al.,* 1970). It is evident, therefore, that if glutamate dehydrogenase contributed significantly to a non-energy-linked equilibration of the NAD and NADP systems, a rapid and useless expenditure of energy would ensue. On these grounds, Krebs and Veech (1969) rejected the possibility of the physiological operation of the energy-linked transhydrogenase in the intact liver. However, this suggestion is unlikely, if only because the total NADPH content of liver is about tenfold higher than the total NADP content, with the mitochondria accounting for up to 50% of the total NADP plus NADPH (Chamalaun, 1971; J. R. Williamson, 1969). An alternative possibility is that glutamate dehydrogenase does not normally function as a non-energy-linked transhydrogenase *in vivo* to equilibrate the mitochondrial NAD and NADP potentials. Hoek *et al.* (1974) reinvestigated the question of the nucleotide specificity of glutamate dehydrogenase in liver mitochondria. They demonstrated an ATPase reaction when ATP and substrates of glutamate dehydrogenase were added to isolated liver mitochondria, attributing this effect to the simultaneous activity of the energy-linked transhydrogenase and the non-energy-linked transhydrogenase catalyzed by glutamate dehydrogenase. The ATPase activity was strongly dependent on the ammonia concentration, and the substrates had a catalytic function. From other experiments in which the kinetics of NAD and NADP changes were measured in the presence of rhein or at low temperatures to inhibit the energy-linked transhydrogenase, it was deduced that reductive amination of α-ketoglutarate to glutamate used mainly NADPH and oxidative deamination of glutamate reacted mainly with NAD, with the energy-linked transhydrogenase being required to convert NADH to NADPH. With high ammonia concentrations or in the energy-deficient state, the redox potentials of the mitochondrial NAD and NADP nucleotides were equalized. It is known from other studies with perfused rat liver that addition of ammonia results in a much larger oxidation of NADPH than NADH (Chamalaun and Tager, 1970; Sies *et al.*, 1974), and a higher than expected utilization of energy for urea synthesis from ammonia has been reported (Hems *et al.*, 1966). Furthermore, recent studies with isolated rat liver cells (Meijer *et al.*, 1975a) have shown large deviations of apparent equilibrium of glutamate dehydrogenase with the mitochondrial NAD system, depending on the ammonia concentration and the energy supply, consistent with an increased reactivity of glutamate dehydrogenase with a pyridine nucleotide pool of lower potential.

Although the data are still insufficient to establish all the details of the complex reactions of glutamate dehydrogenase with NAD, NADH, NADP, and NADPH in the forward and back directions, it appears that the nucleotide specificity of glutamate dehydrogenase in isolated mitochondria and the intact liver is determined by kinetic factors, of which the ammonia concentrations may be the most important. Apparent equilibrium of the glutamate dehydrogenase reaction with the NAD system, but without a transhydrogenase function in relation to a lower potential NADP system, implies a kinetic favoring of the back reaction with NADH rather than NADPH, or nonaccessibility of the latter nucleotide to glutamate dehydrogenase due to compartmentation. Kinetic control could be exerted if the apparent K_m for either α-ketoglutarate or ammonia was lower, perhaps because of allosteric interactions (Goldin and Frieden, 1971), when the enzyme reacts with NADH. Although there is little direct evidence to support this suggestion, it is apparent that NADH rather than NADPH is the first substrate bound to the enzyme (Rydström, 1972). An increased rate of reaction in the forward and back direction with the NADP system would tend to increase the redox potential of the latter system, as observed experimentally, and would cause a deviation of the mitochondrial [Glu]/[α-Kg][NH_4^+] ratio from apparent equilibrium with the NAD system. Under these conditions, the supply of NADPH could be maintained either by NADP-linked isocitrate dehydrogenase or by the energy-linked transhydrogenase, with the latter system involving extra consumption of ATP. The energy expenditure for reduction of NADP by NADH by way of the energy-linked transhydrogenase is equivalent to that required for the synthesis of ATP (Lee and Ernster, 1968), which at equilibrium would result in the redox potential of the NADP system being about 300 mV more negative than the NAD system at an external phosphate potential of 10^4 M^{-1} (Moyle and Mitchell, 1973; Slater, *et al.*, 1973). The activity of the energy-linked transhydrogenase is almost certainly insufficient to maintain a potential difference as large as this against the competing reactions, as evidenced by the fact that ammonia addition readily decreases the NADPH/NADP ratio relative to that of the NADH/NAD ratio. Inhibition of the energy-linked transhydrogenase by fatty acyl-CoA may limit its activity in the intact mitochondria (Rydström, 1972). However, an intermediate value for the mitochondrial NADP redox potential approaching the -400 mV value required to achieve equilibration with the cytosolic NADP redox potential is not unreasonable if the transhydrogenase function of glutamate dehydrogenase is kinetically (or compartmentally) limited. Such a value is in agreement with measured values for the total NADPH/NADP ratio of mitochondria under optimal conditions (Hoek and Tager, 1973; Hoek *et al.*, 1974). Finally, if the redox potentials of the mitochondrial NAD and NADP systems are maintained at different values, equilibrium of glutamate dehydrogenase with the NAD system requires disequilibrium within the NADP system, whereas near equilibrium of mitochondrial NADP-linked isocitrate dehydrogenase with the NADP system requires disequilibrium of the NAD-linked isocitrate dehydrogenase with the NAD system (Greenbaum *et al.*, 1971; J. R. Williamson, 1969). Disequilibrium of NAD-linked isocitrate dehydrogenase is in fact highly probable, since its regulation by ADP, Mg^{2+}, NADH, and NADPH is well documented (Plaut, 1970).

In summary, justifications can be made for the contention that in normal rat liver the redox potentials of the cytosolic NAD and NADP systems are in the regions of -250 and -400 mV, respectively, and the redox potentials of the mitochondrial NAD and NADP systems are in the regions of -300 and -400 mV, respectively. Maintenance of these redox potentials requires energy but is consistent with the postulate of relatively small concentration gradients of α-ketoglutarate, isocitrate, and citrate across the mitochondrial membrane. Transport of these anions by way of non-energy-linked carriers results in equilibration of the NADP potential across the mitochondrial membrane. On the other hand, net transport of reducing equivalents by way of NAD-linked substrates into the mitochondria must involve restricted transport of at least one anion, as discussed below.

VII. REGULATION OF TRANSPORT OF REDUCING EQUIVALENTS FROM CYTOSOL TO MITOCHONDRIA

A. Proposed Substrate-Linked Cycles

A number of substrate-linked cycles have been proposed to account for the transfer of reducing equivalents from cytosol to mitochondria (Borst, 1963; Brunengraber and Lowenstein, 1973; Bücher and Klingenberg, 1958; Whereat, *et al.*, 1969). Net transport of NADH into mitochondria by entry of malate in exchange with oxalacetate (malate-ovalacetate cycle) is depicted by Fig. 2 (page 169). The main objection to this cycle is that since it is theoretically reversible, it would cause an equilibration of the NAD redox potentials on either side of the membrane. In practice this is prevented by the high K_m for oxalacetate transport relative to the physiological concentration of oxalacetate (Gimpel *et al.*, 1973). As previously discussed, the higher NAD/NADH ratio in the cytosolic relative to the mitochondrial space results in a concentration gradient for oxalacetate of less than unity between the mitochondria and cytosol.

Under normal physiological conditions the activity of this cycle will be very small because of the low cytosolic oxalacetate concentration. Experimental verification is lacking to substantiate operation of a proposed fatty acid cycle (Grunnet, 1974; Whereat *et al.*, 1969) or malate-citrate cycle (Brunengraber and Lowenstein, 1973) in the intact cell (Meijer and Van Dam, 1974). The latter cycle in fact is potentially reversible and would tend to equilibrate the NAD redox potential difference. Transport of reducing equivalents from the cytosol to mitochondria has been investigated using either gluconeogenic substrates at a higher state of reduction than glucose (e.g., xylitol, glycerol) which are associated with a net generation of NADH in the cytosol along with glucose, or substrates for cytosolic NAD-linked dehydrogenases such as ethanol. Use of transaminase inhibitors such as aminooxyacetate, cycloserine, or difluoro-oxalacetate with perfused rat livers or isolated rat liver cells (Berry *et al.*, 1973; Werner and Berry, 1974; J. R. Williamson *et al.*, 1971, 1974b) has shown that part of the flux of reducing equivalents into mitochondria occurs by way of the malate-aspartate cycle and part by the rotenone-insensitive α-glycerophosphate cycle (Fig. 1). The latter pathway is relatively inactive at tissue α-glycerophosphate concentrations below the apparent K_m of 0.6 mM for the mitochondrial α-glycerophosphate dehydrogenase (Carnicero, *et al.*, 1972) but can predominate when precursors are available that produce a large increase of the tissue α-glycerophosphate content (Berry *et al.*, 1973; J. R. Williamson *et al.*, 1974b). On the other hand, the malate-aspartate cycle is of special interest because of its general importance and unique features resulting from the characteristics of the glutamate-aspartate translocator.

B. Malate-Aspartate Cycle

If near-equilibrium conditions are assumed for malate dehydrogenase and aspartate aminotransferase on either side of the mitochondrial membrane, the ratio of NAD/NADH in the cytosol (c) and mitochondria (m) may be expressed as:

$$\frac{[\text{NAD}]_c}{[\text{NADH}]_c}\frac{[\text{NADH}]_m}{[\text{NAD}]_m} = \frac{[\text{Mal}]_m}{[\text{Mal}]_c}\frac{[\text{Glut}]_m}{[\text{Glut}]_c}\frac{[\alpha\text{Kg}]_c}{[\alpha\text{Kg}]_m}\frac{[\text{Asp}]_c}{[\text{Asp}]_m}\frac{[\text{H}^+]_c}{[\text{H}^+]_m} \qquad (9)$$

This equation describes the equilibrium conditions for the malate-aspartate cycle, and since all the anion species involved are permeable to the mitochondrial membrane through reaction with specific translocators, an asymmetry of the NAD/NADH ratio across the mitochondrial membrane should be expressed by an asymmetric distribution of one or more anions on the right-hand side of the equation (cf. Chappell, 1969). To a first approximation, the malate gradient may be assumed equal to the α-ketoglutarate gradient, since they are both divalent anions, while the glutamate gradient should be equal to $[\text{H}^+]_c/[\text{H}^+]_m$. Equation 9 therefore reduces to:

$$\frac{[\text{NAD}]_c}{[\text{NADH}]_c}\frac{[\text{NADH}]_m}{[\text{NAD}]_m} = \frac{[\text{Asp}]_c}{[\text{Asp}]_m}\frac{[\text{H}^+]_c^2}{[\text{H}^+]_m^2} \qquad (10)$$

It may be reasoned, therefore, that the difference in redox potentials between the cytosolic and mitochondrial NAD systems is primarily a function of the ΔpH across the mitochondrial membrane and the aspartate concentration gradient. We have already noted that since aspartate transport is electrogenic, its concentration gradient at equilibrium will be controlled by the membrane potential rather than the ΔpH, hence can readily be in the direction from cytosol to mitochondria. According to Eq. 10, the actual gradient will be determined by the ΔpH and the size of the NAD redox potential difference across the mitochondrial membrane.

Studies with perfused rat liver have indicated that the difference between the NAD redox potentials in the cytosol and mitochondria varies greatly with the metabolic state of the liver. Thus in the absence of exogenous substrates, a very low ratio of β-hydroxybutyrate to acetoacetate is observed (J. R. Williamson *et al.*, 1969a) which corresponds to a mitochondrial NAD redox potential of -274 mV and a value of only 6 for the term $[NAD]_c[NADH]_m/[NADH]_c[NAD]_m$. This difference could be accommodated readily by a very small ΔpH and a severalfold concentration gradient of aspartate in favor of the cytosol. Interestingly enough, when flux of reducing equivalents into the mitochondria was increased in the absence of other exogenous substrates by addition of xylitol (Jakob *et al.*, 1971) or ethanol (J. R. Williamson *et al.*, 1969c), the gradient of the NAD/NADH ratio across the mitochondrial membrane decreased to a value of 4 and the aspartate content of the liver decreased by 30%. Addition of substrates, and particularly fatty acids, to the perfused rat liver caused a reduction of the mitochondrial pyridine nucleotide redox state toward values observed *in vivo,* while the degree of reduction of the cytosolic pyridine nucleotide pool varied with the oxidation state of the substrate. With alanine as substrate in the presence of oleate, addition of ethanol increased the NAD/NADH gradient across the mitochondrial membrane from 25 to 38 (J. R. Williamson *et al.*, 1969c). Under both these conditions, flux through malate dehydrogenase and aspartate aminotransferase was increased relative to substrate-free perfusion because of the requirements for glucose and urea synthesis. If near-equilibrium conditions pertain to the system, Eq. 10 predicts a large increase of the ΔpH or a decrease of the aspartate concentration gradient (mitochondria to cytosol). Measurements of the matrix aspartate content of isolated liver mitochondria oxidizing glutamate have consistently produced values of 0.3 to 0.5 mM with very little difference between state 3 and state 4 respiration. The aspartate content of liver *in vitro* (J. R. Williamson, 1969b) or *in vivo* (Greenbaum *et al.*, 1971) provides estimates of the cytosolic aspartate concentration of 1 to 2 mM. In general, the aspartate content of liver increases when aspartate efflux from the mitochondria increases (e.g., on fasting the rat and after oleate or ethanol addition to liver *in vitro*). With alanine as substrate in the presence of ethanol and oleate, the estimated cytosolic aspartate concentration is 6 to 7 mM (J. R. Williamson *et al.*, 1969c). It appears, therefore, that the aspartate gradient across the mitochondrial membrane in the intact liver is likely to be in the range of 0.05–0.5, depending on the flux and the NAD redox potential difference between the cytosol and mitochondria. Even with an aspartate gradient of 0.1, a ΔpH of 0.5 unit would be required from Eq. 10 if the NAD/NADH ratios in the cytosol and mitochondria differ by a factor of 100. In view of the strong buffering action of the bicarbonate-CO_2 system in respiring tissues, together with the effects of anion concentration in diminishing the ΔpH across the mitochondrial membrane (see earlier discussion), a ΔpH as high as 0.5 may be considered somewhat unreasonable. The conclusion is reached, therefore, that Eq. 10 is approximated only when flux of anions across the mitochondrial membrane is low and the gradient of NAD/NADH is small. With increases of flux through mitochondrial malate dehydrogenase and aspartate aminotransferase, deviations from equilibrium may perturb the system to a sufficient extent that calculations based on assumptions of thermodynamic equilibrium become invalid.

C. Energetics

Transport of reducing equivalents into the mitochondria against an NADH/NAD gradient is an energy-consuming process. In the case of the α-glycerophosphate cycle, as pointed out by Borst (1963), energy is derived from the electron transport chain because the mitochondrial flavin-linked α-glycerophosphate dehydrogenase has a midpotential similar to that of cytochrome *b*, and its oxidation is associated with a P/O ratio of 2. Transport of reducing equivalents by way of the malate-aspartate cycle, on the other hand, derives its energy from the energy requirements for aspartate efflux. Since its transport is electrogenic, flux of aspartate from the mitochondria into the cytosol will tend to collapse the membrane potential so that energy from electron transport is required for maintenance of the potential difference in much the same way as that postulated for $ADP^{3-}:ATP^{4-}$ exchange. The actual energy expenditure for aspartate transport is relatively small, however, and although the exact stoichiometry between the rate of aspartate efflux and ATP utilization has not been established, it is certainly greater than 2:1 (K. F. LaNoue and M. Tischler, unpublished observations). An energy utilization of this order of magnitude may also be deduced from maximal values for the free energy change for ATP formation of -300 to -350 mV (Slater, *et al.*, 1973); thus at most transport of aspartate against an NAD redox potential difference of about -50 mV would be expected to diminish the P/O ratio by only 10 to 15% (LaNoue, *et al.*, 1974b). Since aspartate efflux in exchange with glutamate influx is much faster than exchange in the reverse direction (Tischler *et al.*, 1975; La Noue and Williamson, 1971), we are ensured that the NAD redox potential difference across the mitochondrial membrane does not equilibrate when net transport of reducing equivalents is minimal. The unidirectionality of aspartate transport makes the malate-aspartate cycle suitable only for net transport of NADH into the mitochondria. Proposed mechanisms for transport of reducing equivalents from the mitochondria have been discussed earlier (see Fig. 2).

D. Regulation

It is well known that compounds such as xylitol or ethanol produce increased reduction of pyridine nucleotides in the cytosol, and the results of a number of studies have allowed us to conclude that under conditions of suboptimal substrate concentration, the rate-limiting step in the metabolism of these nucleotides is the rate of transport of reducing equivalents to the mitochondria for reoxidation in the electron transport chain (Meijer *et. al.*, 1975b; J. R. Williamson, *et al.*, 1971). Stimulation of flux through the respiratory chain by addition of uncoupling agents has been shown to increase rates of ethanol uptake in fed rat liver cells (Meijer *et al.*, 1975b) and xylitol conversion to glucose in perfused rat liver (J. R. Williamson *et. al.*, 1971); thus a distinction must be made between the possibilities of the rate-limiting step being translocation of reducing equivalents to the mitochondria versus flux in the electron transport chain from NADH to oxygen. In some respects this distinction is artificial, since alterations of flux in the respiratory chain, induced either by uncoupling agents or by changes of the phosphate potential in response to the metabolic load on ATP utilization, will also affect the redox potentials of the NAD system and the electron transport carriers. This in turn is likely to influence flux through the H-transport shuttles, either directly because of the increased concentration of the oxidized component of the redox couple, or indirectly by changes of the concentration gradients of malate, glutamate, and α-ketoglutarate across the mitochondrial membrane. Evidence that the intermediates of the malate-aspartate cycle may be present at suboptimal concentrations in the liver under conditions of limited substrate availability comes from experiments revealing a twofold stimulation of ethanol uptake by isolated liver cells from starved rats after addition of malate, α-ketoglutarate, glutamate, or aspartate (Meijer *et al.*, 1975a; J. R. Williamson *et al.*, 1974a, 1974b). The increased rate of ethanol uptake was inhibited by aminooxyacetate, thereby showing that flux through the malate-aspartate cycle was specifically stimulated. No net utilization of these compounds could be demonstrated, indicating that they had a catalytic role. Furthermore, addition of uncoupling agents to liver cells from fasted rats incubated in the absence of gluconeogenic precursors (Meijer *et al.*, 1975b) caused no increase of ethanol uptake, demonstrating that the rate of reoxidation of NADH in the mitochondria by the respiratory chain was not rate limiting. In contrast, ethanol uptake by liver cells from fed rats was not increased by added intermediates of the malate-aspartate cycle but was stimulated by 80% after addition of uncoupling agents (Meijer *et al.*, 1975b). These data, together with other material reviewed by Meijer and Van Dam (1974), provide strong evidence that the rate of transport of reducing equivalents into mitochondria can be regulated by a kinetic control of the anion translocators.

Details of the regulation are undoubtedly complex because of the interrelationships between the steady-state concentration of metabolites on either side of the mitochondrial membrane. However, two factors appear to be particularly critical for optimal flux through the malate-aspartate cycle. These are the cytosolic malate and glutamate concentrations, which act by regulating the malate:α-ketoglutarate and glutamate:aspartate exchanges, respectively, although it is difficult to separate effects due to one or the other factor because malate influx provides oxalacetate for the intramitochondrial transamination step. Evidence suggesting that the malate-aspartate cycle activity in heart is regulated by a low tissue glutamate content together with inhibition of glutamate:aspartate exchange by a high tissue aspartate content is presented elsewhere (Safer and Williamson, 1973; J. R. Williamson *et al.*, 1973). Recent unpublished data from the author's laboratory illustrate the same point in liver and are presented briefly here. Figure 5 (top trace) shows the rapid increase in the rate of ethanol uptake by the perfused liver from a starved rat upon initiation of infusion of 10 mM aspartate into the arterial fluid flowing through the liver. Perfusion under these conditions was by the flow-through technique (Williamson *et al.*, 1971). Initial rates of ethanol uptake (about 200 μmole per 100 gram body weight

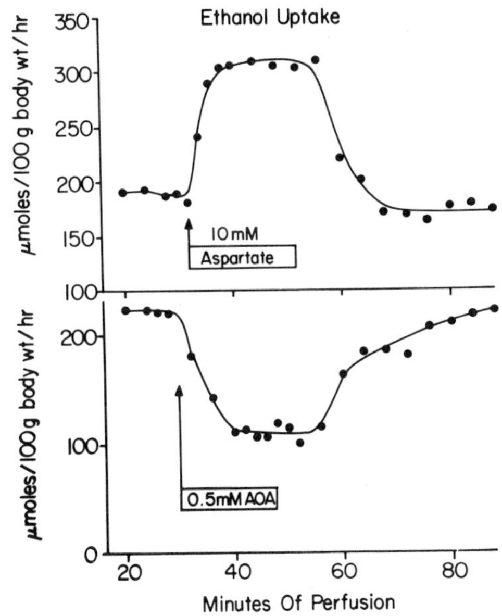

Figure 5. Effects of aspartate and aminooxyacetate (AOA) on ethanol uptake by perfused liver from 24-hours starved rat. The liver was perfused using a flow-through system (J. R. Williamson *et al.*, 1971). Ethanol was continuously infused into the arterial fluid to give a concentration of 1 mM. At the times indicated, infusion of aspartate (10 mM) or aminooxyacetate (0.5 mM) was started and continued for periods indicated by the length of the box. The concentrations refer to final values in the arterial fluid.

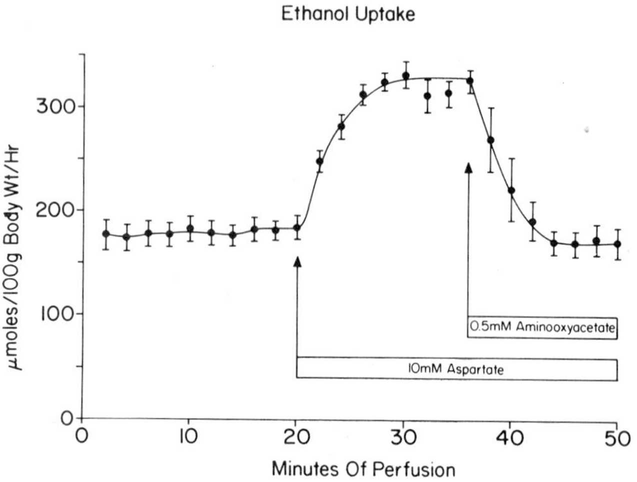

Figure 6. Combined effects of aspartate and aminooxyacetate infusion on ethanol uptake by perfused rat liver. Values shown are means ±SEM of at least 6 livers per point. Other experimental details as in Fig. 5.

per hour) were restored shortly after terminating the aspartate infusion. Figure 5 (bottom trace) illustrates that infusion of 0.5 mM aminooxyacetate reversibly inhibited the control rate of ethanol uptake by about 50% (c.f. Williamson *et al.*, 1974b). Figure 6 pools the results on the rate of ethanol uptake of a number of experiments in which 10 mM aspartate was first infused followed by 0.5 mM aminooxyacetate. The data show that the stimulated rate of ethanol removal was entirely abolished by aminooxyacetate, although the absolute level of inhibition was not as severe as that obtained without prior aspartate perfusion (c.f. Figs. 5 and 6). Livers were freeze clamped under the three conditions of ethanol control (20 minutes of perfusion), ethanol plus aspartate (35 minutes of perfusion), and ethanol plus aspartate plus aminooxyacetate (50 minutes of perfusion). Mean values for the tissue levels of critical metabolites are summarized in Table IV. Ethanol addition produced a high degree of reduction of the cytosolic NAD system but only a moderate degree of reduction of the mitochondrial NAD system. The NAD redox potential difference between the two compartments was 16 mV, corresponding to a value of 3.4 for the term $[NAD]_c[NADH]_m/[NADH]_c[NAD]_m$. Aspartate addition increased ethanol uptake from 185 + 9 to 311 ± 13 µmole/100 g body weight/hour and produced a decrease of the lactate/pyruvate ratio and an increase of the β-hydroxybutyrate/acetoacetate ratio, thereby increasing the NAD redox potential difference to 50 mV. The tissue glutamate content increased twofold and the malate content increased threefold, while the α-ketoglutarate content decreased. Subsequent addition of aminooxyacetate resulted in a decrease of the NAD redox potential difference to 24 mV and reversed the metabolite changes induced by aspartate addition. It may be noted from these data that the malate content of the control ethanol livers is low, giving a concentration in the cytosol of about 0.1 mM. Independent studies with isolated mitochondria have shown that this concentration corresponds to the apparent K_m for malate entry on the α-ketoglutarate carrier; hence malate is in a critical region to regulate flux through the malate-aspartate cycle. The significance of the simultaneous increase of glutamate concentrations (from 2 to 4 mM) after aspartate addition cannot be deduced from these experiments, but may be of less importance than the increase of malate since the apparent K_m for glutamate entry into rat liver mitochondria in exchange with aspartate is below 1 mM.

The effect of ethanol on urea synthesis from ammonia provides a further illustration of the regulatory significance of the cell malate concentration on flux through metabolic pathways (Meijer *et al.*, 1975a). The prevalent pathway for urea formation from added ammonia in the absence of

Table IV. Effects of aspartate and aminooxyacetate in the presence of ethanol on metabolite levels in perfused rat livers[a]

Intermediate	Ethanol control	Ethanol + aspartate	Ethanol + aspartate + AOA
Glutamate	3.91 ±0.34	8.51 ±0.23	5.94 ±0.66
α-Ketoglutarate	0.66 ±0.05	0.38 ±0.11	4.16 ±0.45
Malate	0.25 ±0.02	0.77 ±0.03	0.23 ±0.03
Citrate	1.21 ±0.08	1.23 ±0.11	1.34 ±0.14
ATP/ADP	4.60 ±0.52	4.23 ±0.56	3.81 ±0.37
Lactate/pyruvate	144 ±20	50 ±6	121 ±22
β-Hydroxybutyrate/acetoacetate	1.09 ±0.19	4.79 ±1.08	1.69 ±0.43
Ethanol uptake (μmole/100 g body wt/hr)	185 ±9	311 ±13	168 ±13

[a] Livers from rats starved 18 to 24 hours were perfused by the flow-through technique with 1 mM ethanol. Aspartate infusion (10 mM) was started after 20 minutes and aminooxyacetate (AOA) (0.5 mM) after 35 minutes. Livers were freeze clamped after 20 minutes (ethanol control), after 35 minutes (ethanol + aspartate), or after 50 minutes (ethanol + aspartate + AOA). The experimental protocol is illustrated in Fig. 6. Values, expressed in micromoles per gram of dry weight, are means ±SEM, with five livers in each group.

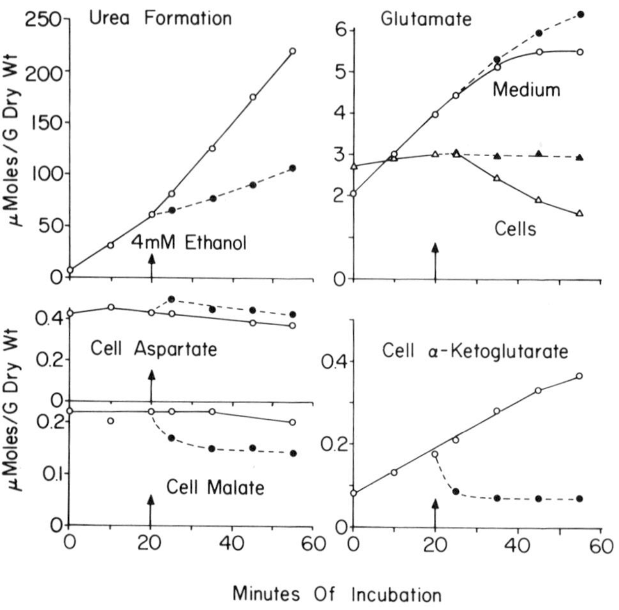

Figure 7. Effect of ethanol addition at 20 minutes on isolated liver cells from starved rats incubated with 8 mM NH_4Cl, 3 mM ornithine, and 1 mM oleate. The cell were centrifuged through silicone oil at the times indicated, and measurements were made of the metabolite contents in the medium and in the cells. Further experimental details are given in Gordon (1973).

gluconeogenic precursors is represented in Fig. 3. As discussed elsewhere (J. R. Williamson *et al.*, 1974a), entry of malate into the mitochondria is a critical rate-limiting step because it supplies oxalacetate for the intramitochondrial transamination of glutamate to aspartate, this being needed in stoichiometric amounts with carbamyl phosphate for urea synthesis. The experiment appearing in Fig. 7 illustrates the usefulness of isolated liver cells for this type of study. Cells obtained from rats fasted from 18 to 24 hours were incubated for 60 minutes with 8 mM NH_4Cl, 3 mM ornithine, and 1 mM oleate, with 4 mM ethanol added to the experimental flask after 20 minutes. At the times indicated, the cells were centrifuged through silicone oil into perchloric acid, to separate the cells from the external medium. Addition of ethanol caused a 70% decrease in the rate of urea formation, an increase of cell glutamate relative to the control, and a decrease of cell malate and α-ketoglutarate. In other experiments addition of β-hydroxybutyrate instead of ethanol caused no inhibition of urea synthesis, indicating that the ethanol effect was caused by an increased reduction of the cytosolic NAD system. Furthermore, addition of ammonia inhibited ethanol uptake, indicative of competition between the pathways of ethanol metabolism and urea synthesis for a common step. A more complete interpretation of the rather complex interactions responsible for the fall of the malate concentration is given elsewhere (Meijer *et al.*, 1975a), but it is evident that the constraint imposed on malate generation by the high cytosolic NADH/NAD ratio causes its steady-state level to fall, hence decreases its rate of entry into mitochondria. This has a feedback effect, since intramitochondrial aspartate formation is thus limited, and the rate of regeneration of malate by way of the urea cycle is also limited.

VIII. CONCLUDING REMARKS

This review of the literature reveals a big gap between many of the studies with isolated mitochondria on anion distribution and the properties of the anion translocators and other studies with intact liver on the regulation and interaction between metabolic pathways. Until recently, in fact, the role of anion translocation and the limitations imposed by the kinetic characteristics of the translocators as a potential site for regulation have been largely ignored. Nonetheless, it is now clear that under special conditions the rate of transport of particular anions of intermediary metabolism across the mitochondrial membrane is an essential feature of the complex array of interactions contributing toward regulation of the overall flux through metabolic pathways. At present perhaps the best illustration of the overriding importance of anion exchange reactions is provided by transport of reducing equivalents into mitochondria by way of the malate-aspartate cycle. The properties of glutamate: aspartate exchange account for its irreversibility and also explain why reducing equivalents can be transported against an NAD redox potential gradient. There is no evidence to suggest that regulation of anion transport in the intact tissue is achieved by alterations of the proton electrochemical gradient across the mitochondrial membrane. Alternatively, control of flux appears to be kinetic, therefore dependent on the relative concentrations of substrates and inhibitors on either side of the mitochondrial membrane for particular translocators. Transport of malate and aspartate are common to many pathways and appear to be particularly important sites of regulation.

Another point emerging from this review is that general assumptions of a uniform distribution of metabolites across the mitochondrial membrane are unjustified. Conversely, any large gradients determined by calculation methods based on assumptions of near equilibrium of reactions occurring in the mitochondria must be regarded with great caution. It is also evident that development in this field requires more knowledge of metabolite distributions in the intact tissue measured under defined conditions, as well as further kinetic measurements of intracellular metabolic changes in relation to known changes of flux through individual steps of a metabolic pathway. It will also be of special importance to relate findings and concepts developed from studies with isolated mitochondria, isolated liver cells, and perfused livers to the liver *in vivo* under normal and abnormal states of metabolism. The author hopes that the chapter will stimulate interest in this area of metabolic regulation.

REFERENCES

Addanki, S., Cahill, F. D., and Sotos, J. F. (1968). *J. Biol. Chem.* **243,** 2337–2348.

Anderson, J. H. and Garfinkel, D. (1971). *Comput. Biomed. Res.* **4,** 43–64.

Anderson, J. H., Nicklas, W. J., Blank, B., Refino, C., and Williamson, J. R. (1971). In *Regulation of Gluconeogenesis: Ninth Conference of the Gesellschaft für Biologische Chemie* (H.-D. Söling and B. Willms, eds.), G. Thieme-Verlag, Stuttgart, pp. 293–315.

Arinze, I. J., Garber, A. J., and Hanson, R. W. (1973). *J. Biol. Chem.* **248,** 2266–2274.

Azzone, G. F., Klingenberg, M., Quagliariello, E., and Siliprandi, N. (eds.). *Proceedings of the International Symposium on Membrane Proteins in Transport and Phosphorylation.* Bressanone, Italy, March 16–19, 1974. In press.

Babior, B. M., Creagan, S., Ingbar, S. H., and Kipnes, R. S. (1973). *Proc. Nat. Acad. Sci. U.S.* **70,** 98–102.

Berry, M. N., Kun, E., and Werner, H. V. (1973). *Eur. J. Biochem.* **42,** 315–324.

Borst, P. (1963). In *Functionelle und Morphologische Organization der Zelle* (P. Karlson, ed.), Springer-Verlag, Berlin, pp. 137–162.

Bradford, N. M. and McGivan, J. D. (1973). *Biochem. J.* **134,** 1023–1029.

Brech, W., Shrago, E., and Wilken, D. (1970). *Biochim. Biophys. Acta* **201,** 145–154.

Brouwer, A., Smits, G. G., Tas, J., Meijer, A. J., and Tager, J. M. (1973). *Biochimie* **55,** 717–725.

Brunengraber, H. and Lowenstein, J. M. (1973). *FEBS Lett.* **36,** 130–132.

Bruni, A., Contessa, A. R., and Luciani, S. (1962). *Biochim. Biophys. Acta* **60,** 301–316.

Bryla, J., Smith, C. M., and Williamson, J. R. (1973). *J. Biol. Chem.* **248,** 4003–4008.

Bücher, T. and Klingenberg, M. (1958). *Angew. Chem.* **70,** 552–

Carnicero, H. H., Moore, C. L., and Hoberman, H. D. (1972). *J. Biol. Chem.* **247,** 418–426.

Chamalaun, R. A. F. M. (1971). M.D. thesis, University of Amsterdam, Academic Service, Amsterdam.

Chamalaun, R. A. F. M. and Tager, J. M. (1970). *Biochim. Biophys. Acta* **222,** 119–134.

Chance, B., DeVault, D., Legallais, V., Mela, L., and Yonetani, T. (1967). In *Nobel Symposium 5: Fast Reactions and Primary Processes in Chemical Kinetics* (S. Claesson, ed.), Wiley-Interscience, New York, pp. 437–468.

Chappell, J. B. (1968). *Brit. Med. Bull.* **24,** 150–157.

Chappell, J. B. (1969). In *Inhibitors: Tools in Cell Research. Twentieth Colloquium of the*

Gesellschaft für Biologische Chemie, April 14–16, 1969, Mosbach/Baden (T. Bücher and H. Sies, eds.), Springer-Verlag, Heidelberg, pp. 335–350.

Chudapongse, P. and Haugaard, N. (1973). *Biochim. Biophys. Acta* **307,** 599–606.

Coty, W. A. and Pedersen, P. L. (1974). *J. Biol. Chem.* **249,** 2593–2598.

Crompton, M. and Chappell, J. B. (1973). *Biochem. J.* **132,** 35–46.

Crompton, M., Palmieri, F., Capano, M., and Quagliariello, E. (1974). *Biochem. J.* **142,** 127–137.

Danielson, L. and Ernster, L. (1963). *Biochem. Z.,* **338,** 188–205.

Davis, E. J. and Lumeng, L. (1975). *J. Biol. Chem.* **250,** 2275–2282.

Davis, E. J., Lumeng, L., and Bottoms, D. (1974). *FEBS Lett.* **39,** 9–12.

DeHaan, E. J., Tager, J. M., and Slater, E. C. (1967). *Biochim. Biophys. Acta* **131,** 1–13.

Donnellan, J. F., Barker, M. D., Wood, J., and Beechy, R. B. (1970). *Biochem. J.* **120,** 467–478.

Elbers, R., Heldt, H. W., Schmucker, P., Soboll, S., and Wiese, H. (1974). *Hoppe-Seyler's Z. Physiol. Chem.* **355,** 378–393.

Engel, P. C. and Dalziel, K. (1967). *Biochem. J.* **105,** 691–695.

Erdelt, H., Weidemann, M. J., Buchholz, M., and Klingenberg, M. (1972). *Eur. J. Biochem.* **30,** 107–122.

Fonyó, A. (1968). *Biochem. Biophys. Res. Commun.* **32,** 624–628.

Fonyó, A., Ligeti, E., Palmieri, F. and Quagliariello, E. (1975). In *Biomembranes, Structure and Function* (G. Gardos and I. Szász, eds.), Adademiai Kiadó, Budapest and North Holland Publishing Co., Amsterdam. In press.

Gamble, J. G. and Lehninger, A. L. (1973). *J. Biol. Chem.* **248,** 610–618.

Gimpel, J. A., DeHaan, E. J., and Tager, J. M. (1973). *Biochim. Biophys. Acta* **292,** 582–591.

Goldin, B. R. and Frieden, C. (1971). *Curr. Top. Cell. Regul.* **4,** 77–117.

Gordon, 1973.

Greenbaum, A. L., Gumaa, K. A., and McLean, P. (1971). *Arch. Biochem. Biophys.* **143,** 617–663.

Greville, G. D. (1969). *Curr. Top. Bioenerg.* **3,** 1–78.

Grunnet, N. (1974). In *Regulation of Hepatic Metabolism. Alfred Benzon Symposium VI* (F. Lundquist and N. Tygstrup, eds.), Academic Press, New York, pp. 520–527.

Gumaa, K. A., McLean, P., and Greenbaum, A. L. (1971). *Essays Biochem.* **7,** 39–86.

Halestrap, A. P. (1975). *Biochem. J.* **148,** 85–96.

Halestrap, A. P. and Denton, R. M. (1974). *Biochem. J.* **138,** 313–316.

Halperin, M. L., Robinson, B. H., and Fritz, I. B. (1972). *Proc. Nat. Acad. Sci. U.S.* **69,** 1003–1007.

Harris, E. J. and Pressman, B. C. (1969). *Biochim. Biophys. Acta* **172,** 66–70.

Harris, R. A., Farmer, B., and Ozawa, T. (1972). *Arch. Biochem. Biophys.* **150,** 199–209.

Heath, D. F. (1968). *Biochem. J.* **110,** 313–335.

Heldt, *et al.*

Heldt, H. W., Klingenberg, M., and Milovancev, M. (1972). *Eur. J. Biochem.* **30,** 434–440.

Hems, R., Ross, B. D., Berry, M. N., and Krebs, H. A. (1966). *Biochem. J.* **101,** 284–292.

Henderson, P. J. F. and Lardy, H. A. (1970). *J. Biol. Chem.* **245,** 1319–1326.

Henley, K. S. and Laughrey, E. G. (1970). *Biochim. Biophys. Acta* **201,** 9–12.

Hoek, J. B. and Ernster, L. (1974). In *Alcohol and Aldehyde Metabolizing Systems,* (R. G. Thurman, T. Yonetani, J. R. Williamson, and B. Chance, eds.), Academic Press, New York, pp. 351–364.

Hoek, J. B. and Tager, J. M. (1973). *Biochim. Biophys. Acta* **325,** 197–212.

Hoek, J. B., Ernster, L., DeHaan, E. J., and Tager, J. M. (1974). *Biochim. Biophys. Acta* **333,** 546–559.

Jakob, A., Williamson, J. R., and Asakura, T. (1971). *J. Biol. Chem.* **246,** 7623–7631.

Johnson, R. N. and Chappell, J. B. (1973). *Biochem. J.* **134,** 769–774.

Julliard, J. H. and Gautheron, D. C. (1973). *FEBS Lett.* **37,** 10–16.

Kaplan, N. O., Colowick, S. P., and Neufeld, E. F. (1953). *J. Biol. Chem.* **205,** 1–15.

Kleineke, J., Sauer, H., and Söling, H. D. (1973) *FEBS Lett.* **29,** 82–86.

Klingenberg, M. (1962). In *Redoxfunktionen Cytoplasmatischer Strukturen* Separatum aus der gemeinsamen Tangung der deutschen Gesellschaft für physiologische Chemie und der österreichischen biochemischen Gesellschaft (T. Bücher, ed.), pp. 163–187.

Klingenberg, M. (1970a). *FEBS Lett.* **6,** 145–154.

Klingenberg, M. (1970b). *Essays in Biochem.* **6,** 119–159.

Klingenberg, M. (1972). In *Mitochondria: Biogenesis and Bioenergetics/Biomembranes: Molecular Arrangements and Transport Mechanisms* Vol. 28, Federation of European Biochemical Societies Proceedings of the Eighth Meeting, Amsterdam, August 1972 (S. G. Van Den Bergh, P. Borst, L. L. M. Van Deenen, J. C. Riemersma, E. C. Slater, and J. M. Tager, eds.), North Holland Publishing Co., Amsterdam, pp. 147–162.

Klingenberg, M. and Buchholz, M. (1970). *Eur. J. Biochem.* **13,** 247–252.

Klingenberg, M. and Buchholz, M. (1973). *Eur. J. Biochem.* **38,** 346–358.

Klingenberg, M., von Haefen, H., and Wenske, G. (1965). *Biochem. Z.* **343,** 452–478.

Klingenberg, M., Buchholz, M., Erdelt, H., Falkner, G., Grebe, K., Kadner, H., Scherer, B., Stengel-Rutkowski, L., and Weidemann, J. (1972). In *Biochemistry and Biophysics of Mitochondrial Membranes* (G. F. Azzone, E. Carafoli, A. L. Lehninger, E. Quagliariello, and N. Siliprandi, eds.), Academic Press, New York, pp. 465–486.

Klingenberg, M., Scherer, B., Stengel-Rutkowski, L., Buchholz, M., and Grebe, K. (1973). In *Mechanisms in Bioenergetics* (G. F. Azzone, L. Ernster, S. Papa, E. Quagliariello, and N. Siliprandi, eds.), Academic Press, New York, pp. 257–284.

Krebs, H. A., Gascoyne, T., and Notton, B. M. (1967). *Biochem. J.* **102,** 275–282.

Krebs, H. A. and Veech, R. L. (1969). In *The Energy Level and Metabolic Control in Mitochondria* (S. Papa, J. M. Tager, E. Quagliariello, and E. C. Slater, eds.), Adriatica Editrice, Bari, Italy, pp. 329–382.

LaNoue, K. F. and Tischler, M. E. (1974). *J. Biol. Chem.* **249,** 7522–7528.

LaNoue, K. F. and Williamson, J. R. (1971). *Metabolism* **20,** 119–140.

LaNoue, K. F., Bryla, J., and Williamson, J. R. (1972). *J. Biol. Chem.* **247,** 667–679.

LaNoue, K. F., Walajtys, E. I., and Williamson, J. R. (1973). *J. Biol. Chem.* **248,** 7171–7183.

LaNoue, K. F., Bryla, J., and Bassett, D. J. P. (1974a). *J. Biol. Chem.* **249,** 7514–7521.

LaNoue, K. F., Meijer, A. J., and Brouwer, A. (1974b). *Arch. Biochem. Biophys.* **161,** 544–550.

Lardy, H. A. (1965). *Harvey Lect.* **60,** 261–278.

Lardy, H. A., Paektau, V., and Walter, P. (1965). *Proc. Nat. Acad. Sci. U.S.* **53,** 1410–1415.

Leblanc, P. and Clauser, H. (1972). *FEBS Lett.* **23,** 107–113.

Lee, C.-P. and Ernster, L. (1964). *Biochim. Biophys. Acta* **81,** 187–190.

Lee, C.-P. and Ernster, L. (1968). *Eur. J. Biochem.* **3,** 385–390.

Lerner, E., Shug, A. L., Elson, C., and Shrago, E. (1972). *J. Biol. Chem.* **247,** 1513–1519.

Longshaw, I. D., Bowen, N. L., and Pogson, C. I. (1972). *Eur. J. Biochem.* **25,** 366–371.

Lowenstein, J. M. (1971). *J. Biol. Chem.* **246,** 629–632.

McGivan, J. D., Bradford, N. M., and Chappell, J. B. (1974). *Biochem. J.* **142,** 359–364.

McGivan, J. D. and Chappell, J. B. (1970). *Biochem. J.* **116,** 37P–38P.

McGivan, J. D. and Klingenberg, M. (1971). *Eur. J. Biochem.* **20,** 392–399.

Meijer, A. J. and Van Dam, K. (1975). *Biochim. Biophys. Acta,* **346,** 213–244.

Meijer, A. J. and Williamson, J. R. (1974). *Biochim. Biophys. Acta* **333,** 1–11.

Meijer, A. J., Groot, G. S. P., and Tager, J. M. (1970). *FEBS Lett.* **8,** 41–44.

Meijer, A. J., Brouwer, A., Reijngoud, D. J., Hoek, J. B., and Tager, J. M. (1972). *Biochim. Biophys. Acta* **283,** 421–429.

Meijer, A. J., DeLeeuw, G., Gimpel, J., and Williamson, J. R. (1975a). *J. Biol. Chem.* **250,** 7728–7738.

Meijer, A. J., Van Woerkom, G. M., Williamson, J. R., and Tager, J. M. (1975b). *Biochem. J.,* **150,** 205–209.

Meisner, H., Palmieri, F., and Quagliariello, E. (1972). *Biochemistry* **11,** 949–955.

Meyer, A. J. and Tager, J. M. (1969). *Biochim. Biophys. Acta* **189,** 136–138.

Meyer, J. and Vignais, P. M. (1973). *Biochim. Biophys. Acta* **325,** 375–384.

Michejda, J. W., Launay, A. N., and Vignais, P. V. (1973). *FEBS Lett.* **32,** 161–165.

Mitchell, P. (1968). *Chemosmotic Coupling and Energy Transduction,* Glynn Research Ltd., Bodmin, England.

Mitchell, P. (1969). In *Mitochondria*—Structure and Function, FEBS Symposia Series, Vol. 17 (L. Ernster and Z. Drahota, eds.), Academic Press, London, pp. 219–232.

Mitchell, P. and Moyle, J. (1969a). *Eur. J. Biochem.* **7,** 471–484.

Mitchell, P. and Moyle, J. (1969b). *Eur. J. Biochem.* **9,** 149–155.

Morel, F., Lauquin, G., Lunardi, J., Duszynski, J., and Vignais, P. V. (1974). *FEBS Lett.* **39,** 133–138.

Mowbray, J. (1975). *Biochem. J.* **148,** 41–47.

Mowbray, J. and Ottaway, J. H. (1973). *Eur. J. Biochem.* **36,** 369–379.

Moyle, J. and Mitchell, P. (1973). *Biochem. J.* **132,** 571–585.

Nakazawa, T., Asami, K., Suzuki, H., and Yukawa, O. (1973). *J. Biochem. (Tokyo)* **73,** 397–406.

Nordlie, R. C. and Lardy, H. A. (1963) *J. Biol. Chem.* **238,** 2259–2263.

Nordlie, R. C., Varricchio, F. E., and Holten, D. D. (1965). *Biochim. Biophys. Acta* **97,** 214–221.

Owen, C. S. and Wilson, D. F. (1974). *Arch. Biochem. Biophys.* **161,** 581–591.

Padan, E. and Rottenberg, H. (1973). *Eur. J. Biochem.* **40,** 431–437.

Palmieri, F., Quagliariello, E., and Klingenberg, M. (1970). *Eur. J. Biochem.* **17,** 230–238.

Palmieri, F., Genchi, G., and Quagliariello, E. (1971a). In *Experientia Suppl.* **18,** 505–512.

Palmieri, F., Prezioso, G., Quagliariello, E., and Klingenberg, M. (1971b). *Eur. J. Biochem.* **22,** 66–74.

Palmieri, F., Quagliariello, E., and Klingenberg, M. (1972a). *Eur. J. Biochem.* **29,** 408–416.

Palmieri, F., Stipani, I., Quagliariello, E., and Klingenberg, M. (1972b). *Eur. J. Biochem.* **26,** 587–594.

Pande, S. V. and Blanchaer, M. C. (1971). *J. Biol. Chem.* **246,** 402–411.

Papa, S. and Paradies G. (1974). *Eur. J. Biochem.* **49,** 265–274.

Papa, S., Tager, J. M., Francavilla, A., and Quagliariello, E. (1969). *Biochim. Biophys. Acta* **172,** 20–29.

Papa, S., Lofrumento, N. E., Quagliariello, E., Meijer, A. J., and Tager, J. M. (1970). *Bioenergetics* **1,** 287–307.

Papa, S., Francavilla, A., Paradies, G., and Meduri, B. (1971). *FEBS Lett.* **12,** 285–288.

Passerella, S., Palmieri, F., and Quagliariello, E. (1973). *FEBS Lett.* **38,** 91–95.

Pfaff, E., Heldt, H. W., and Klingenberg, M. (1969). *Eur. J. Biochem.* **10,** 484–493.

Pfaff, E. and Klingenberg, M. (1968). *Eur. J. Biochem.* **6,** 66–79.

Plaut, G. W. E. (1970). *Curr. Top. Cell. Regul.* **2,** 1–27.

Pressman, B. C. (1965). *Proc. Nat. Acad. Sci. U.S.* **53,** 1076–1083.

Quagliariello, E. and Palmieri, F. (1970). *FEBS Lett.* **8,** 105–108.

Quagliariello, E. and Palmieri, F. (1972). In *Biochemistry and Biophysics of Mitochondrial Membranes* (G. F. Azzone, E. Carafoli, A. L. Lehninger, E. Quagliariello, and N. Siliprandi, eds.), Academic Press, New York, pp. 465–486.

Quagliariello, E., Palmieri, F., Prezioso, G., and Klingenberg, M. (1969). *FEBS Lett.* **4,** 251–254.

Quagliariello, E., Genchi, G., and Palmieri, F. (1971). *FEBS Lett.* **13,** 253–257.

Robinson, B. H. (1971). *FEBS Lett.* **14,** 309–312.

Robinson, B. H. and Chappell, J. B. (1967). *Biochem. Biophys. Res. Commun.* **28,** 249–255.

Robinson, B. H. and Williams, G. R. (1970). *Biochim. Biophys. Acta* **216,** 63–70.

Robinson, B. H., Williams, G. R., Halperin, M. L., and Leznoff, C. C. (1970). *Eur. J. Biochem.* **15,** 263–272.

Robinson, B. H., Williams, G. R., Halperin, M. L., and Leznoff, C. C. (1971a). *Eur. J. Biochem.* **20,** 65–71.

Robinson, B. H., Williams, G. R., Halperin, M. L., and Leznoff, C. C. (1971b). *J. Biol. Chem.* **246,** 5280–5286.

Rognstad, R. and Katz, J. (1970). *Biochem. J.* **116,** 483–491.

Rognstad, R. and Katz, J. (1972). *J. Biol. Chem.* **247,** 6047–6054.

Rottenberg, H. (1970). *Eur. J. Biochem.* **15,** 22–28.

Rottenberg, H. (1973). *J. Membrane Biol.* **11,** 117–137.

Rottenberg, H. and Scarpa, A. (1974). *Biochemistry* **13,** 4811–4817.

Rydström, J. (1972). *Eur. J. Biochem.* **31,** 496–504.

Rydström, J., Teixeira Da Cruz, A. T., and Ernster, L. (1970). *Eur. J. Biochem.* **17,** 56–62.

Safer, B. and Williamson, J. R. (1973). *J. Biol. Chem.* **248,** 2570–2579.

Scott, K. M., Jurkowitz, M., and Brierly, G. P. (1972). *Arch. Biochem. Biophys.* **153,** 682–694.

Shertzer, H. G. and Racker, E. (1974). *J. Biol. Chem.* **249,** 1320–1321.

Shug, A. L. and Shrago, E. (1973). *Biochem. Biophys. Res. Commun.* **53,** 659–665.

Sies, H., Häussinger, D., and Grosskopf, M. (1974). *Hoppe-Seyler's Z. Physiol. Chem.* **355,** 305–320.

Skulachev, V. P. (1971). *Curr. Top. Bioenerg.* **4,** 127.

Slater, E. C., Rosing, J., and Mol, A. (1973). *Biochim. Biophys. Acta* **292,** 534–553.

Sluse, F. E., Sluse-Goffart, C. M., Duyckaerts, C., and Liébecq, C. (1975). *Eur. J. Biochem.* **56,** 1–14.

Sluse, F. E., Meijer, A. J., and Tager, J. M. (1971). *FEBS Lett.* **18,** 149–153.

Sluse, F. E., Ranson, M., and Liébecq, C. (1972). *Eur. J. Biochem.* **25,** 207–217.

Sluse, F. E., Goffart, G., and Liebecq, C. (1973). *Eur. J. Biochem.* **32,** 283–291.

Söling, H.-D. and Seck, A. (1975). FEBS Lett. **51,** 52–59.

Söling, H.-D., Willms, B., Kleineke, J., and Gehlhoff, M. (1970). *Eur. J. Biochem.* **16,** 289–302.

Söling, H.-D., Kleineke, J., Willms, B., Janson, G., and Kuhn, A. (1973). *Eur. J. Biochem.* **37,** 233–243.

Sols, A. and Marco, R. (1970). *Curr. Top. Cell. Regul.* **2,** 227–273.

Souverijn, J. H. M., Huisman, L. A., Rosing, J., and Kemp, A., Jr. (1973). *Biochim. Biophys. Acta* **305,** 185–198.

Srere, P. (1972). *Curr. Top. Cell. Regul.* **5,** 229–283.

Stubbs, M., Veech, R. L., and Krebs, H. A. (1972). *Biochem. J.* **126,** 59–65.

Tager, J. M., Papa, S., DeHaan, E. J., D'Aloya, R., and Quagliariello, E. (1969). *Biochim. Biophys. Acta* **172,** 7–19.

Tischler, M. E., Pachence, J., Williamson, J. R., and LaNoue, K. F. (1975). *Arch. Biochem. Biophys.* in press.

Tulp, A. and Van Dam, K. (1969). *Biochim. Biophys. Acta* **189,** 337–341.

Tyler, D. D. (1969). *Biochem. J.* **111,** 665–678.

Tyler, D. D. and Newton, J. (1970). *FEBS Lett.* **8,** 325.

Vaartjes, W. J., Kemp, A., Jr., Souverijn, J. H. M., and Van den Bergh, S. G. (1972). *FEBS Lett.* **23,** 303–308.

Van Dam, K. and Meijer, A. J. (1971). *Ann. Rev. Biochem.* **40,** 115–160.

Van den Bergh, S. G. and Slater, E. C. (1962). *Biochem. J.* **82,** 362–371.

Veech, R. L., Raijman, L., and Krebs, H. A. (1970). *Biochem. J.* **117,** 499–503.

Vignais, P. M. and Vignais, P. V. (1973). *Biochim. Biophys. Acta* **325,** 357–374.

Vignais, P. V. and Vignais, P. M. (1972). *FEBS Lett.* **26,** 27–31.

Vignais, P. V., Vignais, P. M., and Defaye, G. (1973a). *Biochemistry* **12,** 1508–1519.

Vignais, P. V., Vignais, P. M., Lauquin, G., and Morel, F. (1973b). *Biochimie* **55,** 763–778.

Watson, J. A. and Lowenstein, J. M. (1970). *J. Biol. Chem.* **245,** 5993–6002.

Werner, H. V. and Berry, M. N. (1974). *Eur. J. Biochem.* **42,** 315–324.

Whereat, A. F., Orishimo, M. W., and Nelson, J. (1969). *J. Biol. Chem.* **244,** 6498–6506.

Williamson, D. H., Lund, P., and Krebs, H. A. (1967). *Biochem. J.* **103,** 514–527.

Williamson, J. R. (1969). In *The Energy Level and Metabolic Control in Mitochondria* (S. Papa, J. M. Tager, E. Quagliariello, and E. C. Slater, eds.), Adriatica Editrice, Bari, Italy, pp. 385–400.

Williamson, J. R., Browning, E. T., and Olson, M. S. (1968). *Ad. Enzyme Regul.* **6,** 67–100.

Williamson, J. R., Browning, E. T., and Scholz, R. (1969a). *J. Biol. Chem.* **244,** 4607–4616.

Williamson, J. R., Scholz, R., and Browning, E. T. (1969b). *J. Biol. Chem.* **244,** 4617–4627.

Williamson, J. R., Scholz, R., Browning, E. T., Thurman, R. G., and Fukami, M. H. (1969c). *J. Biol. Chem.* **244,** 5044–5054.

Williamson, J. R., Anderson, J., and Browning, E. T. (1970). *J. Biol. Chem.* **245,** 1717–1726.

Williamson, J. R., Jakob, A., and Refino, C. (1971). *J. Biol. Chem.* **246,** 7632–7641.

Williamson, J. R., Smith, C. M., LaNoue, K. F., and Bryla, J. (1972). In *Energy Metabolism and the Regulation of Metabolic Processes in Mitochondria* (M. A. Mehlman and R. W. Hanson, eds.), Academic Press, New York, pp. 185–210.

Williamson, J. R., Safer, B., LaNoue, K. F., Smith, C. M., and Walajtys, E. (1973). In *Rate Control of Biological Processes* (*Symposium 27 of the Society for Experimental Biology*) (D. D. Davies, ed.), Cambridge University Press, Cambridge, pp. 241–281.

Williamson, J. R., Meijer, A. J., and Ohkawa, K. (1974a). In *Regulation of Hepatic Metabolism* (*Alfred Benzon Symposium VI*) (F. Lundquist and N. Tygstrup, eds.), Munksgaard, Copenhagen, pp. 457–479.

Williamson, J. R., Ohkawa, K., and Meijer, A. J. (1974b). In *Alcohol and Aldehyde Metabolizing Systems* (R. G. Thurman, T. Yonetani, J. R. Williamson, and B. Chance, eds.), Academic Press, New York, pp. 365–381.

Zuurendonk, P. F. (1974). *Abstracts, Ninth FEBS Meeting, August 25–30,* Budapest, Hungary, p. 97.

Zuurendonk, P. F. and Tager, J. M. (1974). *Biochim. Biophys. Acta* **333,** 393–399.

CHAPTER SIX

Regulation of Cell Metabolism by Mitochondrial Transport Systems

EARL SHRAGO, AUSTIN SHUG, AND
CHARLES ELSON

I. INTRODUCTION

The early work on gluconeogenesis emanating from the laboratory of Henry Lardy (1964–1965) elucidated the heretofore unrecognized problem of enzyme compartmentation, an important factor in the overall regulation of the carbon pathway. To account for the subcellular distribution pattern of rat liver phosphoenolpyruvate carboxykinase and pyruvate carboxylase, an amino acid and dicarboxylate shuttle system was proposed as a mechanism for transferring reducing equivalents as well as carbon from the mitochondria to the cytosol (Lardy, *et al.*, 1965; Shrago and Lardy, 1966). More specifically, as illustrated in Fig. 1, oxalacetate synthesized intramitochondrially by pyruvate carboxylase is either converted to aspartate through glutamate oxalacetate transaminase or reduced to malate by malate dehydrogenase. In contrast to oxalacetate, these two compounds are relatively easily transported across the mitochondrial membrane and reconverted to phosphoenolpyruvate, the major carbon precursor of glucose, by appro-

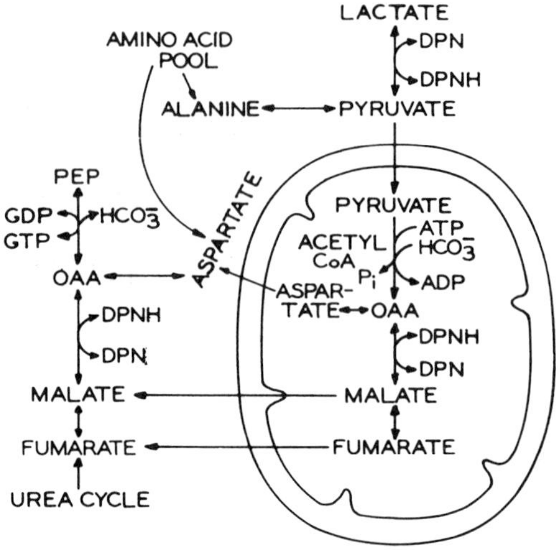

Figure 1. Proposed scheme for phosphoenolpyruvate formation in rat liver from dicarboxylic and amino acids. *Source:* Shrago and Lardy (1966).

priate enzymes in the cytosol. Considerable extension of this work, carried out with a variety of animals in many laboratories, has added new dimensions to the problem (Diesterhaft *et al.*, 1972; Garber and Hanson, 1971; Hanson and Garber, 1972; Johnson *et al.*, 1970; Söling *et al.*, 1970). In particular, it is now recognized that in the majority of animals the distribution of liver phosphoenolpyruvate carboxykinase is bimodal, which would permit synthesis of phosphoenolpyruvate directly in the mitochondria with subsequent diffusion of the substrate into the cytosol (Hanson and Garber, 1972; Utter, 1963). Such biological variation implies a rather complex mechanism for control of gluconeogenesis at the level of phosphoenolpyruvate formation.

The regulation of mitochondrial metabolism through the effect of small molecules on key enzymes is recognized as an important determining factor in the flux of carbon for gluconeogenesis. However, relatively little is known, at least at the physiological level, of the relationship of mitochondrial metabolite transport to control of metabolic pathways. During recent years a series of elegant experiments has been carried out in a number of laboratories delineating the existence of specific carrier systems or permeases that transport dicarboxylate and tricarboxylate anions as well as adenine nucleotides across the inner mitochondrial membrane (Chappell, 1968; Henderson *et al.*, 1970; Pfaff and Klingenberg, 1968; Vignais and Vignais, 1970; Weidemann *et al.*, 1970; Winkler and Lehniger, 1968). In contrast to the outer mitochondrial membrane, the inner membrane is relatively impermeable to any low molecular weight substances except water (Lehninger, 1970). It is now recognized that two distinct pools separated by the inner mitochondrial membrane exist for certain important substances including pyridine nucleotides and coenzyme A (Lehninger, 1970). The most extensively studied inner membrane carrier is that for ADP and ATP, which cannot pass the membrane by simple diffusion.

The delineation and characterization of the adenine nucleotide translocase came about from studies using the inhibitor atractyloside, a toxic plant glycoside that inhibits oxidative phosphorylation of externally added ADP (Bruni *et al.*, 1958). It was concluded that atractyloside inhibits transfer of ATP and ADP across the inner mitochondrial membrane rather than inhibiting oxidative phosphorylation (Bruni *et al.*, 1962; Vignais and Vignais, 1970). Thus the intramitochondrial pool of ATP and ADP is separated from the cytoplasmic pool, and communication occurs by way of an atractyloside-sensitive carrier. Our own interest in this problem arose from the finding that long-chain fatty acyl CoA esters like atractyloside inhibit adenine nucleotide translocation (Lerner *et al.*, 1972; Shrago *et al.*, 1972; Shug *et al.*, 1971). Since long-chain fatty acyl CoA esters must themselves be transported across the inner mitochondrial membrane by the carnitine acyltransferase enzyme(s), it is likely that these esters could act as natural effectors in the translocation of adenine nucleotides. The present discussion is an attempt to develop the proposal that regulation of mitochondrial metabolite transport by long-chain fatty acyl CoA esters plays an important role in the overall control of gluconeogenesis, lipogenesis, and energy-linked respiration. More detailed biochemical accounts of metabolite transport in mitochondria and intracellular membrane function can be found in various excellent review articles (Klingenberg, 1970; Racker, 1970).

II. METHODS OF PROCEDURE

For the most part, the studies described involve measurements on the translocation of adenine nucleotides (ATP, ADP), as well as the anions citrate and phosphoenolpyruvate across the inner mitochondrial membrane.

Experiments in the translocation and binding of ^{14}C-ADP as carried out in this laboratory by the forward exchange reaction have been described (Lerner *et al.*, 1972; Shrago *et al.*, 1974). In general a mixture containing 30 mM tris-HCl pH 7.5, 75 mM KCl, 1.5 mM EDTA, and mitochondrial protein in a volume of 0.9 ml is preincubated with or without inhibitor at ice temperature for 5 minutes. Subsequently, approximately 0.04 μCi ^{14}C-ADP is added, and after 5 minutes the reaction is terminated with 5.0 mM atractyloside. The reaction mixture is then centrifuged for 10 minutes at 4000 rpm, the supernatant is removed by suction, and radioactivity is determined in the pellet following solubilization in Soluene.

In certain experiments ^{14}C-ATP transport was measured by the back exchange technique described by Pfaff and Klingenberg (1968) and Henderson *et al.* (1970). Approximately 200 mg mitochondrial protein is loaded with 0.4 μCi ^{14}C-ATP by incubating at ice temperature for one hour in a medium containing 250 mM sucrose, 40 mM tris-HCl, pH 7.4, and 1.0 mM EDTA. The mitochondria are washed and centrifuged two times at 7000 *g* and finally suspended in the medium at a concentration of 20 mg mitochondrial protein per milliliter. The back exchange is carried out with 3.5 mg of mitochondrial protein containing approximately 9000 dpm preincubated in a medium of 100 mM KCl, 40 mM tris-HCl, pH 7.4, and 1.0 mM EDTA for 5 minutes at ice temperature. Following addition of 0.3 mM ATP, the reaction is stopped after 2 minutes with atracyloside, and the mitochondria are separated by centrifugation at 4000 rpm for 10 minutes. Solubilization and preparation of the precipitate for counting are carried out as previously described.

Procedures for ^{14}C citrate loading and transport are essentially the same as those described by Kleineke *et al.* (1973). Approximately 200 mg of mitochondrial protein is added to 7.0 ml of a mixture containing 125 mM KCl, 20 mM tris-HCl, pH 7.4, 5.6 μg rotenone, 14 μg antimycin A, and 0.5 to 1.0 μCi of 1,5-^{14}C citrate. After incubation at ice temperature for 10 minutes, 90 ml of solution containing 250 mM sucrose, 25 mM tris-HCl, pH 74, and 1.0 mM EGTA is added and the mixture centrifuged 10,000 *g* for 10 minutes. The separated mitochondria are washed twice and resuspended with the above-mentioned buffer. The exchange reaction is carried out in a 1.2 ml volume containing 200 mM manitol, 25 mM sucrose, 25 mM tris-HCl, pH 7.4, 6 mg mitochondrial protein, and approximately 80,000 dpm at ice temperature for 2 minutes, after which the mitochondria are separated by centrifugation at 18,000 *g* for 10 minutes. Solubilization and preparation of precipitate for counting are carried out as before. Determination of radioactivity and calculation of percentage of exchange are made as described by Kleineke *et al.* (1973). Additional details are given in the tables and figures.

III. RESULTS

These studies represent a relatively new concept in the regulation of metabolic pathways. Therefore, presentation of the results in a historical perspective may be advantageous. The energy level of the mitochondria and key enzymes of the citric acid cycle whose control contributes to the direction of carbon flux have been well documented (Veech and Krebs, 1969; Williamson, 1969). As an extension of work carried out while in Henry Lardy's laboratory (Lardy *et al.*, 1965; Shrago and Lardy, 1966), we initiated some studies with liver mitochondria from diabetic and hibernating animals (Shug *et al.*, 1971; Shug *et al.*, 1971a & b). Our interest was drawn to a repeated observation illustrated in Fig. 2. After addition of ADP, the oxygen consumption of hibernating (or diabetic) mitochondria was considerably less than that of normal mitochondria. This finding in itself was not very unusual—the literature is replete with observations showing alterations in mitochondrial structure and function under such metabolic conditions (Harano *et al.*, 1972; Matsubara and Tochino, 1969; South and House, 1967). We were impressed, however, because the low oxygen consumption of the hibernating mitochondria was increased to normal following addition of the uncoupler salicylanilide XIII. The result indicated that these were intact mitochondria in good respiratory control. Our interpretation of the results

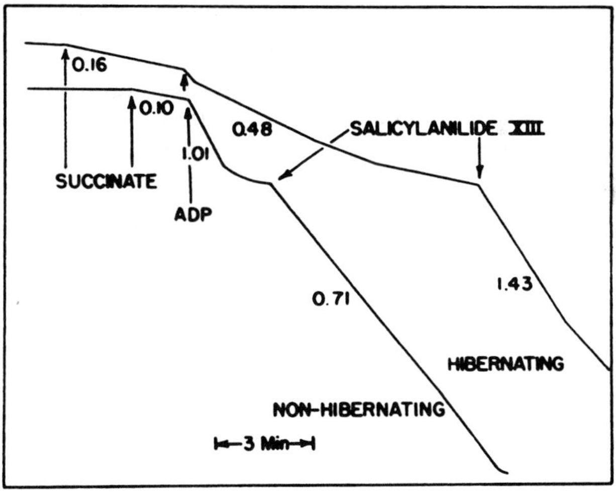

Figure 2. Control of respiration and response to salicylanilide XIII by liver mitochondria from hibernating and nonhibernating ground squirrels. Liver mitochondria (7.0 mg protein) was added to 3.0 ml reaction mixture containing 170 mM KCl, 330 mM sucrose, 17 mM KH_2PO_4, 10 mM $MgCl_2$, 10 mM triethanolamine-HCl, pH 7.4. At the points indicated, 2.0 mM succinate was added followed by the additions of 0.2 mM ADP and 0.2 μM salicylanilide XIII. The numbers in the O_2 tracing represent the respiration rate expressed in micromoles of oxygen per milligram of protein per hour. The oxygen concentration at the time of the succinate addition was 2.4 μM. *Source:* Shug *et al.* (1971).

was that in the livers of hibernating and diabetic animals, ADP cannot penetrate the inner mitochondrial membrane and reach the phosphorylating site.

We then tried to elucidate the mechanism responsible for this effect, and a review of the literature provided some interesting information that subsequently led to the correct conclusion. A previous report by Wojtczak and Zaluska (1967) indicated that fatty acids, in addition to their uncoupling activity at high concentrations, could inhibit translocation of adenine nucleotides through mitochondrial membranes. It therefore seemed logical to assume that increased fatty acid oxidation, characteristic of the hibernating and diabetic liver, would promote the accumulation of such substances and thereby inhibit the translocation of adenine nucleotides. Our own experiments (Lerner *et al.*, 1972; Shug *et al.*, 1971) indicated that oleic acid inhibited adenine nucleotide translocation only when its oxidation was inhibited, and that the actual inhibitory component was not the free fatty acid per se or the acyl carnitine derivative, but the long-chain acyl CoA ester (Table I and Fig. 3). Addition to rat liver mitochondria of a low concentration of oleoyl CoA, or oleic acid plus KCN, to permit endogenous accumulation of oleoyl CoA, produced marked inhibition of adenine nucleotide translocation and state 3 respiration. Transacylation of oleoyl CoA with L-

Table I. Inhibition of ^{14}C-ADP translocation through the mitochondrial membrane by oleic acid and oleoyl CoA, and its reversal by carnitine[a]

Additions	Radioactivity in mitochondria (dpm × 10^2)
Experiment 1:25°C	
None	576
KCN	534
Oleic acid	570
Oleic acid + KCN	274
Experiment 2:15°C	
None	286
Oleic acid	250
Oleoyl CoA	34
Carnitine	280
Oleoyl CoA + carnitine	238

[a] The assay procedure and basic reaction mixture are given in Section II as well as in Lerner *et al.* (1972), Shrago *et al.* (1974), and Shug *et al.* (1971). Additions were 0.3 mM KCl, 0.03 mM oleic acid, 0.03 mM oleoyl CoA, and 5.0 mM D,L-carnitine.

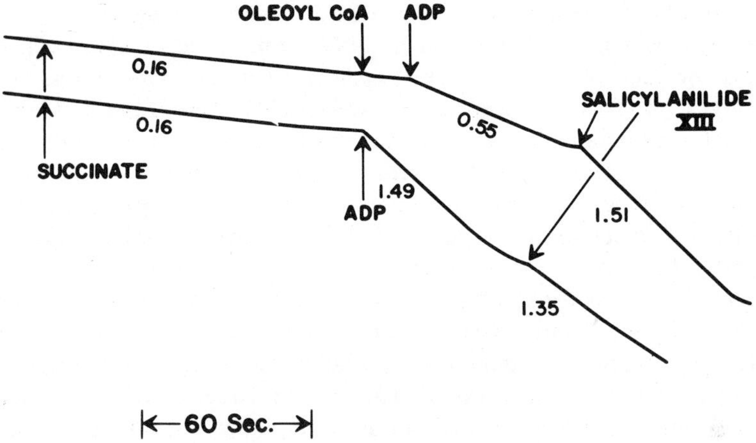

Figure 3. Control of respiration and response to oleoyl CoA by rat liver mitochondria. Liver mitochondria (7.0 mg protein) were added to 2.0 ml reaction mixture containing 20 mM KCl, 225 mM sucrose, 10 mM KH_2PO_4, 5 mM $MgCl_2$, and 20 mM triethanolamine-HCl, pH 7.4. At the point indicated, 2.0 mM succinate was added followed by additions of 0.03 mM oleoyl CoA, 0.35 mM ADP, and 0.2 μM salicylanilide XIII. The numbers in the O_2 tracing represent the respiration reate expressed in micromoles of oxygen per milligram of protein per hour. *Source:* Shug *et al.* (1971).

carnitine to oleoyl carnitine reduced or eliminated the inhibition. Similar conclusions were reached independently by Pande and Blanchaer (1971), who observed a reversible inhibition of electron transport linked phosphorylation of ADP in rat heart mitochondria incubated with palmitoyl CoA, but not with palmitoyl carnitine.

Subsequently, Harris *et al.* (1972) provided more convincing evidence that free fatty acids are not in themselves inhibitory by showing that enzymatic removal of the outer mitochondrial membrane with Nagarse treatment prevented both formation of the acyl CoA ester and inhibition of translocation. Others (Vaartjes *et al.*, 1972; Vaartjes, *et al.* 1972) as well as ourselves immediately became aware of the potential importance of the regulation of adenine nucleotide translocation by long-chain acyl CoA esters to the overall control of cell metabolism.

The adenine nucleotide translocase has been well characterized by elegant studies using the specific inhibitors atractyloside and bongkrekic acid (Henderson *et al.*, 1970; Pfaff and Klingenberg, 1968; Vignais and Vignais, 1970). In contrast to these artificial inhibitors, long-chain fatty acyl CoA esters are natural effectors whose interrelationship with adenine nucleotide metabolism at the inner mitochondrial membrane might offer a likely metabolic control mechanism. A series of experiments characterizing the reversible inhibition of adenine nucleotide translocation by long chain acyl CoA esters was carried out (Lerner *et al.*, 1972; Shrago *et al.*, 1974). Table II reveals the lack of inhibition by the short-chain acyl CoA esters through octanoyl CoA. At the concentration used (5 μM), there is marked inhibition in rat liver mitochondria by the higher chain derivatives except for linoleoyl CoA. Concentration curves for certain of the CoA esters appear in Fig. 4. Octanoyl CoA was completely ineffective up to a concentration of 50 μM. The lack of effectiveness of octanoyl CoA is consistent with evidence that the inhibitory acyl CoA esters are those whose oxidation is carnitine dependent. Linoleoyl CoA, which caused inhibition at concentrations greater than 6 μM, never approached the effectiveness of oleoyl or myristoyl CoA. Under the experimental conditions given, myristoyl CoA was the most inhibitory of the CoA esters tested. Kinetic experiments on the inhibition of the acyl CoA esters were carried out using the lubrol membrane preparation from rat liver mitochondria (Fig. 5). Although no longer able to retain a pool of adenine nucleotides, this preparation does have the capacity to bind adenine nucleotides, and it is atractyloside sensitive (Winkler and Lehninger, 1968). Palmitoyl CoA, which caused a marked inhibition of adenine nucleotide translocation, has a K_i of 0.3 μM. The inhibition by palmitoyl CoA is competitive in type which is similar to that previously described for atractyloside (Pfaff and Klingenberg, 1968; Weidemann *et al.*, 1970). The (Kleineke *et al.*, 1973; Robinson, 1971). In the table, 1, 2, 3-benzene tricarboxylate is shown to cause a strong inhibition of the exchange of ^{14}C-

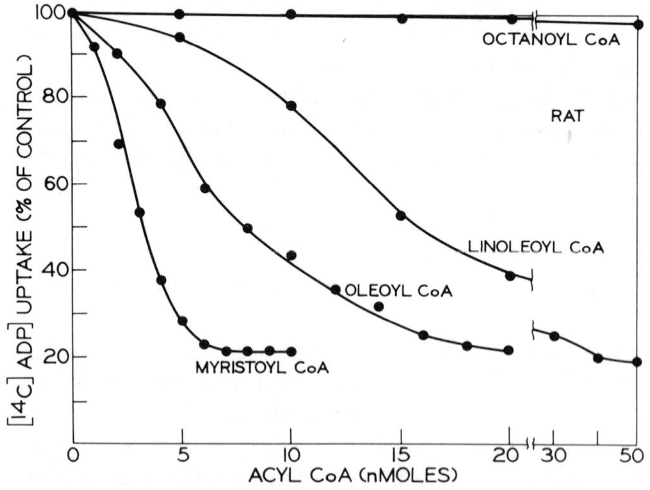

Figure 4. Effect of increasing concentrations of various long-chain fatty acyl CoA esters on the uptake of ^{14}C-ADP by rat liver mitochondria. Experiments were carried out as described in Table *I* and Section II. *Source:* Shrago *et al.* (1974).

Table II. Effect of short- and long-chain fatty acyl CoA esters on ^{14}C-ADP translocation[a]

Fatty acyl CoA ester	Percent of control
None	100
Acetyl CoA	100
Succinyl CoA	100
Butyroyl CoA	98
Octanoyl CoA	103
Lauryl CoA	35
Myristoyl CoA	13
Palmitoyl CoA	3
Stearoyl CoA	10
Oleoyl CoA	35
Linoleoyl CoA	101

[a] Assays were carried out as described in Table I and Section II. Fatty acyl CoA esters were added at a concentration of 5 μM.

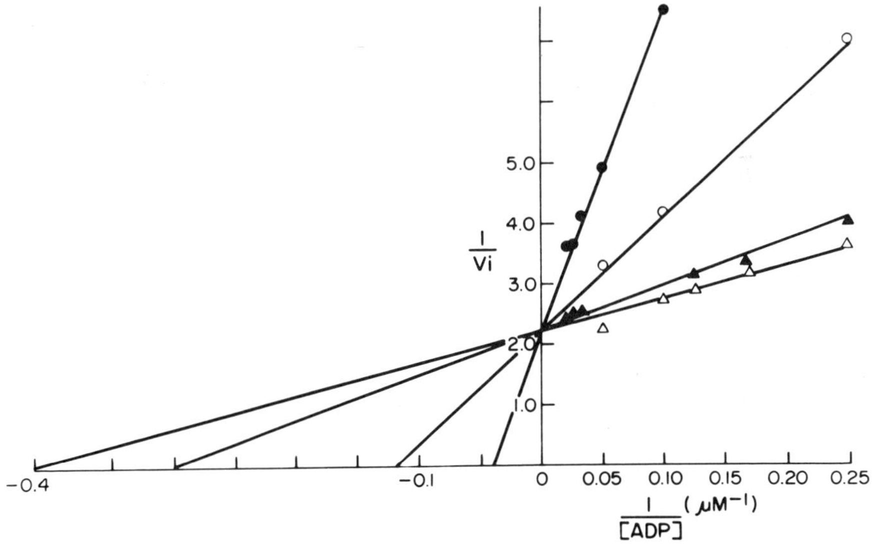

Figure 5. Double reciprocal plot of palmitoyl CoA inhibition of the concentration dependent rate of ^{14}C-ADP uptake by the inner membrane fraction of rat liver mitochondria. Double reciprocal plots are shown for the effect of palmitoyl CoA on the concentration dependent ADP binding. A comparison of a curve for atractyloside is also included. Control △-△-△; palmitoyl CoA 0.1 μmole, ▲-▲-▲; palmitoyl CoA 0.5 μmole O-O-O; atractyloside 5 μmole ●-●-●. *Source:* Shrago *et al.* (1974).

retention of the acyl CoA as well as atractyloside sensitivity by the submitochondrial particle is good evidence for a common action of the inhibitors at the same site on the membrane. It appears likely that the adenine nucleotides and long-chain acyl CoA esters compete for a common binding site on the carrier. This hypothesis is favored because the CoA molecule has an adenine group. Since free CoA and short-chain acyl CoA esters are not effective inhibitors, carbon chain length, which may provide the proper steric configuration, is important.

That changes in adenine nucleotide translocation actually occur during altered metabolic states is additional evidence for the importance of this transport function in cell metabolism (Lerner *et al.*, 1972; Shug *et al.*, 1971). Table III compares the effects of long-chain fatty acyl CoA esters on adenine nucleotide translocation in liver and heart mitochondria prepared from a number of animals. As previously shown, the addition of carnitine to

Table III. Adenine nucleotide translocase activity in heart and liver mitochondria in animals under various experimental conditions[a]

Animal	Adenine nucleotide translocase ($dpm \times 10^2$) Liver	Heart
Rat—normal	290	220
Rat + palmitoyl CoA	30	45
Rat + palmitoyl CoA + carnitine	131	
Rat + octanoyl CoA	280	
Rat—diabetic	157	
Rat—diabetic + carnitine	255	
Guinea pig—normal	250	
Guinea pig + palmitoyl CoA	20	
Ground squirrel—awake	267	
Ground squirrel—hibernating	27	
Ground squirrel—hibernating + carnitine	55	
Chicken embryo—10 days	266	
Chicken embryo + palmitoyl CoA	56	
Host liver	300	
Hepatoma	75	
Dog—normal heart		130
Dog—infarcted heart		70

[a] Assays were carried out as in Table I and Section II. Palmitoyl CoA was added at a concentration of 5 μM and D,L-carnitine at 5 mM.

an experimental system containing the long-chain acyl CoA ester reverses the inhibition of adenine nucleotide translocase activity. Liver mitochondria from diabetic rats and hibernating ground squirrels were found to have lower levels of adenine nucleotide translocase activity, which could be reversed toward normal by addition of carnitine to the reaction mixture. These *in vivo* results complement the *in vitro* studies in which a long-chain fatty acyl CoA ester was added to the experimental system. Embryonic chicken liver, which by necessity is dependent on fatty acid oxidation for energy, contains adenine nucleotide translocase activity susceptible to acyl CoA inhibition. In comparison to host liver tissue, hepatoma mitochondria were found to have rather low adenine nucleotide translocase activity. Of particular interest are the results obtained with mitochondria prepared from the infarcted area of dog hearts. It is well known that the myocardium relies on free fatty acids as a major energy source. A number of observations indicate that during anoxia or ischemia, high concentrations of free fatty acids may be harmful to myocardial function (Henderson *et al.*, 1970; Kurien and Oliver, 1970; Kurien *et al.*, 1971). Previous *in vitro* studies from this laboratory (Shug and Shrago, 1973) demonstrated that impaired mitochondrial oxidation of long-chain fatty acids by cyanide or anaerobiosis with a subsequent accumulation of long-chain acyl CoA esters, caused an inhibition of adenine nucleotide translocation. *In vivo* findings similar to those presented here were observed in the infarcted area of dog hearts. Inhibition of adenine nucleotide translocase could be induced in the ischemic area within 15 minutes after production of the infarct, at which time oxidative phosphrylation and respiratory control were essentially normal (Bittar *et al.*, in press).

Halperin *et al.* (1972) observed an inhibition of citrate efflux from rat liver mitochondria by palmitoyl CoA. This report was of considerable interest because the observation implied a close functional relationship between the tricarboxylate and adenine nucleotide carrier systems. It had already been established that phosphoenolpyruvate is transported across the mitochondrial membrane by the tricarboxylate carrier and that this transport system is rather nonspecific (Robinson, 1971). Additional relevant information, concerned with the inhibitory effect of phosphoenolpyruvate on mitochondrial protein synthesis (McCoy and Doeg, 1972), suggested to us that phosphoenolpyruvate might also be transported by the adenine nucleotide translocase system. A close functional relationship between the adenine nucleotide and tricarboxylate carrier systems has now been established (Shug and Shrago, 1973). Table IV confirms that 1, 2, 3-benzene tricarboxylate is a potent inhibitor of the tricarboxylate transport system citrate with citrate and phosphoenolpyruvate, whereas adenine nucleotide transport is not affected. By contrast, atractyloside and oleoyl CoA, both potent inhibitors of adenine nucleotide translocase, also inhibit the

Table IV. Transport of (^{14}C) ATP and (^{14}C) citrate with various anions and inhibition with 1,2,3-benzene tricarboxylate, atractyloside and oleoyl-CoA[a]

		Percent Exchange	
Anion	Inhibitor	(^{14}C) ATP Exchange	(^{14}C) Citrate Exchange
Citrate	None	2.8	71.9
Citrate	1,2,3-BTCA	(—)	−0.5
Citrate	Atractylate	(—)	−11.1
Citrate	Oleoyl-CoA	(—)	−2.6
Phosphoenolpyruvate	None	67.5	64.4
Phosphoenolpyruvate	1,2,3-BTCA	71.9	−1.2
Phosphoenolpyruvate	Atractylate	3.3	−8.3
Phosphoenolpyruvate	Oleoyl CoA	5.3	−1.1
ADP	None	82.8	(—)
ADP	1,2,3-BTCA	80.8	(—)
ADP	Atractylate	3.4	(—)
ADP	Oleoyl CoA	8.6	(—)

[a] Assays were carried out as described in Section II and Henderson *et al.* (1970), Kleineke *et al.* (1973), and Pfaff and Klingenberg (1968). Results expressed as percent exchange as previously described (Matsubara and Tochino, 1969; Shug and Shrago, (1973) are averages of duplicate incubation from a representative experiment. An absence of exchange activity is expressed by a negative sign (−). Negative values indicate that in the presence of that anion, the counts in the supernatant were lower than when incubated in buffer only. The concentration of externally added citrate was 1.0 mM; ADP, 0.1 mM; and phosphoenolpyruvate 5.0 mM. The concentration of inhibitors were 5 μM oleoyl-CoA, 0.5 mM atractyloside, and 50 mM 1,2,3-benzene tricarboxylate (BTCA).

tricarboxylate carrier system. Since it had been assumed that atractyloside was specific for the adenine nucleotide translocase, these more recent results demonstrating an additional effect on the tricarboxylate carrier strongly suggest a close structural and functional relationship between the two carrier systems.

IV. DISCUSSION

Any enzyme complex or transport system that functionally links one cell compartment with another must play a central role in cellular metabolism.

The adenine nucleotide translocase is especially important in this respect because it is able to control the relative concentrations of ATP and ADP, perhaps the most important chemical species of the cell, in the cytosolic and mitochondrial compartments. The only known physiological effectors of the adenine nucleotide and tricarboxylate carriers are long-chain fatty acyl CoA esters. The importance of such a control mechanism for mitochondrial membrane transport in the overall homeostasis of cell metabolism has been clearly stated by McLean *et al.* (1971). As a consequence of *in vivo* inhibition of adenine nucleotide translocation by long-chain acyl CoA esters, an immediate series of events would occur: (1) a transition from state 3 to state 4 mitochondrial respiration, (2) an increase in the ratio of ATP/ADP $\times$ P_i in the mitochondria, (3) an increase in the mitochondrial NADH/NAD ratio, and (4) a decrease flux through the tricarboxylic acid cycle. A major role for long-chain fatty acyl CoA esters in the regulation of gluconeogenesis, ketogenesis, lipogenesis, and heat production is documented specifically by the references listed in Table V.

The control points at which key mitochondrial enzymatic reactions are modulated subsequent to changes in metabolite transport are diagrammed in Fig. 6. In mitochondria, pyruvate can either be decarboxylated to acetyl CoA or carboxylated to oxalacetate. In the former reaction the pyruvate dehydrogenase complex converts pyruvate to acetyl CoA, leading further to citrate synthesis. In the alternative reaction pyruvate is converted to oxalacetate by pyruvate carboxylase, and oxalacetate may enter the gluconeogenic pathway or condense with acetyl CoA to form citrate, a lipogenic precursor. A potent control of pyruvate dehydrogenase is through phosphorylation-dephosphorylation of the enzyme complex (Linn *et al.*,

Table V. Major metabolic pathways regulated by inhibition of adenine nucleotide translocation

	Effect	References
Gluconeogenesis	+	Lerner *et al.* (1972), McLean *et al.* (1971), Shrago *et al.* (1972)
Ketogenesis	+	Vaartjes *et al.* (1972)
Lipogenesis	−	Halperin *et al.* (1972), Shug and Shrago (1973)
Heat production	+	Christiansen *et al.* (1973)

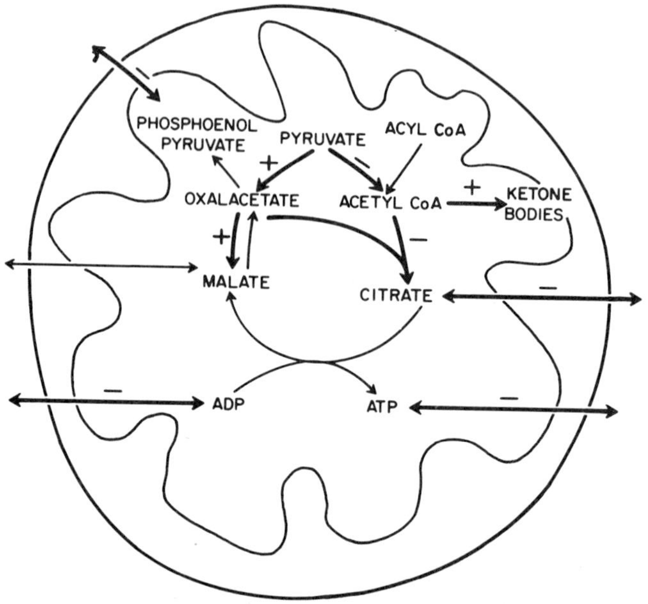

Figure 6. Schematic representation of the positive (+) and negative (−) effects of inhibited adenine nucleotide and tricarboxylic carrier systems on mitochondrial metabolism.

1969). During gluconeogenesis, an increased intramitochondrial ATP level, subsequent to inhibition of adenine nucleotide translocation, would promote the phosphorylation and inactivation of the pyruvate dehydrogenase complex, which in turn would favor an increase in pyruvate carboxylation. It has been convincingly shown that pyruvate carboxylase activity is regulated mainly by the intramitochondrial ATP/ADP ratio (Stucki *et al.*, 1972). Thus lowering mitochondrial ADP through inhibition of adenine nucleotide translocation would also result in the stimulation of pyruvate carboxylase. Insulin and glucagon influence the pyruvate dehydrogenase complex in a reciprocal manner (Taylor *et al.*, 1973; Zahlten *et al.*, 1973). It therefore seems possible that the initial effects of these hormones could occur through changes in the adenine necleotide ratio. The transition from state 3 to state 4 respiration as a consequence of inhibited adenine nucleotide translocation would be conducive to reversal of electron transfer, reduction of NAD, and the intramitochondrial conversion of oxalacetate to malate, a gluconeogenic precursor (Lerner *et al.*, 1972). Since the dicarboxylate carrier system is considerably less sensitive to inhibition by acyl CoA esters than other translocases (Halperin *et al.*, 1972), there would be little hindrance to the transport of malate from the mitochondria. It has been shown (Vaartjes *et al.*, 1972) that palmitoyl CoA inhibition of adenine nucleotide transport in mitochondria is reflected by increased 3-hydroxybutyrate/acetoacetate ratios and in the preferential conversion of acetyl CoA to ketone bodies rather than citrate. Under these conditions, long-chain acyl CoA esters would also inhibit citrate egress from the mitochondria. An inhibition of lipogenesis would then occur (Halperin, *et al.*, 1972), since the availability of citrate in the cytosol as a substrate for fatty acid synthesis and an allosteric activator of actyl CoA carboxylase would be limited. The diagram of Fig. 6 is representative of studies carried out primarily in rat liver. It is interesting to speculate that a similar metabolic response to long-chain acyl CoA esters in animals that synthesize phosphoenolpyruvate in their mitochondria might partially account for the negative effect of fatty acids on gluconeogenesis in these animals (Arinze *et al.*, 1973). Phosphoenolpyruvate synthesized in the mitochondria would be poorly transported on the tricarboxylate and adenine nucleotide carriers, thus would not be available in the cytosol for subsequent conversion to glucose.

More recent studies suggest an ever-widening scope of the influence of mitochondrial metabolite transport on cellular function. We have been focusing our attention on the effect of inhibited adenine nucleotide translocase on myocardial metabolism. The metabolic events following myocardial ischemia could be partly due to the deleterious effect of fatty acids whose impaired oxidation would result in increased levels of long-chain acyl CoA esters and severely inhibited translocase. As demonstrated in dogs, decreased adenine nucleotide translocase activity occurs very rapidly following experimental myocardial infarction. In addition, we have shown that inhibition of translocation is accompanied by a marked increase of long-chain acyl CoA esters in the injured area (Bittar *et al.*, in press). A number of mechanisms can be proposed to show how a disequilibrium of ATP and ADP in the cytosol and mitochondria could contribute to the serious complications of cardiogenic shock and fatal arrhythmias.

Chudapongse and Haugaard (1973) have shown that phosphoenolpyruvate stimulates the egress of calcium from heart mitochondria and that this process is atractyloside sensitive. Based on our own work it appears likely that in the heart, phosphoenolpyruvate enters the mitochondria on the adenine nucleotide carrier with the concurrrent egress of ATP and calcium. Regulation of this metabolic process by long-chain acyl CoA esters would certainly influence calcium effects on myocardial function.

A number of altered mitochondrial functions in ethanol induced fatty livers have been associated with elevated levels of long-chain acyl CoA esters and inhibited adenine nucleotide translocation (Gordon, 1973).

Portnoy *et al.* (1973) have shown a stimulation of adenine nucleotide transport following administration of thyroid hormone and have suggested that the calorigenic action of thyroid hormone may be at the level of the adenine nucleotide translocator. However, a secondary effect subsequent to increased metabolism and lowering of long-chain acyl CoA esters by thyroxin cannot be ruled out.

Considerable documentation must still be provided before any generalizations on the overall significance of the interrelationship of long-chain fatty acyl CoA esters and mitochondrial transport carriers can be unequivocally accepted. A number of the theoretical and experimental problems that remain are clinically as well as biologically important. For effective modulation of liver metabolism, the mitochondrial concentration of long-chain fatty acyl CoA esters and adenine nucleotides must be so coordinated that continuous inhibition of state 3 energy-linked respiration is not permitted. The actual concentrations of long-chain acyl CoA esters and adenine nucleotides within the mitochondria and cytosol must be accurately determined. In attempting to answer these and other questions, we are indeed fortunate to have the continued enthusiastic support and critical expertise of Henry Lardy, who has provided the intellectual atmosphere leading to continued generation of new ideas from many laboratories.

V. SUMMARY AND CONCLUSIONS

The regulation of mitochondrial metabolite transport is of paramount importance to proper cell function. The energy charge and redox state of the cell are dependent on the transport of ATP and ADP between the mitochondria and cytosol. Citrate, malate, and phosphoenolpyruvate, among other anions transported on the di- and tricarboxylate carriers of the inner mitochondrial membrane, are important as substrates and allosteric effectors for key metabolic reactions. The direction of carbon flux reflects resulting changes in the ATP/ADP and NAD/NADH ratios of the mitochondria and cytosol. Alterations in the transport of phosphoenolpyruvate, citrate, and malate will have significant impact on the synthesis of glucose and fatty acids in the liver. It has been shown that long-chain acyl CoA esters act as physiological effectors of cell metabolism by causing a reversible inhibition of the adenine nucleotide and tricarboxylate carrier systems. The interrelationship of metabolite transport and fatty acid metabolism in the inner mitochondrial membrane, therefore, affords a fine control mechanism for the direction and rate of carbon flux through metabolic pathways.

ACKNOWLEDGMENTS

This work was supported by NIH grant GM-14033, the American Diabetes Association and the College of Agriculture and Life Sciences, University of Wisconsin, Madison.

REFERENCES

Arinze, I. J., Garber, A. J., and Hanson, R. W. (1973). *J. Biol. Chem.* **248,** 2266.

Bittar, N., Shug, A., Koke, J., Folts, J. D., and Shrago, E. In *Myocardial Cell Damage, Recent Advances in Cardiac Structure and Metabolism,* Vol. V (A. Fleckenstein, ed.), University Park Press, Baltimore, Md. In press.

Bruni, A., Bistocchi, M., and Contessa, A. R. (1958). *Boll. Soc. Ital. Biol. Sper.* **34,** 459.

Bruni, A., Contessa, A. R., and Luciani, S. (1962). *Biochem. Biophys. Acta.* **60,** 301.

Chappell, J. B. (1968). *Brit. Med. Bull.* **24,** 150.

Christiansen, E. N., Drahota, Z., Duszynski, J., and Wojtczak, L. (1973). *Eur. J. Biochem.* **34,** 506.

Chudapongse, P. and Haugarrd, N., (1973). *Biochem. Biophys. Acta* **307,** 599.

Diesterhaft, M. D., Hsieh, H. C., Elson, C., Sallach, H. J., and Shrago, E. (1972). *J. Biol. Chem.* **247,** 2755.

Garber, A. J. and Hanson, R. W. (1971). *J. Biol. Chem.* **246,** 589.

Gordon, E. R. (1973). *J. Biol. Chem.* **248,** 8271.

Halperin, M. L., Robinson, B. H., and Fritz, I. B. (1972). *Proc. Nat. Acad. Sci. (U.S.)* **69,** 1003.

Hanson, R. W. and Garber, A. J., (1972). *Am. J. Clin. Nutr.* **25,** 1010.

Harano, Y., DePalma, R. G., Lavin, E. L., and Miller, M. (1972). *Diabetes* **21,** 257.

Harris, R. A., Farmer, B., and Oxowa, T. (1972). *Arch. Biochem. Biophys.* **150,** 199.

Henderson, A. H., Most, A. S., Parmley, W. W., Gorlin, R., and Sonnenbleck, E. H. (1970). *Circ. Res.* **26,** 439.

Henderson, P. J. F., Lardy, H. A., and Dorschner, E., (1970). *Biochemistry* **9,** 3453.

Johnson, B. C., Hebert, K. A., and Ray, P. D. (1970). *Biochem. Biophys. Res. Commun.* **39,** 750.

Kleineke, J., Sauer, H., and Söling, H. D. (1973). *FEBS Lett.* **29,** 82.

Klingenberg, M. (1970). In *Essays in Biochemistry,* Vol. VI (P. N. Cambell and F. Dickens, eds.), Academic Press, New York, pp. 119–159.

Kurien, V. A. and Oliver, M. F. (1970). *Lancet* **1,** 813.

Kurien, V. A., Yates, P. A., and Oliver, M. F. (1971). *Eur. J. Clin. Invest.* **1,** 25.

Lardy, H. A. (1964–1965). *The Harvey Lectures,* Series 60, Academic Press, New York, pp. 261–278.

Lardy, H. A., Paetkau, V., and Walter P. (1965). *Proc. Nat. Acad. Sci. (U.S.)* **53,** 1410.

Lehninger, A. L. (1970). *Biochemistry,* Wirth Publishers, New York.

Lerner, E., Shug, A., Elson, C., and Shrago, E. (1972). *J. Biol. Chem.* **247,** 1513.

Linn, T. C., Pettit, F. H., and Reed, L. J. (1969). *Proc. Nat. Acad. Sci. (U.S.)* **69,** 234.

Matsubara, T. and Tochino, Y. (1969). *J. Biochem. Tokyo* **66,** 397.

McCoy, G. D. and Doeg, K. A. (1972). *Biochem. Biophys. Res. Commun.* **46,** 1411.

McLean, P., Gumaa, K. A., and Greenbaum, A. L. (1971). *FEBS Lett.* **17,** 345.

Pande, S. V. and Blanchaer, M. C. (1971). *J. Biochem.* **246,** 402.

Pfaff, E. and Klingenberg, M. (1968). *Eur. J. Biochem.* **6,** 666.

Portnoy, G. I., McClendon, F. D., Bush, J. E., Braverman, L., and Babior, B. M. (1973). *Biochem. Biophys. Res. Commun.* **55,** 17.

Racker, E. (1970). In *Essays in Biochemistry,* Vol. VI (P. N. Campbell and F. Dickens, eds.), Academic Press, New York, pp. 1–22.

Robinson, B. H. (1971). *FEBS Lett.* **14,** 309.

Shrago, E. and Lardy, H. A. (1966). *J. Biol. Chem.* **241,** 663.

Shrago, E., Shug, A., Elson, C., and Lerner, E. (1972). In *The Roles of Membranes in Metabolic Regulation* (M. Mehlman and R. Hanson, eds.), Academic Press, New York, pp. 165–182.

Shrago, E., Shug, A., Elson, C., Spennetta, T., and Crosby, C. (1974). *J. Biol. Chem.* **249,** 5269.

Shug, A., Ferguson, S., Shrago, E., and Burlington, R. (1971). *Biochem. Biophys. Acta.* **226,** 309.

Shug, A., Lerner, E., Elson, C., and Shrago, E. (1971). *Biochem. Biophys. Res. Commun.* **43,** 557.

Shug, A., and Shrago, E. (1973). *J. Lab. Clin. Med.* **81,** 214.

Shug, A., and Shrago, E., (1973). *Biochem. Biophys. Res. Commun.* **53,** 659.

Söling, H. D., Willms, B., Kleineke, J., and Gehlhoff, M. (1970). *Eur. J. Biochem.* **16,** 289.

South, F. E. and House, W. A. (1967). In *Mammalian Hibernation,* Vol. III (K. C. Fisher, A. R. Dawe, C. P. Lyman, E. Schonbaum, and F. E. South, eds.), American Elsevier, New York, pp. 305–331.

Stucki, J. W., Brawand, F., and Walter, P. (1972). *Eur. J. Biochem.* **27,** 181.

Taylor, S. I., Mukherjee, C., and Jungas, R. L. (1973). *J. Biol. Chem.* **248,** 73.

Utter, M. D. (1963). *Iowa State J. Sci.* **38,** 97.

Vaartjes, W. J., Kemp, A., Jr., Souverijn, J. H. M., and van den Bergh, S. G. (1972). *FEBS Lett.* **23,** 303.

Vaartjes, W. J., Lopes-Cardozo, M., and van den Bergh, S. J. (1972). *FEBS Lett.* **26,** 117.

Veech, R. L. and Krebs, H. A. (1969). In *Energy Level and Metabolic Control in Mitochondria* (S. Papa, J. M. Tager, E. Quagliariello, and E. C. Slater, eds.), Adriatica Editrice, Bari, Italy, pp. 329–338.

Vignais, P. V. and Vignais, P. M. (1970). *Biochem. Biophys. Res. Commun.* **38,** 843.

Weidemann, M. J., Erdelt, H., and Klingenberg, M. (1970). *Eur. J. Biochem.* **16,** 313.

Williamson, J. R. (1969). In *Energy Level and Metabolic Control in Mitochondria* (S. Papa, J. M. Tager, E. Quagliariello and E. C. Slater, eds.), Adriatica Editrice, Bari, Italy, pp. 385–398.

Winkler, H. H. and Lehniger, A. L. (1968). *J. Biol. Chem.* **243,** 3000.

Wojtczak, L. and Zaluska, H. (1967). *Biochem. Biophys. Res. Commun.* **28,** 76.

Zahlten, R. N., Stratman, F. W., and Lardy, H. A. (1973). *Proc. Nat. Acad. Sci. (U.S.)* **70,** 3213.

CHAPTER SEVEN

Pyruvate Carboxylase: Intracellular Localization and Regulation

PAUL WALTER

Pyruvate carboxylase catalyzes the conversion of pyruvate to oxalacetate:

$$\text{pyruvate} + \text{ATP} + \text{HCO}_3^- \xrightarrow[\text{acetyl-CoA}]{\text{Mg}^{2+}} \text{oxalacetate} + \text{ADP} + \text{P}_i$$

This reaction is the first step of the pathway of gluconeogenesis in liver and in kidney tissue. Various parameters are known to influence the activity of pyruvate carboxylase, and this enzyme is believed to act as one of the pacemakers for gluconeogenesis (for a review, see Scrutton and Utter, 1968).

This chapter deals with two aspects related to pyruvate carboxylase. In the first part, evidence is presented for an exclusive intramitochondrial localization of the enzyme in rat liver; in the second part several aspects of the regulation of pyruvate carboxylase in intact mitochondria are discussed.

I. INTRACELLULAR DISTRIBUTION OF PYRUVATE CARBOXYLASE IN RAT LIVER

The intracellular distribution of the enzymes participating in gluconeogenesis has been studied by several laboratories. All the enzymes that catalyze the conversion of oxalacetate to glucose-6-phosphate are localized in the cytosol of the cell; however, most of the pyruvate carboxylase activity had been found inside the mitochondria. This enzyme distribution has led to the concept of a partially mitochondrial pathway of gluconeogenesis (Lardy *et al.*, 1965; Haynes, 1965; Walter *et al.*, 1966) and to the proposal that mitochondrial processes such as fatty acid oxidation (Walter *et al.*, 1966), changes in adenine nucleotide concentrations (Stucki *et al.*, 1972), and other parameters (Scrutton and Utter, 1968) can directly influence the rate of gluconeogenesis. This concept has been challenged by the finding that a substantial amount of pyruvate carboxylase is localized outside the mitochondria—therefore the mitochondrial metabolism may not be directly involved in the regulation of gluconeogenesis (Henning and Seubert, 1964; Henning *et al.*, 1966; Seubert *et al.*, 1968; Dugal, 1972).

The most thorough investigation of the intracellular distribution of pyruvate carboxylase in rat liver was carried out by Böttger *et al.* (1969). Their cell fractionation study showed that the intracellular distribution pattern of pyruvate carboxylase activity followed exactly that of the intramitochondrial marker enzymes citrate synthase, 3-hydroxyacyl-dehydrogenase, and glutamate dehydrogenase, and these workers concluded that more than 97% of the pyruvate carboxylase activity had to be localized within the mitochondria. On the other hand, using a different fractionating method, Seubert's group reported that more than 30% of the total pyruvate carboxylase activity was found extramitochondrially, as compared to only 1 to 6% of the total glutamate dehydrogenase activity, indicating that at least 24 to 30% of the total pyruvate carboxylase is localized outside the mitochondria in the liver cell. Moreover, in a more recent publication Dugal (1972) reported that upon prolonged homogenization (10 minutes) of the liver with the media used by Seubert's group, up to 54% of pyruvate carboxylase versus only 11% of glutamate dehydrogenase activity could be measured extramitochondrially.

The object of this work was to try to find the reason for these apparently contradictory reports of Böttger *et al.* (1969) and Seubert's group (1968) and Dugal (1972). The results of our studies lead to the conclusion that under the experimental conditions of Seubert's group, equal percentages of pyruvate carboxylase and glutamate dehydrogenase activities are released by the mitochondria during homogenization, but part of the released glutamate dehydrogenase activity is bound by subcellular fractions such as microsomes and nuclei, whereas this is not the case for pyruvate carboxylase. Since only unbound extramitochondrial enzyme activities are measured by the method of Seubert and Dugal, a higher percentage of pyruvate carboxylase than of glutamate dehydrogenase activity is found outside the mitochondria. With a phosphate-containing homogenization medium in which glutamate dehydrogenase does not bind to other organelles, *equal* percentages of both enzyme activities were found extramitochondrially. These results explain the mentioned discrepancies of results in the literature and support the concept of an exclusive intramitochondrial localization of pyruvate carboxylase. Preliminary accounts of this work were published earlier (Walter and Anabitarte, 1971, 1973).

A. Methodology

We used male rats, either fed or fasted 12 to 24 hours. The animals (CFN COBS) were obtained from the Tierzucht Institut of the University of Zurich and weighing 190 to 240 g. The livers were removed from ether-anesthetized rats and homogenized in the various media indicated in Table I, and the cytosols were prepared according to the fractionation procedure of Henning *et al.* (1966). This procedure involves homogenization in a Potter homogenizer for 40 seconds at a ratio of 1:5, centrifugation at 900 *g* for 8 minutes, rehomogenization of the resulting pellet for 1 minute with the same volume as before, centrifugation at 900 *g* for 8 minutes, and combination of the two supernatants. This combined supernatant is referred to as "900 *g* supernatant" in Tables IV and V. The method of Böttger *et al.* (1969) was followed for the extraction of the total liver activities of pyruvate carboxylase and glutamate dehydrogenase in the respective media. Pyruvate carboxylase activities were measured by the method of Henning and Seubert (1964) by transforming to citrate the oxalacetate, formed from pyruvate, and by measuring the citrate enzymatically (Walter and Stucki, 1970; Moellering and Gruber, 1966). Glutamate dehydrogenase was measured as described by Schmidt (1962) by transforming α-ketoglutarate to glutamate; 0.6 mM ADP (D. H. Williamson *et al.*, 1967) was added when indicated. Sonication of the mitochondria was performed in 0.28 M sucrose at a protein concentration of about 7 mg/ml with a Branson Sonifier model B-12 at setting 5, for four times during 10 seconds. The sonicated mitochondrial preparation was then centrifuged at 100,000 *g* for 10 minutes and the resulting supernatant used for the experiments of Table III.

B. Results

The experiments with the homogenization media I, III, and IV (Table I) were carried out according to the method reported by Henning *et al.* (1966); the only exception concerns the assay system for measuring glutamate dehydrogenase. The system used contained ADP (D. H. Williamson *et al.*, 1967), in contrast to the procedure of Henning *et al.*, (1966), in which this nucleotide was not added. The results show that the main effect

Table I. Pyruvate carboxylase and glutamate dehydrogenase activities in cytosols of rat livers homogenized in various media

Homogenization medium	Metabolic state (number of subjects in parentheses)	Enzyme activities (μmole/min·g liver) (± standard deviation): Total homogenate		Cytosol[a]	
		PC	GDH[b]	PC[c]	GDH[b,c]
I					
0.24 M sucrose, 1 mM EDTA, 1 mM glutathione, 20 mM TEA-HCl pH 7.2	Fed (4)	7.4 ±1.3	187 ±13	0.81 ±0.18(11%)	2.38 ±0.79(1.3%)
	Fasted 24 hours (3)	8.6 ±0.4	253 ±29	1.77 ±0.10(21%)	7.85 ±1.0(3.1%)
II					
0.14 M sucrose, 50 mM potassium phosphate buffer pH 7.2	Fed (5)	7.9 ±1.1	232 ±27	0.52 ±0.1(6.6%)	14.4 ±2.7 (6.2%)
	Fasted 24 hours (1)	11.8	251	1.0 (8.5%)	18.4 (7.4%)
	Fasted 24 hours (1)	10.8	322	1.6 (15%)	40.0 (12%)
III					
0.28 M sucrose, 1 mM EDTA, 1 mM glutathione, 50 mM sodium acetate	Fed (1)	5.8	166	0.44 (7.6%)	0.48 (0.2%)
	Fasted 24 hours (1)	8.2	225	1.52 (19%)	1.09 (0.5%)
IV					
0.30 M sucrose, 1 mM EDTA, 1 mM glutathione	Fed (1)	4.0	94	0.20 (5%)	1.0 (1.1%)
	Fed (1)	5.7	87	0.42 (7.4%)	2.6 (3.0%)

[a] Preparation of cytosol according to Henning *et al.* (1966).
[b] Assayed in the presence of ADP.
[c] Numbers in parentheses refer to percentage of activity of the respective enzymes in the total homogenate.

demonstrated by Seubert *et al.* (1968) could be reproduced—namely, that a higher percentage of pyruvate carboxylase than of glutamate dehydrogenase is measured in the cytosol. Also in agreement with Seubert *et al.* (1968), it was found that this effect was most pronounced with the acetate containing medium III and furthermore that higher percentages of enzyme activities appeared in the cytosols of livers from fasted than from fed animals (Henning *et al.*, 1966). The percentages of enzyme activities found in the cytosol were somewhat smaller than those reported by Seubert's group. They were found to vary from experiment to experiment, and it was observed that the tighter the homogenizer, the more enzyme activity was measured in the cytosol, although this was not statistically proved.

It should be noted that as previously published (Walter and Anabitarte, 1971), all these effects could also be shown when ADP was omitted in the assay system for glutamate dehydrogenase, except that the total glutamate dehydrogenase and the cytosol-glutamate dehydrogenase activities were smaller by a factor of about 3.

Significantly, it was found that equal percentages of both enzyme activities were measured in the cytosol when a phosphate-containing medium was used for the homogenization (medium II). This was true for the fed as well as for the fasted rats. It is obvious that these results cannot be fully explained by assuming an extra- and intramitochondrial pyruvate carboxylase; therefore we must look for other factors to explain the results of Table I.

Whether the various media themselves have an effect on the activities of pyruvate carboxylase and glutamate dehydrogenase was tested in the experiments summarized in Table II. No major effects were observed except with medium III (containing sodium acetate), in which the pyruvate carboxylase activity dropped by more than 75%. As media III and IV differ only in the presence of sodium acetate, it was tested whether pyruvate carboxylase activity also drops in a medium containing only sucrose and sodium acetate without EDTA and glutathione. However, in this experiment (not given here), pyruvate carboxylase activity showed no change. This inactivation effect—which was not further investigated—must therefore be caused by the combined influence of sodium acetate with glutathione and/or EDTA. It should also be noted that this inactivation did not occur when a mitochondrial extract of pyruvate carboxylase was added to the medium III—cytosol, as in Table III.

The results of Table II offer no explanation for the observed dependence between the measured enzyme activities in the cytosol and the compositions of the homogenization media (Table I). Only activation of pyruvate carboxylase or inactivation of glutamate dehydrogenase activity in media I, III, and IV would have been consistent with the results of Table I.

Table II. Stability of crude mitochondrial glutamate dehydrogenase and pyruvate carboxylase in various media[a]

Medium	Activities after incubation (μmole/min)	
	PC	GDH
I	3.8	140
II	3.4	138
III	0.7	126
IV	3.8	136

[a] Liver mitochondria were sonicated in 0.28 M sucrose and the 100,000 *g* supernatant was assayed for glutamate dehydrogenase and pyruvate carboxylase activities. Equal portions of this supernatant containing 3.5 μmole/min pyruvate carboxylase activity and 122 μmole/min glutamate dehydrogenase activity were then added to each incubation medium. Incubations were carried out at 4°C for 2 hours. For composition of media, see Table I. Glutamate dehydrogenase was assayed in the presence of ADP.

The results of Table I could be explained by assuming a differential extractability of the two enzymes from mitochondria in the different media. However, an essentially equal release of both enzymes was observed when the mitochondria were homogenized for different times with medium I. This result is in agreement with previous findings that treatment of mitochondria with sodium deoxycholate or homogenization in a Waring blendor also resulted in an equally gradual release of the two enzyme activities (Seubert *et al.*, 1968; Dugal, 1972). These results further support previous reports that both enzymes are localized within the matrix of the mitochondria (Böttger *et al.*, 1969; Marco *et al.*, 1969). The unequal percentages of total pyruvate carboxylase and glutamate dehydrogenase activities found in the cytosols prepared with certain media are therefore most likely *not* a consequence of a difference in the release of these two enzymes from the mitochondria.

Di Prisco *et al.* (1968) have reported evidence for the existence of a nuclear glutamate dehydrogenase in rat liver. This enzyme was found to be released from sonicated nuclei only in phosphate-containing media, and its activity amounted to up to 18% of the total liver activity. It was also demonstrated that the nuclear enzyme preparation can be activated by about two- to twelvefold if 10 to 200 mM phosphate is added to the assay system for the enzyme. On the basis of these findings, one could postulate that the observed higher activities of glutamate dehydrogenase in cytosols prepared with the phosphate-containing medium II are of mostly nuclear origin. However, our cytosol-glutamate dehydrogenase activity could not be activated by the addition of phosphate (Fig. 1); on the contrary, this activity exhibited exactly the pattern of thc mitochondrial enzyme. This indicates that the glutamate dehydrogenase activity appearing in the cytosol prepared with medium II is not of nuclear origin. Moreover, King and Frieden (1970) have questioned di Prisco's finding of a nuclear glutamate dehydrogenase (see Section I.C).

Tables III to V demonstrate the binding of mitochondrial glutamate dehydrogenase to and the release of bound glutamate dehydrogenase from subcellular fractions. When mitochondrial pyruvate carboxylase and glutamate dehydrogenase activities were added to the homogenates in the various media, the pyruvate carboxylase activity was always fully recovered

Table III. Recovery of pyruvate carboxylase and glutamate dehydrogenase activities added to rat liver homogenates in various media[a]

Homogenization medium	Added enzyme activities to homogenate (μmole/min)		Found enzyme activities in supernatants (μmole/min)		Recovery of added enzyme activities (%)	
	PC	GDH	PC	GDH	PC	GDH
I	—	—	0.52	6.64	—	—
	1.10	41.2	1.79	33.6	115	65
II	—	—	0.43	15.2	—	—
	1.46	41.6	1.76	62.7	91	114
III	—	—	0.37	0.58	—	—
	0.72	25.2	1.16	1.45	110	3.7
IV	—	—	0.43	3.33	—	—
	0.82	32.8	1.34	15.4	111	37

[a] The added enzyme activities were obtained from mitochondria sonicated in 0.28 M sucrose. They were added to the respective homogenization media before the livers were homogenized. Supernatants were obtained by centrifugation of 900 *g* supernatants (prepared according to Henning *et al.*, 1966) at 70,000 *g* for 15 minutes. For composition of media, see Table I. The GDH was assayed in the presence of ADP.
Source: Data from Walter and Anabitarte (1971).

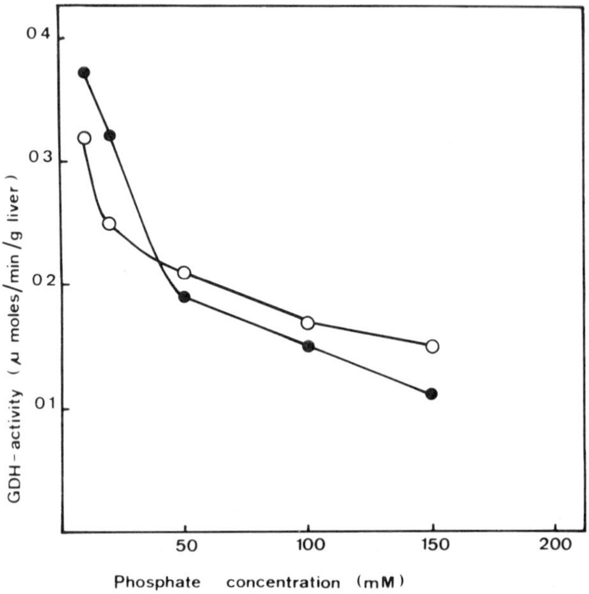

Figure 1. Effect of phosphate on the measurement of glutamate dehydrogenase activity in cytosols prepared with homogenizing medium II. The glutamate dehydrogenase activity was measured according to di Prisco *et al.* (1968) in a medium containing 50 mM tris-HCl buffer, pH 9.0, 25 mM potassium glutamate, and 50 μM NAD. The curves represent two different experiments.

in the cytosol, whereas for the glutamate dehydrogenase activity this was only true with the phosphate-containing medium II (Table III). More than 95% of the glutamate dehydrogenase activity was lost in the acetate-containing medium and between about 40 and 60% in media I and IV. The results in Table IV show that when 900 *g* supernatant fractions prepared in the sucrose-EDTA-glutathione medium (medium IV) were supplemented with the various salts present in the other media, then rehomogenized and centrifuged, the glutamate dehydrogenase activities were somewhat decreased upon addition of triethanolaminehydrochloride. They were strongly decreased in the presence of sodium acetate and dramatically increased when potassium phosphate was added. The pyruvate carboxylase activities, on the other hand, either remained unchanged or showed small increases after rehomogenization. Table V summarizes the effects of phosphate added to various liver fractions. Most glutamate dehydrogenase activity was liberated in the experiment with the acetate-containing medium III. The experiments with medium I indicate that most glutamate dehydrogenase activity was bound to the fraction containing mainly microsomes

(15,000 *g* supernatant) or nuclei (900 *g* sediment), whereas little enzyme activity was released from the mitochondrial fraction (15,000 *g* sediment).

Seubert's group had observed an increased pyruvate carboxylase activity in the cytosols of cortisol-treated rats, and they proposed an induction of the extramitochondrial pyruvate carboxylase by cortisol (Henning *et al.*, 1966). These results had been confirmed by Störmer *et al.* (1970). We repeated the experiments (Table VI and observed similar but somewhat smaller effects of cortisol. The results with media I and II demonstrate that as in livers from fed and fasted rats, cortisol-treated animals show equal percentages of glutamate dehydrogenase and pyruvate carboxylase in the cytosols with medium II, in contrast to the cytosols with medium I. It should be mentioned that the higher pyruvate carboxylase activities in the cytosols with medium I were obtained only when this rather impractical time schedule for fasting and cortisol administration was observed.

Table IV. Effects of various additions to the 900 *g* supernatant fraction on the enzyme activities in the high speed supernatants[a]

		Enzyme activities in 70,000 *g* supernatant (μmole/min·g liver)	
Experiment	Treatment	PC	GDH
1	None	0.58	1.39
	Rehomogenized	0.59	1.59
	Rehomogenized after addition of 20 mM TEA-HCl, pH 7.2	0.59	1.19
2	None	0.35	2.14
	Rehomogenized	0.43	2.22
	Rehomogenized after addition of 50 mM sodium acetate	0.56	0.43
3	None	0.30	2.53
	Rehomogenized	0.34	2.63
	Rehomogenized after addition of 50 mM potassium phosphate, pH 7.2	0.47	15.95

[a] The 900 *g* supernatant fractions were prepared according to Henning *et al.* (1966) in medium IV (Table I). When indicated, rehomogenization was carried out in a Potter type homogenizer for one minute at ice temperature. The supernatant fractions were obtained by centrifugation at 70,000 *g* for 15 minutes.

Source: Data from Walter and Anabitarte (1973).

Table V. The effect of phosphate on the release of glutamate dehydrogenase activity from various liver fractions[a]

Homogenization medium	Liver fraction	GDH-activity released (μmole/min·g liver) Control	Treated with phosphate
I	900 *g* sediment	0.68	5.44
	900 *g* supernatant	0.81	10.93
	15,000 *g* sediment	0.40	2.73
	15,000 *g* supernatant	0.65	8.24
	100,000 *g* supernatant	0.65	0.70
III	900 *g* supernatant	0.60	12.53
	100,000 *g* supernatant	0.60	0.61
IV	900 *g* supernatant	2.72	14.91
	100,000 *g* supernatant	2.72	2.82

[a] Rat livers were homogenized in the media used by Seubert *et al.* (1968) indicated in the table and centrifuged at 900 *g* for 8 minutes. Part of the resulting supernatant was then centrifuged at 15,000 *g* for 10 minutes. Probes were retained of each liver fraction and their GDH activity was measured as follows. (*a*) *Controls.* All 900 *g* and 15,000 *g* "supernatant" fractions were centrifuged at 100,000 *g* and the GDH activities were assayed in the resulting supernatants; "sediment" fractions were resuspended in 10 ml of the respective homogenization media per original gram of liver, then centrifuged at 100,000 *g*, and the GDH activities were determined in the resulting supernatants. (*b*) *Treated with phosphate.* The procedure was the same as with the controls, except that potassium phosphate solution of pH 7.2 was added to all the respective media (final concentration 50 mM) in which the probes were resuspended. In the case of the liver fractions marked as "100,000 *g* supernatants," no recentrifugations were carried out because these solutions were always clear. The control GDH values of the 15,000 and 100,000 *g* supernatant therefore always represent the same measurement. When phosphate was added to the 100,000 *g* supernatant, the solution was thoroughly mixed. Times of centrifugation were 8 minutes for 900 *g*, 10 minutes for 15,000 *g*, and 60 minutes for 100,000 *g*. For composition of media see Table I. The GDH was assayed in the presence of ADP.

C. Discussion

Most studies on the intracellular distribution of an enzyme are performed by comparing the distribution of the enzyme in question with that of a marker enzyme. In this case the glutamate dehydrogenase was assumed to be a marker enzyme for the mitochondria, and the conclusion to be drawn on the localization of the pyruvate carboxylase depends on the validity of this assumption. At the present time there appears to be only one report in conflict with this assumption. As mentioned in Section I.B., di Prisco *et al.* (1968) have postulated the existence of a nuclear glutamate dehydrogenase. Reexamining this question, King and Frieden (1970) have found that in nuclei isolated according to Blobel and Potter (1966), no glutamate dehydrogenase activity could be detected. This and other evidence led King and Frieden to

Table VI. Effect of cortisol treatment on the activities of glutamate dehydrogenase and pyruvate carboxylase in rat liver

		Enzyme activities (μmole/min·g liver)			
		Total homogenate		Cytosol	
Homogenization medium	Metabolic state of animal[a]	PC	GDH	PC	GDH
Experiment A					
I	Fasted (control)	7.6	238	0.87 (12%)	6.36 (3%)
	Fasted + cortisol	8.5	257	1.24 (15%)	6.60 (3%)
II	Fasted (control)	7.5	250	0.68 (9%)	29.4 (10%)
	Fasted + cortisol	9.0	304	0.78 (9%)	31.8 (12%)
Experiment B					
I	Fasted (control)	7.3	250	0.86 (12%)	4.15 (2%)
	Fasted + cortisol	8.1	278	1.05 (13%)	6.60 (2%)
II	Fasted (control)	7.7	258	0.62 (9%)	24.5 (9%)
	Fasted + cortisol	7.8	309	0.71 (8%)	30.0 (10%)

[a] Food was taken away at 10 P.M. and animals were killed at 10 A.M. the following day. When indicated, 5 mg of cortisol (crystalline suspension) per 100 g of body weight was administered intraperitoneally at 4 A.M.; controls received the same volume of physiological saline. Numbers in parentheses are the percentages of activity of the respective enzymes in the total homogenate. For further details see Table I.

Source: Data from Walter and Anabitarte (1973).

conclude that the nuclear sonic extracts prepared by di Prisco *et al.*, who used the method of Hymer and Kuff (1964), must have been contaminated by other cellular elements. Because the existence of an extramitochondrial glutamate dehydrogenase therefore seems to be unlikely, the following discussion is based on glutamate dehydrogenase as mitochondrial marker enzyme.

The results of this chapter have shown that glutamate dehydrogenase and pyruvate carboxylase activities appeared at about equal percentages in the cytosol when prepared with a phosphate-containing medium. For the other media, binding of glutamate dehydrogenase to subcellular fractions can be concluded to be responsible for the lower relative activities of this enzyme measured in these cytosols. A comparison of the results of Tables I, III, IV and V reveals that the binding of glutamate dehydrogenase is strongest in medium III (containing sodium acetate) and somewhat weaker in media I and IV. The presence of potassium phosphate effectively prevents this binding in all three media.

As mentioned in the introduction, Dugal (1972) was also able to measure more pyruvate carboxylase than glutamate dehydrogenase activity in the cytosols of phosphate-free homogenates. Furthermore, when he added deoxycholate after such a prolonged homogenization, 100% of both enzyme activities could be measured. It is now obvious that this treatment with deoxycholate not only destroyed the mitochondria but also released the glutamate dehydrogenase activity bound to the microsomes and nuclei; therefore it is no longer necessary to postulate an extramitochondrial pyruvate carboxylase to explain Dugal's results.

It is important to mention that equal percentages of glutamate dehydrogenase and pyruvate carboxylase were also found in the cytosols of fasted and of cortisol-treated rats. On the basis of these results, an induction of an extramitochondrial pyruvate carboxylase, as has been postulated (Seubert *et al.* 1968), seems to be unlikely.

It can be concluded from these results that the most likely reason for the appearance of pyruvate carboxylase and glutamate dehydrogenase in the cytosol is breakage of some mitochondria during homogenization, with concurrent release of pyruvate carboxylase and glutamate dehydrogenase. When the mitochondrial membranes are more fragile (e.g., under fasting conditions), it is to be expected that higher enzyme quantities will be released by the mitochondria, and this has been confirmed in our experiments.

Ballard and Hanson (1967) found substantial pyruvate carboxylase activities associated with a very crude nuclear fraction (once-washed 600 *g* sediment) or rat liver, however, no marker enzyme study was carried out. As already concluded by Böttger *et al.* (1969), this pyruvate carboxylase activity is most likely due to mitochondrial contamination of the nuclear fraction. Furthermore, Ballard *et al.* (1970) reported that antibodies to the mitochondrial pyruvate carboxylase of rat liver cross-reacted with pyruvate carboxylase found in the cytosols, and they suggested that the cytosol enzyme is of mitochondrial origin. The results of Brech *et al.* (1970) are also in full agreement with this conclusion.

After this work was completed, Godinot and Lardy (1973) published results showing that radioactive glutamate dehydrogenase added to liver homogenates preferentially binds to microsomes, which is in agreement with our findings. These authors also demonstrated that a small fraction of glutamate dehydrogenase activity (< 6%) that does not originate from the mitochondria can be found associated with microsomes. This activity is probably newly synthesized enzyme that is to be transported into the mitochondria. Whether a similar pathway of biosynthesis is also operative in the case of pyruvate carboxylase will have to be investigated. According to the fractionating study of Böttger *et al.* (1969), this nonmitochondrial fraction should amount to less than 3% of the total activity.

Just before this chapter was printed, Seubert's group (Weiss *et al.* 1974) confirmed our published findings (Walter and Anabitarte, 1971, 1973) and agreed with our conclusion of an exclusive intramitochondrial localization of pyruvate carboxylase in rat liver.

II. REGULATION OF PYRUVATE CARBOXYLASE IN INTACT RAT LIVER MITOCHONDRIA

Studies with purified pyruvate carboxylase from rat liver have revealed that the regulation of this enzyme is very complex (McClure *et al.*, 1971; Seufert *et al.*, 1971). Of the various factors that affect the isolated enzyme, we have studied the influence of adenine nucleotides, acetyl-CoA, and calcium on the activity of the enzyme in intact mitochondria; some of the most important results are presented in this section. In many of these experiments, pyruvate dehydrogenase activity was also estimated, and aspects of the regulation of this enzyme are also discussed.

A. Methodology

The product of the pyruvate carboxylate reaction, oxalacetate, does not accumulate in mitochondria but is metabolized to intermediates of the citric acid cycle. The amount of pyruvate carboxylated was therefore estimated by measuring all citric acid cycle intermediates that accumulate in significant amounts. The sum of these products corresponds to the amount of pyruvate carboxylated. The citric acid cycle products were measured partly by enzymatic methods and partly by using labeled pyruvate and by measuring the radioactivity of the various products after electrophoretic separation. The products malate and citrate usually account for about 90% of the carboxylated pyruvate, and in some experiments only the amounts of these products were determined and were taken as a measure of pyruvate carboxylation.

Because pyruvate is only carboxylated or oxidized by liver mitochondria, the amounts of pyruvate oxidized to acetyl-CoA or converted to further metabolites can be estimated by subtracting the amount of pyruvate carboxylated from the measured total amount of pyruvate used in the experiment.

The methods dealing with the measurement of the two enzyme activities have been described in detail and evaluated (Walter and Stucki, 1970; Stucki *et al.*, 1972; Stucki and Walter, 1972). When intramitochondrial intermediates were measured, the mitochondria were separated from the incubation medium by a molecular sieve filtration technique described by Brawand and Walter (1974). It was shown that the metabolic state did not change during the filtration process by this technique.

Unless otherwise indicated, the incubation medium used had the following composition: 4 mM ATP; 10 mM magnesium sulfate; 6.6 mM potassium phosphate buffer, pH 7.4; 6.6 mM triethanolamine buffer, pH 7.4; 10 mM potassium bicarbonate; 10 mM potassium pyruvate, and 0.25 M mannitol–0.07 M sucrose solution to make the reaction medium about isotonic. Mitochondria isolated from 0.5 g of liver (12–15 mg mitochondrial protein) were added in 0.5 ml mannitol-sucrose solution. The final volume was 3 ml. Incubations were carried out in 25 ml Erlenmeyer flasks in a shaking waterbath for 10 minutes at 37°C. For further methods see legends or references cited in the text.

B. Regulation of Pyruvate Metabolism by Adenine Nucleotides

Previous work has shown that mitochondria that were incubated in the presence of saturating amounts of pyruvate (10 mM) oxidized and carboxylated about equal amounts of pyruvate (see also Table VII). Addition of creatine and creatine kinase, which caused a lowering of the ATP:ADP ratio in the incubation medium, led to strong inhibition of pyruvate carboxylation, whereas pyruvate oxidation changed relatively little. In the presence of octanoate, oleate, oleoyl-CoA, or atractylate, the inhibition of pyruvate carboxylation was partially released, whereas the addition of uncouplers such as dinitrophenol (DNP) increased this inhibition. It could be shown that in most cases the rate of pyruvate carboxylation correlated with the intramitochondrial ATP:ADP ratios, and that fatty acids,

Table VII. Pyruvate metabolism in mitochondria incubated with various pyruvate concentrations[a]

	Pyruvate (μmole/3 ml)				
System	Added	Remaining	Used	Carboxylated	Oxidized
Control I	31.5	9.5	22.0	11.4	10.6
+ Creatine kinase	31.5	17.7	13.8	4.7	9.1
+ Creatine kinase and oleoyl-CoA	31.5	13.7	17.8	7.6	10.2
Control II	7.7	0.3	7.4	3.7	3.6
+ Creatine kinase	7.7	0.4	7.3	1.7	5.6
+ Creatine kinase and oleoyl-CoA	7.7	0.3	7.4	2.6	4.8

[a] To the standard medium for all incubations was added 13 mM creatine. In the experiment with limiting pyruvate, 0.6 μmole of pyruvate was added at the beginning and 7.1 μmole was added by infusion during the 10 minute incubation.

long-chain acetyl-CoA esters, atractylate, and DNP regulated the enzyme activity indirectly by causing changes in the intramitochondrial ATP:ADP ratios by various mechanisms. This result is in agreement with the work on the isolated enzyme that has been found to the regulated by this ratio over a similar concentration range (McClure and Lardy, 1971). In these experiments, the levels of acetyl-CoA did not correlate with the rate of pyruvate carboxylation and were 2 to 4 times higher than the K_a values reported for the isolated enzyme. For details of the experiments summarized in this paragraph, see Walter and Stucki (1970) and Stucki *et al.* (1972). the results of the work of Wojtczak *et al.* (1972) and of Lopez-Cardozo *et al.* (1972) are also compatible with our conclusions.

The concentrations of several components in our incubation mixture do not correspond to those of the extramitochondrial medium in the liver cell. This is especially true for the pyruvate concentration, which was 10 mM or higher in our experiments but is only around 0.1 mM in the liver cell (Hohorst *et al.*, 1959). The pyruvate concentration is of critical importance because McClure and Lardy (1971) have shown that the isolated pyruvate carboxylase is much less sensitive to the ATP:ADP ratio in presence of low pyruvate than of high pyruvate concentrations. Furthermore, Linn *et al.* (1969) and Wieland *et al.* (1973) have reported that more pyruvate dehydrogenase is in the active form at a low ratio than at a high ratio of ATP:ADP. However, this regulation occurs only at pyruvate concentrations below 0.5 mM pyruvate, whereas at higher pyruvate concentrations, the enzyme is insensitive to the adenine nucleotide ratio. To evaluate the importance of pyruvate concentration for our mitochondrial system, we have carried out experiments in which pyruvate was infused at a constant rate during the 10 minutes incubation. By this technique the pyruvate concentration in the incubation medium was in the physiological range and was shown to remain constant during these experiments (Walter *et al.*, 1974). The results in Table VII demonstrate that the same pattern for the rate of pyruvate carboxylation was found with 0.1 mM pyruvate as with saturating pyruvate. However, a different pattern was observed for the rates of pyruvate oxidation, which increased when the rates of pyruvate carboxylation were decreased. As a result, the amounts of pyruvate used and of pyruvate remaining stayed constant for all three experimental conditions when 0.1 mM pyruvate was used, whereas there was great variation in the presence of saturating pyruvate concentrations. This opposite pattern of the two enzyme activities was shown to correlate with the intramitochondrial ATP:ADP ratio (Walter *et al.*, 1974) as is demonstrated in Fig. 2. The ATP:ADP ratio in incubations 1 to 4 were varied by increasing the concentration of creatine in the medium. The incubation with octanoate is discussed later.

Because the concentration of pyruvate is in the K_m region of both enzymes, the observed pattern of enzyme activities is to be expected if only one enzyme was regulated by the nucleotide ratio or if both were. From our results (Walter *et al.*, 1974) and the recent work of Wieland and Portenhauser (1974), it can be concluded that pyruvate dehydrogenase activity is regulated by the adenine nucleotide ratio. Whether pyruvate carboxylase activity is also directly ragulated by the ATP:ADP ratio or indirectly through the changed availability of pyruvate cannot be concluded from these experiments.

When octanoate was added (Fig. 2), the results clearly gave no correlation between the adenine nucleotide ratio and the enzyme activities. However, a distinct increase in acetyl-CoA was noted. A regulating role of acetyl-CoA as allosteric co-factor of pyruvate carboxylase was at first not considered to be very likely because the concentrations of acetyl-CoA within the mitochondria can be calculated to be in the range of 4 to 6 times higher than the K_a of 60 μM reported for the isolated enzyme by McClure and Lardy (1971) and of 15 to 40 μM by Seufert *et al.* (1971). This calculation is based on the value of 1 μl of intramitochondrial water per milligram of protein (Pfaff *et al.*, 1968). On the other hand, other experiments in our laboratory indicated that such high intramitochondrial acetyl-CoA concentrations might have a role in the regulation of pyruvate carboxylase, and we therefore decided to study this aspect more thoroughly.

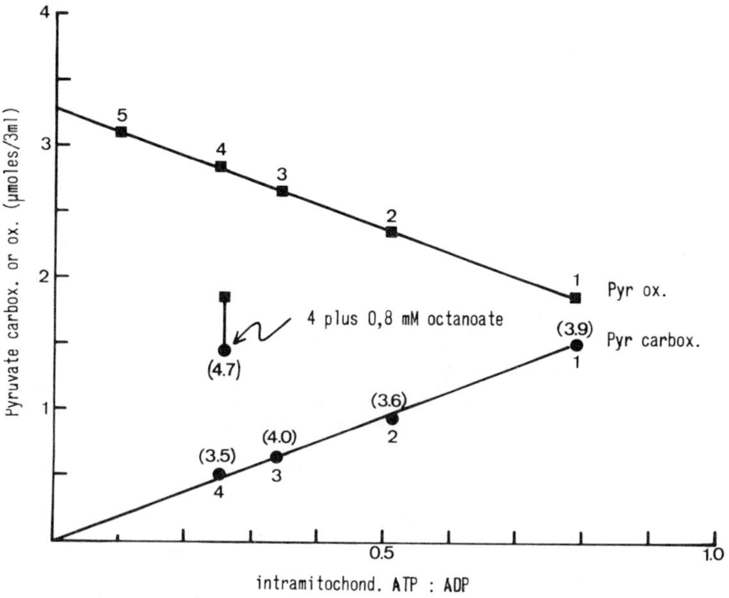

Figure 2. Regulation of pyruvate metabolism by intramitochondrial ATP:ADP ratios in the presence of limiting pyruvate concentrations. The standard medium was used throughout. Point 1, no additions; point 2, 4 mM creatine; point 3, 7 mM creatine; point 4, 15 mM creatine; 250 μg creatine kinase was added to the incubations of points 2, 3, and 4. The adenine nucleotides were measured in the pellet after millipore separation. Initially 0.56 μmole of pyruvate was added and 2.81 μmole was infused during the 10 minute incubation. The pyruvate remaining at the end amounted to an average of 0.07 μmole/3 ml.

C. Regulation of Pyruvate Carboxylase by Acetyl-CoA in Mitochondria

To study the apparent K_a' of acetyl-CoA for pyruvate carboxylase in mitochondria, we had to find a system in which acetyl-CoA levels could be varied at constant ATP:ADP ratios. As described in Fig. 3, the acetyl-CoA production from pyruvate could be varied by adding acetylcarnitine and arsenite. Furthermore, the addition of 3-hydroxybutyrate resulted in a more or less constant ATP:ADP ratio. A K_a' of about 170 μM for acetyl-CoA can be calculated from the curve. This value is about 3 to 4 times higher than the value obtained for the isolated enzyme. It was observed in additional experiments that the enzyme activity is less sensitive to the ATP:ADP ratio at low concentrations as compared to high levels of acetyl-CoA. Furthermore, the K_a' values increased slightly with decreasing ATP:ADP ratios (e.g., at a ratio of 0.25 the K_a' value was about 230 μM). The latter effect is in agreement with the results of Seubert *et al.* (1971),

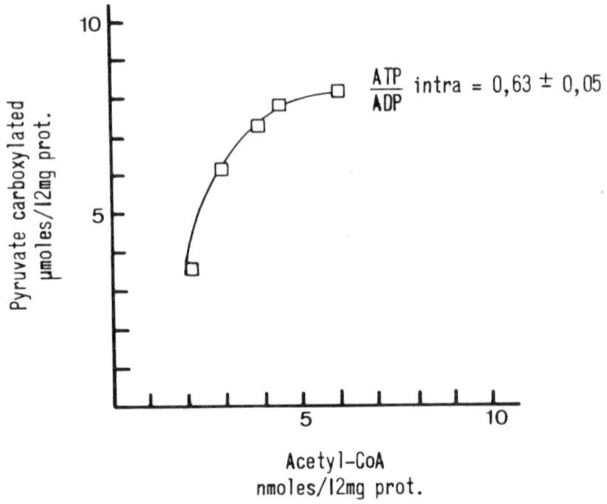

Figure 3. Relation between the rate of pyruvate carboxylation and intramitochondrial acetyl-CoA levels. The incubations corresponding to the points from right to left contained, in addition to the standard medium, the following: 10 mM L-acetylcarnitine in point 1, no additions in point 2, 1.67 μM arsenite in point 3, 3.33 μM arsenite in point 4, and 8.33 μM arsenite in point 5. In all incubations 10 mM 3-hydroxybutyrate and 12 mg mitochondrial protein were present. Acetyl-CoA was measured in the mitochondrial pellets after millipore filtration.

who reported that free magnesium and/or the $MgATP^{2-}$ complex exerted a positive heterotropic effect with respect to the binding of acetyl-CoA to the isolated enzyme.

Based on the results presented in Fig. 3, the experiment in Fig. 2 with octanoate can now be explained by an activation of pyruvate carboxylase by the increased acetyl-CoA levels. It should be noted that the effects of octanoate could also be explained by an inhibition of pyruvate oxidation which is independent of the adenine nucleotide level. Taylor *et al.* (1974) have recently reported such an inhibition and have suggested that this effect of octanoate may be mediated by elevated NADH or acetyl-CoA levels in the mitochondria.

The concentration of pyruvate in the experiment of Fig. 2 was only 25 μM, and one might ask what the acetyl-CoA level would be when the more physiological level of 0.1 mM pyruvate was used. In several experiments with 0.1 mM pyruvate, acetyl-CoA levels between 240 and 300 μM have been measured; this is within the regulating range for pyruvate carboxylation in mitochondria.

How can the difference between the K_a and the K_a' for acetyl-CoA be explained? Based on published work with the isolated enzyme, it seems rather unlikely that changes in ATP, Mg^{2+}, or pH are responsible for this large difference. A more likely explanation is that acetyl-CoA is in some way compartmented. The compartmentation might be within the mitochondrion (e.g., by nonspecific binding of the co-factor to proteins in the matrix space). There may also be different types of mitochondria having different enzyme activities and pool sizes of acetyl-CoA. We have studied this problem and have concluded (see Section II. D) that acetyl-CoA is indeed compartmented.

D. Compartmentation of Acetyl-CoA in Mitochondria

An incubation system was chosen in which the mitochondria formed acetyl-CoA from pyruvate as well as from (1-^{14}C) palmitic acid. Furthermore, oxalacetate was provided by pyruvate carboxylation, and sufficient malonate was added to block the citric acid cycle at the site of the conversion of succinate to fumarate. Without any compartmentation, we would expect the ratio of the specific radioactivities of 3-hydroxybutyrate: citrate to be 2. The results in Table VIII, however, reveal that ratios of about 3 were observed in this type of experiment. A comparison of the results at 7.5 and 15

Table VIII. Specific radioactivities of 3-hydroxybutyrate and citrate in incubations of rat liver mitochondria in the presence of pyruvate and (1-^{14}C) palmitate[a]

	Metabolite changes (μmole/3 ml)	
Acids	7.5 min	15 min
Pyruvate used	15.9	30.4
Pyruvate carboxylated	6.7	13.4
Pyruvate oxidized	9.2	17.0
Malate found	4.1	7.9
Citrate found	2.6	5.5
Citrate, specific radioactivity (cpm/μmole)	56,000	43,000
3-OH-butyrate found	0.3	0.8
3-OH-butyrate, specific radioactivity (cpm/μmole)	164,000	141,600
Quotient of specific radioactivity 3-OH-butyrate: citrate	3.0	3.2

[a] To the standard medium for all incubations were added 40 μM CoA, 1 mM L-carnitine, and 0.3 mM (1-^{14}C), palmitate. The radioactivities of 3-hydroxybutyrate and citrate were measured after separation on paper electrophoresis.

minutes indicates that the system was more or less at a steady state. Additional information for compartmentation was obtained in the experiments summarized in Table IX. Here, the production of acetyl-CoA from pyruvate was blocked by arsenite. The results show that acetyl-CoA derived from acetylarnitine, palmityl-carnitine (not shown), and octanoate stimulated pyruvate carboxylation, whereas leucine caused an increase in acetyl-CoA but no increase in enzyme activity. All these experiments clearly demonstrate compartmentation of acetyl-CoA. However, it cannot be decided from these results whether acetyl-CoA is compartmented within each mitochondrion or whether different types of mitochondria are present in the liver cell. Strong evidence for the second possibility was obtained in an experiment in which mitochondria were divided by centrifugation into a light and a heavy fraction. The quotient of the specific radioactivities of 3-hydroxybutyrate:citrate was 3.3 ± 0.1 for the light fraction and 3.7 ± 0.06 for the heavy fraction. The difference between the two ratios was statistically significant ($P < 0.01$).

E. Regulation of Pyruvate Carboxylation by Calcium Ions

Studies with isolated pyruvate carboxylase have shown that calcium is a potent inhibitor of this enzyme (McClure and Lardy, 1971; Wimhurst and Manchester, 1970). Kimmich and Rasmussen (1969) reported that addition of 100 μM calcium to mitochondrial incubations caused an inhibition of

Table IX. Effect of acetyl-CoA produced from various precursors on the carboxylation of pyruvate in rat liver mitochondria[a]

	Metabolite changes		
Additions (mM)	Pyruvate carboxylated (μmole/3 ml)	Acetyl-CoA (nmole/12 mg protein)	ATP:ADP
Control	1.0	0.7	0.35
1.6 Octanoate	5.1	6.2	0.25
10 Leucine	1.0	2.3	0.29
10(−)-Acetylcarnitine	1.8	2.4	0.27

[a] To the standard medium were added 10 mM 3-hydroxybutyrate and 0.17 mM arsenite. In the experiment with leucine 0.1 mM 2-oxoglutarate was present. Acetyl-CoA, ATP, and ADP were measured in the mitochondrial pellet after millipore filtration.

Table X. Incubation of low calcium containing mitochondria in the presence of limiting pyruvate concentrations and 0.1 μM calcium[a]

Additions	Calcium content (nmole/ mg protein)	Pyruvate (μmole/3 ml) Used	Carboxylated	Oxidized
Control	5	5.8	2.4	3.4
0.1 μM Ca^{2+}	27	5.8	2.3	3.5

[a] The reactions were carried out in the standard medium. Calcium was measured in the mitochondrial pellet after millipore filtration.
Source: Data from Mörikofer-Zwez *et al.* (1973).

57% of pyruvate carboxylation. In the course of our investigations (Mörikofer-Zwez *et al.*, 1973), we were able to reproduce the results of Kimmich and Rasmussen (1969); however, we noted that the inhibition was strongly dependent on the sucrose concentration in the incubation medium and that sucrose itself was also inhibitory. Pyruvate carboxylation was maximal at sucrose concentrations below 60 mM and was inhibited by 50% at 200 mM sucrose when the tonicity of the medium was kept constant with mannitol. Upon addition of 100 μM calcium, the inhibition was less than 10% at low sucrose concentrations, whereas in the presence of high sucrose concentrations we observed a relative inhibiton by calcium of up to 74%. However, the absolute decrease of pyruvate carboxylase activity in the presence of calcium never exceeded 15% of the enzyme activity measured under optimal conditions.

The mechanism of the action of sucrose and calcium has not been elucidated. The fact that pyruvate oxidation was also inhibited by sucrose as well as by calcium indicates that a change may have occurred in the transport capacity of the mitochondrial membrane for pyruvate. On the other hand, decreases in acetyl-CoA content were observed as a result of sucrose addition and a regulatory role of this co-factor also must be considered.

The mitochondria used in these experiments had a calcium content of about 24 nmole/mg protein. Based on the calcium content of fresh liver, a maximal content of about 10 to 12 nmole/mg mitochondrial protein is possible, indicating that the isolation medium (sucrose-mannitol solution) contained calcium. It was of interest to examine whether mitochondria having a calcium content below 10 nmole/mg protein behaved differently from those with a higher content. For the experiment in Table X, mitochondria were isolated in a mannitol-sucrose solution from which calcium had been removed by ion exchange. The experiment was carried out in the presence of low sucrose concentration, and the pyruvate concentration was kept below 0.1 mM by constant infusion. The results showed absolutely no effect of added calcium on either the oxidation or the carboxylation of pyruvate.

The results with calcium can be summarized by saying that the calcium had almost no effect on pyruvate metabolism when tested under optimal conditions. In the presence of sucrose concentrations that inhibited pyruvate metabolism by itself, an additional inhibiting effect of calcium on the carboxylation and oxidation of pyruvate was observed.

F. Discussion

The first half of this chapter states that it is now generally accepted that pyruvate carboxylase is exclusively localized in the mitochondria of rat liver. Because the pyruvate dehydrogenase system is also a mitochondrial enzyme, this organelle is the site at which regulation of pyruvate metabolism must occur. Experiments with isolated mitochondria therefore represent a useful way for the elucidation of regulatory mechanisms that may also be operative in the whole cell.

It has been well established that when pyruvate is needed for gluconeogenesis in the liver cell, less pyruvate is converted to acetyl-CoA. It is therefore likely that regulatory mechanisms exist by which pyruvate oxidation and carboxylation are effected in an opposite manner. In principle, any cellular inhibitor of one of the two enzymes may have an opposite effect on the other enzyme, since the physiological concentration of pyruvate in the cell is in the nonsaturating range for pyruvate carboxylase as well as for the pyruvate dehydrogenase system. Our results demonstrate that the ATP:ADP ratio inside the mitochondria regulates the two enzyme activities in such an opposite manner. A high ratio stimulated pyruvate carboxylation and inhibited pyruvate oxidation; at low ratios of ATP:ADP, the effects were exactly reversed.

Another effector stimulating pyruvate carboxylation and inhibiting pyruvate oxidation was octanoate. It is well known that fatty acids inhibit pyruvate oxidation in liver mitochondria (Bremer, 1966; Walter *et al.*, 1966), but the exact mechanism has never been elucidated. Although a shift of the ATP:ADP ratio has recently been proposed (Wieland and Portenhauser, 1974), other factors such as acetyl-CoA and NADH may also be important (Taylor *et al.*, 1974). Our experiments allow no conclusion about the exact mechanism, but they demonstrate that the ATP:ADP ratio is not the only regulating factor. Acetyl-CoA was found to be increased in the experiment with octanoate, and stimulation of pyruvate carboxylase or (and) inhibition of pyruvate oxidation by this compound are likely mechanisms for the observed effects.

Our results clearly show that the K_a' of acetyl-CoA for pyruvate carboxylase in mitochondria is 3 to 4 times higher than the K_a for the isolated enzyme. Moreover, the acetyl-CoA levels in experiments with 0.1 mM pyruvate were in the regulating range of this co-factor for the enzyme in mitochondria. Our findings in the experiments with 1-^{14}C-palmitic acid demonstrate a compartmentation of acetyl-CoA that is likely to be responsible for the observed difference in the K_a values. Such a compartmentation had been previously postulated by several workers (e.g., Fritz, 1968; Müllhofer and Kuntzen, 1972). It is noteworthy that pyruvate carboxylase could be stimulated by acetyl-CoA supplied by the oxidation of pyruvate or of fatty acids but not of leucine. Different ratios of specific radioactivities of 3-hydroxybutyrate: citrate were found in the light as compared to the heavy mitochondrial fraction, and Weiss *et al.* (1974) have reported a heterogeneous distribution of pyruvate carboxylase activity in mitochondria—these data support the conclusion that different mitochondrial populations exist with varying enzyme patterns and pool sizes for acetyl-CoA.

Calcium ions are known to inhibit purified pyruvate carboxylase (McClure and Lardy, 1971) and to activate the pyruvate dehydrogenase activity (Pettit *et al.*, 1972). Under our experimental conditions, however, the addition of calcium in the physiological range showed no effect on either enzyme activity in intact mitochondria. Walajtys *et al.* (1974) also concluded from their work on the effects of calcium on pyruvate dehydrogenase in rat liver mitochondria that the normal content of magnesium and calcium in isolated mitochondria is sufficient to provide optimal activation of the enzyme.

What is the relevance of these experiments with isolated mitochondria for the whole cell? This question cannot be answered conclusively, but some comments and speculations are worthwhile. We have observed that regulation of pyruvate metabolism is effected by ATP:ADP ratios below 0.7, whereas the cellular ratio is about 3 to 8. Table XI presents an experiment in which the extramitochondrial ATP:ADP ratios in our incubations were measured. The observed large differences between the extra- and intramitochondrial ratios are in agreement with the findings of Heldt *et al.* (1972). Furthermore, Elbers *et al.* (1974) have isolated mitochondria from freezed-stopped liver tissue in a nonaqueous medium and have reported ATP:ADP ratios of about 6 for the extramitochondrial space and 0.12 to 0.89 for the mitochondrial space. In other words the ratios, found in our experiments appear to be similar to those for the mitochondria of whole cells.

Table XI. Relation between pyruvate carboxylation and extra- and intramitochondrial ATP/ADP ratios in the presence of limiting pyruvate concentrations[a]

	ATP/ADP Ratio		
System	Extramitochondrial	Intramitochondrial	Pyruvate carboxylated (μmole/3 ml)
1	91	0.55	1.24
2	56	0.31	0.78
3	24	0.22	0.53
4	7.8	0.17	0.27

[a] The following amounts of creatine were added to the standard incubation media: 3.3 mM in system 1, 6.6 mM in system 2, 13.3 mM in system 3, and 46.6 mM in system 4. The amount of 250 μg of creatine kinase was added to all flasks. Intra- and extramitochondrial adenine nucleotides were measured in the same incubation according to Brawand and Walter (1974).

For acetyl-CoA, only measurements of the whole liver content are available. Tubbs and Garland (1964) reported values of 17 nmole/g liver for normal rats and 39 nmole/g liver for starved animals. For perfused livers, Söling *et al.* (1968) reported 25 nmole of acetyl-CoA per gram of liver, Toews *et al.* (1970) 26, and Williamson, J. R. *et al.* (1968) 60. If all acetyl-CoA were mitochondrial, the values of Tubbs would be in the range of 300 to 650 μM (based on 60 mg mitochondrial protein/g liver and 1 μl water/mg mitochondrial protein). Based on our K_a' for acetyl-CoA in mitochondria of 170 to 230 μM, it can be concluded that if up to half the total liver acetyl-CoA were intramitochondrial, we would still be in the regulating range of acetyl-CoA for pyruvate carboxylase. Furthermore, Toews *et al.* (1970) were able to measure an apparent K_a of acetyl-CoA for pyruvate carboxylase in the perfused liver under experimental conditions in which pyruvate carboxylation had become limiting. Their value of 18 μM for whole liver would correspond approximately to the K_a' found in our mitochondrial experiments under the assumption that half the liver acetyl-CoA is localized extramitochondrially.

The work of Dambach and Friedmann (1974) has demonstrated that addition of glucagon to perfused livers results in changes in sodium, potassium, and calcium concentrations in the liver cell, and it is possible that some of these changes have an effect on gluconeogenic enzymes. The potassium concentration in our medium is about 30 mM, whereas the extramitochondrial concentration in the cell is probably above 100 mM. We therefore replaced some of the mannitol solution in our incubation medium by 40 mM potassium chloride, no change in pyruvate metabolism occurred, however. This result is in contrast to the findings of Schaefer and Veneziale (1973), who reported lower activity of pyruvate carboxylation when the potassium concentration was raised.

Another major difference between our incubation medium and the extramitochondrial space in the liver cell is that the mitochondria are incubated in the presence of mannitol-sucrose in the *in vitro* experiments instead of potassium salts of proteins and of various acids that are present in the cell. If sucrose was used instead of mannitol-sucrose in our incubations, pyruvate metabolism was severely inhibited, and we observed a relative large effect of calcium that was absent in the sucrose-free medium. In other words, it is possible that calcium and other effectors react differently on pyruvate metabolism in a medium that is identical with the extramitochondrial space in the cell as compared to the media used in our experiments.

In closing, let us touch on some aspects of the regulation of pyruvate metabolism that we have not yet discussed. Halestrap and Denton (1974) recently found a specific inhibitor of pyruvate transport in mitochondria. It is therefore very likely that a specific carrier for pyruvate transport through the inner membrane of mitochondria exists. Adam and Haynes (1969) had proposed earlier that this transport may be limiting under certain conditions because they had observed an increased uptake of pyruvate by mitochondria from rats treated with epinephrine, glucagon, or cyclic-AMP. Furthermore, Borle (1974) has reported a stimulating effect of cyclic-AMP on calcium release from isolated rat liver mitochondria. To our knowledge, this is the first report of an effect of cyclic-AMP on the metabolism of rat liver mitochondria. All these results show that the link(s) between hormone action and gluconeogenesis are still not known and that an interesting and challenging field of research lies ahead.

ACKNOWLEDGMENTS

I wish to thank my collaborators who participated in this work: M. Anabitarte, F. Brawand, G. von Glutz, and S. Mörikofer-Zwez. This research was supported by grants from the Schweizerischer Nationalfonds and from the Emil Barell-Stiftung (Hoffmann-La Roche, Basel).

REFERENCES

Adam, P. A. and Haynes, R. C., Jr. (1969). *J. Biol. Chem.* **244,** 6444.

Ballard, F. J., and Hanson, R. W. (1967). *Biochem. J.* **104,** 866.

Ballard, F. J., Hanson, R. W., and Reshef, L. (1970). *Biochem. J.* **119,** 735.

Blobel, G., and Potter, V. R. (1966). *Science* **154,** 1662.

Borle, A. B. (1974) *J. Membrane Biol.* **16,** 221.

Böttger, I., Wieland, O., Brdiczka, D., and Pette, D. (1969). *Eur. J. Biochem.* **8,** 113.

Brawand, F. and Walter, P. (1974). *Anal. Biochem.* **62,** 485.

Brech, W., Shrago, E., and Wilken, D. (1970). *Biochem. Biophys. Acta* **201,** 145.

Bremer, J. (1966). *Biochim. Biophys. Acta* **116,** 1.

Dambach, G. and Friedmann, N. (1974). *Biochim. Biophys. Acta* **332,** 374.

Dugal, S. (1972). *FEBS Lett.* **26,** 97.

Elbers, R. Heldt, H. W., Schmucker, P., Soboll, S., and Wiese, H. (1974). *Hoppe-Seyler's Z. Physiol. Chem.* **355,** 378.

Fritz, J. B. (1968). In *Cellular Compartmentalization and Control of Fatty Acid Metabolism,* Vol. 6 (J. Bremer and F. C. Gran, eds.), FEBS Symposium No. 39, Academic Press, London.

Godinot, C. and Lardy, H. A. (1973). *Biochemistry* **12,** 2051.

Halestrap, A. P. and Denton, R. M. (1974). *Biochem. J.* **138,** 313.

Haynes, R. C., Jr. (1965). *J. Biochem.* **240,** 4103.

Heldt, H. W., Klingenberg, M., and Milovancev, M. (1972). *Eur. J. Biochem.* **30,** 434.

Henning, H. V. and Seubert, W. (1964). *Biochem. Z.* **340,** 160.

Henning, H. V., Stumpf, B., Ohly, B., and Seubert, W. (1966). *Biochem. Z.* **344,** 274.

Hohorst, H. J., Kreutz, F. H., and Bücher, T. (1959) *Biochem. Z.* **332,** 18.

Hymer, W. C. and Kuff, E. L. (1964). *J. Histochem. Cytochem.* **12,** 359.

Kimmich, G. A. and Rasmussen, H. (1969). *J. Biol. Chem.* **244,** 190.

King, K. S. and Frieden, C. (1970). *J. Biol. Chem.* **245,** 4391.

Lardy, H. A., Paetkau, V., and Walter, P. (1965). *Proc. Nat. Acad. Sci. U.S.* **53,** 1410.

Lopez-Cardozo, M. L., Vaartjes, W. J., and van den Bergh, S. G. (1972). *FEBS Lett.* **28,** 265.

Marco, R., Sebastian, J., and Sols, A. (1969). *Biochem. Biophys. Res. Commun.* **34,** 725.

McClure, W. R., Lardy, H. A., and Kneifel, H. P. (1971). *J. Biol. Chem.* **246,** 3569.

McClure, W. R. and Lardy, H. A. (1971). *J. Biol. Chem.* **246,** 3591.

Moellering, H. and Gruber, W. (1966). *Anal. Biochem.* **17,** 369.

Mörikofer-Zwez, S., Kunin, A. S., and Walter, P. (1973). *J. Biol. Chem.* **248,** 7588.

Müllhofer, G. and Kuntzen, O. (1972). *Hoppe-Seyler's Z. Physiol. Chem.* **353,** 1461.

Pettit, F. H., Roche, T. E., and Reed, L. J. (1972). *Biochem. Biophys. Res. Commun.* **49,** 563.

Pfaff, E., Klingenberg, M., Ritt, E., and Vogel, W. (1968). *Eur. J. Biochem.* **5,** 222.

di Prisco, G., Banay-Schwartz, M., and Strecker, H. J. (1968). *Biochem. Biophys. Res. Commun.* **33,** 606.

Schaefer, P. C. and Veneziale, C. M. (1973). *Eur. J. Biochem.* **35,** 18.

Schmidt, E. (1962). In *Methoden der enzymatischen Analyse* (H. U. Bergmeyer, ed.), Verlag Chemie, Weinheim, p. 752.

Scrutton, M. C. and Utter, M. F. (1968). *Ann. Rev. Biochem.* **37,** 249.

Seubert, W., Henning, H. V., Schoner, W., and l'Age, M. (1968). In *Advances in Enzyme Regulation,* Vol. 6, (G. Weber, ed.), Pergamon Press, New York, p. 153.

Seufert, D., Herlemann, E.-M., Albrecht, E., and Seubert, W. (1971). *Hoppe-Seyler's Z. Physiol. Chem.* **352,** 459.

Söling, H. D., Willms, B., Friedrichs, D., and Kleineke, J. (1968). *Eur. J. Biochem.* **4,** 364.

Störmer, B., Janssen, W., Reinauer, H., Staib, W., and Hollmann, S. (1970). *Hoppe-Seyler's Z. Physiol. Chem.* **351,** 296.

Stucki, J. W., Brawand, F., and Walter, P. (1972). *Eur. J. Biochem.* **27,** 181.

Stucki, J. W. and Walter, P. (1972). *Eur. J. Biochem.* **30,** 60.

Taylor, S. I., Mukherjee, G., and Jungas, R. L. (1974). *Fed. Proc.* **33,** 1426.

Toews, C. J., Lowy, C., and Rudermann, N. B. (1970). *J. Biol. Chem.* **245,** 818.

Tubbs, P. K. and Garland, P. B. (1964). *Biochem. J.* **93,** 550.

Walajtys, E. I., Gottesman, D. P., and Williamson, J. R. (1974). *J. Biol. Chem.* **249,** 1857.

Walter, P., Paetkau, V., and Lardy, H. A. (1966). *J. Biol. Chem.* **241,** 2523.

Walter, P. and Stucki, J. W. (1970). *Eur. J. Biochem.* **12,** 508.

Walter, P. and Anabitarte, M. (1971). *FEBS Lett.* **12,** 289.

Walter, P. and Anabitarte, M. (1973). *FEBS Lett.* **37,** 170.

Walter, P., Mörikofer-Zwez, S., and Brawand, F. (1974). In *Regulation of Hepatic Metabolism,* Vol. 6, *Alfred Benzon Symposium* (F. Lundquist and N. Tygstrup, eds.), Munksgaard, Copenhagen, p. 79.

Weiss, G., Lamartiniere, C. A., Müller-Ohly, B., and Seubert, W. (1974). *Eur. J. Biochem.* **43,** 391.

Wieland, O. H., Löffler, G., Patzelt, C., and Portenhauser, R. (1974). In *Regulation of Hepatic Metabolism,* Vol. 6, *Alfred Benzon Symposium* (F. Lundquist and N. Tygstrup, eds.). Munksgaard, Copenhagen, p. 62.

Wieland, O. H. and Portenhauser, R. (1974). *Eur. J. Biochem.* **45,** 577.

Williamson, D. H., Lund, P., and Krebs, H. A. (1967). *Biochem. J.* **103,** 514.

Williamson, J. R., Browning, E. T., and Olson, M. S. (1968). In *Advances in Enzyme Regulation,* Vol. 6, (G. Weber, ed.), Pergamon Press, New York, p. 67.

Williamson, J. R., Browning, E. T., and Scholz, R. (1969). *J. Biol. Chem.* **244,** 4617.

Wimhurst, J. M. and Manchester, K. L. (1970). *Biochem. J.* **120,** 79.

Wojtczak, A. B., Lenartowicz, E., Rodionova, M. A., and Duszynski, J. (1972). *FEBS Lett.* **28,** 253.

PART III

Experimental Approach to Gluconeogenesis in Liver and Kidney

CHAPTER EIGHT

Interrelations Between Gluconeogenesis and Urea Synthesis

HANS A. KREBS, PATRICIA LUND, AND MARION STUBBS

I. INTRODUCTION

Interrelations between gluconeogenesis and urea synthesis have two roots. First, glucose synthesis shares intermediate stages with urea synthesis. Second, both syntheses depend on, and compete for, ATP.

As for the sharing of intermediary stages, it has recently become clear that earlier formulations of the pathways of the two processes must be supplemented by reactions that transfer intermediates and, in the case of gluconeogenesis from certain substrates, reducing equivalents, through the inner mitochondrial membrane. The present concepts of the pathways and compartmentation of gluconeogenesis and urea synthesis are illustrated in Diagrams 1, 2, 3, and 4.

It follows from Diagram 1 that aspartate is an obligatory intermediate in gluconeogenesis from lactate, as a carrier of the carbon skeleton of oxaloacetate from the mitochondria to the cytosol; it is also clear that glutamate

Diagram 1. Intermediate steps and compartmentation of glucose formation from lactate in rat liver

Compartment		Intermediate steps
		Lactate
Cytosol	(1)	↓ + NAD^+
		Pyruvate + NADH
	(2)	↓ (possibly carrier mediated)
		Pyruvate
	(3)	↓ + HCO_3^- + ATP
Mitochondria		Oxaloacetate
	(4)	↓ + glutamate
		Aspartate + α-oxoglutarate
	(5)	↓ (carrier mediated)
		Aspartate + α-oxoglutarate
	(6)	↓
		Oxaloacetate + glutamate
	(7)	↓ + GTP (ATP)
		Phosphoenolpyruvate + GDP (ADP) + CO_2
	(8)	↓
		2-Phosphoglycerate
Cytosol	(9)	↓
		3-Phosphoglycerate + ATP
	(10)	↓
		1,3-Diphosphoglycerate + ADP
	(11)	↓ + NADH
		Glyceraldehyde-3-phosphate + HPO_4^{2-} + NAD^+
		↓
		$\frac{1}{2}$ Glucose

The transfer of pyruvate into the mitochondria is probably carrier mediated in exchange for OH^- (Papa *et al.*, 1971). The transfer of aspartate out of the mitochondria (step 5) is taken to be coupled with the transfer of glutamate formed by step 6 into the mitochondria, where it is used in step 4 (see Azzi *et al.*, 1967). The α-oxoglutarate required in step 5 in the cytosol is that formed in step 4 in the mitochondria. The transfer of α-oxoglutarate from the mitochondria is assumed to occur in exchange for malate (Quagliariello *et al.*, 1969; Papa *et al.*, 1970). The NADH need at step 11 in the cytosol is that generated in step 1 in the cytosol. Thus no transfer of reducing equivalents to the cytosol is required in gluconeogenesis from lactate in contrast to that from pyruvate. When pyruvate accumulates in the cell (as occurs on incubation with high lactate concentrations), another set of reactions must take place because for each molecule of pyruvate that accumulates, one molecule of NADH must be disposed of by transfer to the mitochondria and oxidation by the respiratory chain. This is taken to involve the reversal of steps 5b to 4a of Diagram 2 (see Borst, 1963).

In some species (e.g., pigeon), step 7 occurs in the mitochondria, making steps 4, 5, and 6 unnecessary.

Diagram 2. Early steps and compartmentation of gluconeogenesis from alanine or pyruvate in rat liver

Compartment		Intermediate steps
		Alanine
Cytosol or mitochondria	(1a)	↓ + α-oxoglutarate
		Pyruvate + glutamate
	(3)	↓ + HCO_3^- + ATP
		Oxaloacetate
Mitochondria	(4a)	↓ + NADH
		Malate
	(5a)	↓ (carrier mediated)
		Malate
Cytosol	(5b)	↓ + NAD
		Oxaloacetate + NADH

Steps 1a, 4a, 5a, and 5b replace 1, 4, and 5 of Diagram 1. The NADH needed for step 4a is generated within the mitochondria by NAD-linked dehydrogenase reactions in the mitochondria. The NADH generated in step 5b is used in step 11 of Diagram 1. When propionate, glutamate, glutamine, histidine, proline, or arginine are glucogenic precursors, the pathways join that from alanine at the stage after mitochondrial malate formation (step 5a). The exchange of malate out of the mitochondria is carrier mediated in exchange for various tricarboxylic acids (Robinson *et al.*, 1970), α-oxoglutarate (Quagliariello *et al.*, 1969; Papa *et al.*, 1970), and phosphate (Robinson and Chappell, 1967).

When ethanol is added together with high concentrations of alanine or pyruvate, transfer of reducing equivalents from mitochondria to cytosol may be unnecessary because ethanol readily causes reduction of cytosolic NAD. In this situation oxaloacetate formed from the carbon skeleton of alanine or pyruvate in the mitochondria may be transferred to the cytosol in the form of aspartate, as in the case of gluconeogenesis from lactate (Diagram 1).

and oxoglutarate are obligatory participants. The mitochondrial glutamate needed for transfer (reaction 4) is that generated at step 6 in the cytosol, which enters the mitochondria by exchange with the aspartate in step 5. Aspartate is not an intermediate in gluconeogenesis from alanine or pyruvate (Diagram 2); instead malate is an intermediate. Aspartate functions as a carrier of the C_4-carbon skeleton of oxaloacetate, while malate carries both the C_4-carbon skeleton and reducing equivalents. As shown in Diagrams 3 and 4, aspartate, oxoglutarate, and glutamate are also involved in urea synthesis.

As for the sharing of the requirement for ATP, it is relevant that the quantities of ATP needed in gluconeogenesis and urea synthesis can be high

Diagram 3. Intermediary stages and compartmentation of synthesis of urea from ammonium ions

Compartment Intermediary steps Compartment

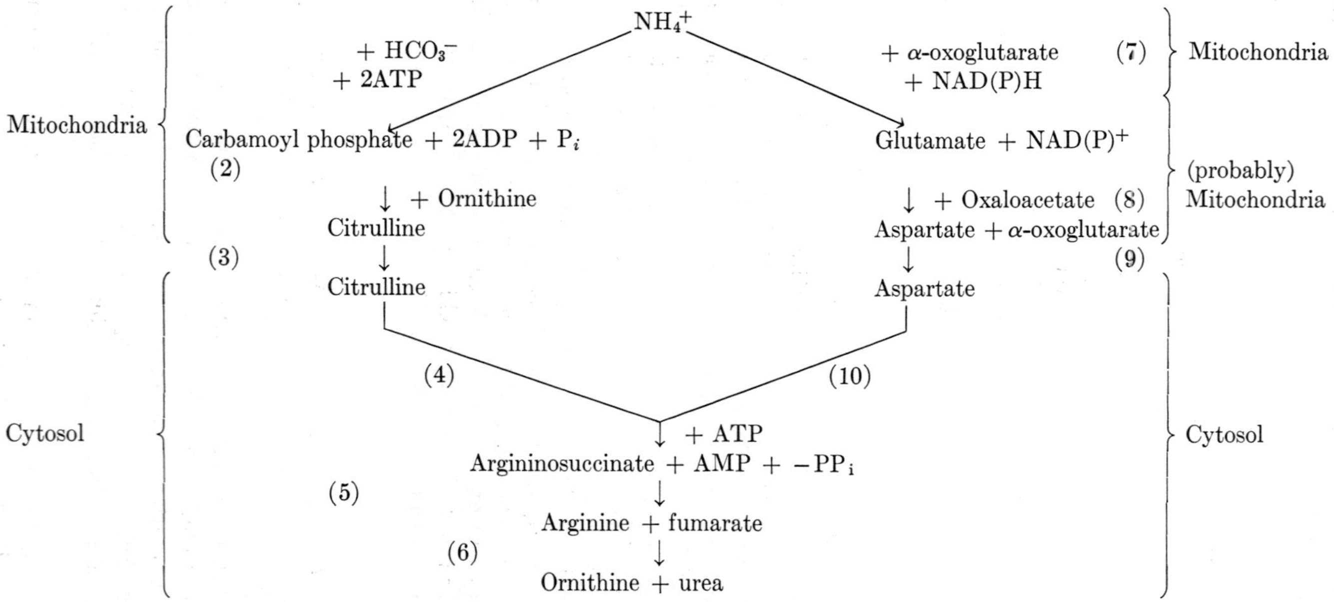

Since glutamate arises in step 7 in the mitochondria, the formation of aspartate by transamination probably also occurs in the mitochondria; this is not certain, however, because aspartate aminotransferase is also present in the cytosol. The fumarate formed together with arginine in step 5 is reconverted by way of malate into oxaloacetate to be available for step 8 in either cytosol or mitochondria. Ornithine formed in the cytosol by step 6 becomes available in the mitochondria for step 2. The transfer of citrulline from mitochondria to cytosol (step 3) and the entry of ornithine into the mitochondria involve special carrier mechanisms (Gamble and Lehninger, 1973).

Diagram 4. Initial intermediary steps of urea synthesis from some amino acids

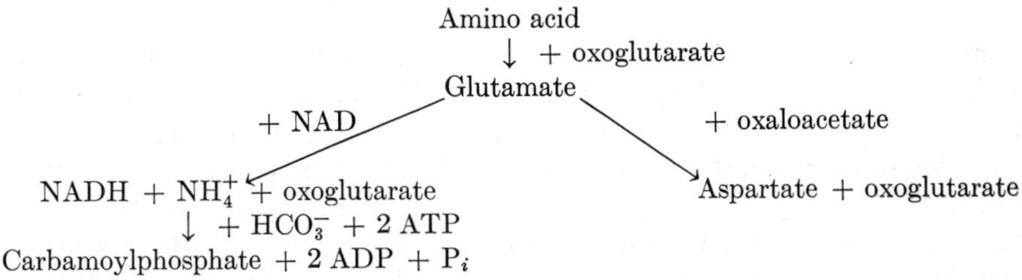

The scheme applies to amino acids that transaminate with oxoglutarate or form glutamate in the course of their degradation (histidine, proline, arginine, ornithine, glutamine).

in relation to the basal rate of ATP synthesis. Under conditions that are optimal for the two synthetic processes, the O_2 uptake of liver cells rises by a factor of up to 4, and since the O_2 uptake depends on the rate of ATP utilization, we can assume that ATP consumption may exceed by a factor of up to 4 the basal ATP production. The synthesis of one molecule of glucose from lactate is known to require six molecules of ATP, and the synthesis of one molecule of urea from either amino acids or ammonium ions requires four molecules of ATP.

One of the problems arising from the interrelations between gluconeogenesis and urea synthesis is that of the competition of the two processes for ATP and for the joint intermediary reactants, such as aspartate and oxoglutarate. Competition is most likely to occur when the conditions for the two processes are optimal. It is thus necessary to establish the conditions of maximal rates of gluconeogenesis and to test how the two processes affect each other under optimal conditions for both.

A problem connected with the optimal rate of urea synthesis is that of the nature of the control mechanisms which regulate the stoichiometric formation of aspartate and carbamoylphosphate, a requirement evident in Diagram 3.

Experiments bearing on the problems just mentioned are reported in this chapter. Most of the experiments were carried out on isolated rat hepatocytes (a more convenient material for serial experiments that the perfused liver) after it had been ascertained that the metabolic behavior of such cells is, under suitable conditions, indistinguishable from that of the perfused liver.

II. MATERIALS AND METHODS

A. Rats

Female Wistar rats (180–220 g) were obtained from Charles River U.K. Ltd., Manston Road, Margate, Kent, U.K. The normal diet was Oxoid pasteurized breeding diet for rats and mice (Oxoid Ltd., London S.E. 1.). Changes in physiological state were brought about by (1) starvation for 48 hours, (2) feeding a pure protein diet (boiled egg white) for the 4 days preceding the experiment, and (3) feeding a high protein diet (a mixture of raw minced pig heart, sheep kidney, and ox liver from which all adhering fat was removed) for 7 days.

B. Preparation of Isolated Liver Cells

Cells were prepared essentially as described by Berry and Friend (1969) with the modifications described by Cornell *et al.* (1973) and Krebs *et al.* (1974). The main modification of the method of Berry and Friend (1969) was the use of a medium containing physiological concentrations of bicarbonate. This is essential in experiments involving CO_2-fixation reactions such as gluconeogenesis and urea synthesis. Usually 4 ml cell suspensions were incubated in 25 ml vessels. The cell concentration was about 40 mg wet weight/4 ml when rates were high and 80 mg/4 ml otherwise. Within a given experiment, the concentration was constant.

C. Oxygen Consumption

Oxygen uptake in the presence of 5% v/v CO_2 was measured in special manometer vessels as described by Krebs *et al.* (1974).

D. Isolated Liver Perfusion

The method of perfusion was that described by Hems *et al.* (1966) and Krebs *et al.* (1969). The substrates were added as 0.2 M neutral solutions, except for oleate, which was 0.1 M and was added together with albumin as described by Krebs *et al.* (1969). A period of 38 minutes perfusion was allowed before addition of substrates.

E. Reagents

Urease tablets (type III-20) and urease powder (type VI) were obtained from Sigma London Chemical Co. Ltd., London S.W. 6. Other purified enzymes and coenzymes were products of Boehringer Corp. (London) Ltd., London, W. 5.

F. Determination of Metabolites

Glucose was determined by the method of Slein (1963). Urea and NH_3 were determined successively in the same cuvette in the assay system described by Bernt and Bergmeyer (1970) for urea. The "reagent blank" in the urea determination was considerably decreased by the use of NH_3-free urease (Sigma type VI). Glutamate and alanine also were determined in the same cuvette by successive addition of glutamate dehydrogenase in glycerol and alanine dehydrogenase, respectively. The conditions were essentially as described for the determination of glutamate by Bernt and Bergmeyer (1963). Aspartate was determined by the method of Pfleiderer (1963) and oxoglutarate by that of Bernt and Bergmeyer (1963) using glutamate dehydrogenase in glycerol and excess of $(NH_4)_2SO_4$.

III. RESULTS

A. Optimal Conditions for Urea Synthesis from Ammonium Ions in Isolated Hepatocytes

Earlier work on urea synthesis in liver slices (Krebs and Henseleit, 1932; Krebs, 1934) indicated that there are three substances that accelerate the rate of urea synthesis upon addition to the medium: NH_4^+ ions, ornithine, and lactate (or substances metabolically closely related to lactate, such as pyruvate or fructose); it was also learned that optimal rates of urea synthesis require physiological concentrations of HCO_3^- (26 mM) and CO_2 (1 mM). It was clear that ammonia and bicarbonate can be rate limiting because they are the starting materials for the synthesis of urea. That ornithine can be rate limiting became obvious when it was recognized that the substance is an intermediary reactant. It was not properly understood, however, why lactate accelerates the rate of urea synthesis. Its ready oxidation in the liver led Krebs and Henseleit to suspect that lactate might act more effectively than endogenous substrates as a source of the energy required for the synthesis of urea.

When these observations were made in 1932 it was not yet known that half the ammonia molecules participating in the synthesis of urea pass through the stages of glutamate and aspartate (Ratner and Petrack, 1951). The need for lactate thus can be explained by the need for a precursor of the carbon skeleton of these amino acids. Lactate, by way of the pyruvate carboxylase and tricarboxylic acid cycle reactions, can supply both α-oxoglutarate and oxaloacetate.

For the systematic investigation of optimal conditions for hepatic urea synthesis, isolated hepatocytes of rats fasted for 48 hours were used. As far as a carbon source of α-oxoglutarate and aspartate (lactate, pyruvate, or amino acids) is concerned, 2 mM was found to give maximum rates of urea synthesis. At this concentration the rate of gluconeogenesis from these precursors is low. Oleate (1 mM) was added as a fuel of respiration, since fatty acids are the normal fuel in the fasted liver.

Both ornithine and lactate accelerate urea synthesis from NH_4Cl, ornithine being more effective, and the effects are additive as already reported by Krebs and Henseleit (1932) for liver slices (Table I). Oleate alone also increases the rate of urea synthesis, but much less than lactate—presumably because it is not a precursor of aspartate. The highest rates are obtained when lactate, ornithine, and oleate are present together with NH_4Cl (Table I).

Lactate can be replaced by glutamine, asparagine, proline, pyruvate, or

Table I. Optimal conditions for urea synthesis from ammonium ions in isolated hepatocytes from fasted rats[a]

Additions	Urea formed (μmole/min/g)	O_2 Uptake (μmole/min/g)
NH_4Cl	0.51	2.12
NH_4Cl; lactate	1.42	3.28
NH_4Cl; orn	1.80	2.05
NH_4Cl; orn; lactate	2.65	2.86
NH_4Cl; oleate	0.95	2.33
NH_4Cl; oleate; orn	1.17	2.35
NH_4Cl; oleate; orn; lactate	5.73	6.39
NH_4Cl; oleate; orn; asn	6.30	7.37
NH_4Cl; oleate; orn; pro	4.47	4.92
NH_4Cl; oleate; orn; gln	6.11	5.54
NH_4Cl; oleate; orn; pyruvate	5.91	7.48
NH_4Cl; oleate; orn; ala	6.09	6.70
None	0.23	2.27
Oleate	0.21	3.28

[a] The initial concentrations of the added substrates were NH_4Cl, 10 mM; ornithine, 2 mM; oleate, 1 mM, and all other substrates 2 mM. Figures are from a representative experiment. Incubations were for 40 minutes; rats fasted for 48 hours.

Table II. Optimal conditions for urea synthesis from ammonium ions in isolated hepatocytes from fed rats[a]

Additions	Urea formed (μmole/min/g)	O_2 Uptake (μmole/min/g)
None	0.33	2.01
NH_4Cl	1.06	3.34
NH_4Cl; lactate	0.87	4.22
NH_4Cl; orn	2.40	3.92
NH_4Cl; orn; lactate	4.95	6.48
NH_4Cl; oleate	1.36	3.88
NH_4Cl; oleate; orn	4.19	5.64
NH_4Cl; oleate; orn; lactate	4.30	6.13

[a] The experimental conditions were as described for Table I except that the lactate concentration was 10 mM.

alanine (Table I)—that is, by substances that serve as precursors of glutamate and aspartate. Added glutamate and aspartate were not effective (not shown in table) because their rate of entry into hepatocytes is very slow.

The increases in the rates of urea synthesis are roughly paralleled by increases in the rates of oxygen consumption. When oleate, ornithine, and lactate were added, the increased rate of oxygen consumption over the blank was 4.12 μmole/min/g. This corresponds to the rate of ATP synthesis of 24.7 μmole/min/g on the assumption that one O_2 yields 6 ATP. The increase in the rate of urea synthesis was 5.50 μmole/min/g, requiring 22.0 molecules of ATP/min/g. In other experiments we found that at a lactate concentration of 2 mM, glucose is synthesized at the rate of 0.5 μmole/min/g, which requires 3 μmole ATP/min/g. Thus the expected additional ATP requirement for urea synthesis and gluconeogenesis is 25 μmole/min/g, and the additional ATP supply calculated from the extra O_2 uptake was 24.7 μmole/min/g. This good agreement between calculated and expected extra oxygen uptake does not apply to all experimental conditions.

Hepatocytes from well-fed rats differ from those of starved rats in that the maximal rates of urea synthesis are about 20% lower (Table II). Lower rates are expected because the enzymes of urea synthesis are known to increase on fasting (Schimke, 1962). Lactate or oleate has no major effects when added alone, but in the presence of ornithine, lactate more than doubles the rate and oleate also causes a substantial increase. The result when oleate plus lactate were added was the same as that of lactate alone. Thus the main limiting factor of urea synthesis in the fed liver is the concentration of ornithine. When addition of ornithine removes this limitation, the capacity of argininosuccinate synthetase appears to become rate limiting rather than the supply of aspartate, because lactate and oleate were about equally effective. When rats were fed a high protein diet (boiled egg white or raw meat; see Section II.A the rates of urea synthesis in the presence of ornithine, oleate, and lactate were about 20% higher than in cells from rats given the standard diet, and of the same order as in fasted rats.

B. Rate of Urea Formation from Amino Acids

The rate of urea synthesis from amino acids is always lower than that from NH_4Cl (Table III), indicating that the rate of supply of NH_4^+ is a limiting factor. The amino acids listed in Table III are those which give the highest rate of urea formation. Glutamine and asparagine are the most effective precursors because they release ammonia directly by the glutaminase and asparaginase reactions. In the case of asparagine the formation of aspartate in the cytosol by asparaginase may contribute to the high rate.

Addition of ornithine and oleate did not accelerate urea synthesis except when asparagine was the precursor. This is not unexpected if ammonia release is the main rate-limiting factor. Both urea synthesis and gluconeogenesis from proline and glutamine are somewhat decreased by oleate, possibly because of inhibition of substrate oxidation. In every case, oleate somewhat increased oxygen consumption, regardless of whether it inhibited or accelerated urea synthesis and gluconeogenesis.

Some ammonia accumulates during incubation with amino acids, especially with glutamine and asparagine. At 40 minutes the concentration was 0.4 mM in the cell suspension when glutamine was the urea precursor, compared with 0.11 mM with alanine. This concentration was unchanged by the presence of ornithine. With asparagine alone the ammonia concentration was somewhat higher (0.74 mM). In this case it was decreased to 0.4 mM by ornithine. Of the total amino acid nitrogen recovered as urea plus ammonia, the latter contributed no more than 18% (i.e., with asparagine) in the absence of ornithine, and less than 10% in the presence of ornithine.

A stoichiometric relationship between urea formation and gluconeogenesis is expected when glucogenic amino acids are the added precursors and when the formation of glucose is quantitative. Alanine, serine, and proline should form equal amounts of urea and glucose, whereas asparagine and glutamine should form twice as much urea as glucose. The stoichiometry is reasonably good with alanine and asparagine, provided oleate is present. Lack of stoichiometry is due to the formation of products other than glucose. In the absence of oleate, alanine forms much lactate

Table III. Urea synthesis from amino acids[a]

Added amino acids	Additional substrates	Urea formed (μmole/min/g)	Glucose formed (μmole/min/g)	O_2 Uptake (μmole/min/g)
None	None	0.31	0.23	2.67
None	Orn	0.51	0.26	2.65
Alanine	None	1.06	0.50	4.09
	Orn	1.23	0.48	4.29
	Oleate	1.09	0.97	5.60
	Orn, oleate	1.12	0.92	5.64
Glutamine	None	1.85	0.78	4.73
	Orn	1.87	0.74	4.58
	Oleate	1.49	0.61	5.33
	Orn, oleate	1.46	0.56	5.41
Asparagine	None	1.68	0.53	4.10
	Orn	2.19	0.67	4.23
	Oleate	2.12	0.92	5.97
	Orn, oleate	2.30	1.01	5.98
Proline	None	0.61	0.48	3.85
	Orn	0.71	0.44	3.89
	Oleate	0.54	0.33	4.34
	Orn; oleate	0.66	0.30	4.20
Serine	None	0.89	0.47	3.68
	Orn	0.93	0.56	3.82
	Oleate	0.84	0.57	4.10
	Orn, oleate	0.98	0.65	4.64

[a] The conditions were as described in Table I. The rats were starved for 48 hours. The initial concentrations of the added amino acids were 10 mM; 4 ml suspension contained 88 mg of cells. The experiment is representative.

(Krebs *et al.*, 1973), and glutamine and asparagine form some other amino acids (see Cornell *et al.*, 1973).

C. Maximum Rates of Gluconeogenesis from Lactate

Gluconeogenesis from high concentrations of lactate (10 mM) has a lag period in freshly prepared hepatocytes, probably because the concentration of glutamate is too low for the full functioning of the aspartate shuttle, owing to loss during the preparation of the cells (see Cornell *et al.*, 1974). Several amino acids and NH_4Cl can overcome the lag, L-lysine being most effective. Moreover, lysine accelerates gluconeogenesis beyond the lag period for reasons that are not yet understood. Thus lysine is a requirement for maximum rates of gluconeogenesis from lactate. As shown in Table IV, addition of oleate or dibutyryl cyclic AMP causes further increases of the rate (see Exton and Park, 1968; Ross *et al.* 1967). In the absence of lysine, the effects of these two accelerators are fully additive (Table IV), but in the presence of lysine (when the flux from lactate to glucose is near-maximum) the effects are not fully additive.

D. Inhibition of Urea Synthesis by L-Lysine

In the course of experiments designed to test whether the substances favoring maximum rates of gluconeogenesis from lactate interfere with urea synthesis, it was found that lysine can inhibit urea synthesis, presumably by competing with ornithine for the binding site of ornithine carbamoyl transferase (Table V). Lysine at 10 mM decreased by 83% the stimulation of urea synthesis that occurs with 0.2 mM ornithine. Increasing the ornithine concentration to 2 mM almost doubled the rate of urea synthesis and decreased the inhibitory effect of lysine from 83 to 14%. Lysine is already known to be a competitive inhibitor of ornithine carbamoyl transferase in *Streptococcus faecalis* where the enzyme plays a role in the biosynthesis of

Table IV. Maximum rates of gluconeogenesis from lactate[a]

Additions	Glucose formed (μmole/min/g wet wt)
Lactate	0.67
Lactate; oleate	1.40
Lactate; lysine	1.47
Lactate; c-AMP	0.86
Lactate; oleate; lysine	2.19
Lactate; c-AMP; lysine	1.86
Lactate; oleate; c-AMP	1.67
Lactate; oleate; c-AMP; lysine	2.35

[a] Liver cells were prepared from rats fasted for 48 hours. Initial concentrations were: lactate, 10 mM; L-lysine, 2 mM; oleate, 2 mM, and dibutyryl cyclic AMP (c-AMP) 0.1 mM. Incubation was for 60 minutes at 37°C. Data are from a representative experiment; see also Table I of Cornell *et al.* (1973).

Table V. Inhibition of urea synthesis by L-lysine[a]

Additional substances				
Ornithine (mM)	Lysine (mM)	Glucose formed (μmole/min/g wet wt)	Urea formed (μmole/min/g wet wt)	Inhibition of ornithine effect by lysine (%)
—	—	1.25	0.42	
—	2	1.20	0.26	38
—	10	1.03	0.20	52
0.2	—	1.42	2.40	
0.2	2	1.32	1.57	35
0.2	10	1.20	0.76	68
2.0	—	1.43	4.57	
2.0	2	1.39	4.59	0
2.0	10	1.40	3.99	13

[a] Liver cells were prepared from rats fasted for 48 hours. All suspensions contained initially lactate, 10 mM; pyruvate, 1 mM; oleate, 1 mM, and NH_4Cl, 10 mM. Incubation was for 40 minutes at 37°C. Data are from a representative experiment.

arginine (Nakamura and Jones, 1970). The formation of homocitrulline in the presence of excess lysine also indicates a reaction of lysine with ornithine carbamoyl-transferase in the liver (Ryan *et al.*, 1968).

E. Interrelations Between Gluconeogenesis and Urea Synthesis

Because of the inhibitory effect of lysine on urea synthesis, lysine could not be used in experiments designed to test interrelations between urea synthesis and gluconeogenesis when rates of both processes are high. In the absence of lysine, the highest rates of gluconeogenesis from lactate were found when lactate and pyruvate were present in the physiological ratio of 10:1 (10 mM lactate, 1 mM pyruvate) together with 1 mM oleate and 2 mM ornithine. These conditions essentially remove the lag, and as the data in Tables IV and VI reveal, the rates of glucose synthesis were of the same order as with lysine and oleate (in the absence of NH_4Cl). Rates of glucose and urea formation under various conditions appear in Table VI. Special tests in which the lactate concentration was varied between 2.5 to 20 mM in the presence of 10 mM NH_4Cl, 2 mM ornithine, and 1 mM oleate showed that the rate of urea synthesis does not change when the cells synthesize glucose at high rates. By contrast, glucose synthesis is inhibited between 32 and 39%

Table VI. Interrelations between gluconeogenesis and urea synthesis in isolated liver cells[a]

Experimental conditions	Rates without added NH_4Cl (μmole/min/g wet wt)			Rates with added NH_4Cl (μmole/min/g wet wt)			Inhibition of gluconeogenesis in presence of NH_4Cl (%)
	Glucose formed	Urea formed	O_2 Uptake	Glucose formed	Urea formed	O_2 Uptake	
1. Lactate; pyruvate	0.95±0.07	0.16±0.02	3.56±0.17	1.04±0.09	1.02±0.15	4.19±0.35	
2. Lactate; pyruvate; oleate	1.87±0.21	0.13±0.02	5.36±0.43	1.33±0.35	0.56±0.08	6.08±0.43	32
3. Lactate; pyruvate; orn	1.20±0.12	0.29±0.03	4.00±0.27	0.72±0.07	2.87±1.04	5.85±0.59	39
4. Lactate; pyruvate; oleate; orn	2.50±0.41	0.25±0.03	6.38±0.73	1.65±0.32	4.64±0.35	8.80±1.25	35
5. Oleate	0.09±0.03	0.20±0.03	3.21±0.18	0.06±0.02	0.78±0.18	3.21±0.22	
6. Oleate; orn	0.11±0.02	0.32±0.02	3.46±0.22	0.05±0.02	1.01±0.59	3.61±0.24	

[a] Liver cells were prepared from rats fasted for 48 hours. The initial concentrations were for lactate, 10 mM; for pyruvate, 1 mM; for ornithine 2 mM; for oleate, 1 mM, and for NH_4Cl, 10 mM. Incubation was for 60 minutes in special manometer vessels designed for measurement of O_2 uptake in the presence of 95% O_2:5% CO_2 (see Krebs *et al.*, 1974). In each experiment the flasks contained only about 50 mg wet weight of cells. Rates are means ±SEM of three experiments.

when NH_4Cl is added (Table VI). In the presence of ornithine (conditions 3 and 4), addition of NH_4Cl causes high rates of urea synthesis, and this might be taken to suggest that in the competition for ATP or aspartate the synthesis of urea is favored against the synthesis of glucose. However this interpretation does not explain why the addition of NH_4Cl also inhibits gluconeogenesis in the absence of ornithine when the rate of urea synthesis is low (condition 2).

Thus other factors must play a role in the inhibition of gluconeogenesis in the presence of NH_4Cl.

One such factor is the *presence* of 10 mM NH_4Cl, rather than a *high rate* of urea synthesis. The NH_4Cl does not inhibit gluconeogenesis under all conditions (see condition 1 of Table VI). Only when oleate or ornithine or both are present is NH_4Cl inhibitory. The effect of oleate with NH_4Cl is to cause a large accumulation of aspartate (Table VII), which is diverted to urea synthesis by the presence of ornithine. This implies inhibition of the aspartate aminotransferase in the cytosol, since aspartate is an intermediate in gluconeogenesis from lactate. Such an inhibition can be accounted for by a fall of the concentration of α-oxoglutarate that occurs in the presence of oleate and NH_4Cl (Table VII).

As discussed in Section III.A, the extra oxygen consumption above the

Table VII. Metabolites found in hepatocyte suspensions after incubation under the conditions described in Table VI[a]

Substrates added	Metabolite found (μmole/4 ml of suspension containing 263 mg cells)			
	α-Oxo-glutarate	Aspartate	Glutamate	Alanine
Lactate; pyruvate	0.206	0.074	0.52	0.40
Lactate; pyruvate; orn	0.147	0.145	1.06	1.28
Lactate; pyruvate; NH_4Cl	0.066	0.360	1.10	3.18
Lactate; pyruvate; orn; NH_4Cl	0.091	0.130	0.53	1.09
Lactate; pyruvate; oleate	0.164	0.136	0.60	0.27
Lactate; pyruvate; oleate; orn	0.118	0.502	1.16	0.67
Lactate; pyruvate; oleate; NH_4Cl	0.044	5.99	2.31	3.51
Lactate; pyruvate; oleate; orn; NH_4Cl	0.073	1.84	1.47	2.31

[a] Concentrations of substrates were as described in Table VI. Incubation was for 10 minutes at 37°C.

Table VIII. Calculated and observed O_2 requirements for gluconeogenesis and urea synthesis[a]

Experimental conditions	Rates without added NH_4Cl		Rates with added NH_4Cl			
	ΔO_2 Needed for glucose synthesis	ΔO_2 Observed	ΔO_2 Needed for glucose synthesis	ΔO_2 Needed for urea synthesis	Total O_2 calculated	ΔO_2 Observed
Lactate; pyruvate	0.95	1.29	1.04	0.68	1.72	2.07
Lactate; pyruvate; oleate	1.87	2.15	1.33	0.37	1.70	2.87
Lactate; pyruvate; ornithine	1.20	1.73	0.72	1.92	2.64	3.73
Lactate; pyruvate; oleate; ornithine	2.50	3.17	1.65	3.10	4.75	5.59

[a] The calculations are based on the data in Table VI. The assumption is made that 6 molecules of ATP are generated for each molecule of O_2 consumed in excess of the appropriate basal rates when no substrate or NH_4Cl (Table I) or oleate (Table VI) was present. The ATP requirements are taken to be 6 molecules for one molecule of glucose and 4 molecules for one molecule of urea. The ΔO_2 refers to the extra O_2 for the synthesis of glucose and urea expressed as micromoles per minute per gram.

basal rate is about the same as that calculated for the ATP requirements of urea synthesis and gluconeogenesis when the rate of urea synthesis is high and that of gluconeogenesis is low. When the rate of glucose synthesis is high (Table VI), the observed extra oxygen consumption is rather greater than the calculated values (Table VIII). This holds when ammonia is absent as well as in its presence. It suggests that under the experimental conditions described in Table VI, additional biosyntheses or "futile" cycles (Newsholme and Gevers, 1967; Friedmann *et al.*, 1971; Clark *et al.*, 1974) take place.

IV. STOICHIOMETRIC FORMATION OF ASPARTATE AND CARBAMOYL PHOSPHATE IN UREA SYNTHESIS

It follows from Diagrams 1 and 2 that half the nitrogen in urea synthesis passes through the stage of carbamoylphosphate and half through the stage of aspartate. Normally (*in vivo*) most of the nitrogen originates from amino acids and arises in the form of glutamate, as a result of the transamination of various amino acids with α-oxoglutarate. Smaller amounts of nitrogen arise primarily in the form of ammonia from the amide group of glutamine and asparagine, from serine, threonine, histidine, from aminopurines, nucleotides, and a few other precursors.

Since glutamate may form either aspartate or ammonia, there must be a mechanism by which stoichiometry between carbamoylphosphate and aspartate is achieved. A satisfactory explanation of the mechanism is the following, which is based on the fact that the value of the mass action ratio of the aspartate aminotransferase system in the normal liver corresponds very nearly to the equilibrium constant. This has been noted in many experiments (Williamson *et al.*, 1967; Brosnan and Williamson, 1974), and it means that the given steady-state concentrations of glutamate, α-oxoglutarate, and oxaloacetate in the liver fix the concentration of aspartate in the region of 0.7 mM. Appreciable amounts of aspartate, it should be noted, can be formed in animal tissues only by the aminotransferase reaction (minor amounts can arise by proteolysis). Thus whenever aspartate is removed in the course of the synthesis of urea by the formation of argininosuccinate from citrulline and aspartate, the tendency to reestablish equilibrium automatically replaces the aspartate by the aminotransferase reaction and maintains it at a semiconstant concentration. The steady-state concentration of aspartate changes only under exceptional experimental conditions (Sections V.B and V.D).

Any excess of ammonia formed either by the glutamate dehydrogenase reaction or from other precursors then enters the carbamoylphosphate pathway leading to the formation of citrulline. Thus stoichiometry is achieved by the thermodynamic limitation of aspartate formation and the disposal of the excess nitrogen by the formation of carbamoylphosphate. The regulatory device is thus exceedingly simple. No regulatory mechanism controlling the carbamoylphosphate synthetase is required, although there must be a mechanism to prevent tissue ammonia from being removed completely. In the liver a steady-state concentration of about 0.7 mM ammonia is maintained, and it is not yet clear why carbamoylphosphate synthetase does not lower this concentration.

V. DISCUSSION

A. Comparison of Rates of Urea Synthesis in Isolated Liver Cells and in the Perfused Rat Liver

The rates of urea synthesis given in Table I are higher in general than those reported for the perfused liver by Hems *et al.* (1966). However recent unpublished work on the perfused liver shows that in the experiments of Hems *et al.* the rate of blood flow may have been a limiting factor in urea synthesis. When the flow rate of the perfusion medium containing 2.5 g hemoglobin/100 ml was raised to about 40 ml/min, the rates of urea synthesis in the presence of 10 mM NH_4Cl, 10 mM lactate, 2 mM oleate, and 2.7 mM ornithine rose to 5 μmole/min/g. In the experiments of Hems *et al.* the venous blood still contained considerable amounts of oxygen. Whether the rate of release of O_2 from hemoglobin, diffusion of O_2 into the liver cell, or the diffusion of ammonia into the cell, was rate limiting is still an open question.

B. The Mechanism of Inhibition of Gluconeogenesis by Ammonium Ions

The following facts bear on the interrelations between gluconeogenesis and the synthesis of urea.

1. In the presence of NH_4Cl, gluconeogenesis from lactate is inhibited to about the same extent when oleate, ornithine, or oleate and ornithine together are added (Table VI). The inhibition of gluconeogenesis occurs not only when the rate of urea synthesis is very high but also when the rate is low (as in the absence of ornithine).
2. An accumulation of aspartate is caused by NH_4Cl when lactate and oleate are also present. The accumulation of aspartate is largely prevented by the addition of ornithine (Table VII).

3. In the presence of lactate and oleate, NH_4Cl causes a fall in the concentration of α-oxoglutarate (Table VII).
4. Oleate accelerates glucose synthesis from lactate (Table IV) and (at low lactate concentrations) urea synthesis from NH_4Cl in the presence of ornithine (Table I).
5. Aspartate, oxoglutarate, glutamate, and ATP are involved in both glucose synthesis from lactate and urea synthesis from NH_4Cl (Diagrams 1 and 3).

The important point is that high rates of urea synthesis as such do not inhibit gluconeogenesis. It is the presence of NH_4Cl at high, unphysiological concentrations, in combination with oleate, ornithine, or both, that causes the inhibition of gluconeogenesis. The inhibition in the presence of lactate, oleate, and NH_4Cl can be explained as follows: oleate increases the rate of formation of oxaloacetate by its activation of pyruvate carboxylase through the supply of acetyl CoA. The high concentration of NH_4Cl, together with the high rate of NADH supply from the degradation of oleate, facilitates the rate of glutamate synthesis from oxoglutarate. The high activity of the mitochondrial aspartate aminotransferase causes oxaloacetate and glutamate to react promptly. The oxoglutarate formed, instead of being transferred to the cytosol in stoichiometric amounts with aspartate, is reconverted to glutamate in the mitochondria. This causes a shortage of oxoglutarate in the cytosol and decreases the rate of supply of cytosolic oxaloacetate, resulting in a decreased rate of gluconeogenesis. The decrease in the accumulation of aspartate on addition of ornithine is explained by the increased rate of argininosuccinate synthesis in the presence of ornithine.

This explanation for the inhibition of gluconeogenesis cannot account for the inhibition that occurs in the absence of oleate when lactate, ornithine, and NH_4Cl are present together. In this situation aspartate does not accumulate, and the rate of urea synthesis is low. The reasons for the inhibition of gluconeogenesis under these conditions are still obscure. The inhibition suggests that NH_4Cl has an additional effect, possibly acting at other stages of metabolism, perhaps by activating phosphofructokinase (Passonneau and Lowry, 1964).

It is evident that a phenomenon that at first may be interpreted as an inhibition of gluconeogenesis by urea synthesis is in fact an inhibition independent of the rate of urea synthesis but dependent on the presence of NH_4Cl. Although competition for ATP, or for aspartate, can be expected when the rates of gluconeogenesis and urea synthesis are both high, the experiments give no indication that ATP is a limiting factor under any of the conditions tested. The maximum rates of the two processes thus do not appear to be limited by the supply of ATP but rather by the capacity of the slowest intermediate steps—the argininosuccinate synthetase reaction in urea synthesis and a reaction between pyruvate and phosphoenolpyruvate in gluconeogenesis.

C. Formation of Alanine

The formation of alanine in the presence of lactate and NH_4Cl can be accounted for by the equilibrium relations in the glutamate dehydrogenase and alanine aminotransferase systems:

$$\text{oxoglutarate} + NH_4^+ + \text{NADH} \rightleftharpoons \text{glutamate} + \text{NAD} \quad (1)$$

$$\text{glutamate} + \text{pyruvate} \rightleftharpoons \text{alanine} + \text{oxoglutarate} \quad (2)$$

$$\text{pyruvate} + NH_4^+ + \text{NADH} \rightleftharpoons \text{alanine} + \text{NAD}$$

The equilibrium constant for Eq. 1 is 3.87×10^{-6} M and for Eq. 2 it is 1.52. It follows that at equilibrium

$$[\text{ala}] = [\text{pyruvate}] \times [NH_4^+] \times \frac{[\text{NADH}]}{[\text{NAD}]} \times 3.93 \times 10^5\ M^{-1}$$

In the present experiments (Table VI) [pyruvate] was initially about 10^{-3} M, $[NH_4^+]$ about 10^{-2} M, and [NADH]/[NAD] (in the cytosol) about 10^{-3}. Hence the equilibrium concentration of [ala] was of the order of 3.93 mM. Thus alanine is expected to be formed because of the tendency to establish equilibrium in reactions 1 and 2.

D. Accumulation of Aspartate in the Presence of Ethanol

Aspartate also accumulates (in the perfused rat liver) when ethanol is added to the perfusion medium together with either alanine or with lactate plus NH_4Cl (Krebs *et al.*, 1973). Stubbs and Krebs (1975) showed that the basic mechanism of accumulation of aspartate on addition of ethanol is the same as in the experiments reported in this chapter on addition of oleate, a common factor being the introduction of reducing equivalents by oleate or ethanol.

VI. SUMMARY

1. Gluconeogenesis from lactate and urea synthesis are interconnected because they share the need for ATP and therefore compete for that source of energy. Furthermore, aspartate is an intermediate in both processes, and

oxoglutarate and glutamate participate in both. Interference by one process with the other may be expected, especially when rates are high.

2. High rates of gluconeogenesis from lactate are obtained in isolated hepatocytes from fasted rats in the presence of lysine (2 mM), oleate (1 mM), and dibutyryl cyclic AMP. The rates are of the same order as those obtained in the isolated, perfused liver.

3. The highest rates of urea synthesis from ammonium chloride are obtained in the presence of lactate (2 mM), ornithine (2 mM), and oleate (1 mM). Lactate is needed as a precursor of the carbon skeleton of aspartate as well as of glutamate. It can be replaced by various other precursors. Maximum rates in hepatocytes from fed rats were about 20% lower than those from rats fasted for 48 hours.

4. Lysine inhibits urea synthesis from ammonia by competition with ornithine.

5. The maximum rates of urea synthesis from amino acids are much lower than those from ammonium chloride.

6. When the rates of gluconeogenesis are relatively high, the rates of urea synthesis are not decreased when 10 mM lactate is added. By contrast, gluconeogenesis is inhibited when the rates of urea synthesis are high under the conditions described in paragraph 3 (i.e., in the presence of oleate, ornithine, and ammonium chloride).

7. Inhibition of gluconeogenesis in the presence of ammonium chloride occurs also when either ornithine or oleate is omitted (but not both) and when the rate of urea synthesis is relatively low. This indicates that the presence of ammonia rather than a high rate of urea synthesis is responsible for the inhibition of gluconeogenesis.

8. In the presence of lactate, oleate, and ammonium chloride, large amounts of aspartate are formed and the concentration of oxoglutarate falls.

9. The accumulation of aspartate and the inhibition of gluconeogenesis from lactate in the presence of ammonium chloride and oleate can be explained by the assumption that the oxoglutarate formed by the intramitochondrial aspartate aminotransferase reaction is not translocated to the cytosol because glutamate dehydrogenase rapidly reconverts it to glutamate, the degradation of oleate providing the required NADH. Aspartate on the other hand is translocated to the cytosol, where it accumulates because of the lack of stoichiometric amounts of oxoglutarate. Since the cytosolic aspartate aminotransferase reaction is an obligatory step in gluconeogenesis from lactate, this process is inhibited.

10. When ornithine is present under the conditions specified in paragraph 9, much less aspartate accumulates because it is removed by the argininosuccinate synthetase reaction.

11. No explanation can be offered yet for the inhibition of gluconeogenesis in the presence of lactate, ornithine, and ammonium chloride when aspartate accumulation is slight and the rate of urea synthesis relatively low.

12. Alanine is always formed by liver cells in the presence of lactate and ammonium chloride. This is accounted for by the tendency to establish equilibrium in the interlinked glutamate dehydrogenase and alanine aminotransferase systems.

13. Urea synthesis and gluconeogenesis are accompanied by increased oxygen consumption. This increase is of the expected order on the assumption that one molecule of O_2 energizes the synthesis of 6 molecules of ATP, and that urea synthesis requires 4 ATP, and glucose synthesis from lactate 6 ATP per molecule.

14. Stoichiometric formation of aspartate and carbamoylphosphate required in the course of urea synthesis is achieved by the limit set to the formation of aspartate because of the equilibrium constant of the aspartate aminotransferase reaction. At the normal steady-state concentrations of glutamate, oxoglutarate, and oxaloacetate, the formation of aspartate stops when its concentration is about 0.7 mM. Removal of aspartate by the interaction with citrulline is followed automatically by replacement. Any excess nitrogen becomes available for the carbamoylphosphate synthetase reaction.

ACKNOWLEDGMENTS

The authors are indebted to Philip Gregory for preparing the isolated liver cells and to David Wiggins for technical assistance.

The work was supported by grants from the Medical Research Council, the U.S. Public Health Service (grant no. AM11748) and from Schering AG, Berlin.

REFERENCES

Azzi, A., Chappell, J. B., and Robinson, B. H. (1967). *Biochem. Biophys. Res. Commun.* **29**, 148–152.

Bernt, E. and Bergmeyer, H. U. (1963). In *Methods of Enzymatic Analysis* (H. U. Bergmeyer, ed.), Academic Press, New York, pp. 324–327.

Bernt, E. and Bergmeyer, H. U. (1963). In *Methods of Enzymatic Analysis* (H. U. Bergmeyer, ed.), Academic Press, New York, pp. 384–388.

Bernt, E. and Bergmeyer, H. U. (1970). In *Methoden der enzymatischen Analyse* Vol. 2, 2nd ed. (H. U. Bergmeyer, ed.), Verlag Chemie, Weinheim, pp. 1741–1745.

Berry, M. N. and Friend, D. S. (1969). *J. Cell Biol.* **43,** 506–520.

Borst, P. (1963). In *Funktionelle und morphologische Organisation der Zelle,* Springer-Verlag, New York, pp. 137–158.

Brosnan, J. T. and Williamson, D. H. (1974). *Biochem. J.* **138,** 453–462.

Clark, M. G. Kneer, N. M., Bosch, A. L., and Lardy, H. A. (1974). *J. Biol. Chem.* **249,** 5695–5703.

Cornell, N. W., Lund, P., Hems, R., and Krebs, H. A. (1973). *Biochem. J.* **134,** 671–672.

Cornell, N. W., Lund, P., and Krebs, H. A. (1974). *Biochem. J.* **142,** 327–337.

Exton, J. H. and Park, C. R. (1968). *J. Biol. Chem.* **243,** 4189–4196.

Friedmann, B., Goodman, E. H., Saunders, H. K., Kostos, V., and Weinhouse, S. (1971). *Metabolism* **20,** 2–12.

Gamble, J. G. and Lehninger, A. L. (1973). *J. Biol. Chem.* **248,** 610–618.

Hems, R., Ross, B. D., Berry, M. N., and Krebs, H. A. (1966). *Biochem. J.* **101,** 284–292.

Krebs, H. A. (1934). *Ergebn. Enzymforschung.* **3,** 247–264.

Krebs, H. A. and Henseleit, K. (1932). *Hoppe-Seyler's Z. Physiol. Chem.* **210,** 33–66.

Krebs, H. A., Wallace, P. G., Hems, R., and Freedland, R. A. (1969). *Biochem. J.* **112,** 595–600.

Krebs, H. A., Hems, R., and Lund, P. (1973). *Biochem. J.* **134,** 697–705.

Krebs, H. A., Cornell, N. W., Lund, P., and Hems, R. (1974). In *Regulation of Hepatic Metabolism,* Vol. 6, *Alfred Benzon Symposium* (F. Lundquist and N. Tygstrup, ed.), Munksgaard, Copenhagen, pp. 726–750.

Nakamura, M. and Jones, M. E. (1970). In *Methods in Enzymology,* Vol. 17A (H. Tabor and C. W. Tabor, eds.), Academic Press, New York, pp. 286–294.

Newsholme, E. A. and Gevers, W. (1967). *Vitam. Horm.* **25,** 1–87.

Papa, S., Lofrumento, N. E., Quagliariello, E., Meijer, A. J., and Tager, J. M. (1970). *J. Bioenerg.,* **1,** 287–307.

Papa, S., Francavilla, A., Paradies, G., and Meduri, B. (1971). *FEBS Lett.* **12,** 285–288.

Passonneau, J. V. and Lowry, O. H. (1964). *Adv. Enzyme Regul.* **2,** 265–274.

Pfleiderer, G. (1963). In *Methods of Enzymatic Analysis* (H. U. Bergmeyer, ed.), Academic Press, New York, pp. 381–383.

Quagliariello, E., Papa, S., Meijer, A. J., and Tager, J. M. (1969). In *Mitochondria—Structure and Function* (L. Ernster and Z. Drahota, eds.), Academic Press, London, pp. 335–346.

Ratner, S. and Petrack, B. (1951). *J. Biol. Chem.* **191,** 693.

Robinson, B. H. and Chappell, J. B. (1967). *Biochem. Biophys. Res. Commun.* **28,** 249–255.

Robinson, B. H., Williamson, J. R., Halperin, M. L., and Leznoff, C. C. (1970). *Eur. J. Biochem.* **15,** 263–272.

Ross, B. D., Hems, R., Freedland, R. A., and Krebs, H. A. (1967). *Biochem. J.* **105,** 869–875.

Ryan, W. L., Barak, A. J., and Johnson, R. J. (1968). *Arch. Biochem. Biophys.* **123,** 294–297.

Schimke, R. T. (1962). *J. Biol. Chem.* **237,** 1921–1924.

Slein, M. W. (1963). In *Methods of Enzymatic Analysis* (H. U. Bergmeyer, ed.). Academic Press, New York, pp. 117–123.

Stubbs, M. and Krebs, H. A. (1975). *Biochem J.* **150,** 41–45.

Williamson, D. H., Lopes-Vieira, O., and Walker, B. (1967). *Biochem. J.* **104,** 497–502, 1967.

CHAPTER NINE

Hepatic Gluconeogenesis in the Rabbit

PAUL D. RAY

I. INTRODUCTION

The long-established facts that (*a*) the central nervous system and other tissues such as the erythrocytes are highly dependent on blood glucose as an energy source and (*b*) most mammals maintain their blood glucose levels relatively constant, even after lengthy caloric deprivation, demonstrate not only that glucose can be made but that its synthesis can be regulated. The last 20 years have accordingly seen intensive efforts made by numerous investigators to unravel the intricacies that control glucose formation. The results of these efforts have provided us with a general concept of pathways whereby noncarbohydrate can be converted into carbohydrate under regulation by various means of controls.

Gluconeogenesis is known to occur mainly in liver (because of that organ's greater mass) but also in kidney tissue. If we consider gluconeogenesis as being the synthesis of glucose and other hexoses by way of molecules containing five or fewer carbon atoms (Krebs, 1964), then substrates providing the carbon are primarily lactate from glycolysis, glucogenic amino acids (mainly alanine), and the glycerol moeity of triglycerides, and the contribution from each substrate(s) varies with the metabolic situation.

Earlier Krebs (1954) pointed out the unlikelihood that under physiological conditions, 3-carbon fragments would be converted to glucose by a direct reversal of glycolytic reactions. Since reviews of the process of glycolysis and its reversal are readily available, (Krebs, 1954; Lardy, 1961; Lardy *et al.*, 1964; Utter, 1963), this process is not elaborated on here. Suffice it to state that although certain of the glycolytic reactions are currently considered to be reversible, other of these catabolic reactions are thought to be circumvented for purposes of glucose synthesis. Attention has been focused primarily on four enzymes: pyruvate carboxylase (EC 6.4.1.1), P-enolpyruvate carboxykinase (EC 4.1.1.32), fructose-1,6-diphosphatase (EC 3.1.3.11), and glucose 6-phosphatase hydrolase (EC 3.1.3.9), which catalyze the conversions of pyruvate to oxalacetate, of oxalacetate to P-enolpyruvate, of fructose-1,6-diphosphate to fructose-6-phosphate, and of glucose-6-phosphate to glucose, respectively.

The subcellular locations of fructose-1,6-diphosphatase (Gomori, 1943) and glucose-6-phosphatase (Hers *et al.*, 1951) in hepatic cytosolic and microsomal fractions, respectively, were established more than two decades ago. On the other hand, the existence and subcellular location of pyruvate carboxylase were not so easily or quickly arrived at. Initial reports suggested a mitochondrial location primarily (Utter and Keech, 1960; Henning and Seubert, 1964), yet subsequent work indicated the existence of cytosolic pyruvate carboxylase activity as well (Seubert and Huth, 1965; Henning *et al.*, 1966). Even though more recent results of Böttger *et al.* (1966) and of Walter and Anabitarte (1971) provided strong evidence that pyruvate carboxylase is an exclusively mitochondrial enzyme in rat liver, controversial reports still appeared (Dugal, 1972). Very recently, Seubert's group (Weiss *et al.*, 1974) has also published data confirmatory to those of Walter and Anabitarte (1971) which indicate that preferential extraction of pyruvate carboxylase, as compared to the marker enzyme glutamate dehydrogenase, can occur under various conditions, thus negating their previous studies, which provided the bases for projecting an extramitochondrial location for pyruvate carboxylase. These authors, on the basis of these recent findings, now agree that pyruvate carboxylase is located solely within the mitochondrion. Thus essentially all the evidence now available supports the contention that hepatic pyruvate carboxylase, fructose-1,6-diphosphatase, and glucose-6-phosphatase are primarily found in the mitochondrial, cytosolic, and microsomal fractions, respectively, of all species.

In the early 1960s Nordlie and Lardy (1963) published data showing that hepatic P-enolpyruvate carboxykinase activity can be found in mitochondrial and/or cytosolic fractions of liver depending on the species of animal under investigation. They reported, for example, that P-enolpyruvate carboxykinase is a mitochondrial enzyme in rabbit liver, nearly equally distributed between the mitochondrial and cytosolic fractions in guinea pig liver, and found essentially only in cytosol of rat, mouse, and hamster liver.

The mitochondrial location of pyruvate carboxylase coupled with the cytosolic location of P-enolpyruvate carboxykinase in rat liver might suggest that to form glucose from pyruvate, pyruvate could be carboxylated to oxalacetate, which could presumably pass into cytosol to be converted to P-enolpyruvate by way of P-enolpyruvate carboxykinase.

However Lardy *et al.* (1965) showed that rat liver mitochondria incubated with pyruvate, ATP, and HCO_3^- do not generate oxalacetate in the medium, rather, malate and citrate accumulate. Furthermore, addition of NH_4^+ or glutamate to the incubation medium leads to accumulation of aspartate. Additional data were soon published by Walter *et al.* (1966), and corroborative data were published by Haynes (1965). On the basis of these observations, Lardy and his group projected the scheme shown in Fig. 1 for the conversion of pyruvate to glucose in rat liver. Pyruvate from glycolysis or from amino acids such as alanine are carboxylated by way of mitochondrial pyruvate carboxylase to oxalacetate. The oxalacetate is converted to malate or aspartate, which exits from the mitochondria to the cytosol to be reconverted to oxalacetate by way of cytosolic NAD-malate dehydrogenase

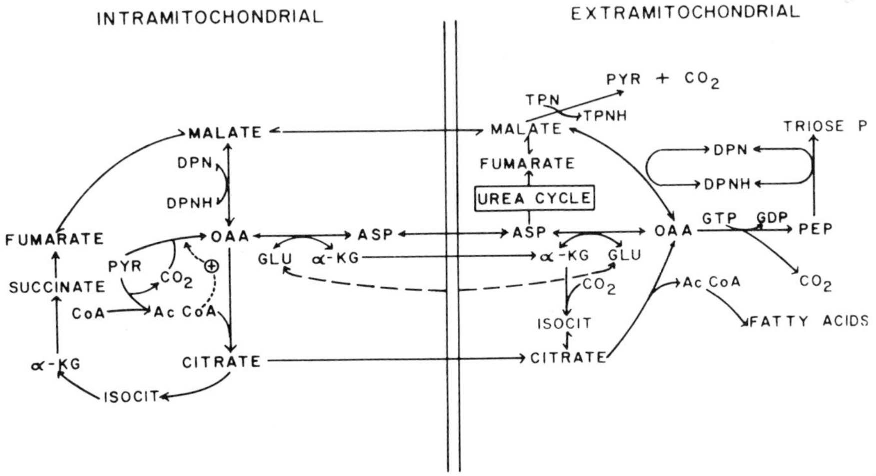

Figure 1. Paths of carbon supply for gluconeogenesis and lipogenesis in liver. *Source:* Lardy *et al.* (1965), courtesy of the National Academy of Sciences.

and aspartate aminotransferase, respectively. Aspartate can also be converted to oxalacetate by a combination of reactions involving urea cycle enzymes, fumarase, and malate dehydrogenase. This metabolic scheme can thus account for movement of carbon and reducing equivalents (if needed) from mitochondria to cytosol, where both are known to be necessary for glucose formation.

Until 1967, the bulk of our knowledge of the process of gluconeogenesis was based mainly on data obtained with the intact rat or with rat-derived tissues; thus a number of laboratories, including ours, initiated studies into carbohydrate formation as it occurs in other species.

II. RATIONALE FOR STUDY

Our studies of glucose formation in rabbits were begun on the basis of the observation by Nordlie and Lardy (1963) that P-enolpyruvate carboxykinase is a mitochondrial enzyme in rabbit liver. On this basis alone, one might predict that the process of gluconeogenesis in the rabbit could deviate considerably from that projected to occur in rats. Additionally, comparison of this process and its regulation as it occurs in various species other than the rat appeared necessary prior to assuming generalities regarding gluconeogenesis that might or might not exist.

III. ORDER OF PRESENTATION AND CONTENT

This chapter has been written to conform with the chronological development of our own studies, starting with observations made with isolated mitochondrial systems, followed by studies related to the intracellular location of P-enolpyruvate carboxykinase, the effects or lack thereof of various dietary and hormonal treatments on activities of this and several other enzymes potentially involved in gluconeogenesis, and finally some results obtained utilizing the isolated perfused rabbit liver. Each of these aspects of our research is discussed in a separate major section, complete with a résumé of earlier work (if applicable), followed by our results, those of others, and a summary of the results. A section of conclusions integrates all the material covered into a current overview of glucose formation as it is thought to occur in rabbit liver.

Data and references contained herein intentionally pertain primarily to rabbits, although information from other species is also included and referenced for background and comparative purposes.

IV. STUDIES WITH ISOLATED MITOCHONDRIA

A. Background

In the early 1950s Mudge *et al.* (1954) reported on the ability of rabbit liver mitochondria to make P-enolpyruvate from α-ketoglutarate incubated with P_i under aerobic conditions in the absence of any added acceptor and in the presence of the uncoupling agent, dinitrophenol (DNP). Interestingly, these investigators observed that DNP had a biphasic effect on P-enolpyruvate production, enhancing its formation at low (10^{-5} M) and high concentrations (10^{-3} M) while decreasing P-enolpyruvate formation at an intermediate concentration of 5×10^{-5} M. The biphasic effects of DNP observed by Mudge *et al.* (1954) were subsequently confirmed and rationalized by Nordlie and Lardy (1963; see below). In retrospect, an important aspect of the work of Mudge *et al.* (1954) was the use of DNP as an uncoupler that not only promoted oxidation of α-ketoglutarate but also permitted substrate level generation of GTP to be used in phosphorylation of oxalacetate to P-enolpyruvate by way of mitochondrial P-enolpyruvate carboxykinase. However, these data of Mudge *et al.* were not particularly helpful in attempting to speculate about a net production of P-enolpyruvate for glucose formation from readily available substrates nor was it shown in these studies that P-enolpyruvate could actually exit from mitochondria.

Bartley (1954) showed the formation of P-enolpyruvate from pyruvate plus HCO_3^- incubated with particles (mainly mitochondrial) derived from sheep kidney cortex and in fact suggested that this mitochondrially derived P-enolpyruvate might be involved in the synthesis of glucose. Yet Bandurski and Lipmann (1956) were the first to demonstrate hepatic formation of P-enolpyruvate from pyruvate aerobically incubated with mitochondria isolated from guinea pig liver. Interestingly, these workers also revealed the inability of mitochondria to convert any substrate other than oxalacetate to P-enolpyruvate under anaerobic conditions (N_2), even in the presence of ATP. Thus, again in retrospect, these early observations provided support for involvement of both P-enolpyruvate carboxykinase and pyruvate carboxylase in the mitochondrial conversion of pyruvate to P-enolpyruvate, and the subsequent demonstration of the presence of these two enzyme activities in guinea pig liver mitochondria explains why a net formation of P-enolpyruvate from pyruvate was found to occur with mitochondria from this species. Once more however, no attempt was made in these studies to measure the exit of P-enolpyruvate or other metabolites from mitochondria.

Nordlie and Lardy (1963a) reported data indicating that guinea pig liver mitochondria can convert α-ketoglutarate to P-enolpyruvate and that this conversion is enhanced in the presence of DNP, thus confirming the observations that Mudge *et al.* (1954) had made using rabbit liver mitochondria. They also confirmed the biphasic effects of DNP in these systems and subsequently suggested that these could be attributed to (*a*) the ability of DNP at 10^{-4} M to enhance the concentration of ADP, which can compete with oxalacetate as an acceptor for P_i from GTP and (*b*) the ability of DNP at 10^{-3} M to inhibit ATPase, thus maintaining ADP at a very low level. Additionally, these authors reported that a correlation exists between the rates of P_i uptake and P-enolpyruvate formation with mitochondria from various species with the levels of P-enolpyruvate carboxykinase activity present in the mitochondria (Nordlie and Lardy, 1963, 1963a).

In 1965 Lardy and his group first reported their work (Lardy *et al.*, 1965; Walter *et al.*, 1966) alluded to earlier, leading to the elaborate scheme for gluconeogenesis (Fig. 1) whereby pyruvate-derived carbon and reducing equivalents egress from mitochondria to cytosol as malate or aspartate, which are then converted back to oxalacetate, thus providing substrate for cytosolic P-enolpyruvate carboxykinase. Haynes (1965) followed with confirmatory observations. A salient feature of the work of Lardy and his group was the observation that rat liver mitochondria, under proper conditions, could generate malate and aspartate from pyruvate at rates sufficient to account for the rate of gluconeogenesis from pyruvate (based on CO_2 fixation) previously observed in intact fasted rats—that is, formation of 0.5 μmole glucose/min/g liver (Solomon *et al.*, 1941; Lardy *et al.*, 1965). A two- to threefold greater rate for lactate's conversion to glucose has since been obtained in the isolated, perfused rat liver (Exton and Park, 1967); the isolated, perfused guinea pig liver (Söling *et al.*, 1970), and with homogenates of pigeon liver (Krebs *et al.*, 1964). Because of these observations, the capacities of the various systems described below to generate glucogenic intermediates have been expressed, where possible, in micromoles per minute per gram of liver for comparative purposes, on the assumption that rabbit liver possesses a somewhat similar capacity to make glucose from lactate.

Gamble and Mazur (1967) demonstrated the ability of rabbit liver mitochondria to convert various Krebs cycle intermediates (not pyruvate) to P-enolpyruvate in the absence of P_i. However, synthesis of P-enolpyruvate from α-ketoglutarate was increased by addition of P_i. The synthesis of P-enolpyruvate from citrate, α-ketoglutarate, fumarate, and malate was enhanced in each case by DNP. These authors found a rate of conversion of α-ketoglutarate to P-enolpyruvate (at 23°C and in the presence of 10^{-4} M DNP) of approximately 0.02 μmole/min/mg protein nitrogen (Gamble and Mazur, 1967, Fig. 2), which is probably equivalent to about 0.20 μmole P-enolpyruvate formed per minute per gram of liver (assuming a content of 65 mg mitochondria protein/g liver).[1] Movement of P-enolpyruvate from mitochondria into the surrounding medium was also demonstrated in these studies. Gamble and Mazur thus concluded that "since pyruvate carboxylase, P-enolpyruvate carboxykinase and P-enolpyruvate are present in rabbit liver mitochondria, the first steps of the sequence (of gluconeogenesis) must occur within the particles in this animal." However, no consideration was given to the quantity of P-enolpyruvate that might be needed to meet the rabbits' requirements for glucose nor that the quantity of P-enolpyruvate generated in these experiments would likely be insufficient to meet those needs.

About the same time, Davis and Gibson (1967) showed that a net conversion of pyruvate and other TCA cycle intermediates to P-enolpyruvate by rabbit liver mitochondria is enhanced up to a point not only by DNP but also by oleate. In these studies, a rate of conversion of α-ketoglutarate to P-enolpyruvate, at 30°C and in the presence of 10^{-3} M DNP, of 0.01 μmole/min/mg protein or about 0.7 μmole/min/g liver was observed, which is approximately double the rate they observed in the presence of 4×10^{-4} M oleate (Davis and Gibson, 1967, Table I). These authors also observed a rate of conversion of pyruvate to P-enolpyruvate in the presence of 8×10^{-3} M DNP of 0.004 μmole/min/mg protein or about 0.24 μmole/min/g liver (Davis and Gibson, 1967, Fig. 1). Formation of P-enolpyruvate from pyruvate in the absence of any uncoupler was 0.05 to 0.08 μmole/min/g liver (Davis and Gibson, 1967, Fig. 1, Table II). The net formation of citrate and malate from pyruvate was also demonstrated in the absence as well as in the presence of DNP, and the ratio of P-enolpyruvate to citrate and malate increased with increasing concentrations of DNP. These and additional observations led these authors to project that fatty acids, functioning as uncouplers of mitochondrial respiration, may be important to *in vivo* regulation of gluconeogenesis in species possessing mitochondrial P-enolpyruvate carboxykinase activity. Again, movement of metabolites from mitochondria into the medium was not demonstrated, and the quantity of P-enolpyruvate produced was not related to the animals' needs. Many of these observations were subsequently reiterated and expanded on by Davis and Gibson (1969).

Recently Garber and Ballard (1969; 1970) and Garber and Hanson (1971) have reported on the ability of guinea pig liver mitochondria and also rabbit liver mitochondria (Garber and Hanson, 1971a) to metabolize various substrates to P-enolpyruvate, malate, and aspartate. With rabbit liver mitochondria, they observed a conversion of α-ketoglutarate, at 30°C in the

[1] We find a mitochondrial protein content of 64.6 + 7.6 mg/g liver for fed rabbits (13 observations).

presence of either DNP or ATP, to P-enolpyruvate at a rate of about 0.07 μmole/min/mg mitochondrial nitrogen, which is equivalent to about 0.75 μmole/min/g liver (Garber and Hanson, 1971a, Table III). These workers also reported that in both guinea pigs and rabbits, the hepatic intramitochondrial $NAD^+/NADH$ ratio shifts to a more oxidized state during fasting, whereas the opposite is known to occur in rat liver (Williamson *et al.*, 1967). If the $NAD^+/NADH$ ratios observed in fasted guinea pig or rabbit livers *in vivo* are artificially reproduced (by use of β-hydroxybutyrate and acetoacetate) with isolated mitochondria incubated with various substrates including pyruvate, P-enolpyruvate formation is favored over malate and aspartate synthesis. As the $NAD^+/NADH$ ratio is incrementally shifted *in vitro* from that of the fed to that of the fasted rabbit, the quantity of P-enolpyruvate generated from 3 mM α-ketoglutarate (Garber and Hanson, 1971a, Fig. 1) or from a mixture of 1 mM pyruvate, 1 mM aspartate, 1 mM α-ketoglutarate, and 1 mM glutamate (Garber and Hanson, 1971a, Fig. 3) increases from about 0.25 to 0.70 μmole/min/g liver. Thus these authors conclude that with regard to mitochondrial P-enolpyruvate formation in the rabbit, the interaction of the $NAD^+/NADH$ ratio at the malate dehydrogenase equilibrium appears to be an important determinant.

B. Studies of Ray *et al.*

Our own studies (Johnson and Ray, 1969; Johnson *et al.*, 1970; 1973), also initiated in 1967, were formulated on the basis of Lardy's work with rat liver mitochondria in which he and his group suggested the movement of pyruvate-derived carbon and reducing equivalents from mitochondria to cytosol (Lardy *et al.*, 1965; Walter *et al.*, 1966). Even though Nordlie and Lardy (1963) had reported that P-enolpyruvate carboxykinase was a mitochondrial enzyme in rabbit liver and some of the work just reviewed had shown that P-enolpyruvate could be synthesized and liberated by rabbit liver mitchondria, it was not clear whether carbon for gluconeogenesis might leave the mitochondria in a form other than that of P-enolpyruvate. This appeared to be a reasonable question because up to this time the amount of P-enolpyruvate obtained in most of the studies from various substrates, especially from pyruvate, seemed inadequate if one were assuming that the rabbit possesses a capacity to convert pyruvate to glucose that is similar to that measured in the intact rat liver (Solomon *et al.*, 1941), in the isolated, perfused rat liver (Exton and Park, 1967), or in guinea pig liver (Söling *et al.*, 1970). We also thought that some means should exist for movement of reducing equivalents to the cytosol to facilitate gluconeogenesis from substrates such as alanine. Of course if such a "reducing equivalent shuttle" did exist, some means had to be available to deal with the metabolic product(s) of the shuttle.

*1. **Definition of our System and Conditions.*** Our first studies were designed to estimate the capacity of rabbit liver mitochondria to effect a net conversion of pyruvate to P-enolpyruvate and/or other metabolites. We chose the same system with rabbit liver mitochondria (except for omission of ATP) used by Lardy *et al.* (1965), that is, 6.7 mM sodium pyruvate, 6.7 mM potassium phosphate (pH 7.4), 6.7 mM triethanolamine (pH 7.4), 7.5 mM $MgSO_4$, and 10 mM $KHCO_3$ made up to 2.5 ml with isotonic sucrose. Incubations, initiated by addition of 0.5 ml of mitochondria (5–7 mg mitochondrial protein) in isotonic sucrose, were for 10 minutes at 37° C, and reactions were stopped by placing reaction vessels in ice at 0° C. Mitochondria were removed from the media by centrifugation, and the supernatant solutions plus washes were utilized for determination of substrate utilization and metabolite production. Removal of mitochondria from the medium by centrifugation allows the assumption that metabolites had been formed in mitochondria and had egressed into the medium. Even though the experimental conditions employed were those used by Lardy *et al.* (1965), we verified experimentally that these conditions were at or near optimum in terms of substrate utilization and total metabolite production (Johnson *et al.*, 1973).

We used pyruvate as a substrate because under *in vivo* conditions, where extensive glucose formation is taking place, pyruvate becomes an obligatory intermediate in the conversion to glucose of lactate and amino acids such as alanine, glycine, threonine, and serine. Again, our metabolite data are expressed in terms of micromoles of substrate used or product generated per unit time per liver, for reasons mentioned earlier.

*2. **Production of Malate, Citrate, Aspartate, and P-enolpyruvate from Pyruvate.*** Data presented in Fig. 2 demonstrate that rabbit liver mitochondria incubated under the above-described conditions can utilize pyruvate to effect a net synthesis and liberation of substantial quantities of malate and citrate together with much smaller amounts of P-enolpyruvate and aspartate. The inclusion of NH_4Cl in the medium leads to an extensive increase in aspartate production, apparently at the expense of citrate and especially of malate; again, the amount of P-enolpyruvate generated is small in comparison to total metabolite production. It is assumed that the increased production of aspartate in the presence of NH_4^+ is due to the formation of catalytic quantities of glutamate from α-ketoglutarate (via glutamate dehydrogenase) sufficient to transaminate with oxalacetate

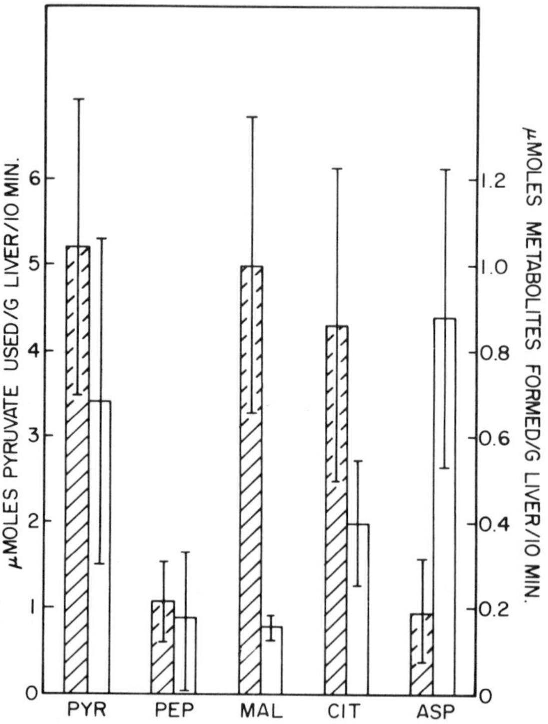

Figure 2. Pyruvate utilization and metabolite production in the absence and in the presence of 3 mM NH_4CL. Other incubation conditions are described in the text. The results depicted are averages of data from six experiments, and the vertical lines represent one standard deviation. Symbols: ▨ basic reaction medium; □ basic reaction medium plus 3 mM NH_4Cl; PYR, pyruvate; PEP, P-enolpyruvate; MAL, malate; CIT, citrate; ASP, aspartate. *Source:* Johnson *et al.* (1973), courtesy of the American Society of Biological Chemists.

(generated by carboxylation of pyruvate) to form aspartate, thereby concomitantly decreasing the amount of oxalacetate available for production of malate and citrate.

On the basis of data presented in Fig. 2, it is apparent that in the absence of NH_4^+, mitochondria from fed rabbits generate 0.02 μmole P-enolpyruvate, 0.1 μmole malate, 0.09 μmole citrate, and 0.02 μmole aspartate for a total of 0.23 μmole metabolites/min/g liver. Since we lose 40 to 45% of mitochondria during preparation,[2] the corrected value for total metabolite production is about 0.40 μmole/min/g liver. Of this total

[2] With our fractionation procedures, we lose about 40 to 45% of mitochondria into the nuclear pellet based on P-enolpyruvate carboxykinase activity in this pellet. This loss is comparable to those reported by Nordlie and Lardy (1963) and by Garber and Hanson (1971a).

amount of metabolites formed and liberated by mitochondria, only about 10% is accounted for by P-enolpyruvate. Thus our data on P-enolpyruvate production from pyruvate are quantitatively rather similar to those of Davis and Gibson (1967, 1969). However, as indicated earlier, they had shown that P-enolpyruvate production could be enhanced by uncouplers and had concluded that naturally occurring uncouplers such as fatty acids could be important *in vivo* regulators of gluconeogenesis in the rabbit. They arrived at these conclusions on the assumption that P-enolpyruvate carboxykinase was exclusively a mitochondrial enzyme in this species. However, since P-enolpyruvate production and liberation by our system appeared to be quite limited, we considered the possibility that the nature and quantity of the other metabolites produced and liberated (i.e., malate, citrate, and, under proper conditions, aspartate) were more important. These are, of course, metabolites formed by rat liver mitochondria, and they are the metabolites thought to be involved in the transfer of carbon and reducing equivalents across the mitochondrial membrane(s) to the cytosol. These results led to the experiments (discussed in Section V.B) whereby we showed that a P-enolpyruvate carboxykinase activity, capable of responding to gluconeogenic stimuli, exists in rabbit liver cytosol, and all our observations taken together led first to our report (Johnson and Ray, 1969) and subsequently to our published proposal (Johnson *et al.*, 1970) that gluconeogenesis could occur in rabbits by a process similar to that projected to occur in rats.

3. ***Influence of Exogenous ATP on Metabolite Production.*** Since we did observe a net synthesis of metabolites from pyruvate which presumably requires an ATP-dependent fixation of CO_2 (Utter and Keech, 1963), exogenous ATP was added to our basic system in an attempt to enhance metabolite production. Instead, however, exogenous ATP was found to exert detrimental effects (Fig. 3). Generally speaking, increasing the concentration of ATP in small increments from 0 to 6 mM decreases pyruvate utilization and metabolite production. The production of P-enolpyruvate does increase somewhat with concentrations of ATP of 0.75 and 1.5 mM, but it declines sharply thereafter. Production of malate and citrate are severely decreased by the addition of ATP. Although the ratio of P-enolpyruvate produced to that of total metabolites increases substantially with concentrations of ATP up to 1.5 mM, the actual amount of P-enolpyruvate produced in the system is not substantially increased, at least not to an amount meaningful in terms of the rabbit's projected requirements for glucose. These effects are being observed in the presence of, in some cases, near-physiological concentrations of ATP, since Garber and Hanson (1971a) have reported a level of 2.4 μmole ATP/g liver in fed rab-

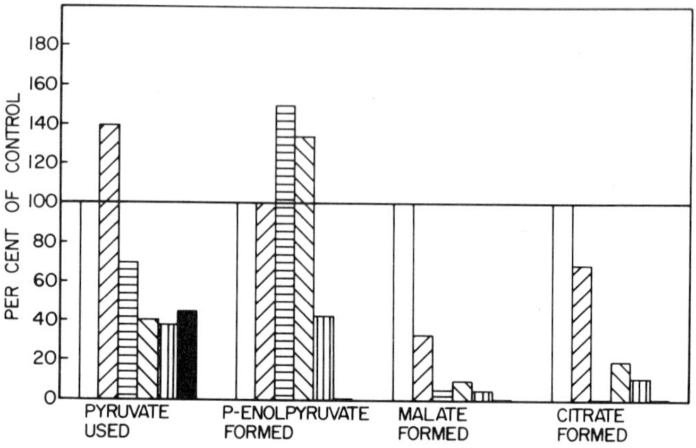

Figure 3. The effects of various concentrations of ATP on pyruvate utilization and metabolite production. Other incubation conditions are described in the text. These data are reported as percentages of control, since all concentrations of ATP were not tested in the same experiments, thus were not compared to identical controls. Nonetheless, data reported for each concentration of ATP are representative, and the results depicted are averages of data from four experiments. Symbols: □, no ATP; ▨, 0.37 mM ATP; ▤, 0.75 mM ATP; ▧, 1.5 mM ATP; ▥, 3.0 mM ATP; ■, 6.0 mM ATP. *Source:* Johnson (1970).

bits. Generally speaking, exogenous ATP suppresses both substrate utilization and total metabolite production; thus we assume that endogenous production of ATP by this system is adequate to meet the ATP requirement for CO_2 fixation. Nonetheless, 0.75 and 1.5 mM ATP specifically enhances P-enolpyruvate production somewhat (Fig. 3), which means that to some extent the pool of ATP may be specifically important for P-enolpyruvate formation, most likely due to an increase in GTP production by way of nucleoside diphosphokinase. This conclusion would be consistent with that of Garber and Hanson (1971a), who concluded on the basis of experiments with uncouplers, various nucleotides, and combinations of both that transphosphorylation and substrate level phosphorylation are important for P-enolpyruvate formation in rabbit liver (Garber and Hanson, 1971a; Table III). However the nucleoside diphosphokinase activity in rabbit liver is much less than that found in either guinea pig or rat liver (Ishihara and Kikuchi, 1968; Garber and Hanson, 1971a), and this observation suggests that P-enolpyruvate formation in rabbit liver is likely to be more dependent on substrate level phosphorylation than on transphosphorylation. An explanation for the suppression of the production of citrate and especially malate by exogenous ATP is currently lacking, but even if the level of GTP available enhances the generation of P-enolpyruvate from oxalacetate rather than its conversion to malate or citrate, the decreased production of the latter two metabolites is not nearly accounted for by the amount of P-enolpyruvate produced. Perhaps related is the observation of Krebs *et al.* (1964) that addition of ATP reduces the formation, by pigeon liver homogenates, of glucose from lactate.

4. Influence of Other Nucleotides on Metabolite Production. Since production of P-enolpyruvate by our system was so limited, and since GTP is the *in vivo* co-factor required by P-enolpyruvate carboxykinase (Utter and Kurahashi, 1954), its addition to the system at a concentration of 1 mM was also tested but found to be without effect, perhaps simply because of its inability to enter the mitochondria (Fig. 4). However ITP, which can substitute for GTP as a co-factor for P-enolpyruvate carboxykinase *in vitro* (Utter and Kurahashi, 1954), was found to exert a detrimental effect on the system decreasing substrate utilization to about the same extent as did ATP. Additionally, ITP decreases P-enolpyruvate formation as well, although it appears to exert somewhat less effect than does ATP on formation of citrate and especially malate. The addition of 1 mM ADP results in a greater decrease in both substrate utilization and metabolite production than does the addition of any other nucleotide tested (Fig. 4). Exogenous GTP exerts no effect on P-enolpyruvate production, presumably because of its inability to enter mitochondria (Heldt, 1966). Yet ITP, which apparently

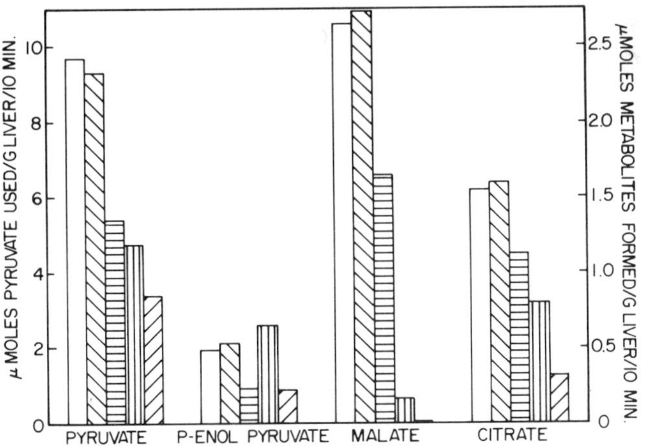

Figure 4. The effects of exogenous nucleotides on pyruvate utilization and metabolite production. Other incubation conditions are described in the text. The results depicted are representative of a minimum of three experiments. Symbols: □, no added nucleotides; ⧅, 1 mM GTP; ⊟, 1 mM ITP; ▥, 1 mM ATP; ▨, 1 mM ADP. *Source:* Johnson *et al.* (1973), courtesy of the Society of Biological Chemists.

does enter mitochondria because it is known to exert effects on both substrate utilization and metabolite production, still does not enhance but rather depresses P-enolpyruvate production. It may be that exogenous ADP so severely decreases both substrate utilization and metabolite production because of its inhibition of pyruvate carboxylation (Walter *et al.*, 1966; Walter and Stucki, 1970), although such an effect has been questioned (Davis and Gibson, 1969). Such inhibition of pyruvate carboxylation could decrease production of ATP by oxidative processes by limiting the formation of sparker quantities of oxalacetate necessary for oxidation of pyruvate.

5. ***Influence of Endogenously Generated GTP on Metabolite Production.*** Since P-enolypyruvate production by our system was minimal and was not particularly enhanced by addition of exogenous ATP, GTP, or ITP, the influence of GTP produced by substrate level phosphorylation was tested by adding 3 mM α-ketoglutarate to the basic system (Fig. 5). Its addition was found to significantly suppress pyruvate utilization and to significantly increase citrate formation. Production of P-enolpyruvate was generally increased and malate production generally decreased, but neither to a statistically significant degree. Figure 5 also reveals the influence of 3 mM glutamate added to the system. Glutamate obviously causes a highly significant increase in aspartate production; but again, the increase observed in P-enolpyruvate formation is only of marginal significance. The production of malate is significantly decreased. These data indicate that the availability of either α-ketoglutarate or glutamate, in addition to pyruvate, does not substantially enhance the total amount of P-enolpyruvate production by the system nor the ratio of P-enolpyruvate produced to the total amount of metabolites formed. On the other hand, an increase in P-enolpyruvate is consistently observed and although not statistically significant, this increase is probably real and is probably due to greater availability of GTP generated by way of substrate level phosphorylation. This observation is again consistent with the conclusion of Garber and Hanson (1971a) based on their more detailed experimentation into the requirements for conversion of oxalacetate to P-enolpyruvate by rabbit liver.

6. ***Influence of Fasting and Alloxan Diabetes on Metabolite Production.*** Since our system utilizing mitochondria from fed rabbits generated only small amounts of P-enolpyruvate in relationship to other metabolites, we decided to evaluate the capacity of mitochondria isolated from fasted or alloxan diabetic rabbit livers to produce P-enolpyruvate and other metabolites from pyruvate. These data are presented in Fig. 6. Fasting

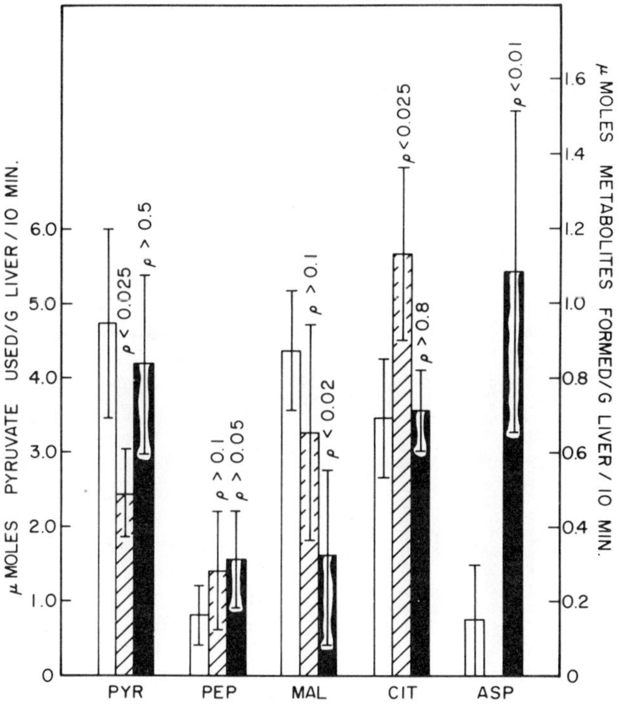

Figure 5. The effects of 3 mM α-ketoglutarate (▨) or 3 mM glutamate (■) on pyruvate (PYR) utilization and metabolite production. Other incubation conditions are described in the text. Aspartate (ASP) was not measured in incubations carried out in the presence of α-ketoglutarate. Values represent averages of data from five experiments. The vertical lines represent one standard deviation. Significance of differences between sample means; data from incubations of pyruvate in the presence of α-ketoglutarate or glutamate tested against data obtained with pyruvate alone. *Source:* Johnson *et al.* (1973), courtesy of the American Society of Biological Chemists.

generally causes an increase in both substrate utilization and metabolite production, although only the production of malate is significantly increased. Mitochondria from fed diabetic rabbits not only utilize more substrate but produce significantly greater amounts of malate and P-enolpyruvate. Even though changes in hepatic glycogen and lipid content caused by fasting or diabetes could influence data expressed per grams of liver (Fig. 6), similar changes of equal or greater statistical significance are seen if the same data are expressed in terms of either mitochondrial protein or body weight. From the data presented in Fig. 6, we can calculate that mitochondria from fasted and diabetic rabbits, respectively, generate 0.54

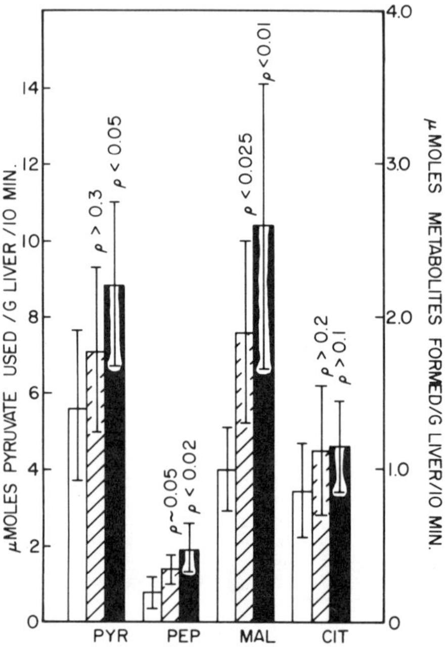

Figure 6. A comparison of pyruvate (PYR) utilization and metabolite production by mitochondria prepared from 14 fed versus 4 48-hour fasted rabbits, and 14 fed versus 5 fed, alloxan diabetic rabbits. Other incubation conditions are described in the text. The vertical lines represent one standard deviation. Significance of differences between sample means; data from fasted and from fed, diabetic rabbits tested against data from fed controls. Symbols: □, fed; ▨, 48-hour fasted; ■, fed, diabetics; PEP, P-enolpyruvate; MAL, malate; CIT, citrate. *Source:* Johnson *et al.* (1973), courtesy of the American Society of Biological Chemists.

and 0.69 μmole metabolites/min/g liver (corrected for mitochondrial loss) as compared to about 0.34 μmole/min/g liver generated by mitochondria from fed rabbits.

C. Summary

From the studies performed with rabbit liver mitochondria, it is apparent that these mitochondria, like those from guinea pig liver and unlike those from rat liver, are capable of generating P-enolpyruvate from a number of Krebs cycle intermediates and pyruvate.

The majority of these studies, including our own, indicate that rabbit liver mitochondria synthesize and liberate rather small amounts of P-enolpyruvate related to the quantity of P-enolpyruvate needed to satisfy the amount of glucose that this species can presumably make.

Studies of Davis and Gibson (1967, 1969) and of Garber and Hanson (1971a) demonstrate that rabbit liver mitochondria given various substrates including pyruvate can generate malate and citrate along with P-enolpyruvate and that uncouplers such as DNP or oleate can enhance P-enolpyruvate production at the expense of malate, citrate, and aspartate formation.

The work of Garber and Hanson (1971a) also shows that rabbit liver mitochondria can generate substantial amounts of P-enolpyruvate at the expense of other metabolites if the ratio of $NAD^+/NADH$ (as measured *in vivo* in fasted rabbits) is artificially maintained *in vitro*. To explain this observation, they suppose that the increase in the $NAD^+/NADH$ ratio induced by fasting influences the equilibrium of the mitochondrial malate dehydrogenase in favor of oxalacetate to the extent that more oxalacetate is available for mitochondrial P-enolpyruvate formation. Thus they suggest that the mitochondrial $NAD^+/NADH$ ratio and its subsequent influence on the malate dehydrogenase equilibrium is an important determinant in P-enolpyruvate production in rabbit liver. Their data also suggest an alternate explanation for the ability of uncouplers to enhance mitochondrial production of P-enolpyruvate rather than other metabolites; that is, that uncouplers cause an increase in the $NAD^+/NADH$ ratio in mitochondria.

Our own studies (Johnson and Ray, 1969; Johnson *et al.*, 1970, 1973), like the majority of the earlier work, suggest that the capacity of rabbit liver mitochondria to generate P-enolpyruvate from pyruvate is rather limited compared to their capacity for the formation and liberation of malate, citrate, and aspartate. Although mitochondria from fasted or diabetic rabbits can generate greater quantities of metabolites, the proportion of P-enolpyruvate to total metabolites does not change appreciably.

Generally speaking then, data of Davis and Gibson, Garber and Hanson, and ourselves all demonstrate that rabbit liver mitochondria presented with a variety of substrates can produce P-enolpyruvate, malate, citrate, and aspartate (under proper conditions) in combined quantities that in some cases approach the capacity for glucose production projected for this species. The major differences in much of the data reported thus far are in the amounts of each of the various metabolites generated from the given substrate. Variations in the proportions of the metabolites to one another obviously can be influenced by uncouplers. However, uncoupler-mediated influences observed *in vitro* are difficult to extrapolate to physiological significance.

The reason(s) for the $NAD^+/NADH$ ratio becoming more oxidized in fasted rabbits and guinea pigs as opposed to the same ratio going more reduced in rats is not yet apparent. Nonetheless, there is little doubt that *in vitro* at least, the $NAD^+/NADH$ ratio can extensively influence the quantitative relationship of the various metabolites to one another.

Although much emphasis has been placed on the capacity of rabbit liver mitochondria to synthesize and liberate P-enolpyruvate, it becomes evident from the following section that inducible cytosolic P-enolpyruvate carboxykinase activity exists in rabbit liver, and under various conditions conducive to gluconeogenesis, this activity can negate the need for mitochondria to bear sole responsibility for P-enolpyruvate production.

V. ACTIVITIES OF ENZYMES INVOLVED IN GLUCONEOGENESIS IN RABBIT LIVER

A. Background

As mentioned in the introduction, it became apparent some time ago that the reactions catalyzed by four enzymes—pyruvate carboxylase, P-enolpyruvate carboxykinase, fructose-1,6-diphosphatase, and glucose-6-phosphatase—were likely candidates for consideration as regulatory sites in glucose formation because they circumvent the need for an undesirable (thermodynamically speaking) reversal of certain enzymatic reactions involved in glycolysis. Subsequently, a large volume of data has been accumulated primarily from rats, but from other species too, on the abilities of various dietary and hormonal treatments to influence the activities of some or all of these enzymes. For the most part, the alterations in these enzymes' activities have been consistent with the direction of the changes observed in glucose formation. However, relatively few such studies have been made with rabbits. Thus it appeared worthwhile to investigate the activities of these enzymes and their responses to various types of gluconeogenic stimuli.

As mentioned earlier, Nordlie and Lardy (1963) published data indicating that in livers from fed rabbits, P-enolpyruvate carboxykinase activity was demonstrable in nuclear and mitochondrial fractions but undetectable in cytosolic and microsomal fractions. Even though Ilyin *et al.,* (1966) reported indications that an inducible P-enolpyruvate carboxykinase activity was also present in rabbit liver cytosol, this work apparently went without notice or mention.

We too were unaware of the work of Ilyin *et al.* (1966) at the time, and based on our observations with isolated rabbit liver mitochondria, it appeared to us that reinvestigation of the subcellular location of P-enolpyruvate carboxykinase activity was especially necessary.

B. P-Enolpyruvate Carboxykinase

1. Intracellular Location. In 1970 and 1973 we reported (Johnson *et al.* 1970, 1973), in contrast to the findings of Nordlie and Lardy (1963), that a measurable amount of P-enolpyruvate carboxykinase activity could be demonstrated in hepatic cytosol from fed rabbits, although most of the activity was in the mitochondrial fraction as they had reported. Our data (Table I) indicate that although the specific activity of the cytosolic enzyme is only about one-eighth that of the mitochondrial enzyme, the cytosol does contain about 8 to 10% of the total hepatic carboxykinase activity even

Table I. Activity of P-enolpyruvate carboxykinase in rabbit liver

	Mitochondrial activity[a]		Cytosolic activity[a]	
Metabolic condition (number of rabbits in parentheses)	Specific (mμmole P-enolpyruvate formed/min/mg protein	Total (μmole P-enolpyruvate formed/min/total hepatic subcellular fraction	Specific (mμmole P-enolpyruvate formed/min/mg protein)	Total (μmole P-enol-pyruvate formed/min/total hepatic subcellular fraction)
Fed (16)	39.5 ±19	77.7 ±39	5.2 ±3.3	25.4 ±22
Fasted, 48 hours (15)	78.7 ±24	58.9 ±33	35.3 ±15	115 ±51
	$P < 0.001$	$P > 0.1$	$P < 0.001$	$P < 0.001$
Fed, diabetic (11)	93.2 ±23 $P < 0.001$	116 ±58 $P > 0.05$	38.4 ±13 $P < 0.001$	154 ±66 $P < 0.001$
Fasted, 48 hours plus refeeding 48 hours (12)	53.7 ±13	127 ±28	4.2 ±2.3	23.9 ±18
	$P > 0.025$	$P > 0.001$	$P > 0.3$	$P > 0.8$

[a] Standard deviations from the mean represented by ± figures; P = significance of differences between sample means; data from fasted, fed diabetic, and fasted, refed rabbits tested against data from fed controls.

after correction for mitochondrial loss (40–45%—see footnote 2) and leakage during preparation (according to data in Figs. 7, and 8).

2. ***Effects of Starvation and Diabetes.*** With the demonstrated presence of cytosolic as well as mitochondrial P-enolpyruvate carboxykinase activity in rabbit liver, it became important to know whether either one or both of these activities could respond to fasting or diabetes, both conditions known to be conducive to gluconeogenesis. As is apparent from data also presented in Table I, both fasting and alloxan diabetes cause highly significant four- and sixfold (7.7- and 9-fold when corrected) increases in total cytosolic carboxykinase activity. When these data are corrected for both mitochondrial loss and leakage, amounts of about 46 and 33% of the total carboxykinase are found in the hepatic cytosol of fasted and diabetic rabbits, respec-

tively. On the other hand, neither of these conditions causes a statistically significant alteration in total mitochondrial carboxykinase content. Other data also presented in Table I indicate that refeeding of 48-hour-fasted rabbits for 48 hours causes the elevated cytosolic carboxykinase activity to return toward normal. These findings of elevated cytosolic carboxykinase activity in rabbits with fasting are similar to those made with rats and guinea pigs (Lardy *et al.*, 1964) but are in contrast to what occurs in sheep (Filsell *et al.*, 1969) and in cattle (Ballard, *et al.*, 1968), where no changes in carboxykinase activity are observed with fasting. The effect of alloxan diabetes on the cytosolic carboxykinase of rabbit liver is consistent with data obtained with rats (Lardy *et al.*, 1964) and also with sheep, where interestingly, both mitochondrial and cytosolic carboxykinase activities are increased (Filsell *et al.*, 1969; Taylor *et al.*, 1971).

As pointed out previously, Ilyin *et al.* (1966) published data indicating the existence of cytosolic carboxykinase in livers of fed rabbits and also the enhancement of this cytosolic activity by starvation. Nearly simultaneous to publication of our data (Johnson *et al.*, 1970), Usatenko (1970) published similar data and subsequently, Garber and Hanson (1971a) also reported confirmatory data that were also similar.

*3. **Evidence for** in vivo **existence in rabbit liver cytosol.*** Since the publications of Ilyin *et al.* (1966) and Usatenko (1970) as well as our own (Johnson *et al.*, 1970) provided no conclusive evidence to substantiate the *in vivo* presence of carboxykinase in rabbit liver cytosol (as opposed to its being an artifact of cellular fractionation), we decided to establish the intracellular location of this enzyme activity making use of the mitochondrial "marker" enzymes, citrate synthase, and cytochrome oxidase, these data are presented in Figs. 7 and 8.

The data regarding the appearance of citrate synthase (Fig. 7) in fractions other than mitochondria indicate that about 10% of the total hepatic citrate synthase activity measured is found as a contaminant in cytosolic plus microsomal fractions prepared by our procedures from livers of fed rabbits. Rather minor increases in the amounts of contaminating citrate synthase activity to 14 and 16% are observed in the same two fractions prepared from livers of fasted and diabetic rabbits, respectively. Relatively similar data are obtained when cytochrome oxidase is used as the marker enzyme (Fig. 8). In other words, when cytosolic carboxykinase activity is extensively increased, little increase is seen in the combined cytosolic and microsomal contents of these exclusively mitochondrial enzymes. Thus cytosolic carboxykinase activity apparently does exist *in vivo*. Our data, when published, were in agreement with data published by Garber and Hanson (1971a), who also concluded on the basis of a comparison of the in-

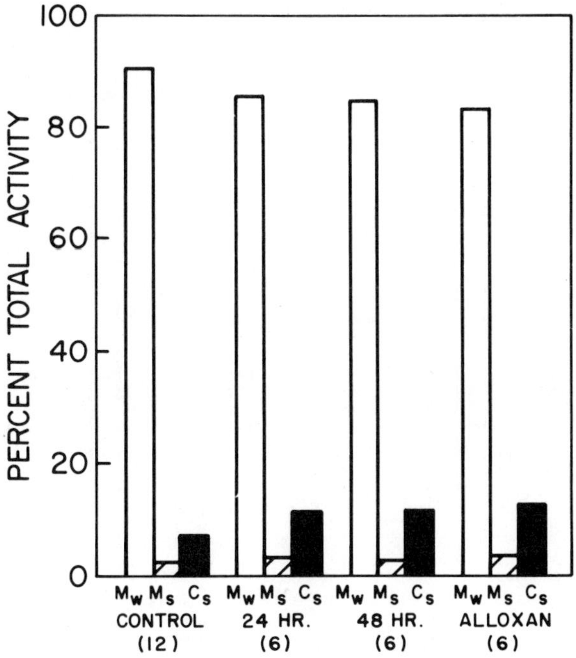

Figure 7. A comparison of the citrate synthase activity in the subcellular fractions of livers from fed, 24-, and 48-hour-fasted, and alloxan diabetic rabbits. The activity is expressed as a percentage of the total activity measured in all three fractions. The numbers in parentheses indicate the number of animals tested for citrate synthase activity. Abbreviations: M_w, mitochondria; M_s, microsomes; C_s, cytosol.

tracellular locations of P-enolpyruvate carboxykinase, pyruvate carboxylase, glutamate dehydrogenase, and β-hydroxybutyrate dehydrogenase activities that P-enolpyruvate carboxykinase is a normal constituent of rabbit liver cytosol. Thus data based on a number of different marker enzymes' activities lead to the same conclusion.

The presence in rabbit liver cytosol of P-enolpyruvate carboxykinase activity that responds to conditions conducive to enhanced glucose formation suggests, circumstantially at least, that this activity must have a role in gluconeogenesis. Fasting and alloxan diabetes have been shown to cause marked increases in cytosolic carboxykinase activity, but the changes take place over a relatively lengthy period of time and such data do not provide any evidence for the ability of the cytosolic carboxykinase to respond to acute conditions such as those caused by more rapid alterations in hormone levels. We have accordingly used several means of determining whether this cytosolic activity can respond in a shorter period of time.

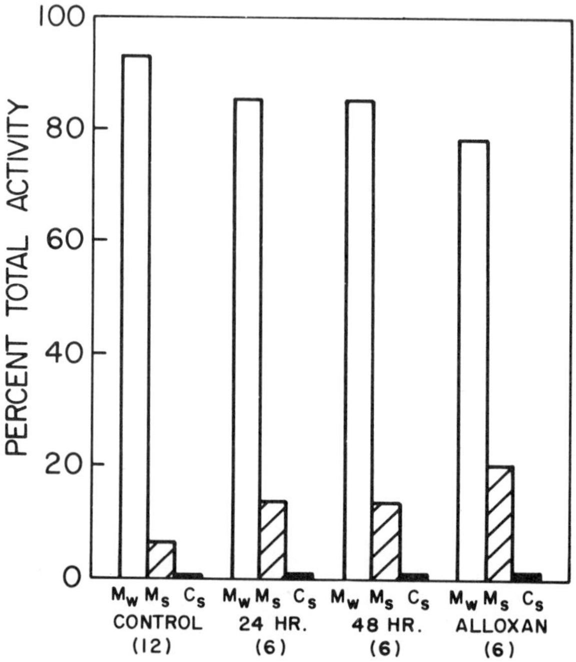

Figure 8. A comparison of the cytochrome oxidase activity in the subcellular fractions of livers from fed, 24-, and 48-hour-fasted, and alloxan diabetic rabbits. The activity is expressed as percentage of the total activity measured in all three fractions. The numbers in parentheses indicate the number of animals tested for cytochrome oxidase activity. Abbreviations: M_w, mitochondria; M_s, microsome; C_s, cytosol.

4. Effect of administration of mannoheptulose. Mannoheptulose temporarily suppresses pancreatic release of insulin (Coore *et al.*, 1963) and/or enhances the release of glucagon (Simon *et al.*, 1972), causing an "instant diabetes" of sorts. Therefore we have investigated the influence of its administration to fed rabbits on the animals' hepatic carboxykinase activity. Some of the data appear in Fig. 9. Within 3 hours of receiving mannoheptulose, the cytosolic activity is increased approximately twofold and it remains near this level even after 6 hours. In this instance then, cytosolic carboxykinase responds to mannoheptulose as does the cytosolic carboxykinase activity in both rat and guinea pig livers (Shrago *et al.*, 1963; Lardy *et al.*, 1964; Foster *et al.*, 1966). On the other hand, the mitochondrial carboxykinase activity in rabbit liver is not significantly altered by administration of mannoheptulose. The response of only the cytosolic carboxykinase of rabbit liver to mannoheptulose is consistent with the data obtained from guinea pigs after similar treatment (Lardy *et al.*, 1964).

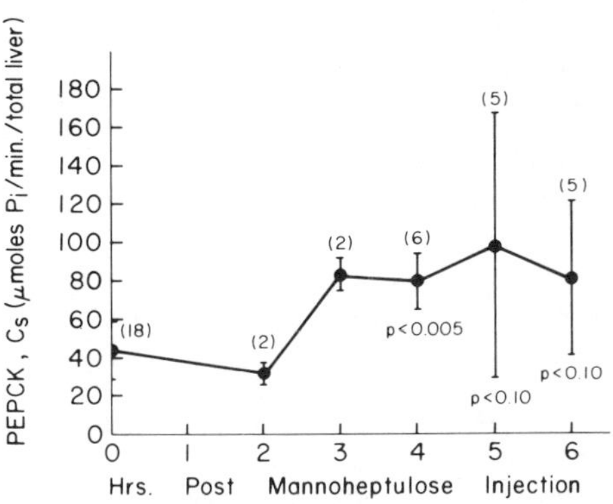

Figure 9. Hepatic cytosolic P-enolpyruvate carboxykinase activity of fed rabbits after subcutaneous administration of a single 2 g/kg dose of mannoheptulose. The number of animals contributing to each point appears in parentheses, and the vertical lines represent one standard deviation. Values of significance (*P*) were calculated by testing the mean values from treated animals against the mean control value.

5. ***Effect of Administration of Hydrocortisone.*** Data presented in Fig. 10 reveal that administration of a single dose of hydrocortisone to fed rabbits also significantly enhances the cytosolic carboxykinase activity within 12 hours. Here again, the cytosolic activity in rabbit liver responds just as does that of rat liver, although the response is somewhat slower (Foster *et al.*, 1966). As observed in the case with mannoheptulose, hydrocortisone administration exerts little effect on the mitochondrial carboxykinase activity. In sharp contrast to these data, neither the mitochondrial nor cytosolic carboxykinase activities of guinea pig liver (Lardy *et al.*, 1964) or of sheep liver (Filsell *et al.*, 1969) respond to hydrocortisone.

We have not tested the effect of a single dose of hydrocortisone on carboxykinase activity beyond 16 hours, nor have we tested the effect of

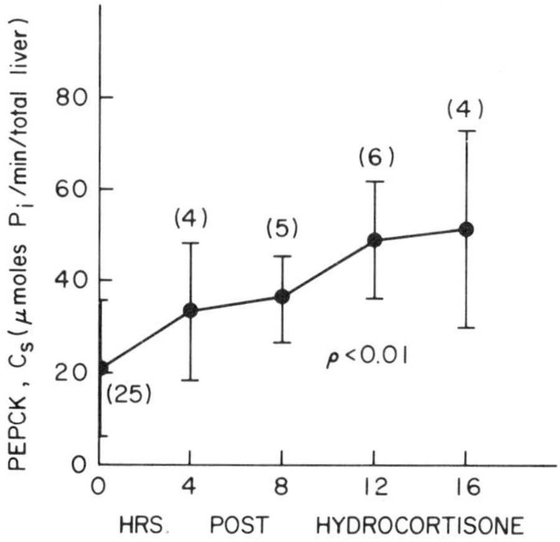

Figure 10. Hepatic cytosolic P-enolpyruvate carboxykinase activity of fed rabbits after subcutaneous administration of a single 2.5 mg/kg dose of hydrocortisone. The number of animals contributing to each point appears in parentheses, and the vertical lines represent one standard deviation. Values of significance (*P*) were calculated by testing the mean values from treated animals against the mean control value.

multiple doses of hydrocortisone. Since fasting per se extensively enhances cytosolic carboxykinase activity, we have not tested the effect of hydrocortisone in the fasted rabbit. However, Ilyin *et al.*, (1966) reported that repeated doses of hydrocortisone given to fed rabbits over a period of 66 hours do significantly increase cytosolic carboxykinase activity.

To date, we have not observed meaningful increases in the cytosolic carboxykinase activity over a short period of time (less than 8 hours) following administration of a single dose of triamcinolone (2.5 mg/kg), glucagon (2 mg/kg), theophylline (20 mg/kg), or glucagon and theophylline in combination. Interestingly enough, a large dose of tryptophan (750 mg/kg) that inhibits hydrocortisone-induced gluconeogenesis and also causes a paradoxical increase in the assayable activity of P-enolpyruvate carboxykinase in rats (Foster *et al.*, 1966a; Ray *et al.*, 1966), exerts neither of these effects when administered to fed or fasted rabbits. Nonetheless, the abovementioned observations made with mannoheptulose and hydrocortisone indicate that the cytosolic carboxykinase activity can respond with varying degrees of rapidity depending on the particular conditions.

6. ***Developmental Aspects.*** With respect to the developmental aspects of heptatic P-enolpyruvate carboxykinase in rabbits, Ilyin et al. (1966) and Usatenko (1970) have reported that this activity, which first appears in mitochondria of fetal rabbits after 25 days of embryonic development, gradually increases to 59% (on the basis of specific activity) of adult levels by term (31 days). In the first day following birth, the mitochondrial activity sharply increases to levels somewhat above those found in adult rabbits, remaining there for 15 to 30 days of postnatal life. In a somewhat similar fashion, cytosolic carboxykinase activity is also essentially negligible until about the 25th day of embryonic development, at which time low levels of activity appear and progressively increase to 45% of adult levels at term. Cytosolic carboxykinase activity also rapidly increases to somewhat supra-adult levels within the first postnatal day and stabilizes at normal adult levels at several weeks after birth. It should be mentioned that none of these studies contained any totally definitive experiments to preclude the possibility that some fraction of the "cytosolic" carboxykinase activity might be of mitochondrial origin. The time at which the cytosolic carboxykinase of the neonatal rabbit becomes responsive to gluconeogenic stimuli has not yet been established.

The appearance of hepatic P-enolpyruvate carboxykinase has also been studied in developing rats (Ballard and Hanson, 1967; Yeung and Oliver, 1967; Yeung *et al.*, 1967) and in swine (Swiatek *et al.*, 1970; Mersmann, 1971). Ballard and Hanson (1967) report that 90% of the carboxykinase activity in 17-day fetal rat livers is in mitochondria and nuclei. However cytosolic activity increases some 25-fold within 2 days after birth, while mitochondrial activity remains relatively constant. Data on the enzyme's intracellular location are not provided, but Swiatek *et al.* (1970) and Mersmann (1971) discovered that although carboxykinase in swine liver is easily measurable in the late term fetal liver, it normally increases some three- to fourfold in the first few days after delivery and is even more responsive following starvation from birth. Thus hepatic P-enolpyruvate carboxykinase seems to appear relatively late in fetal development in the various species studied thus far. Species differences obviously exist with respect to the intracellular location of hepatic carboxykinase, and differences also exist in enzyme distribution in the fetus compared to the post-term animal of the same species (as, e.g., in the rat).

7. ***Purification and Isolation.*** P-enolpyruvate carboxykinase has been isolated from rat liver cytosol (Ballard and Hanson, 1969) and from pig liver mitochondria (Chang *et al.*, 1966; Lane *et al.*, 1969). Ballard and Hanson (1969) have shown by immunological means that the P-enolpyruvate carboxykinases found in rat liver mitochondria and cytosol are two distinct enzymes. Nonetheless essentially no differences, based on various kinetic and physical parameters, were found to exist between the mitochondrial and cytosolic carboxykinases from rat liver (Ballard and Hanson, 1967) or from guinea pig liver (Holten and Nordlie, 1965). To date, neither the mitochondrial nor cytosolic carboxykinases(s) in rabbit liver has been isolated, nor have their properties been compared.

C. Pyruvate Carboxylase

1. ***Cellular Location and Regulation.*** Pyruvate carboxylase activity was reported by Keech and Utter (1963) to be present in mitochondria of rabbit liver. Their findings have been confirmed by Garber and Hanson (1971a). Data of Garber and Hanson also seem to preclude extramitochondrial existence of this activity in rabbit liver consistent with recent evidence (Böttger *et al.*, 1969; Walter and Anabitarte, 1971; Weiss *et al.*, 1974), which suggests that the pyruvate carboxylase activity in rat liver and sheep liver (Taylor *et al.*, 1971) is also limited to mitochondria. Garber and Hanson (1971a) have reported that a 48 hour fast does not significantly alter the mitochondrial content of pyruvate carboxylase activity in rabbit liver (on the basis of units activity per gram of liver, wet weight). Our own unpublished data, generally speaking, concur with the above-mentioned results of others indicating that negligible amounts of pyruvate carboxylase appear in cytosol. We also find, consistent with the observations of Garber and Hanson, that fasting actually seems to cause a decrease of the total hepatic activity of this enzyme. We also find that total hepatic pyruvate carboxylase activity is not significantly altered during a period up to 16 hours following a single dose of hydrocortisone, or by diabetes induced 10 days previously with alloxan. in contrast to these observations, both fasting and diabetes increase hepatic pyruvate carboxylase activity in sheep (Filsell *et al.*, 1969) and also in rats and mice according to (Freedman and Kohn, 1964; Henning *et al.*, 1966). However, other investigators report that neither starvation nor diabetes causes any change in hepatic pyruvate carboxylase activity of rats (Filsell *et al.*, 1969; Krebs, 1966).

These data taken as a whole suggest that pyruvate carboxylase activity is confined to mitochondria in rabbit liver as it is in other species and that this activity does not fluctuate significantly under several conditions in which carbohydrate synthesis is known to vary extensively.

2. ***Developmental Aspects.*** The development of pyruvate carboxylase activity has been studied in embryonic rats (Ballard and Hanson, 1967; Yeung *et al.*, 1967) and pigs (Mersmann, 1971), but no such studies have been conducted concerning the development of this activity in fetal or newborn rabbits.

3. ***Purification and Isolation.*** Pyruvate carboxylase has been extensively purified from chicken liver (Scrutton and Utter, 1965; Scrutton *et al.*, 1969) and isolated from rat liver (McClure *et al.*, 1971; Seufert *et al.*, 1971). Yet to the author's knowledge, this enzyme has not been extensively purified from rabbit liver.

D. Fructose-1,6-diphosphatase

1. ***Cellular Location and Regulation.*** Gomori (1943) first identified fructose-1,6-diphosphatase activity in rabbit liver cytosol. However very few studies have been made of its *in vivo* regulation in this species. Mokrasch *et al.* (1956) reported that this activity (expressed per liver or per body weight) was increased in rabbit liver by feeding diets high in protein or fructose and also by prolonged, daily administration (3–7 days) of cortisone acetate. We find (Fig. 11) that a 24 or 48 hour fast does not alter the total hepatic activity of the enzyme, although alloxan diabetes does cause a doubling of its total hepatic activity. We also find no change in the total hepatic activity of fructose-1,6-diphosphatase up to 16 hours following administration of a single dose of hydrocortisone, or after administration of mannoheptulose, triamcinolone, glucagon, theophylline, or a combination of the latter two agents. Starvation increases this particular enzyme's activity in sheep liver, but the same activity in rat liver remains essentially unaffected (Filsell *et*

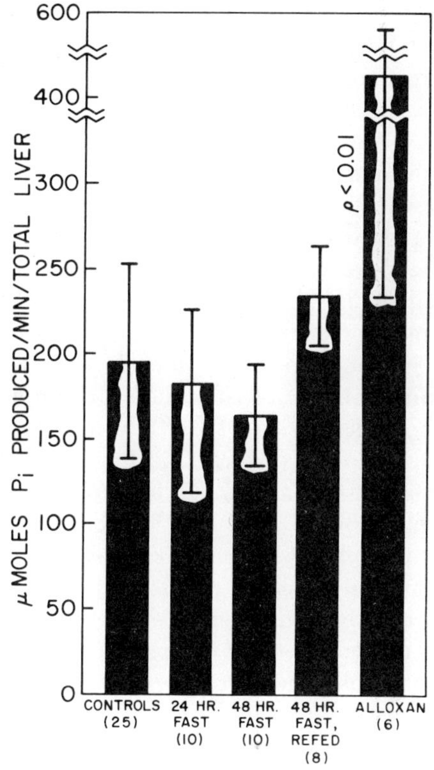

Figure 11. Hepatic cytosolic fructose-1,6-diphosphatase activity from fed, 24-, and 48-hour-fasted, 48-hour-fasted and 48-hour-refed, and alloxan diabetic rabbits. Data are expressed in terms of total hepatic activity. The numbers in parentheses indicate the number of animals tested with each experimental condition. The vertical lines represent one standard deviation from the mean. Values of significance (*P*) were calculated by testing the mean values from the fasted, refed, and diabetic animals against the mean control value.

al., 1969). Kvam and Parks (1960) and Weber *et al.* (1964) have shown that this activity can be significantly increased in rat liver by administration of glucocorticoids and also by alloxan diabetes (Weber *et al.*, 1964).

2. ***Developmental Aspects.*** Mokrasch *et al.* (1956) have reported the presence of fructose-1,6-diphosphatase activity in fetal rabbit livers at 20 days. This activity, which is only about 30% of adult levels at 20 days (expressed per gram of liver), is increased to about 70% of adult levels in 29-day-old fetuses.

The development of fructose-1,6-diphosphatase activity has also been studied in rats (Yeung *et al.*, 1967; Wallace and Newsholme, 1967), pigs (Swiatek *et al.*, 1970; Mersmann, 1971), sheep (Wallace and Newsholme, 1967), and chickens (Wallace and Newsholme, 1967). Generally speaking, it appears that hepatic fructose-1,6-diphosphatase activity in rabbits, rats, and pigs is relatively low during early fetal development and increases until delivery, at which time enzyme activity continues to increase to adult levels.

On the other hand, the activity of this enzyme remains relatively constant in fetal and adult sheep liver and is decreased in adult chicken livers compared to chick embryo livers.

3. ***Purification and Isolation.*** Unlike the other enzymes involved in gluconeogenesis in rabbit liver, fructose-1,6-diphosphatase has been isolated and extensively studied *in vitro*. It was first isolated in crystalline form from rabbit liver by Pontremoli *et al.*, (1965) and has recently been isolated from beef liver as well (Byrne *et al.*, 1971). The properties of the enzyme are discussed in detail in reviews by Pontremoli and Grazi (1968) and Pontremoli and Horecker (1970) and are not extensively elaborated on here.

It should be briefly mentioned, however, that the enzyme has a K_m of the order of 10^{-6} M (for fructose-1,6-diphosphate), is inhibited by substrate [at a concentration of 0.1 mM or more (Taketa and Pogell, 1965)] and by cyclic AMP (Taketa and Pogell, 1965), and is inactivated by physiological concentrations of ADP and ATP (Taketa *et al.*, 1971). The inactivation induced by ADP or ATP can be reversed by physiological concentrations of 3-phosphoglycerate (Taketa *et al.*, 1971). It has also recently been shown that various fatty acids, and especially oleate, can stimulate rabbit liver fructose-1,6-diphosphatase activity (Baxter *et al.*, 1972). Thus it is obvious that the activity of this particular enzyme can be regulated by a wide variety of effectors at least *in vitro* and most likely *in vivo* as well.

E. Glucose-6-Phosphatase Hydrolase

1. ***Cellular Location and Regulation.*** Glucose-6-phosphatase activity was first reported in rabbit liver extracts by Fantl and Rome (1945). Hers *et al.* (1951) fractionated liver tissue homogenates from various species including rabbits and concluded that glucose-6-phosphate hydrolase activity was located mainly in the microsomal fraction of livers from all species studied. A measurable amount of hydrolase activity has recently been reported to reside in the nuclear membrane of liver cells from various species, including the rabbit, although the physiological significance of this activity remains unknown (Gunderson and Nordlie, 1973).

Very few studies concerning the regulation of glucose-6-phosphatase activity have been performed with rabbit liver. It has been reported by Verne *et al.* (1961) that glucagon (250–7500 μg) does not modify the glucose-6-phosphate hydrolase pattern of rabbit liver viewed histologically, although Devrim (1961) has reported that a large dose of glucagon (50 mg/kg body weight) causes a rapid and extensive increase in the activity of the enzyme in this species. It has also been stated by Ilyin and Stepanova (1964) that administration of cortisone acetate or hydrocortisone acetate increases glucose 6-phosphate hydrolase activity in adult rabbit liver.

We have investigated the influence of various conditions known to affect cytosolic P-enolpyruvate carboxykinase on the total activity of hepatic microsomal glucose-6-phosphate hydrolase. Figure 12 reveals that a 24 or 48 hour fast exerts no effect on the total activity of this enzyme, although alloxan diabetes causes a more than twofold increase in total hepatic hydrolase activity. We also find the total hydrolase activity to be increased significantly within 12 hours following a single dose of hydrocortisone (Fig. 13). However, administration of mannoheptulose, glucagon, theophylline, or a combination of these two agents does not significantly affect the total activity of the enzyme.

Starvation also does not exert any influence on total hepatic glucose-6-phosphate hydrolase activity of rats (Ashmore and Weber, 1959), although fasting does increase this activity in the livers of pig (Swiatek *et al.*, 1970)

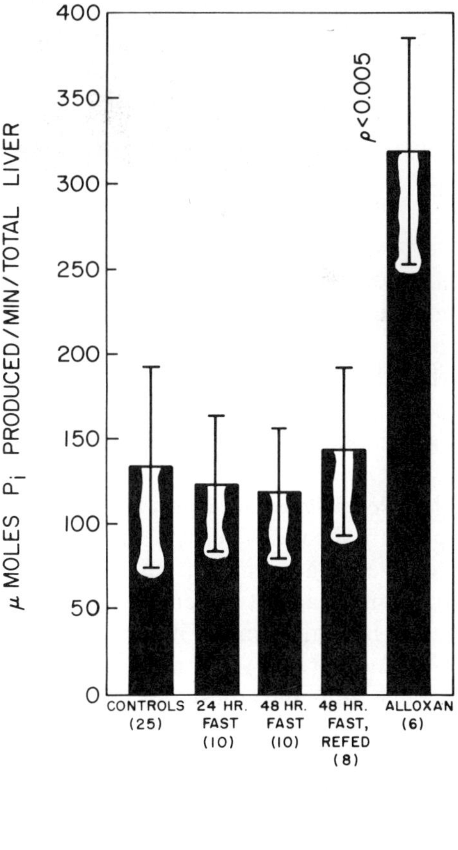

Figure 12. Hepatic microsomal glucose-6-phosphate hydrolase activity from fed, 24-, and 48-hour-fasted, 48-hour-fasted and 48-hour-refed, and alloxan diabetic rabbits. Data are expressed in terms of total hepatic activity. The numbers in parentheses indicate the number of animals tested with each experimental condition. The vertical lines represent one standard deviation from the mean. Values of significance (*P*) were calculated by testing the mean values from the fasted, refed, and diabetic animals against the mean control value.

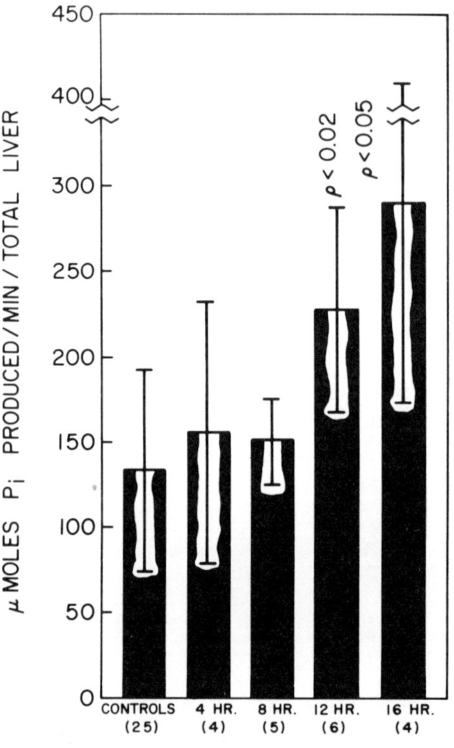

Figure 13. Hepatic microsomal glucose-6-phosphate hydrolase activity from fed rabbits after subcutaneous administration of a single 2.5 mg/kg dose of hydrocortisone. Data are expressed in terms of total hepatic activity. The numbers in parentheses indicate the number of animals contributing to each point, and the vertical lines represent one standard deviation. Values of significance (*P*) were calculated by testing the mean values from treated animals against the mean control value.

and sheep (Filsell *et al.,* 1969). Administration of hydrocortisone and alloxan-induced diabetes have each been shown to increase glucose-6-phosphate hydrolase activity in rat liver, but certain evidence presented suggests that the increase in hydrolase activity brought about in rat liver by glucocorticoids is due to enzyme activation rather than to enzyme synthesis (Arion and Nordlie, 1967).

2. ***Developmental Aspects.*** The developmental pattern of glucose-6-phosphate hydrolase activity has been studied in rabbits, and activity is reportedly absent until the 24th to 26th day of fetal life, when a slight amount of activity becomes detectable; activity continues to increase until term, by which time it has reached adult levels (Stepanova, 1964). Ilyin and Stepanova (1964) report that embryonic hepatic hydrolase activity is also increased by intramuscular injection of glucocorticoids into fetal rabbits.

The developmental pattern of glucose-6-phosphate hydrolase activity has also been studied in rats (Burch *et al.,* 1963; Yeung *et al.,* 1967), sheep (Wallace and Newsholme, 1967), and swine (Swiatek *et al.,* 1970;

Mersmann, 1971). Generally speaking, this activity becomes apparent rather late in embryonic development and slowly increases until birth, when activity usually undergoes a further increase to levels in some cases exceeding those found in adults of the same species.

*3. **Purification and Isolation.*** Glucose-6-phosphate hydrolase has not, to date, been isolated or even extensively purified from any species. Many of the purification attempts and the associated problems have been discussed by Nordlie (1971).

F. Cytosolic Malate Dehydrogenase, Aspartate Aminotransferase, and Fumarase Activities

Since our data suggest that potential glucose carbon may exit from mitochondria in the forms of malate and aspartate, and since considerable carboxykinase activity appears in cytosol under conditions conducive to gluconeogenesis, we measured the activities of cytosolic malate dehydrogenase, aspartate aminotransferase, and fumarase activities because these enzymes would be instrumental in converting malate and aspartate into oxalacetate.

The cytosolic activities of malate dehydrogenase, aspartate aminotransferase, and fumarase were found to be 3150, 845, and 1736 μmole of product formed per minute per total liver. Fasting exerts very little effect on the activities of any of these enzymes. Alloxan diabetes does cause a significant increase in the activity of aspartate aminotransferase, but malate dehydrogenase and fumarase activities remain unaffected.

G. Summary

The preceding information regarding enzyme activities in rabbit liver discloses a number of interesting observations. First of all, there is no doubt that P-enolpyruvate carboxykinase activity is present in both mitochondrial and cytosolic fractions of liver—even in the fed rabbit (Garber and Hanson, 1971a; Johnson *et al.*, 1973; Section V.B). Yet only cytosolic carboxykinase activity, as opposed to the mitochondrial activity, responds significantly to conditions such as fasting (and refeeding) or alloxan diabetes, or, also, to administration of hydrocortisone or mannoheptulose (Ilyin *et al.*, 1966; Garber and Hanson, 1971a; Johnson *et al.*, 1970, 1973; Section V.B). Thus cytosolic carboxykinase in the rabbit responds to a number of dietary and hormonal manipulations in a manner consistent with alterations in glucose synthesis, and the response can be a rapid one, as occurs following administration of mannoheptulose. Generally speaking, hepatic cytosolic carboxykinase of rabbits responds to various stimuli in a manner more analogous to the response of hepatic carboxykinase in rats (as opposed to the responses of this activity in various other species—Section V.B). In some cases, however, the response is slower (e.g., after hydrocortisone), and in some cases it is absent (e.g., after glucagon). It is of interest that carboxykinase activity appears in both the mitochondrial and cytosolic fractions of near-term fetal rabbit liver as well as adult rabbit liver (Ilyin *et al.*, 1966; Usatenko, 1970). However, the point at which the rabbit can first synthesize glucose from three carbon fragments remains unknown at this time, as does the point at which the cytosolic carboxykinase activity first becomes responsive to gluconeogenic stimuli.

Among the enzymes usually considered to be potential points of regulation in gluconeogenesis, cytosolic carboxykinase activity in rabbit liver clearly seems to be more responsive to more conditions than do any of the other enzymes investigated (Sections V.B to V.E). On the basis of our data, a severalfold excess of carboxykinase activity resides in rabbit liver mitochondria under all conditions, and in cytosol too, during starvation or diabetes, to catalyze formation of P-enolpyruvate in amounts sufficient to support a rate of glucose synthesis of 0.5 μmole/min/g liver.

As far as the activities of the other enzymes in rabbit liver are concerned, pyruvate carboxylase activity appears to be relatively unresponsive to glucogenic stimuli (Garber and Hanson, 1971a) and the activities of fructose-1,6-diphosphatase and glucose-6-phosphate hydrolase, while individually responsive to some stimuli (that elicit a response of cytosolic carboxykinase activity) are unresponsive to others (Sections V.D, V.E). The total hepatic activities of pyruvate carboxylase and mitochondrial carboxykinase are about equivalent (Garber and Hanson, 1971a), whereas total fructose-1,6-diphosphatase and glucose-6-phosphate hydrolase activities are roughly twofold that of total hepatic carboxykinase activity in terms of potential glucose formation from their reaction products. (Section V.B to V.E).[3]

VI. STUDIES WITH PERFUSED LIVERS

A. Background

From time to time throughout this chapter reference has been made to the rate at which the intact, fasted rat or livers isolated from rats or guinea pigs

[3] Although total P-enolpyruvate carboxykinase activity is roughly equivalent to fructose-1,6-diphosphatase and glucose-6-phosphatase activities in terms of micromoles of product per minute per gram of liver, two equivalents of P-enolpyruvate, as compared to one equivalent of fructose-6-phosphate or glucose 6-phosphate, is needed to make one equivalent of glucose.

can make glucose from lactate; and it has been suggested here that the rabbit might synthesize glucose at a somewhat comparable rate. To the author's knowledge, however, no one has provided data obtained with intact rabbits from which the glucose synthesizing capability for this species can be calculated. To at least make an estimate of this capacity, we have measured the conversion of L-lactate to glucose by the isolated, perfused rabbit liver. In addition, some data have been obtained with isolated, perfused rabbit livers by Arinze *et al.* (1973) and by Rothschild *et al.* (1968).

B. Rate and Regulation of Glucose Synthesis

In our own studies, the methods and apparatus employed are similar to those referenced or described by Veneziale *et al.* (1967) for perfusion of isolated rat livers. In our system, a perfusate consisting of washed rabbit erythrocytes and 3% bovine serum albumin (Fraction V, fatty acid poor—Miles Laboratories) in oxygenated Krebs-Ringer bicarbonate buffer is recycled through the liver for the duration of the experiment. Rabbits fasted for 48 hours are used as liver donors, and the liver is flushed with 50 to 75 ml of perfusate prior to being coupled into the recycling system.

Data presented in Fig. 14 depict the ability of fasted rabbit livers to synthesize glucose from 10 mM L-lactate. It is apparent that in the absence of added substrate, very little glucose formation occurs. However, addition

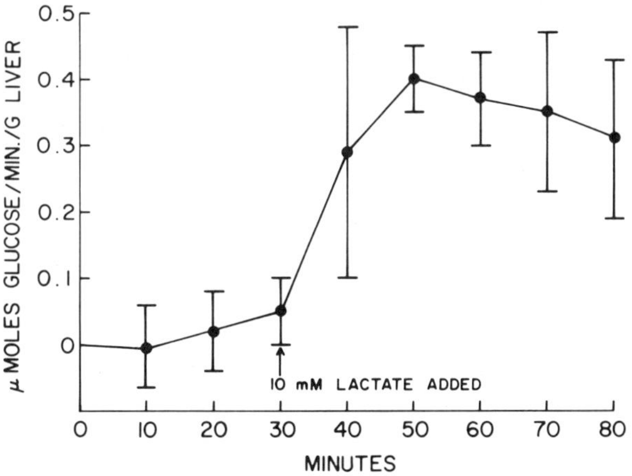

Figure 14. Glucose synthesis from 10 mM lactate by perfusion of livers isolated from 48-hour-fasted rabbits. After 30 minutes of perfusion in the absence of exogenous substrate, lactate was added to a final concentration of 10 mM (subsequently maintained by infusion of additional lactate). Points plotted are the average of data from three perfusions; the vertical lines represent one standard deviation from the mean.

of lactate to a final concentration of 10 mM (and its maintenance by subsequent infusion) results in a rapid initiation of glucose synthesis which reaches a maximum rate of 0.40 μmole/min/g liver (uncorrected for glucose utilization by erythocytes). Arinze *et al.* (1973) have reported a similar maximal rate of glucose synthesis from 2 mM lactate presented (in a nonrecycling system) to livers isolated from fasted rabbits. Interestingly, these authors observe nearly identical maximal rates of glucose synthesis from lactate with livers isolated from either fed or fasted rabbits. Rothschild *et al.* (1968) have also demonstrated the conversion of lactate to glucose by isolated rabbit livers, although the rate of glucose formation calculated from their studies is considerably less than that found in our own studies or in those of Arinze *et al.* (1973). No data on the ability of rabbit livers to form glucose from substrates other than lactate are currently available.

Arinze *et al.* (1973) also report that aminooxyacetate[4] inhibits gluconeogenesis from lactate by 55 to 60% in fasted rabbit livers but exerts no inhibition of glucose formation from lactate with livers from fed rabbits. They find that aminooxyacetate also inhibits gluconeogenesis from lactate by 50 to 55% in guinea pig livers. Infusion of guinea pig livers with octanoate or β-hydroxybutyrate results in inhibition of gluconeogenesis from lactate and cessation of their infusion results in a return of glucose formation to its original rate (Arinze *et al.*, 1973).

C. Summary

Only a small amount of data is currently available on glucose synthesis by isolated perfused rabbit livers. The data demonstrate, however, that rabbit liver can synthesize glucose from lactate at a rate of at least 0.4 μmole/min/g liver (Fig. 14; also Arinze *et al.*, 1973). Although livers from fed and fasted rabbits reportedly synthesize glucose from lactate at relatively similar rates, glucose formation in the fasted rabbit appears to rely substantially on cytosolic P-enolpyruvate carboxykinase, since aminooxyacetate inhibits glucose formation from lactate by nearly 60% in livers from fasted rabbits (Arinze *et al.*, 1973). Nonetheless, if glucose formation occurs at relatively similar rates in both fed and fasted rabbits, and if the mechanism of action of aminooxyacetate is correctly interpreted, mechanisms must exist for regulating P-enolpyruvate production by the individual or combined actions of mitochondrial and cytosolic P-enolpyruvate carboxykinase activities.

Infusion of rat livers with octanoate or β-hydroxybutyrate enhances glucose production from lactate, whereas it inhibits the formation of lactate-

[4] Aminooxyacetate has been reported by Rognstad and Katz (1970) to inhibit gluconeogenesis from lactate in rat kidney cortex slices, and these authors suggest that such inhibition is exerted at the level of aspartate aminotransferase.

derived glucose by guinea pig livers (Arinze *et al.*, 1973). The effects of octanoate or β-hydroxybutyrate on glucose formation by rabbit livers have not been reported, but one might assume that these substances would exert inhibitory effects similar to those observed in guinea pig livers. The ability of octanoate and β-hydroxybutyrate to inhibit gluconeogenesis in guinea pigs (and presumably rabbits) suggests an influence of the mitochondrial $NAD^+/NADH$ ratio on (*a*) mitochondrial production of P-enolpyruvate and (*b*) the availability of mitochondrially derived malate and aspartate, which ultimately provide substrate for cytosolic P-enolpyruvate carboxykinase activity. (The ability of $NAD^+/NADH$ ratio to influence substrate production by isolated rabbit and guinea pig liver mitochondria is also discussed in Sections IV.A and IV.C).

VII. FINAL CONCLUSIONS

Devotion of a chapter to gluconeogenesis in rabbits needs to be considered in light of the relatively limited number of studies into this process as it occurs in species other than the rat. Furthermore, evidence is accumulating to support the use of species such as the guinea pig and rabbit, as opposed to the rat, for models of gluconeogenesis as it occurs in a number of different species including man.

A number of earlier studies performed with rabbit liver mitochondria were interpreted on the assumption that P-enolpyruvate carboxykinase was confined to mitochondria (Section IV.A). A reinterpretation of certain conclusions drawn from some of these studies is required, however, because of the recent demonstration of cytosolic P-enolpyruvate carboxykinase activity in rabbit liver (Section V.B).

It has recently been shown by several groups that given various substrates, rabbit liver mitochondria can generate and liberate P-enolpyruvate, malate, citrate, and aspartate, although the quantity of each metabolite varies with the substrate and the experimental conditions used (Sections IV.A, IV.B). In certain cases (Davis and Gibson, 1967; Garber and Hanson, 1971a; Section IV.A), the production of P-enolpyruvate from certain substrates was nearly sufficient to account for glucose formation at the rate observed to occur in isolated rabbit livers (Section VI.B).

One of the interesting observations often made with mitochondria from rabbits and guinea pigs is that an uncoupler such as DNP or a fatty acid is capable of increasing the quantity of P-enolpyruvate produced compared to the production of other metabolites. These effects of uncouplers have been attributed by some investigators to enhanced production of oxalacetate and/or substrate level production of GTP (Nordlie and Lardy, 1963a; Davis and Gibson, 1967; Section IV.A). On the other hand, Jomain-Baum and Hanson (1973) have found that 0.2 mM octanoate decreases the ability of either guinea pig liver or human liver mitochondria to convert malate or α-ketoglutarate to P-enolpyruvate, and Garber and Hanson (1970), using guinea pig liver mitochondria, have found that when the oxidation-reduction potential is held relatively constant, uncouplers cause a decrease in the rate of P-enolpyruvate formation from various substrates.

More recently, it has been observed that the *in vivo* $NAD^+/NADH$ ratio in hepatic mitochondria of rabbits and guinea pigs is increased by starvation, whereas the opposite occurs in fasted rats (Section IV.A). Furthermore, if one artificially regulates the $NAD^+/NADH$ ratio of mitochondria isolated from guinea pigs, rabbits, or humans (Jomain-Baum and Hanson, 1973), the production of P-enolpyruvate increases as the $NAD^+/NADH$ ratio increases (Section IV.A). Finally, it has been observed that perfusion of livers from guinea pigs (Arinze *et al.*, 1973) or cats (Arinze and Hanson, 1973) with β-hydroxybutyrate or oleate inhibits glucose formation from lactate. These observations have all been interpreted as being consistent with the theory that any situation that causes an increase in the $NAD^+/NADH$ ratio will affect primarily the mitochondrial malate dehydrogenase equilibrium, such that less oxalacetate is reduced to malate and more oxalacetate is available for P-enolpyruvate formation (Garber and Hanson, 1971a). Additionally, it is suggested that alterations in the $NAD^+/NADH$ ratio will affect mitochondrial levels of α-ketoglutarate and glutamate and ultimately P-enolpyruvate, by influencing glutamate dehydrogenase as well as α-ketoglutarate oxidation. The ability of uncouplers to preferentially enhance mitochondrial production of P-enolpyruvate has likewise been attributed to their ability to cause an increase in the $NAD^+/NADH$ ratio.

The reason(s) for the shift toward oxidation in the $NAD^+/NADH$ ratio in mitochondria of fasted rabbits and guinea pigs is not readily apparent (Arinze *et al.*, 1973; Hanson and Garber, 1972). The foregoing observations, however, appear to be consistent among a number of species that possess both mitochondrial and cytosolic P-enolpyruvate carboxykinase activities. In fact, it has been suggested that the presence of carboxykinase in mitochondria may be a key factor in altering the control mechanisms of oxalacetate metabolism in these species (Garber and Salganicoff, 1973).

All the observations that demonstrate or imply an increase in mitochondrial production of P-enolpyruvate with conditions conducive to gluconeogenesis appear to present an anomaly if one assumes that gluconeogenesis (e.g., from amino acids) requires an enhanced flow of reducing equivalents, by way of malate, from mitochondria to cytosol. Furthermore an increase in cytosolic carboxykinase activity under some particular condition would also seem to be paradoxical if the same condition simultaneously enhanced mitochondrial production of P-enolpyruvate rather than production of potential substrates for cytosolic carboxykinase. Thus much of the data and their interpretations pointing to the involvement of mitochondrial carboxykinase in gluconeogenesis tend to leave the function of cytosolic carboxykinase undefined.

It has been pointed out that no evidence exists that the induction of cytosolic carboxykinase is a prerequisite for gluconeogenesis (Garber and Hanson, 1971a). Yet this activity responds to a number of conditions and treatments that are known to enhance glucose synthesis (Ilyin *et al.*, 1966; Johnson *et al.*, 1970, 1973; Section V.B). Additionally, this activity has been found to be capable of responding rapidly under certain conditions (Section V.B). Finally, more direct evidence that cytosolic carboxykinase contributes substantially to gluconeogenesis in the fasted rabbit comes from the observations that aminooxyacetate inhibits by 55 to 60% the formation of lactate-derived glucose by livers isolated from fasted rabbits (Arinze *et al.*, 1973). Similar results have been obtained with livers isolated from fasted guinea pigs (Arinze *et al.*, 1973). Additionally, 2.4 mM quinolinate, which presumably inhibits only cytosolic carboxykinase, also inhibits by about 55% the conversion of alanine to glucose by isolated guinea pig livers (Söling *et al.*, 1970).

On the basis of available data on the enzymatic activities of pyruvate carboxylase, P-enolpyruvate carboxykinase, fructose-1,6-diphosphatase, and glucose-6-phosphate hydrolase, cytosolic carboxykinase seems to be responsive to a greater variety of conditions and treatments than does mitochondrial carboxykinase or any of these other enzymes (Sections V.B to V.E). Since none of these enzymes in rabbit liver has been isolated or even extensively studied (except for fructose-1,6-diphosphatase), one cannot speak directly to this aspect of regulation of gluconeogenesis in this species. One can say only that on the basis of total assayable hepatic activities, all these enzymes appear to be present in amounts severalfold greater than necessary to catalyze glucose formation from lactate at the rate observed in perfusion experiments (i.e., 0.4 μmole glucose/min/g liver). Nonetheless, it is unlikely that any of these enzymes operate at their maximal capacity under physiological conditions.

Perfusion studies with isolated rabbit livers are underway in several laboratories, but relatively little information is currently available. To date, we have obtained a rate of glucose formation from 10 mM lactate of 0.4 μmole glucose formed/min/g liver (uncorrected). Arinze *et al.* (1973) have observed a relatively similar rate of glucose formation from 2 mM lactate using livers from either fed or fasted rabbits. This rate of glucose formation seems low, since isolated rat and guinea pig livers can form glucose from lactate at rates of 1 and 1.5 μmole/min/g liver, respectively. However, optimal conditions for glucose formation by rabbit livers may not yet have been achieved.

Based on information presented here, it appears that the process of glucose formation, as it occurs in rabbits, utilizes P-enolpyruvate originating from the actions of both mitochondrial and cytosolic P-enolpyruvate carboxykinase (under certain conditions such as fasting), although the glucose formation that occurs in the fed rabbit most likely depends primarily on P-enolpyruvate of mitochondrial origin.

Many questions remain to be answered, and what seem to be anomalies must be resolved before the process of glucose formation in rabbits can be clearly defined. Nonetheless, it appears that the rabbit can be included in a group containing an increasing number of other species (including man) that make glucose by relatively similar processes under the influence of quite similar regulatory mechanisms. As these similarities become more apparent, the rabbit may become more important as a general model for this most important metabolic process.

VIII. ACKNOWLEDGMENTS

The author wants to acknowledge the creative research efforts and contributions of Dr. Douglas C. Johnson, Mrs. Kathleen (Ebert) Zawasky, Robert A. Brunsvold, and Charles A. Huibregtse.

Research support by the National Institutes of Health (AM 12705) and the American Diabetes Association is also acknowledged. The American Heart Association provided an Established Investigatorship to the author from 1967 to 1972.

IX. EPILOGUE

The author spent five most enjoyable years in the stimulating atmosphere of Professor Lardy's laboratory. During this time a number of new and exciting observations were developed regarding glucose metabolism and its regulation, and it seems totally fitting that this book be dedicated to him for his pioneering efforts and achievements in this area of research.

On a more personal note, the author is grateful to have shared in these scientific adventures with this man, whom he has been priviledged to know as scientist, teacher, gentleman, and friend.

REFERENCES

Arinze, I. J., Garber, A. J., and Hanson, R. W. (1973). *J. Biol. Chem.* **248,** 2266–2274.

Arinze, I. J. and Hanson, R. W. (1973). *FEBS Lett.* **31,** 280–282.

Arion, W. J. and Nordlie, R. C. (1967). *J. Biol. Chem.* **242,** 2207–2210.

Ashmore, J. and Weber, G. (1959). *Vitam. Horm.* **17,** 91–132.

Ballard, F. J. and Hanson, R. W. (1967). *Biochem. J.* **104,** 866–871.

Ballard, F. J. and Hanson, R. W. (1969). *J. Biol. Chem.* **244,** 5625–5630.

Ballard, F. J., Hanson, R. W., Kronfeld, D. S., and Raggi, F. (1968). *J. Nutr.* **95,** 160–172.

Ballard, F. J. and Oliver, I. T. (1962). *Nature* **195,** 498–499.

Bandurski, R. S. and Lipmann, F. (1956). *J. Biol. Chem.* **219,** 741–752.

Bartley, W. (1954). *Biochem. J.* **56,** 387–390.

Baxter, R. C., Carlson, C. W., and Pogell, B. M. (1972). *J. Biol. Chem.* **247,** 2969–2971.

Böttger, I., Wieland, O., Brdiczka, D., and Pette, D. (1969). *Eur. J. Biochem.* **8,** 113–119.

Burch, H. B., Lowry, O. H., Kuhlman, A. M., Skerjance, J., Diamant, E. J., Lowry, S. R., and Dippe, P. V. (1963). *J. Biol. Chem.* **238,** 2267–2273.

Byrne, W. L., Rajagopalan, G. T., Griffin, L. D., Ellis, E. H., Harris, T. M., Hochachka, P., Reid, L., and Geller, A. M. (1971). *Arch. Biochem. Biophys.* **146,** 118–133.

Chang, H. C., Maruyama, H., Miller, R. S., and Lane, M. D. (1966). *J. Biol. Chem.* **241,** 2421–2430.

Coore, H. G., Randle, P. J., Simon, E., Kraicer, P. F., and Shelesnyak, M. C. (1963). *Nature* **197,** 1264–1266.

Davis, E. J. and Gibson, D. M. (1967). *Biochem. Biophys. Res. Commun.* **29,** 815–821.

Davis, E. J. and Gibson, D. M. (1969). *J. Biol. Chem.* **244,** 161–170.

Devrim, S. (1961). *Türk Tip Cemiyeti Mecmuasi* **27,** 43; *C.A.* **55,** 20225a, (1961).

Dugal, B. S. (1972). *FEBS Lett.* **26,** 97–101.

Exton, J. H. and Park, C. R. (1967). *J. Biol. Chem.* **242,** 2622–2636.

Fantl, P. and Rome, N. M. (1945). *Aust. J. Exp. Biol. Med. Sci.* **23,** 20–27.

Filsell, O. H., Jarrett, I. G., Taylor, P. H., and Keech, D. B. (1969). *Biochim. Biophys. Acta* **184,** 54–63.

Foster, D. O., Ray, P. D., and Lardy, H. A. (1966). *Biochemistry* **5,** 555–562.

Foster, D. O., Ray, P. D., and Lardy, H. A. (1966a). *Biochemistry* **5,** 563–569.

Freedman, A. D. and Kohn, L. (1964). *Science* **145,** 58–60.

Gamble, J. L. and Mazur, J. A. (1967). *J. Biol. Chem.* **242,** 67–72.

Garber, A. J. and Ballard, F. J. (1969). *J. Biol. Chem.* **244,** 4696–4703.

Garber, A. J. and Ballard, F. J. (1970). *J. Biol. Chem.* **245,** 2229–2240.

Garber, A. J. and Hanson, R. W. (1971). *J. Biol. Chem.* **246,** 589–598.

Garber, A. J. and Hanson, R. W. (1971a). *J. Biol. Chem.* **246,** 5555–5562.

Garber, A. J. and Salganicoff, L. (1973). *J. Biol. Chem.* **248,** 1520–1529.

Gomori, G. (1943). *J. Biol. Chem.* **148,** 139–149.

Gunderson, H. M. and Nordlie, R. C. (1973). *Biochem. Biophys. Res. Commun.* **52,** 601–607.

Hanson, R. W. and Garber, A. J. (1972). *Am. J. Clin. Nutr.* **25,** 1010–1021.

Haynes, R. C., Jr. (1965). *J. Biol. Chem.* **240,** 4103–4106.

Heldt, H. W. (1966). In *Regulation of Metabolic Processes in Mitochondria,* Vol. 7 (J. M. Tager, S. Papa, E. Quagliariello, and E. C. Slater, eds.), Elsevier Publishing Co., Amsterdam, pp. 51–64.

Henning, H. V. and Seubert, W. (1964). *Biochem. Z.,* **340,** 160–170.

Henning, H. V., Stumpf, B., Ohly, B., and Seubert, W. (1966). *Biochem. Z.* **344,** 274–288.

Hers, H. G., Berthet, J., Berthet, L., and deDuve, C. (1951). *Bull. Soc. Chim. Biol.* **33,** 21–41.

Holten, D. D. and Nordlie, R. C. (1965). *Biochemistry* **4,** 723–731.

Ilyin, V. S. and Stepanova, N. G. (1964). *Vopr. Med. Khim.* **10,** 576–584; *C.A.* **62,** 6772b, (1965).

Ilyin, V. S., Usatenko, M. S., and Evstratove, L. A. (1966). *Zh. Evol. Biokhim. Fiziol.* **2,** 185–191.

Ishihara, N. and Kikuchi, G. (1968). *Biochim. Biophys. Acta* **153,** 733–748.

Johnson, D. C. and Ray, P. D. (1969). Abstracts, 158th American Chemical Society National Meeting, Division of Biological Chemistry, No. 130.

Johnson, D. C., Ebert, K. A., and Ray, P. D. (1970). *Biochem. Biophys. Res. Commun.,* **39,** 750–756.

Johnson, D. C., Brunsvold, R. A., Ebert, K. A., and Ray, P. D. (1973). *J. Biol. Chem.* **248,** 763–770.

Jomain-Baum, M. and Hanson, R. W. (1973). *FEBS Lett.* **29,** 145–148.

Keech, D. B. and Utter, M. F. (1963). *J. Biol. Chem.* **238,** 2609–2614.

Krebs, H. A. (1954). *Bull. Hopkins Hosp.* **95,** 19–33.

Krebs, H. A. (1964). *Proc. R. Soc.* **B159,** 545–564.

Krebs, H. A. (1966). *Adv. Enzyme Regul.,* **4,** 339–353.

Krebs, H. A., Dierks, C., and Gascoyne, T. (1964). *Biochem. J.* **93,** 112–121.

Kvam, D. C. and Parks, R. E., Jr. (1960). *Am. J. Physiol.* **198,** 21–24.

Lane, M. D., Chang, H. C., and Miller, R. S. (1969). *Meth. Enzymol.* **13,** 270–277.

Lardy, H. A. (1961). *Proc. Robert A. Welch Found. Conf. Chem. Res.* **5,** 267–276.

Lardy, H. A., Foster, D. O., Shrago, E., and Ray, P. D. (1964). *Adv. Enzyme Regul.* **2,** 39–47.

Lardy, H. A., Paetkau, V., and Walter, P. (1965). *Proc. Nat. Acad. Sci. U.S.* **53,** 1410–1415.

McClure, W. R., Lardy, H. A., and Kneifel, H. P. (1971). *J. Biol. Chem.* **246,** 3569–3578.

Mersmann, H. J. (1971). *Am. J. Physiol.* **220,** 1297–1302.

Mokrasch, L. C., Davidson, W. D., and McGilvery, R. W. (1956). *J. Biol. Chem.* **222,** 179–184.

Mudge, G. H., Neuberg, H. W., and Stanbury, S. W. (1954). *J. Biol. Chem.* **210,** 965–979.

Nordlie, R. C. and Lardy, H. A. (1963). *J. Biol. Chem.* **238,** 2259–2263.

Nordlie, R. C. and Lardy, H. A. (1963a). *Biochem. Z.* **338,** 356–363.

Nordlie, R. C. (1971). *Enzymes* **4,** 543–610.

Pontremoli, S. and Horecker, B. L. (1970). *Curr. Top. Cell Regul.* **2,** 173–199.

Pontremoli, S. and Grazi, E. (1968). In *Carbohydrate Metabolism and its Disorders,* Vol. 1 (F. Dickens, P. F. Randle, and W. J. Whelan, eds.) pp. 259–295.

Pontremoli, S., Traniello, S., Luppis, B., and Wood, W. A. (1965). *J. Biol. Chem.* **240,** 3459–3463.

Ray, P. D., Foster, D. O., and Lardy, H. A. (1966). *J. Biol. Chem.* **241,** 3904–3908.

Rognstad, R. and Katz, J. (1970). *Biochem. J.* **116,** 483–491.

Rothschild, M. A., Oratz, M., Mongelli, J., and Schreiber, S. S. (1968). *J. Clin. Invest.* **47,** 2591–2599.

Scrutton, M. C., Olmsted, M. R., and Utter, M. F. (1969). *Meth. Enzymol.* **13,** 235–249.

Scrutton, M. C. and Utter, M. F. (1965). *J. Biol. Chem.* **240,** 1–9.

Scrutton, M. C. and Utter, M. F. (1968). *Ann. Rev. Biochem.* **37,** 249–302.

Seubert, W. and Huth, W. (1965). *Biochem. Z.* **343,** 176–191.

Seufert, D., Herlemann, E. M., Albrecht, E., and Seubert, W. (1971). *Z. Physiol. Chem.* **352,** 459–478.

Shrago, E., Lardy, H. A., Nordlie, R. C. and Foster, D. O., (1963). *J. Biol. Chem.* **238,** 3188–3192.

Simon, E., Frenkel, G., and Kraicer, P. F. (1972). *Isr. J. Med. Sci.* **8,** 743–752.

Söling, H. D., Willms, B., Kleineke, J., and Gehlhoff, M. (1970). *Eur. J. Biochem.* **16,** 289–302.

Solomon, A. K., Vennesland, B., Klemperer, F. W., Buchanan, J. M., and Hastings, A. B. (1941). *J. Biol. Chem.* **140,** 171–182.

Stepanova, N. G. (1964). *Vopr. Med. Khim.* **10,** 64–70; *C.A.* **60,** 14909a (1964).

Swiatek, K. R., Chao, K. L., Chao, H. L., Cornblath, M., and Tildon, S. T. (1970). *Biochim. Biophys. Acta* **222,** 145–154.

Taketa, K. and Pogell, B. M. (1965). *J. Biol. Chem.* **240,** 651–662.

Taketa, K., Sarngadharan, M. G., Watanabe, A., Aoe, H., and Pogell, B. M. (1971). *J. Biol. Chem.* **246,** 5676–5683.

Taylor, P. H., Wallace, J. C., and Keech, D. B. (1971). *Biochim. Biophys. Acta* **237,** 179–191.

Usatenko, M. S. (1970). *Biochem. Med.* **3,** 298–310.

Utter, M. F. (1963). *Iowa State J. Sci.* **38,** 97–113.

Utter, M. F. and Keech, D. B. (1960). *J. Biol. Chem.* **235,** PC17–PC18.

Utter, M. F. and Keech, D. B. (1963). *J. Biol. Chem.* **238,** 2603–2608.

Utter, M. F. and Kurahashi, I. (1954). *J. Biol. Chem.* **207,** 787–802.

Veneziale, C. M., Walter, P., Kneer, N., and Lardy, H. A. (1967). *Biochemistry* **6,** 2129–2138.

Verne, J., Vergman, R., Guha, S., and Petkov, P. E. (1961). *C. R. Acad. Bulgar. Sci.* **14,** 863–866; *C.A.,* **57,** 1422e (1962).

Wallace, J. C. and Newsholme, E. A. (1967). *Biochem. J.* **104,** 378–384.

Walter, P. and Anabitarte, M. (1971). *FEBS Lett.* **12,** 289–292.

Walter, P. and Stucki, J. W. (1970). *Eur. J. Biochem.* **12,** 508–519.

Walter, P., Paetkau, V., and Lardy, H. A. (1966). *J. Biol. Chem.* **241,** 2523–2532.

Weber, G., Singhal, R. L., Stamm, N. B., Fisher, E. A., and Mentendiek, M. A. (1964). *Adv. Enzyme Regul.* **2,** 1–38.

Weiss, G., Lamartiniere, C. A., Müller-Ohly, B., and Seubert, W. (1974). *Eur. J. Biochem.* **43,** 391–403.

Williamson, D. H., Lund, P., and Krebs, H. A. (1967). *Biochem. J.* **103,** 514–527.

Yeung, D. and Oliver, I. T. (1967). *Biochem. J.* **108,** 325–331.

Yeung, D., Stanley, R. S., and Oliver, I. T. (1967). *Biochem. J.* **105,** 1219–1227.

CHAPTER TEN

Phosphoenolpyruvate Carboxykinase and Renal Gluconeogenesis

CHRISTOPHER I. POGSON, IAN D. LONGSHAW, ANNE ROOBOL, STEPHEN A. SMITH, AND GEORGE A. O. ALLEYNE

I. INTRODUCTION

For every ten papers devoted to aspects of gluconeogenesis, it is doubtful whether more than one, or at most, two, have been concerned with the contribution made by the kidney. This is presumably because the part played by this organ in glucose homeostasis is very limited in most physiological states, renal gluconeogenic flux being rarely more than 10% of total body glucose synthesis (although it may rise to as much as 50% in prolonged starvation: Owen *et al.*, 1969). The present authors contend, however, that kidney gluconeogenesis is of more physiological significance than might be suggested by this one fact and that the kidney offers particular advantages for the experimenter in the field of carbohydrate metabolism.

The capacity of the kidney cortex to synthesize glucose from various substrates has been known since the studies of Benoy and Elliott (1937). These were extended quantitatively and qualitatively particularly by Krebs and his group (Krebs *et al.*, 1963). The early work was excellently reviewed by Krebs himself in a Croonian lecture (Krebs, 1964); subsequent studies have been considered mostly in reviews devoted to general aspects of kidney physiology rather to gluconeogenesis per se (e.g., Orloff and Burg, 1971; Relman, 1972; Goodman, 1972).

In this chapter we consider the aspects of phosphoenolpyruvate carboxykinase (PEP-CK) involvement in kidney metabolism that (*a*) illustrate differences between this tissue and liver and (*b*) may throw further light on the basic mechanisms whereby the expressed activity of this enzyme is regulated. In view of the necessary restriction on space, attention is focused on the following topics:

1. The significance of the biphasic distribution of PEP-CK in the kidney of the guinea pig, and other species.
2. The relationship between PEP-CK activity and acid-base balance.
3. The changes in PEP-CK activity occasioned by disturbances in hormonal status and by other putative physiological effectors.

Before entering a detailed discussion of these specific points, it is perhaps useful initially to give brief consideration to some of the more general aspects of renal metabolism.

Table I lists several of the nutritional, hormonal, and other factors known to affect gluconeogenesis in liver and kidney. Although not exhaustive, the list is sufficient to show that substantial similarities and differences in response exist between the two tissues.

A further marked difference between liver and kidney exists in respect of the permeability of the plasma membrane to potential substrates. The liver cell membrane effectively excludes citric acid cycle intermediates, which therefore yield negligible or zero rates of glucose synthesis even under optimal conditions (Ross *et al.*, 1967). Kidney cortex, on the other hand, has a reasonable gluconeogenic capacity with such cycle intermediates, which seem to penetrate without difficulty (Goodman *et al.*, 1966; Goorno *et al.*, 1967; Krebs *et al.*, 1963; Rutman *et al.*, 1965). Although this is of use experimentally, it is not clear whether and to what extent such observations are relevant to the *in vivo* situation. The major physiological precursors of glucose in kidney are lactate, glutamine, and, to a lesser extent, glycerol. Lactate is primarily derived from glycolytic activity in skeletal muscle, but a small amount comes also from the renal medulla (Hems and Gaja, 1972). Glutamine is a major end product of nitrogen metabolism in muscle

(London *et al.*, 1965; Marliss *et al.*, 1971; Pozefsky *et al.*, 1969; Ruderman and Lund, 1972). Some may also arise from the liver through the action of glutamine synthetase (Addae and Lotspeich, 1968a, 1968b; Lund *et al.*, 1970; Marliss *et al.*, 1971). In kidney cortex glutamine is readily deaminated and deamidated to yield ammonia; this ammonia neutralizes fixed acid in the urine and aids in conserving sodium (Pitts, 1964). The remaining carbon skeleton is converted with varying degrees of efficiency (depending on the prevailing metabolic state) into glucose, a process whereby three-fifths of the original carbon is conserved.

It is relevant at this juncture to draw attention to two points. First, it is important to take note that there are quite wide quantitative differences in kidney metabolism from species to species. Thus consideration of glutamine metabolism is complicated by the observation that glutamine synthetase is present in the kidney cortex of some species (rat, guinea pig, rabbit, and sheep) but not in that of others (dog, cat, and pig) (Krebs, 1935; Lyon and Pitts, 1969). Again, the response of the kidney to acidosis varies from one species to another, particularly in respect of the relative rates of glucose and ammonia production (Klahr, 1971); this has been correlated with the type of nutrition occurring in the various animals examined (Richterich and Goldstein, 1958; Klahr *et al.*, 1972).

Second, the kidney cortex is not histologically homogeneous (Longley, 1969). Not surprisingly, the capacity for ammoniagenesis and gluconeogenesis is not present in all structures and cell types. Recent studies by Curthoys and Lowry (1973a) have shown that the phosphate-dependent glutaminase is normally low in the proximal convoluted tubule and relatively higher in the medulla. In acidosis, however, the glutaminase in the proximal convoluted tubule alone is increased twentyfold. This agrees with the demonstration that glucose-6-phosphatase (Jacobson *et al.*, 1967; Pretlow *et al.*, 1974), PEP-CK (Guder and Schmidt, 1974) and, by inference, gluconeogenesis are very largely associated only with the proximal convoluted tubule. These structures are also known to contain the γ-glutamyl cycle enzymes of amino acid transport (Meister, 1973).

The effect of this intercellular compartmentation can be seen in comparisons of the rates of glucose production from various kidney preparations. Thus isolated tubules from rat kidney, containing a high proportion of gluconeogenic cells, are approximately twice as active in this respect as whole cortical slices (Guder and Wieland, 1971), which are weight for weight more active than the whole perfused organ (Nishiitsutsuji-Uwo *et al.*, 1967). This contrasts with the situation in rat liver, where isolated cell preparations produce amounts of glucose similar to or smaller than those synthesized by the whole organ perfused under standard conditions (e.g., Johnson *et al.*, 1972; Garrison and Haynes, 1973). Although the isolated tubule system, and, in the future, separated convoluted tubule cells, offer

Table I. Effects of various stimuli on gluconeogenesis in liver and kidney cortex[a]

	Liver		Kidney	
Stimulus	Effect	References	Effect	References
Starvation, short term	+	Exton *et al.* (1970)	+	Krebs *et al.* (1963)
Starvation, long term	–	Owen *et al.* (1969)	+	Owen *et al.* (1969); Cahill (1970) (human)
Insulin	–	Cahill *et al.* (1966); Felig and Wahren (1971)	0	Exton (1972)
			–	Joseph and Subrahmanyam (1968)
Glucagon	+	Exton (1972)	0	Exton (1972); Bowman (1970); Nishiitsutsuji-Uwo *et al.* (1967)
Steroids	+	Exton *et al.* (1970)	+	Roobol and Alleyne (1973)
			+	Stumpf *et al.* (1972)
Catecholamines	+	Exton *et al.* (1971)	0	Bowman (1970)
			+	Kurokawa and Massry (1973)
Cyclic AMP	+	Exton and Park (1968)	+	Nagata and Rasmussen (1971a); Pagliara and Goodman (1969)

exciting possibilities, the preparation most commonly employed is the cortex slice. Slices exhibit good metabolic viability, are simply prepared, and allow a large number of separate experiments to be performed simultaneously. Again, there is a contrast with liver slices, which are either too thick for efficient gas diffusion or so thin that a high proportion of their relatively large parenchymal cells are damaged, the preparation thus effectively approximating to a homogenate.

II. PATHWAYS OF RENAL GLUCONEOGENESIS IN RAT AND GUINEA PIG

Lardy and his collaborators several years ago drew attention to the means whereby gluconeogenic intermediates in the liver might pass between the mitochondrial and cytosol compartments (Lardy *et al.*, 1965; Walter *et al.*,

Table I. (Continued)

	Liver		Kidney	
Stimulus	Effect	References	Effect	References
Growth hormone	?	Exton (1972)	+	Joseph and Subrahmanyam (1968)
Parathyroid hormone	(?)		+	Nagata and Rasmussen (1971a)
Exercise	+	Exton (1972)	+	Krebs and Yoshida (1963)
Alcohol	–	Krebs *et al.* (1969); Kaden *et al.* (1969)	0	Krebs *et al.* (1969)
Vasopressin	+	Hems *et al.* (1974)	0 +	Bowman (1970) Stumpf *et al.* (1972)
Fatty acids	+ ?	Ruderman *et al.* (1969) Exton (1972)	+	Krebs *et al.* (1965)
Acidosis	0	Exton (1972)	+	Goodman *et al.* (1966)
Tryptophan	–	Foster *et al.* (1966); Veneziale *et al.* (1967)	0	Pogson *et al.* (1976)
K^+ depletion	(?)		+	Goodman *et al.* (1966)
Ca^{2+}	+	Roobol and Alleyne (1972)	+	Krebs *et al.* (1963); Alleyne *et al.* (1973)

[a] Key: + = stimulation of glucose synthesis; – = inhibition; 0 = no, or very slight effect; ? = effect unknown or ill defined. Data refer to the rat unless otherwise specified. Other species may show different responses (e.g., fatty acids inhibit gluconeogenesis in the perfused guinea pig liver).

1966). Subsequent work has served to confirm and extend these observations, especially in respect of the specific mitochondrial anionic porter systems involved (Chappell, 1968). It was also shown in Lardy's laboratory that the distribution of PEP-CK between cell compartments varies from one species to another (Nordlie and Lardy, 1963). The rat, like the mouse and the hamster, is atypical in that at least 90 to 95% of the total PEP-CK activity is cytosolic; most species exhibit a ratio of cytosolic to mitochondrial enzymes of approximately 1:1, a few, such as the rabbit and the pigeon, are at the other extreme, having more than 90% of the activity associated with the mitochondrial fraction (Hanson and Garber, 1972). Early experiments with cytosolic and mitochondrial PEP-CKs from the guinea pig suggested that the two activities might be associated with distinct protein species (Holten and Nordlie, 1965). This has been subsequently confirmed in other species by both immunological (Ballard and Hanson, 1969) and chromatographic methods (Ballard, 1971; Diesterhaft *et al.*, 1971).

But to what extent does the situation in liver reflect that in kidney? Measurements both of metabolite pathways and PEP-CK distributions (Flores and Alleyne, 1971) indicate that qualitatively at least, the answer is "to a considerable degree." In guinea pig, for example, kidney PEP-CK was found to be approximately 60% cytoplasmic (Longshaw *et al.*, 1972a), in good agreement with the earlier liver data of Nordlie and Lardy (1963). Better agreement still appeared in an experiment where anti-rat-liver cytosolic PEP-CK antibody (Ballard and Hanson, 1969) was titrated with rat kidney cytosol; the similar titers of this and the original antigenic activity strongly suggests that the same protein is present in both tissues (Longshaw and Pogson, 1972).

We have investigated the role of the two PEP-CKs in guinea pig kidney cortex, making use of the specific transaminase inhibitor, aminooxyacetate. Referring to Fig. 1, it is apparent that when L-lactate is provided as substrate to a cell type synthesizing carbohydrate in which the active PEP-CK is cytosolic, sufficient reducing power is already present in the cytosol following lactate oxidation, and oxaloacetate may pass from the mitochondrion to the cytosol as aspartate (Rognstad and Katz, 1970). If lactate is replaced by pyruvate, or an equivalent "oxidized" substrate, reducing equivalents must pass from the mitochondrion to the cytosol; thus oxaloacetate is "transported" as malate, which can generate cytosolic NADH.

Conversely, where the active PEP-CK is mitochondrial, lactate gives rise to mitochondrial PEP, which may leave the mitochondrion by way of a citrate-dependent porter system (Robinson, 1971a, 1971b). When an oxidized substrate is provided, however, the need for cytosolic NADH implies that a portion of the oxaloacetate pool must undergo reduction to malate and enter the cytosol to generate NADH; the excess oxaloacetate in the cytosol then must return to the mitochondrion by way of the transaminase systems and the aspartate porter. Since aminooxyacetate blocks the glutamate-oxaloacetate transaminase, it follows that if the scheme in Fig. 1 holds good, the inhibitor should be effective when lactate is the substrate and the active PEP-CK is cytosolic, and also when pyruvate or equivalent substrates are present in a system dependent on mitochondrial PEP-CK. Conversely, no such inhibition should be seen with cytosolic PEP-CK and oxidized substrates or with mitochondrial PEP-CK and lactate.

We have confirmed the earlier data of Rognstad and Katz (1970) with rat kidney, and we find substantially similar effects with the guinea pig (Longshaw *et al.*, 1972a, Table II). This is in agreement with the proposition that gluconeogenesis under these conditions depends on the cytosolic

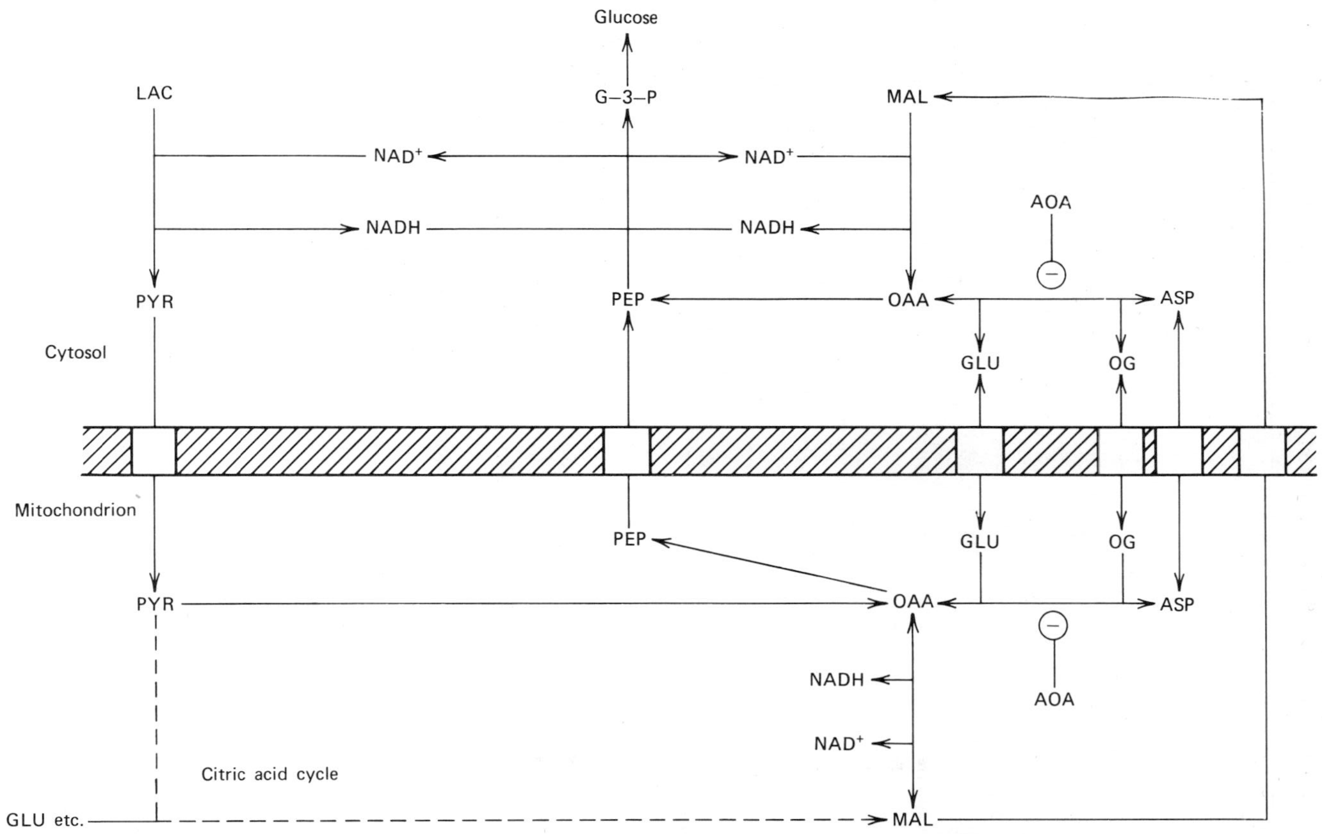

Figure 1. Pathways of gluconeogenesis in kidney cortex. Reactions are for both intra- and extramitochondrial PEP-CK sequences. Arrows represent physiological net fluxes, not equilibria. "Boxes" in the mitochondrial membrane represent specific porter systems, and ⊖ signifies inhibition. Abbreviation: AOA, aminooxyacetate; ASP, aspartate; GLU, glutamate; G-3-P, glyceraldehyde-3-phosphate; LAC, lactate; MAL, malate; OAA, oxaloacetate; OG, oxoglutarate; PEP, phosphoenolpyruvate; PYR, pyruvate.

Table II. Effects of the transaminase inhibitor aminooxyacetate (AOA) on gluconeogenesis in kidney cortex slices[a]

		Glucose production (nmole/hr/mg dry wt)		
		Control	Plus AoA (0.1 mM)	Inhibition
L-Lactate	Rat	65.8 ±4.5	9.2 ±1.8	86%[b]
L-Lactate	Guinea pig	23.0 ±3.9	6.4 ±0.7	72%[b]
Pyruvate	Rat	169 ±10	168 ±6	—
	Guinea pig	30.8 ±3.3	31.1 ±4.5	—
L-Glutamate	Rat	60.7 ±5.5	64.0 ±4.9	—
L-Malate	Guinea pig	53.6 ±7.5	53.6 ±6.0	—

[a] All substrates were initially 10 mM. Rats were starved overnight, guinea pigs were allowed to feed *ad libitum*. Rates, corrected for endogenous activity, are the means ±SEM for 5 or 6 determinations.
[b] $P < 0.001$.
Source: Data from Longshaw *et al.* (1972a).

PEP-CK activity. It is also apparent that the inhibition by aminooxyacetate cannot be "relieved" by an increase in the activity (i.e., catalytic efficiency) of the mitochondrial enzyme. Again, these results, which have recently been confirmed elsewhere (Peng *et al.*, 1973), correlate with the "inducibility" of the kidney cytosolic enzyme in the rat under conditions favoring gluconeogenesis (Longshaw and Pogson, 1972; Sanchez-Medina *et al.*, 1971); whether the same holds true for the guinea pig, as perhaps might be surmised, depends for a definitive answer on the purification of guinea pig cytosol PEP-CK and the preparation of a suitable corresponding antibody.

In an apparent partial contradiction of this proposition, Arinze *et al.* (1973) have shown that the degree of aminoxyacetate inhibition in the perfused guinea pig liver depends on the redox state of the preparation. These authors contend that the data reported in Table II can be said to arise from the relatively reduced condition of the slices during the 60 minute incubation and that both PEP-CKs may come into play when the tissue is in a more oxidized state. Manipulation of the kidney slice intramitochondrial redox state by suitable addition of acetoacetate and/or β-hydroxybutyrate, however, does not, in our hands, alter the extent of the effectiveness of aminooxyacetate (N. Stamper and C. I. Pogson, unpublished results).

A possible resolution of these differences may lie in the fact that different tissues are under study, and in this respect liver may differ quantitatively from kidney in animals with biphasic PEP-CK distributions. It is difficult, however, to side finally with either "extreme" view: if the cytosolic enzyme alone is concerned with glucose synthesis, what is the function of the mitochondrial activity? If the converse, what is the function of the cytosolic enzyme and, teleologically, why is it "induced" by gluconeogenic conditions? It seems more probable, therefore, that both enzymes play important roles in gluconeogenesis in the animal as a whole and that the biphasic distribution may be a reflection of the overall dietary requirement (Hanson and Garber, 1972).

III. ACIDOSIS AND RENAL GLUCONEOGENESIS

Changes in the acid-base status of the rat and other species are associated with simultaneous alterations in renal glutamine metabolism. In response to increasing acidity, the kidney increases ammonia production. This ammonia is derived from glutamine by at least two pathways. It is generally considered that the phosphate-dependent glutaminase I system is responsible for most of the ammoniagenesis (Orloff and Burg, 1971); the level of this enzyme in many species, although not in the dog, is increased in metabolic acidosis by an actinomycin D-sensitive mechanism (Goldstein, 1965). The glutaminase II system, which may occur in both mitochondrial and extramitochondrial compartments (Cooper and Meister, 1974), seems to play a minor role only in the normal dog but may increase to approximately 10% when the animal is acidotic (Bourke *et al.*, 1971c). In man under optimal conditions, a 10% overall contribution has been noted (Bourke *et al.*, 1971a, 1971b). This system may also account for the major proportion of the alanine produced by the rat kidney, which has negligible glutamate-pyruvate transaminase (Szepesi *et al.*, 1970); at least one of the four glutamine aminotransferase activities described in rat kidney reacts *in vitro* with pyruvate as keto—acceptor (Cooper and Meister, 1974).

Two other potential pathways have been described. The first, proposed by Whelan *et al.*, (1969), involves the formation and subsequent deamination of γ-aminobutyric acid; although some ammonia is indeed formed from added γ-aminobutyrate, the overall rate is low and insensitive to changes in the acid-base status (Vavatsi-Manos *et al.*, 1973). The second postulates that glutamine may condense with itself to form glutamyl-glutamine with concomitant release of NH_3 (Welbourne, 1974). This activity, glutamine glutamyltransferase, has been detected in rat kidney, possibly as a side activity of glutamine synthetase (Welbourne, 1974). It is not clear, however, whether the resulting dipeptide is subsequently hydrolyzed and whether it may eventually contribute to renal glucose production; in any case it is unlikely to be of major consequence in the acid-stressed animal (Welbourne, 1974).

The major pathways of ammoniagenesis and gluconeogenesis in rat kidney are shown in Fig. 2. Goodman *et al.* (1966) proposed that ammoniagenesis is secondary to an acidotic increase in gluconeogenesis. Acidosis increases PEP-CK activity within 4 to 6 hours of the administration of NH_4Cl; this increased activity, it was suggested, might cause a drop in the 2-oxoglutarate concentration, which in turn would lead to a drop in glutamate. Since glutamate is an inhibitor of glutaminase I (Goldstein, 1966), the activity of the latter should increase, thereby generating increased amounts of ammonia. This hypothesis, although probably no longer tenable, has proved fruitful in that it has stimulated a number of studies.

The correlation between ammonia release and glucose production is good after 48 hours in the rat but is not apparent in the first 6 hours (Alleyne, 1970) after administration of an acid load. Indeed such a correlation may not obtain at all in species other than the rat and the dog (Klahr *et al.*, 1072). The changes in glutamate concentration in acidosis are not consistent with the postulated deinhibition of glutaminase I (Pagliara and Goodman, 1970; Goodman, 1972; Irias and Greenberg, 1972), nor are they of sufficient magnitude to produce a significant change in enzyme activity. Although this latter point may be evaded by invoking either inter- or intracellular compartmentation of glutamate, Curthoys and Lowry (1973b) have shown that changes in this metabolite in the proximal tubule reflect those noted earlier for the whole organ. Furthermore, experiments with various inhibitors of the conversion of 2-oxoglutarate to malate (Preuss, 1968, 1969; Churchill and Malvin, 1970), although open to some criticism of experimental design, tend to indicate that ammoniagenesis and gluconeogenesis may be dissociated, at least under artificial conditions. It has yet to be shown, however, that there is no connection between the two processes under physiological conditions. It still remains an attractive hypothesis that gluconeogenesis serves as a trap for the carbon skeleton of glutamine after its deamination and deamidation.

What are the factors involved in acidosis? There are a number of possible candidates. If acidosis acts "directly," then H^+ ions, HCO_3^- ions, and pCO_2 are all possible factors. If the effect is indirect, a number of other factors may come into play, including inorganic ions such as Ca^{2+}, plasma protein components, and other hormones known to have renal actions.

At high concentrations (2.5 mM) of calcium, a reduction of pH in the incubation medium does not stimulate gluconeogenesis in rat kidney slices *in vitro* (Alleyne *et al.*, 1973). When the calcium concentraion is lowered, two

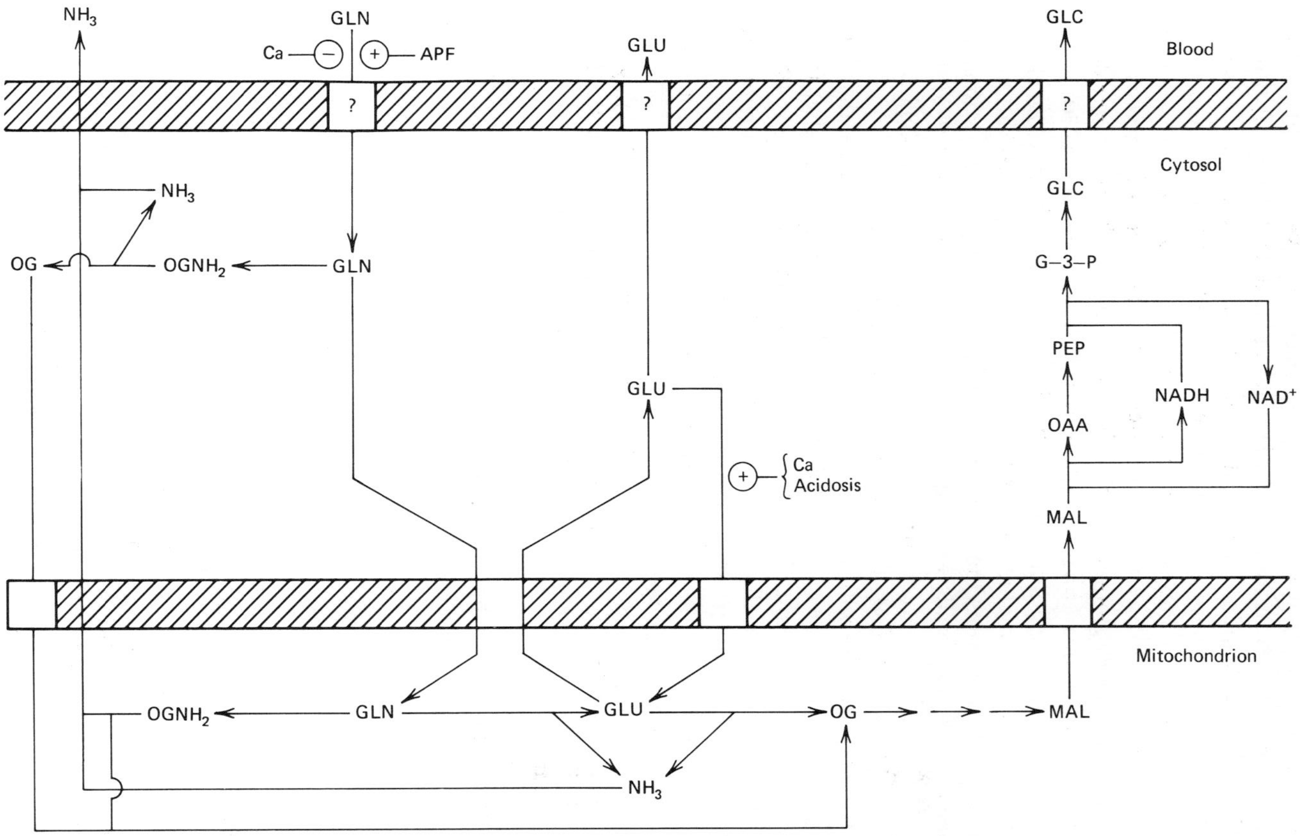

Figure 2. Pathways of ammoniagenesis and gluconeogenesis from glutamine in rat kidney cortex. Arrows represent physiological net fluxes, not equilibria. "Boxes" in the mitochondrial and plasma membranes represent specific transport systems (those marked ? are not yet characterized); ⊕ and ⊖ represent stimulation and inhibition, respectively. Abbreviations: APF, acidotic plasma factor; GLC, glucose, GLN, glutamine; GLU, glutamate; G-3-P, glyceraldehyde-3-phosphate; MAL, malate; OAA, oxaloacetate; OG, oxoglutarate; $OGNH_2$, oxoglutaramate; PEP, phosphoenolpyruvate.

effects may be observed; (1) the gluconeogenic rate drops as the calcium level decreases, and (2) gluconeogenesis becomes sensitive to the pH of the medium, being highest at more acid pH values (Table III). This effect may be attributed to a drop in $\{HCO_3^-\}$ rather than to an increase in pCO_2, since only the former produces consistent increases in glucose production from glutamine, glutamate, lactate, pyruvate, and citric acid cycle metabolites (Alleyne *et al.*, 1973). This agrees well with the marked increase in gluconeogenesis from glutamine and other substrates in the isolated kidney perfused with low bicarbonate medium at pH 7.23 (Bowman, 1970). It is perhaps pertinent to mention that most workers who have observed an effect of acidosis *in vitro* on gluconeogenesis have used Ca^{2+} concentrations of about 1 mM, and not those concentrations of Ca^{2+} which stimulate gluconeogenesis maximally (Kamm *et al.*, 1967). Alleyne *et al.* (1973) have shown that with succinate as substrate, the stimulatory effect of acidosis *in vitro* disappears as the Ca^{2+} concentration rises above 1 mM.

Similar experiments with isolated rat kidney tubules also agree that low $\{HCO_3^-\}$ is more effective than increased pCO_2, but they indicate that these agents have similar qualitative effects (Kurokawa and Rasmussen, 1973). These authors conclude that $\{H^+\}$ itself may be the true regulator: on the basis of measurements of citric acid cycle intermediates, they propose that an increase of $\{H^+\}$ inhibits citrate synthase and activates both isocitrate and succinate dehydrogenases.

Both Kurokawa and Rasmussen (1973) and Alleyne *et al.*, (1973) agree, however, that HCO_3^- may have a specific effect as an inhibitor of PEP-CK, a finding first reported by Chang *et al.* (1966). There is further agreement that

Table III. Effects of calcium and *in vitro* acidosis on gluconeogenesis in rat kidney cortex slices[a]

	Ca^{2+} Concentration		
	0	1 mM	
pH 7.4	13.6 ±2.1	28.8 ±1.4	$P < 0.001$
pH 7.0	24.9 ±2.7	31.0 ±1.4	NS
	$P < 0.005$	NS	

[a] The substrate was 2 mM-glutamine, $n = 14$ throughout. The pH 7.0 buffer was produced by lowering the $NaHCO_3$ concentration from 25 to 10 mM. Glucose synthesis rates are expressed as nanomoles per hour per milligram of dry weight; NS = not significant.

Source: Data from Roobol and Alleyne (1974).

the interaction between pH changes and HCO_3^- and Ca^{2+} concentrations is complex and may involve a number of separate regulatory loci, possibly including mitochondrial transport systems (Simpson and Angielski, 1973).

Analysis of the effect of acute (2 hour) acidosis on glutamine metabolism in rat kidney cortex slices reveals that ammoniagenesis and gluconeogenesis may be effectively separated by appropriate manipulation of the medium calcium concentration. In the presence of 1 mM Ca^{2+}, acidotic slices utilize more glutamine and form more ammonia, glucose, and glutamate than do slices from normal control rats. When Ca^{2+} is omitted, glutamine uptake and ammoniagenesis are relatively similar in both acidotic and control slices; the rate of gluconeogenesis, however, remains higher in the acidotic slice, and glutamate output is correspondingly lower (Table IV). When slices from normal animals are incubated at pH 7.4 or 7.0 (low HCO_3^-, *in vitro* acidosis) addition of 1 mM Ca^{2+} in both cases depresses glutamine uptake but has no effect on ammonia production. *In vitro* acidosis in calcium-free medium leads to stimulation of gluconeogenesis, but ammoniagenesis is still unchanged. When slices from normal or acidotic rats are incubated with glutamate as substrate, both gluconeogenesis and ammoniagenesis are higher in the acidotic slices. *In vitro* acidosis also stimulates glucose and ammonia production from glutamate; this effect, however, is observed only in the absence of calcium (Roobol and Alleyne, 1974).

The difference between results obtained with slices derived from acutely (2 hour) acidotic rats and slices rendered acidotic *in vitro* by incubation with low bicarbonate medium implies that other factors may be operative *in vivo*. When slices from normal animals are preincubated with plasma from acidotic animals, their response in the subsequent incubation with glutamine is similar to that seen with *in vivo* acidotic slices. Control experiments (i.e., preincubation with normal plasma or acidified Krebs bicarbonate buffer) demonstrate that this is a specific effect. When the acidotic plasma is neutralized with $NaHCO_3$, ammoniagenesis remains significantly increased, although the stimulation of gluconeogenesis is reduced to a barely significant level. Acidotic plasma remains effective both after storage at 5°C for 24 hours and dialysis against standard incubation medium (Table V; Alleyne and Roobol, 1974). Figure 2 is a summary of the reactions and control points of ammoniagenesis and gluconeogenesis based on data presently available.

IV. ACIDOSIS AND KIDNEY PEP-CK

Although gluconeogenesis can be shown in increase within 2 hours of the administration of NH_4Cl, the assayable activity of PEP-CK does not show

Table IV. Effects of calcium and *in vivo* acidosis on gluconeogenesis in rat kidney cortex slices[a]

		Net metabolic changes			
	Medium (Ca^{2+})	Glutamine uptake	Glutamate formed	Ammonia formed	Glucose formed
Control	1.0 mM	231 ±6	47.9 ±5.4	298 ±11	27.6 ±1.4
Acidotic	1.0 mM	275 ±9	63.4 ±4.5	361 ±12	34.1 ±2.0
Control	0	275 ±11	73.0 ±7.4	293 ±9	13.3 ±1.4
Acidotic	0	274 ±12	55.0 ±3.4	295 ±7	19.8 ±0.9
Control vs. acidotic (1 mM Ca^{2+})		$P < 0.001$	$P < 0.05$	$P < 0.001$	$P < 0.02$
Control vs. acidotic (no Ca^{2+})		NS	$P < 0.05$	NS	$P < 0.001$
Ca^{2+} (1 mM vs. none, controls)		$P < 0.005$	$P < 0.02$	NS	$P < 0.001$
Ca^{2+} (1 mM vs. none, acidotic)		NS	NS	$P < 0.001$	$P < 0.001$

[a] The substrate was 2 mM-glutamine, $n = 17$ throughout. Metabolic changes are expressed as nanomoles per hour per milligram of dry weight; NS = not significant. *Source:* Data from Roobol and Alleyne (1974).

significant change until the fourth to sixth hours. After this lag period activity rises rapidly to reach a near-maximal value at between 24 and 48 hours after treatment (Longshaw *et al.*, 1972b). This is accompanied by a sustained increase in the rate of gluconeogenesis in slices incubated *in vitro* (Alleyne and Scullard, 1969).

Of a number of enzymes tested, only PEP-CK and glutaminase I (Goldstein, 1965) exhibit increased activity in acidosis. This contrasts with the action of the synthetic glucocorticoid, triamcinolone, which increases both glucose-6-phosphatase and PEP-CK in kidney. A further contrast is seen between the responses of the same enzymes in liver and kidney (Table VI). Glucocorticoid reproducibly *increases* kidney PEP-CK but has variable effects on the liver enzyme, being reported variously to increase activity

(Shrago *et al.*, 1963; Henning *et al.*, 1966; Longshaw and Pogson, 1972), decrease activity (Reshef *et al.*, 1969), or have little significant effect (S. A. Smith and C. I. Pogson, unpublished results). Such results may depend rather on the strain of rat and the constitution of the diet and other environmental factors than on individual experimental expertise.

Acidosis and triamcinolone at maximally effective dosages produce additive effects on kidney PEP-CK, suggesting that the points of action of the two stimuli may be different. A further indication of this is obtained from related experiments with actinomycin D (Fig. 3). Steroid induction of rat kidney PEP-CK is strongly inhibited by actinomycin D at relatively low concentrations; that due to acidosis is relatively insensitive under the same conditions (Flores and Alleyne, 1971; Longshaw and Pogson, 1972). In animals treated with both triamcinolone and NH_4Cl, actinomycin D produces partial inhibition, presumably indicative of the blocking of the steroid-mediated portion of the response. A simple interpretation of these experiments is, however, precluded by the demonstration that actinomycin D at similar dosages is an effective blocker of the *early* rise of PEP-CK

Table V. Effects of preincubation of rat kidney cortex slices in rat plasma on gluconeogenesis and ammoniagenesis[a]

Preincubation medium	Glucose formed	Ammonia formed
Normal rat plasma	27.6 ±1.5(17)	258 ±8(27)
Plasma from acutely acidotic rats	35.6 ±2.0(18) ($P < 0.005$)	329 ±10(27) ($P < 0.001$)
Plasma from acutely acidotic rats (brought to pH 7.48 with $NaHCO_3$)	30.2 ±2.0(12) (NS, $P < 0.1$)	315 ±19(12) ($P < 0.005$)
Normal rat plasma, 5°C for 24 hours	28.2 ±1.3(6)	213 ±8(6)
Plasma from acutely acidotic rats, 5°C for 24 hours	35.2 ±1.1(6) ($P < 0.001$)	262 ±12(6) ($P < 0.005$)
Normal rat plasma, dialyzed	28.5 ±1.9(6)	225 ±9(6)
Plasma from acutely acidotic rats, dialyzed	36.1 ±2.4(6) ($P < 0.05$)	267 ±11(6) ($P < 0.005$)

[a] The substrate was 2 mM-glutamate; the number of determinations is shown in parentheses. Rates are expressed as nanomoles per hour per milligram of dry weight; NS = not significant.
Source: Data from Alleyne and Roobol (1973).

Table VI. Induction of gluconeogenic enzymes in liver and kidney by steroid and acidosis[a]

	Enzyme activities (μmole/min/g wet wt)					
	Control	Triamcinolone	% Increase	Control	Acidosis	% Increase
Liver						
Glucose-6-phosphatase	7.5 ±0.3	8.1 ±0.4	8 (NS)	10.5 ±0.8	9.0 ±0.4	−14 (NS)
Fructose-1,6-diphosphatase	9.9 ±0.9	10.4 ±0.3	5 (NS)	12.7 ±0.3	12.2 ±0.3	−4 (NS)
PEP-CK	2.1 ±0.2	4.4 ±0.5	110 ($P < 0.01$)	6.1 ±0.3	7.4 ±0.7	21 (NS)
Kidney						
Glucose-6-phosphatase	3.6 ±0.4	8.2 ±0.3	128 ($P < 0.005$)	4.2 ±0.1	4.3 ±0.2	2 (NS)
Fructose-1,6-diphosphatase	11.3 ±0.1	12.4 ±0.8	10 (NS)	11.9 ±0.3	10.2 ±0.7	−14 (NS)
PEP-CK	2.6 ±0.2	11.3 ±0.4	335 ($P < 0.001$)	7.4 ±0.5	24.9 ±1.0	236 ($P < 0.001$)

[a] Triamcinolone (2.5 mg/100 g rat) and NH_4Cl (5 ml 0.4 M/200 g rat) were given at 12 hour intervals. Rats (150–200 g) were killed at 48 hours. Steroid and acidosis experiments were performed with different rat populations; results are means ±SEM, $n = 3$ (steroid) or 4 (acidosis) in each case.

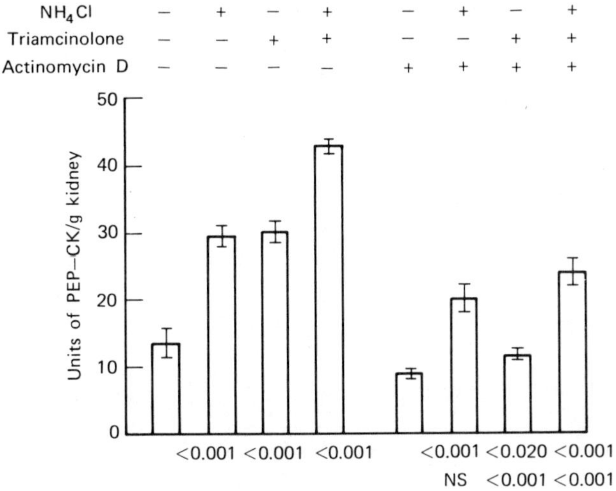

Figure 3. Effects of steroid, acidosis, and actinomycin D on renal PEP-CK activity. Animals were killed 48 hours from the start of the experiment. Maximally stimulating doses of triamcinolone and NH_4Cl were given 4 times, 12 hours apart. Actinomycin D was given 30 minutes before each treatment at a dosage of 15 μg/100 g rat. Results are means ± SEM of six determinations; **P*, significance of the differences between the means in treated animals and those in appropriate control groups; ***P*, significance of the differences between the mean increments in animals treated with actinomycin D the mean increments in animals similarly treated but not given actinomycin D. *Source:* Longshaw and Pogson (1972).

activity—that is, during the first 12 hours after induction of acidosis (C. I. Pogson and R. W. Hanson, unpublished results). Furthermore, the use of actinomycin D has been much criticized in regard to the specificity of its action (e.g., Georgiev *et al.*, 1963), and there is at present too little information about the *in vivo* time course of action and clearance of such inhibitors. It appears, however, that a substantial proportion of injected actinomycin D remains in the rat after 24 hours (Philips *et al.*, 1960) and that the inhibitory action may persist beyond clearance of the molecule (Sawacki and Godman, 1972).

Experiments with translational inhibitors, also, are less clear-cut than one would like. Ethionine appears to block the steroid-induced increase in activity but not that due to acidosis (Flores and Alleyne, 1971). Cycloheximide, on the other hand, a more potent and specific inhibitor (Farber, 1971; Harris *et al.*, 1968), blocks both steroid and acidotic effects completely (Table VII).

It thus appears that there are both quantitative and qualitative differences between the induction processes due to steroid and to acidosis. For further

Table VII. Inhibition of the increase in rat kidney PEP-CK activity following acidosis by cycloheximide

Experimental conditions	Enzyme activity (μmole/min/g tissue) "Backward"	"Forward"
Experiment A[a]		
Control	4.3 ±0.5 (7)	5.9 ±0.9 (7)
Cycloheximide	4.0 ±0.5 (6)	5.7 ±0.8 (6)
Acidosis	7.7 ±0.4 (7)[b]	12.2 ±0.7 (7)[b]
Acidosis plus cycloheximide	4.1 ±0.9 (6)	6.3 ±1.2 (6)
Experiment B[c]		
Control	5.2 ±0.2 (4)	
Acidosis plus cycloheximide	0 hours 4.7 ±0.6 (4)	
	1 hour 3.8 (mean of 2)	
	2 hours 4.2 (mean of 2)	
	4 hours 5.0 ±0.2 (4)	
	5 hours 6.3 ±0.7 (4)	
	6 hours 7.4 ±0.5 (4)	
	7 hours 8.0 ±0.3 (4)	
Acidosis alone	9.2 ±0.9 (4)	

[a] Male rats weighing 200 g each were given two doses *per os* of 5 ml of a 0.4 M solution of either NaCl (control) or NH_4Cl (acidosis) at zero time and 12 hours. Cycloheximide (200 μg) was simultaneously administered intraperitoneally in 0.9% NaCl with appropriate controls. Animals were killed at 24 hours; kidney PEP-CK activities were assayed in both the "backward" (Ballard and Hanson, 1967) and "forward" (Seubert and Huth, 1965) directions. Figures in parentheses indicate the number of determinations in each case.

[b] P vs. control < 0.001.

[c] Male rats weighing 400 g each were given an equivalent dose of either NaCl or NH_4Cl and were killed 7.5 hours later. Cycloheximide (100 μg/100 g rat in 0.9% NaCl) was administered intraperitoneally at the times stated.

clarification of this problem, however, we must determine the relative *in vivo* effectiveness in terms of biological half-lives of the various effectors employed in such studies. It seems clear, for example, that the time dependence of triamcinolone clearance from the plasma, binding to tissues, and subsequent elimination (Florini *et al.*, 1961) is very distinct from that of the pH changes due to NH_4Cl administration (Fig. 4).

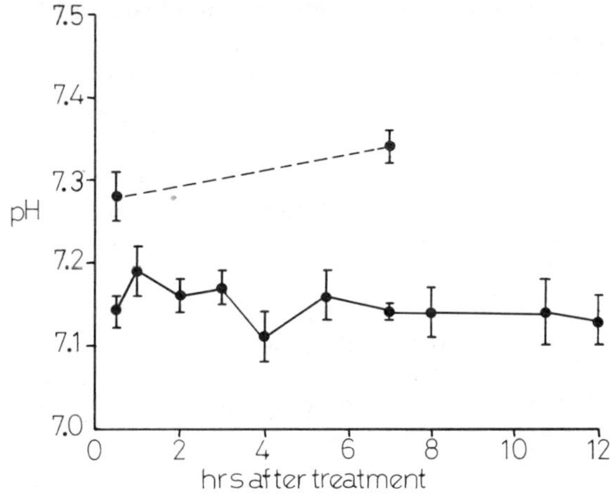

Figure 4. Time course of blood of pH changes after a single dose of NH_4Cl *per os* (400 g rats received 10 ml 0.4 M NH_4Cl). The dotted line joins values obtained from control animals given equivalent amounts of NaCl. Results are means ± SEM for four or five determinations.

V. THE INFLUENCE OF OTHER HORMONAL AND DIETARY FACTORS

It is doubtful whether insulin has any great activity on renal cortex. Acute 48 hour diabetes produced by administration of alloxan (60 mg/kg, i.v.) produces either no significant change or a slight rise only in kidney PEP-CK. In no case is this small effect comparable to that in the liver of the same animal (Tables VIII and XII). Experiments with animals in which diabetes was induced in a less acute form show somewhat larger increases in kidney PEP-CK (Anderson and Stowring, 1973; Murphy and Anderson, 1974, Table IX). The extent to which these longer term changes are due to hormonal factors other than insulin, altered in the diabetic state, is not clear.

Short-term starvation produces a marked increase in the activities of both liver and kidney PEP-CK (Hanson and Garber, 1972). This does not seem to be mediated through changes in circulating steroid levels, since the increment is similar in both normal and adrenalectomized animals (Table IX). It is similarly unlikely that insulin is involved, for the reasons given in the preceding paragraph. Kamm and Cahill (1969) have suggested that the enhancement of renal gluconeogenesis in starvation may be at least partly a consequence of the accompanying acidosis. Although such acidosis may play a role in the "fine" controls involved in the increase, it seems improba-

Table VIII. Effects of alloxan diabetes and acidosis on PEP-CK activity in rat liver and kidney[a]

Experimental conditions	Enzyme activity (μmole/min/g tissue) Liver	Kidney	Blood glucose (mg %)	n
Experiment A[b]				
Control	1.8 ±0.1	7.3 ±0.6	109 ±4	6
Acidosis	1.6 ±0.2	21.7 ±0.7[c]	107 ±2	6
Diabetes	8.8 ±0.6[c]	7.9 ±0.6	633 ±89[c]	7
Diabetes plus acidosis	10.4 ±0.6[c]	24.9 ±1.0[c]	661 ±43[c]	9
Experiment B[d]				
Control		7.4 ±0.8		11
Diabetes		15.1 ±1.0[c]		16

[a] All rats were male and weighed between 175 and 250 g.
[b] Acidosis was induced by administration *per os* of 2.5 ml of 0.4 M NH_4Cl per 100 g rat at 12 and 24 hours before killing. Controls received similar volumes of 0.4 M NaCl. Diabetes was induced by intravenous injection of 60 mg/kg rat of fresh alloxan in 0.9% NaCl, 44 hours before killing; PEP-CK activity was determined by the method of Ballard and Hanson (1967).
[c] P versus appropriate control < 0.001; all other differences not significant.
[d] These data have been recalculated from Anderson and Stowring (1973). Diabetes was induced by intraperitoneal injection of alloxan (100–125 mg/kg rat) after a 60 hour fast. Animals were killed after 25 to 35 days. PEP-CK was assayed substantially as described by Ballard and Hanson (1967).

ble that the concomitant PEP-CK increases are a consequence of this; animals simultaneously starved and made acidotic with NH_4Cl exhibit a partial additive response to these two stimuli (Table IX).

VI. EFFECTS OF L-TRYPTOPHAN AND ITS METABOLITES ON KIDNEY PEP-CK AND GLUCONEOGENESIS

Tryptophan administration to rats *in vivo* increases the activity of liver PEP-CK as measured by the formation of PEP (Foster *et al.*, 1967). This effect is biphasic. The initial rapid rise is maximal at 30 to 60 minutes and is insensitive to actinomycin D; the second phase is slower and is inhibited

Table IX. Effects of adrenalectomy, starvation, acidosis, and refeeding on PEP-CK activities in rat liver and kidney[a]

	Enzyme activity[b] (μmole/min/g tissue)	
Experimental conditions	Liver	Kidney
Experiment A		
Control	1.05 ±0.09(4)	4.2 ±0.1(8)
18-hour-starved	7.1 ±0.2(8)***	12.3 ±0.8(8)***
Experiment B		
48-hour-starved	6.7 ±0.2(6)	13.1 ±0.3(6)
48-hour-starved, 24 hours, refed	0.61 ±0.05(6)***	7.7 ±0.5(6)***
Experiment C		
1. Control	0.72 ±0.04(5)	2.9 ±0.1(5)
2. 18-hour-starved	4.7 ±0.2(5)***	8.1 ±0.3(5)***
3. Adrenalectomized	0.98 ±0.09(7)*	2.3 ±0.1(7)**
4. Adrenalectomized, 18-hour-starved	6.2 ±0.3(7) (vs. 2**, vs. 3***)	8.1 ±0.3(7) (vs. 2 NS, vs. 3***)
Experiment D		
1. Adrenalectomized	1.19 ±0.12(7)	4.6 ±0.2(7)
2. Adrenalectomized, 18-hour-starved	8.0 ±0.2(7)***	14.6 ±0.3(7)***
3. Adrenalectomized acidosis	2.1 ±0.2(7)**	13.7 ±0.5(7)***
4. Adrenalectomized 18-hour-starved, acidosis	8.7 ±0.7(6) (vs. 2 NS, vs. 3***)	19.8 ±0.7(6) (vs. 2***, vs. 3***)

[a] Male rats weighing 175–200 g were used throughout. In experiment B refed animals received a diet of 70% casein, 30% commercial defatted dried milk *ad libitum*. Adrenalectomized animals were given 1% NaCl to drink and were used 5 to 9 days after operation. Twenty-four hour-acidosis was induced as described in footnote *b*, Table VIII. PEP-CK activities were measured by the method of Ballard and Hanson (1967).

[b] Significance (against controls unless otherwise specified): * $P < 0.05$, ** $P < 0.005$, *** $P < 0.001$; NS = not significant.

by actinomycin D. The transcription-independent" phase has been attributed to the formation of quinolinate (Veneziale *et al.*, 1967), a metabolite that accumulates under conditions of tryptophan loading (McDaniel *et al.*, 1972, 1973; Snoke *et al.*, 1971). In an apparently paradoxical manner, tryptophan also inhibits hepatic gluconeogenesis both *in vivo* and *in vitro* (Ray *et al.*, 1966; Veneziale *et al.*, 1967), and the locus of action has been identified as the PEP-CK-catalyzed step (Ray *et al.*, 1966). Again, a series of tryptophan metabolites also mediates this effect, and quinolinate has been proposed as the direct mediator, possibly through chelation of Fe^{2+}, which activates PEP-CK (Foster *et al.*, 1967; Snoke *et al.*, 1971).

We have determined PEP-CK activities in both forward and backward reactions in the livers and kidneys of fed rats treated with tryptophan for 60 minutes. Although we were able to confirm a 150% increase in "forward" activity in the liver, there was no similar change in kidney (Pogson and Smith, 1975; Table X). In *in vitro* experiments with kidney cortex slices, L-tryptophan exerted no significant effect on gluconeogenesis from lactate, even at 2.5 mM (Table XI). This compares with the pronounced inhibition seen in isolated rat liver perfusion experiments (Veneziale *et al.*, 1967) and with isolated mouse liver cells. Quinolinate, however, is markedly more effective in this system than it is with liver, producing 50% inhibition, with lactate as substrate, at approximately 0.1 mM (Table XI). This may be attributed to the greater permeability of the kidney cortex, as opposed to the liver parenchymal, plasma membrane (Hagino *et al.*, 1968). The observed potency is consistent with the reported K_i for quinolinate inhibition of PEP-CK (McDaniel *et al.*, 1972). Quinolinate is less effective with certain citric acid cycle intermediates; this is again consistent with a pattern of inhibition competitive with oxaloacetate (Pogson *et al.*, 1976). We failed to find any significant effect of quinolinate on either mitochondrial electron transport or glutamate dehydrogenase (cf. Klahr and Schoolwerth, 1972), but we cannot exclude the mitochondrial metabolite porter systems as possible further points of action.

A further contrast with earlier studies on mouse liver cells is that kidney gluconeogenesis in the rat is almost totally insensitive toward metabolites of the indole-3-acetate pathway (Pogson *et al.*, 1976).

Since tryptophan oxygenase is absent from kidney cortex, circulating tryptophan cannot give rise directly to metabolites of the kynurenine-quinolinate pathway. Kidney cortex does, however, contain appreciable levels of kynureninase and kynurenine-3-hydroxylase (De Castro *et al.*, 1957), a finding that gives rise to the speculation that tryptophan may be converted to kynurenine in the liver and that a portion of this is subsequently metabolized in the kidney (Rose, 1972). If the action of tryp-

Table X. Effects of L-tryptophan administration on PEP-CK activities in the liver and kidneys of normal rats[a]

Tissue	Experimental conditions	PEP-CK activity[b] (μmole/min/g tissue)			Ratios[b]	
		A	B	C	A/B	A/C
Liver	Control	0.89 ±0.11	0.18 ±0.02	1.34 ±0.23	4.9 ±0.2	0.71 ±0.07
	Tryptophan treated	1.22 ±0.06*	0.45 ±0.03***	3.22 ±0.33***	3.1 ±0.2***	0.39 ±0.02**
Kidney	Control	7.02 ±0.23	0.76 ±0.08	7.55 ±0.50	9.6 ±0.8	0.95 ±0.06
	Tryptophan treated	6.50 ±0.31	0.78 ±0.06	8.92 ±0.25*	8.5 ±0.7	0.73 ±0.02*

[a] Male rats weighing 200 g each were injected intraperitoneally at 8 A.M. with 0.8 g/kg L-tryptophan; controls received a similar volume of suspending medium. All animals were killed 1 hour later, and PEP-CK activities were measured in three assay systems: A, Ballard and Hanson (1967); B, ^{32}P incorporation from GT-^{32}P into PEP, a modification from Walsh and Chen (1971); C, Seubert and Huth (1965). All values are the means ±SEM for six determinations.

[b] Significances * $P < 0.05$, ** $P < 0.005$, *** $P < 0.001$, other differences not significant.

Source: Data from Pogson and Smith (1975).

Table XI. Effects of tryptophan, indole-3-acetate, and quinolinate on gluconeogenesis in rat kidney cortex slices[a]

Substrate (10 mM)	Additions	Rates of gluconeogenesis (μmole glucose/g dry wt/hr) Control	Plus addition	n	% Inhibition	P
Experiment A						
L-Lactate	L-tryptophan (2.5 mM)	61 ±8	57 ±7	6	—	NS
	Quinolinate (2.5 mM)	—	7 ±1	6	89	< 0.001
Experiment B						
L-Lactate	Indole-3-acetate (0.5 mM)	72 ±13	77 ±6	4	—	NS
	Indole-3-acetate (0.5 mM)	—	55 ±7	4	24	NS
	Quinolinate (0.1 mM)	—	35 ±7	4	51	< 0.05
	Quinolinate (0.5 mM)	—	17 ±5	4	76	< 0.01
	Quinolinate (1.0 mM)	—	5 ±4	4	93	< 0.005
Experiment C						
L-Lactate	Quinolinate (1.0 mM)	72 ±4	14 ±1	24	81	< 0.001
Pyruvate	Quinolinate (1.0 mM)	199 ±11	76 ±2	6	62	< 0.001
Citrate	Quinolinate (1.0 mM)	45 ±3	24 ±4	6	47	< 0.005
2-Oxoglutarate	Quinolinate (1.0 mM)	149 ±4	109 ±4	12	27	< 0.001
Succinate	Quinolinate (1.0 mM)	205 ±8	150 ±5	6	27	< 0.001
L-Malate	Quinolinate (1.0 mM)	202 ±13	166 ±8	6	17	< 0.05
L-Proline	Quinolinate (1.0 mM)	25 ±3	7 ±4	6	72	< 0.005
L-Glutamate	Quinolinate (1.0 mM)	73 ±5	22 ±4	12	70	< 0.001
L-Glutamine	Quinolinate (1.0 mM)	40 ±3	−8 ±3	12	100	< 0.001
Glycerol	Quinolinate (1.0 mM)	48 ±8	54 ±10	6	—	NS
D-Fructose	Quinolinate (1.0 mM)	262 ±13	252 ±21	6	—	NS

[a] Male rats weighing between 150 and 200 g were starved overnight before use. The conditions of incubation, and so on, were as described by Longshaw *et al.* (1972a); NS = not significant.

tophan on PEP-CK activity is indeed attributable exclusively to its conversion into quinolinate, as has been proposed, the failure to observe any increase in assayable activity 1 hour after the administration of the amino acid is evidence that quinolinate production in kidney is substantially less than that in liver. It is unlikely, therefore, that formation of quinolinate from tryptophan plays a major role in the regulation of kidney gluconeogenesis.

VII. PHYSICAL AND KINETIC PROPERTIES OF KIDNEY PEP-CK

Ballard and Hanson (1969) have shown that rat liver and adipose tissue cytosolic PEP-CKs are immunologically identical. Using antibody generously provided by Dr. Hanson, we have confirmed that kidney cortex cytosol PEP-CK is also indistinguishable from that in the liver (Longshaw and Pogson, 1972). We have further shown that increases in total tissue activity, whether elicited by steroid or acidosis, are parallelled by the increases in antibody titer.

The K_M values of cytosolic PEP-CK for oxaloacetate display variation over a considerable range (5–300 μM; Ballard, 1970; Chang *et al.*, 1966; Foster *et al.*, 1967; Walsh and Chen, 1971). There are fewer reports of the K_M for the nucleotide substrate, but once more the apparent divergence is considerable (Foster *et al.*, 1967; Guder and Schmidt, 1974). These large discrepancies are less likely to have arisen from simple experimental error; more probably, they are a result of the several different assay systems employed. In view of the ratio of the rate of gluconeogenic flux to the maximum measureable activity of PEP-CK, and allowing for some recycling through the pyruvate kinase "energetically wasteful" ATPase cycle (Rognstad and Katz, 1972), it seems probable that the lower values may more truly represent the *in vivo* condition.

Several of the PEP-CK assay systems used measure the rate of the "back" reaction (i.e., the reaction in the direction of oxalacetate production). Values so obtained frequently have been converted to give forward rates by multiplying by the appropriate factor (8.3 to 18.5) as determined for the homogeneous pig liver mitochondrial enzyme by Chang *et al.*, (1966). In the course of our studies it was noted that the ratios of the rates of the backward and forward reactions in the same sample are consistently at variance with the expected values. In a more systematic study (Pogson and Smith, 1975), we have measured PEP-CK activities in liver and kidney under a range of dietary and hormonal conditions. Representative data are shown in Table XII. In general, variations in hormonal status do not affect the ratio for either of the two tissues. The only conditions effective in this respect are

Table XII. The effects of diabetes and starvation on PEP-CK activities in rat liver and kidney[a]

	PEP-CK activity (μmole/min/g tissue)[b]			Ratios[b]		
	A	B	C	A/B	A/C	C/B
Experiment I						
Normal fed liver	1.47 ±0.08(70)	0.58 ±0.08(70)	2.39 ±0.18(35)	3.7 ±0.24(70)	0.60 ±0.03(35)	6.6 ±0.41(35)
Normal fed, kidney	5.33 ±0.15†(89)	0.74 ±0.03*(82)	5.86 ±0.24†(54)	7.3 ±0.21†(82)	0.91 ±0.03†(54)	8.7 ±0.29†(47)
Experiment II						
Control, liver	1.54 ±0.19(8)	0.44 ±0.05(8)	2.92 ±0.47(8)	3.8 ±0.4(8)	0.56 ±0.04(8)	6.7 ±0.6(8)
Diabetic liver	7.22 ±0.84†(8)	1.70 ±0.19†(8)	12.8 ±1.5†(8)	4.5 ±0.1(8)	0.57 ±0.02(8)	8.0 ±0.3(8)
Control, kidney	3.30 ±0.20(8)	0.44 ±0.02(8)	4.83 ±0.37(8)	7.5 ±0.3(8)	0.69 ±0.02(8)	10.8 ±0.6(8)
Diabetic kidney	5.37 ±0.67**(8)	0.71 ±0.08***(8)	8.25 ±1.2*(8)	7.6 ±0.6(8)	0.65 ±0.01(8)	11.6 ±0.6(8)

Experiment III				
Control, liver	1.15 ±0.09(4)	0.73 ±0.18(4)		1.7 ±0.3(4)
18-hour-starved liver	7.1 ±0.2†(8)	1.21 ±0.09*(8)		6.1 ±0.5†(8)
Control kidney	4.2 ±0.1(8)	0.58 ±0.04(8)		7.6 ±0.6(8)
18-hour-starved kidney	12.3 ±0.8†(8)	1.02 ±0.11***(8)		12.6 ±1.0†(8)

[a] Male rats weighing between 150 and 250 g were used throughout. PEP-CK was assayed as follows: A, method of Ballard and Hanson (1967), "backward"; B, method modified from Walsh and Chen (1971), "forward"; C, method of Seubert and Huth (1965), "forward." Results are means ± SEM; the number of determinations is given in parentheses. Differences (against appropriate controls in experiments II and III, and for liver vs. kidney in experiment I) are insignificant unless marked.

[b] Significances = * $P < 0.05$, ** $P < 0.01$, *** $P < 0.005$, † $P < 0.001$.

Source: Data from Pogson and Smith (1975).

tryptophan administration and, less reproducibly, starvation. It is not apparent why the ratios for liver and kidney should be distinct. Values for such ratios are remarkably reproducible in any given experiment (see Table XII, experiments II and III), but exhibit wider variation from day to day and from one group of animals to another (Table XII, experiment I). Since the enzyme activities arise from apparently identical proteins, it is conceivable that some effector with actions analogous to those of quinolinate may be involved in one or both of these tissues. The relevance of such findings to the *in vivo* situation is similarly a matter of conjecture.

Although we now have strong evidence that the reaction catalyzed by PEP-CK is a rate-limiting step, determinations of PEP-CK composition and structure have generally and unexpectedly indicated that the protein is a monomer of molecular weight 55,000 to 75,000 daltons (Ballard and Hanson, 1969; Chang and Lane, 1966; Foster *et al.*, 1967). Data derived from sedimentation experiments with sucrose gradients and gel filtration are in agreement with a molecular weight also in this region for rat kidney PEP-CK (Longshaw and Pogson, 1972). These methods involve dilution of the sample with water or buffer, and the possible introduction of nonphysiological effects. Using the press-juice method of Amberson *et al.* (1964), however, we have obtained an undiluted cytosol fraction from rat kidney cortex. When this is subjected to sedimentation in the moving-partition cell designed by Yphantis and Waugh (1956) for the analytical ultracentrifuge, proteins are separated on the basis of their molecular characteristics. From measurements of the activities of marker and the "unknown" enzymes in the top compartment of the cell and in the original extract, first approximation values for the apparent molecular weight can be obtained (Fig. 5). We have shown that PEP-CK from the acidotic kidney has a molecular weight of 128,000, compared with 83,000 for PEP-CK from control kidneys. These values are consistent with the presence of a proportion of dimeric, or tetrameric, species of PEP-CK in acidosis. Some indication of this is also discernible in the earlier sucrose density gradient experiments with diluted extracts: in these the PEP-CK activity is "skewed" because of an extended "leading edge" (Longshaw and Pogson, 1972).

VIII. CONTROL OF PEP-CK IN KIDNEY

Exton *et al.* (1971) have demonstrated that hormones producing an increase in liver intracellular cyclic AMP also increase the rate of glucose synthesis from C_3 and C_4 precursors. This increase is accompanied by a metabolic crossover between malate and phosphoenolpyruvate (Exton *et al.*, 1970, 1971; Shrago *et al.*, 1963). Although mitochondrial porter systems cannot

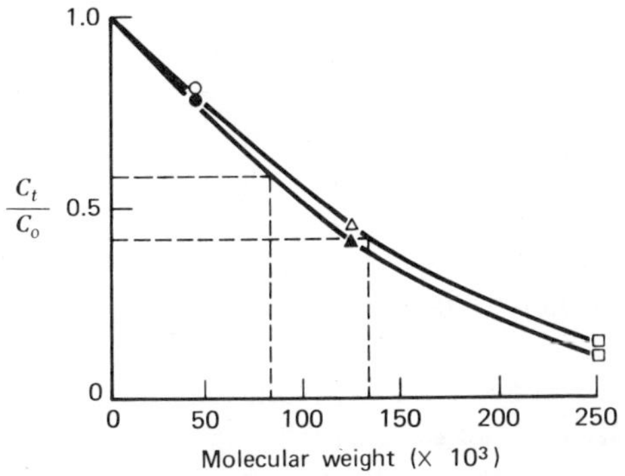

Figure 5. Typical partition cell distribution pattern obtained with undiluted extracts of acidotic (open symbols) and normal (closed symbols) rat kidney cortices. Dotted lines indicate C_t/C_o (ratio of enzyme activity in top compartment to that in the original extract) obtained for PEP-CK activities in the two cases. For further discussion see text and Longshaw and Pogson (1972), from which this figure is taken. Symbols: ○●, triosephosphate isomerase; △▲, lactate dehydrogenase; □■, catalase.

be excluded as possible control sites, these and related experiments have been taken as evidence of the regulatory role of PEP-CK.

Rasmussen has shown analogously for kidney that parathyroid hormone increases both gluconeogenesis and cyclic AMP, although the site of action is less clear (Nagata and Rasmussen 1970a, 1970b; Rasmussen, 1970). Since a direct action of cyclic AMP on PEP-CK has not been demonstrable (Exton *et al.*, 1971), one is drawn to the possibility of interconvertible enzymic forms. In this context a phosphorylation-dephosphorylation mechanism, analogous to that known for phosphorylase and glycogen synthetase, is the most obvious, although other systems, such as the adenylylation-deadenylylation and uridylylation-deuridylylation cycles of *E. coli* glutamine synthetase interconversion should be considered (Mangum *et al.*, 1973). Since the two molecular weight species of PEP-CK in the acidotic kidney are clearly not in free equilibrium with each other, the experiments reported in this chapter are at least consistent with the foregoing hypothesis. In other experiments, however, we have been unable to detect any differences in PEP-CK activities extracted with, or incubated in, buffer variously containing EDTA, fluoride, ATP plus Mg^{2+} or cyclic AMP (I. D. Longshaw and C. I. Pogson, unpublished results). Wicks and Walsh (quoted in Wicks *et al.*, 1972) were equally unsuccessful with attempts to phosphorylate PEP-CK with $AT^{32}P$ in the presence of active protein kinase.

It is probable that at least two factors affect kidney PEP-CK during the onset of acidosis. The gluconeogenic flux is increased during the first 3 hours of acidosis, whereas total PEP-CK activity increases only after a longer lag period. It is tempting to relate the molecular weight changes to the short-term increase (and possibly to the plasma factors involved), although there is no direct evidence that this is the case.

Total PEP-CK activity is increased by three distinct mechanisms—namely, starvation, steroids, and acidosis. Induction by steroid involves both transcription of new messenger RNA and its translation; that due to acidosis may also involve transcription but seems to be quantitatively distinct, particularly in terms of sensitivity to actinomycin D; the mechanisms whereby activity is increased in starvation remain for the present a matter of conjecture.

These different mechanisms of induction afford a further interesting contrast between liver, kidney and adipose tissue PEP-CK. We hope that subsequent studies on the kidney enzyme will throw further light on the general mechanisms of PEP-CK regulation in all tissues.

ACKNOWLEDGMENTS

This work was supported by grants from the British Diabetic Association and the Medical Research Council to Professor P. J. Randle and Dr. C. I. Pogson, and grants by the Medical Research Council and the Wellcome Trust to Professor G. A. O. Alleyne.

REFERENCES

Addae, S. K. and Lotspeich, W. D. (1968a). *Am. J. Physiol.* **215,** 269.

Addae, S. K. and Lotspeich, W. D. (1968b). *Am. J. Physiol.* **215,** 278.

Alleyne, G. A. O. (1970). *J. Clin. Invest.* **49,** 943.

Alleyne, G. A. O. and Roobol, A. (1974). *J. Clin. Invest.* **53,** 117.

Alleyne, G. A. O. and Scullard, G. H. (1969). *J. Clin. Invest.* **48,** 364.

Alleyne, G. A. O., Flores, H., and Roobol, A. (1973). *Biochem. J.* **136,** 445.

Amberson, W. R., Bauer, A. C., Philpott, D. E., and Roisen, F. (1964). *J. Cell. Comp. Physiol.* **63,** 7.

Anderson, J. W. and Stowring, L. (1973). *Am. J. Physiol.,* **224,** 930.

Arinze, I. J., Garber, A. J., and Hanson, R. W. (1973). *J. Biol. Chem.* **248,** 2266.

Ballard, F. J. (1970). *Biochem. J.* **120,** 809.

Ballard, F. J. (1971). *Biochim. Biophys. Acta* **242,** 470.

Ballard, F. J. and Hanson, R. W. (1967). *Biochem. J.* **104,** 866.

Ballard, F. J. and Hanson, R. W. (1969). *J. Biol. Chem.* **244,** 5625.

Benoy, M. P. and Elliott, K. A. C. (1937). *Biochem. J.* **31,** 1268.

Bourke, E., Fine, A., and Scott, J. M. (1971a). *Biochem. J.* **125,** 94P.

Bourke, E., Fine, A., and Scott, J. M. (1971b). *Nat. New Biol.* **233,** 249.

Bourke, E., Frindt, G., Rubio-Paez, D., and Schremer, G. E. (1971c). *Am. J. Physiol.* **220,** 1033.

Bowman, R. H. (1970). *J. Biol. Chem.* **245,** 1604.

Cahill, G. F., Jr., Herrera, M. G., Morgan, A. P., Soeldner, J. S., Steinke, J., Levy, P. L., Reichard, P. A., Jr., and Kipnis, D. M. (1966). *J. Clin. Invest.* **45,** 1751.

Cahill, G. F., Jr. (1970). **282,** 668.

Chang, H.-C. and Lane, M. D. (1966). *J. Biol. Chem.* **241,** 2413.

Chang, H.-C., Maruyama, H., Miller, R. S., and Lane, M. D. (1966). *J. Biol. Chem.* **241,** 2421.

Chappell, J. B. (1968). *Brit. Med. Bull.* **24,** 150.

Churchill, P. C. and Malvin, R. L. (1970). *Am. J. Physiol.* **218,** 353.

Cooper, A. J. L. and Meister, A. (1974). *J. Biol. Chem.* **249,** 2554.

Curthoys, N. P. and Lowry, O. H. (1973a). *J. Biol. Chem.* **248,** 162.

Curthoys, N. P. and Lowry, O. H. (1973b). *Am. J. Physiol.* **224,** 884.

De Castro, F. T., Brown, R. R., and Price, J. M. (1957). *J. Biol. Chem.* **228,** 777.

Diesterhaft, M. D., Shrago, E., Sallach, H. J. (1971). *Biochem. Med.* **5,** 297.

Exton, J. H. (1972). *Metabolism* **21,** 945.

Exton, J. H. and Park, C. R. (1968). *J. Biol. Chem.* **243,** 4189.

Exton, J. H., Mallette, L. E., Jefferson, L. S., Wong, E. H. A., Friedmann, N., Miller, T. B., Jr., and Park, C. R. (1970). *Rec. Prog. Horm. Res.* **26,** 411.

Exton, J. H., Ui, M., Lewis, S. B., and Park, C. R. (1971). In *Regulation of Gluconeogenesis, Ninth Conference of the Gesellschaft für Biologische Chemie,* H.-D. Söling and B. Willms, (eds.), Academic Press, New York, p. 160.

Farber, E. (1971). *Ann Rev. Pharmacol.* **11,** 71.

Felig, P. and Wahren, J. (1971). *J. Clin. Invest.* **50,** 1702.

Flores, H. and Alleyne, G. A. O. (1971). *Biochem. J.* **123,** 35.

Florini, J. R., Peets, E. A., and Buyske, D. A. (1961). *J. Pharmacol. Exp. Ther.* **131,** 287.

Foster, D. O., Lardy, H. A. Ray, P. D., and Johnston, J. B. (1967). *Biochemistry* **6,** 2120.

Garber, A. J. and Hanson, R. W. (1971). *J. Biol. Chem.* **246,** 589.

Garrison, J. C. and Haynes, R. C. (1973). *J. Biol. Chem.* **248,** 5333.

Georgiev, G. P., Samarina, O. P., Lerman, M. I., Smirnov. M. N., and Severtzov, A. N. (1963). *Nature (Lond.)* **200,** 1291.

Goldstein, L. (1965). *Nature (Lond.)* **205,** 1330.

Goldstein, L. (1966). *Am. J. Physiol.* **210,** 661.

Goodman, A. D. (1972). *Isr. J. Med. Sci.* **8,** 285.

Goodman, A. D., Fuisz, R. E., and Cahill, G. F., Jr. (1966). *J. Clin. Invest.* **45,** 612.

Goorno, W. R., Rector, F. C., and Seldin, D. W. (1967). *Am. J. Physiol.* **213,** 969.

Guder, W. G. and Wieland, O. (1971). In *Regulation of Gluconeogenesis, Ninth Conference of the Gesellschaft für Biologische Chemie,* H.-D. Söling, B. Willms, (eds.), Academic Press, New York, p. 226.

Guder, W. G. and Schmidt, U. (1974). *Hoppe-Seyler's Z. Physiol. Chem.* **355,** 273.

Hagino, Y., Lan, S. Y., Ng, C. Y., and Henderson, L. M. (1968). *J. Biol. Chem.* **243,** 4980.

Hanson, R. W. and Garber, A. J. (1972). *Am. J. Clin. Nutr.* **25,** 1010.

Harris, C., Grady, H. and Svoboda, D. (1968) *J. Ultrastruct. Res.* **22,** 240.

Hems, D. A. and Gaja, G. (1972). *Biochem. J.* **128,** 421.

Hems, D. A. and Whitton, P. D. (1973). *Biochem. J.* **136,** 705.

Henning, H. V., Stumpf, B., Ohly, B., and Seubert, W. (1966). *Biochem. Z.* **344,** 274.

Holten, D. D. and Nordlie, R. C. (1965). *Biochemistry,* **4,** 723.

Irias, J. J. and Greenberg, R. E. (1972). *Am. J. Physiol.* **223,** 750.

Jacobson, N. O., Jorgensen, F., and Thomsen, A. C. (1967). *J. Histochem. Cytochem.* **15,** 456.

Johnson, M. E. M., Das, N. M., Butcher, F. R., and Fain, J. M. (1972). *J. Biol. Chem.* **247,** 3229.

Joseph, P. K. and Subrahmanyam, K. (1968). *Arch. Biochem. Biophys.* **127,** 288.

Kaden, M., Oakley, N. W., and Field, J. B. (1969). *Am. J. Physiol.* **216,** 756.

Kamm, D. E. and Cahill, G. F., Jr. (1969). *Am. J. Physiol.* **216,** 1207.

Kamm, D. E., Fuisz, R. E., Goodman, A. D., and Cahill, G. F., Jr. (1967). *J. Clin. Invest.* **46,** 1172.

Klahr, S. (1971). *Am. J. Physiol.* **221,** 69.

Klahr, S. and Schoolwerth, A. C., (1972). *Biochim. Biophys. Acta* **279,** 157.

Klahr, S., Schoolwerth, A. C., and Bourgoignie, J. J. (1972). *Am. J. Physiol.* **222,** 813.

Krebs, H. A. (1935). *Biochem. J.* **29,** 1951.

Krebs, H. A. (1964). *Proc. R. Soc. B.* **159,** 545.

Krebs, H. A. and Yoshida, T. (1963). *Biochem. Z.* **338,** 241.

Krebs, H. A., Speake, R. N., and Hems, R. (1965). *Biochem. J.* **94,** 712.

Krebs, H. A., Freedland, R. A., Hems, R., and Stubbs, M. (1969). *Biochem. J.* **112,** 117.

Krebs, H. A., Bennett, D. A. H., de Gasquet, P., Gascoyne, T., and Yoshida T. (1963). *Biochem. J.* **86,** 22.

Kurokawa, K. and Massry, S. G. (1973). *J. Clin. Invest.* **52,** 961.

Kurokawa, K. and Rasmussen, H. (1973). *Biochim. Biophys. Acta* **313,** 42.

Lardy, H. A., Paetkau, V., and Walter, P. (1965). *Proc. Nat. Acad. Sci. U.S.* **53,** 1410.

London, D. R., Foley, T. H., and Webb, C. G. (1965). *Nature (Lond.)* **208,** 588.

Longley, J. B. (1969). *The Kidney, Morphology, Biochemistry, Physiology,* Vol. 1 C. Rouiller and A. F. Muller, (eds.), Academic Press, New York and London. p. 157.

Longshaw, I. D. and Pogson, C. I. (1972). *J. Clin. Invest.* **51,** 2277.

Longshaw, I. D., Bowen, N. L., and Pogson, C. I. (1972a). *Eur. J. Biochem.* **25,** 366.

Longshaw, I. D., Alleyne, G. A. O., and Pogson, C. I. (1972b). *J. Clin. Invest.* **51,** 2284.

Lund, P., Brosnan, J. T., and Eggleston, L. V. (1970). *Essays in Cell Metabolism* (W. Bartley, H. L. Kornberg, and J. R. Quayle, eds.), John Wiley, New York and London. p. 167.

Lyon, M. L. and Pitts, R. F. (1969). *Am. J. Physiol.* **216,** 117.

McDaniel, H. G., Reddy, W. J., and Boshell, B. R. (1972). *Biochim. Biophys. Acta* **276,** 543.

McDaniel, H. G., Boshell, B. R., and Reddy, W. J. (1973). *Diabetes* **22,** 713.

Mangum, J. H., Magni, G., and Stadtman, E. R. (1973). *Arch. Biochem. Biophys.* **158,** 514.

Marliss, E., Aoki, T. T., Pozefsky, T., Most, A. S., and Cahill, G. F., Jr. *J. Clin, Invest.* **50,** 814.

Meister, A. (1973). *Science* **180,** 33.

Murphy, E. D. and Anderson, J. W. (1974). *Endocrinology* **94,** 27.

Nagata, N. and Rasmussen, H. (1970a). *U.S. Proc. Nat. Acad. Sci. U.S.* **65,** 368.

Nagata, N. and Rasmussen, H. (1970b). *Biochim. Biophys. Acta* **215,** 1.

Nishiitsutsuji-Uwo, J. M., Ross, B. D., and Krebs, H. A. (1967). *Biochem. J.* **103,** 852.

Nordlie, R. C. and Lardy, H. A. (1963). *J. Biol. Chem.* **238,** 2259.

Orloff, J. and Burg, M. (1971). *Ann. Rev. Physiol.* **33,** 83.

Owen, O. E., Felig, P., Morgan, A. P., Wahren, J., and Cahill, G. F., Jr. *J. Clin. Invest.* **48,** 574.

Pagliara, A. S. and Goodman, A. D. (1969). *J. Clin. Invest.* **48,** 1408.

Pagliara, A. S. and Goodman, A. D. (1970). *J. Clin. Invest.* **49,** 1967.

Peng, Y.-S., Brooks, M., Elson, C., and Shrago, E. (1973). *J. Nutr.* **103,** 1489.

Philips, F. S., Schwartz, H. S., Sternberg, S. S., and Tan, C. T. C. (1960). *Ann. N.Y. Acad. Sci.* **89,** 348.

Pitts, R. F. (1964). *Am. J. Med.* **36,** 720.

Pogson, C. I. and Smith, S. A. (1975). *Biochem. J.* **152,** 401.

Pogson, C. I., Crisp, D. M., and Smith, S. A. (1976).

Pozefsky, T., Felig, P., Tobin, J. D., Soeldner, J. S., and Cahill, G. F., Jr. (1969). *J. Clin. Invest.* **48,** 2273.

Pretlow, T. G., Jones, J., and Dow, S. (1974). *Am. J. Pathol.* **74,** 275.

Preuss, H. G. (1968). *J. Lab. Clin. Med.* **72,** 370.

Preuss, H. G. (1968). *Nephron* **6,** 235.

Rasmussen, H. (1970). *Science* **170,** 404.

Ray, P. D., Foster, D. O., and Lardy, H. A. (1966). *J. Biol. Chem.* **241,** 3904.

Relman, A. S. (1972). *Kidney Int.* **1,** 347.

Reshef, L., Ballard, F. J., and Hanson, R. W. (1969). *J. Biol. Chem.* **244,** 5577.

Richterich, R. W. and Goldstein, L. (1958). *Am. J. Physiol.* **195,** 316.

Robinson, B. H. (1971a). *FEBS Lett.* **14,** 309.

Robinson, B. H. (1971b). *FEBS Lett.* **16,** 267.

Rognstad, R. and Katz, J. (1970). *Biochem. J.* **116,** 483.

Rognstad, R. and Katz, J. (1972). *J. Biol. Chem.* **247,** 6047.

Roobol, A. and Alleyne, G. A. O. (1972). *Biochem. J.* **129,** 231.

Roobol, A. and Alleyne, G. A. O. (1973). *Biochem. J.* **134,** 157.

Roobol, A. and Alleyne, G. A. O. (1974). *Biochim. Biophys. Acta.* **362,** 83.

Rose, D. P. (1972). *J. Clin. Pathol.* **25,** 17.

Ross, B. D., Hems, R., and Krebs, H. A. (1967). *Biochem. J.* **102,** 942.

Ruderman, N. B. and Lund, (1972). *Isr. J. Med. Sci.* **8,** 295.

Ruderman, N. B., Toews, C. J., and Shafrir, E. (1969). *Arch. Intern. Med.* **123,** 299.

Rutman, J. Z., Meltzer, L. E., Kitchell, J. R., Rutman, R. J., and George, P. (1965). *Am. J. Physiol.* **208,** 841.

Sanchez-Medina, R., Sanchez-Urrutia, L., Medina, J. M. and Mayor, F. (1971). *FEBS Lett.* **19,** 128.

Sawicki, S. G. and Godman, G. C. (1972). *J. Cell. Biol.* **55,** 299.

Seubert, W. and Huth, W. (1965). *Biochem. Z.* **343,** 176.

Shrago, E., Lardy, H. A., Nordlie, R. C., and Foster, D. O. (1963). *J. Biol. Chem.* **238,** 3188.

Simpson, D. P. and Angielski, S. (1973). *Biochim. Biophys. Acta* **298,** 115.

Snoke, R. E., Johnston, J. B., and Lardy, H. A. (1971). *Eur. J. Biochem.* **24,** 342.

Söling, H.-D., Willms, B., Kleineke, J., and Gehlhoff, M. (1970). *Eur. J. Biochem.* **16,** 269.

Stumpf, B., Boie, A., and Seubert, W. (1972). *Biochem. Biophys. Res. Commun.* **49,** 164.

Szepesi, B., Avery, E. H., and Freedland, R. A. (1970). *Am. J. Physiol.* **219,** 1627.

Vavatsi-Manos, O., Roxe, D. M., Schreiner, G. E., and Preuss, H. G. (1973). *Am. J. Physiol.* **224,** 154.

Veneziale, C. M., Walter, P., Kneer, N., and Lardy, H. A. (1967). *Biochemistry* **6,** 2129.

Walsh, D. A. and Chen, L.-J. (1971). *Biochem. Biophys. Res. Commun.* **45,** 669.

Walter, P., Paetkau, V., and Lardy, H. A. (1966). *J. Biol. Chem.* **241,** 2523.

Welbourne, T. C. (1974). *Am. J. Physiol.* **226,** 544.

Whelan, D. T., Scriver, C. R., and Mohyuddin, F. (1969). *Nature (Lond.)* **224,** 916.

Wicks, W. D., Lewis, W., and McKibbin, J. B. (1972). *Biochim. Biophys. Acta* **264,** 177.

Yphantis, D. A. and Waugh, D. F. (1956). *J. Phys. Chem.* **60,** 630.

CHAPTER ELEVEN

Species Dependent Regulation of Hepatic Gluconeogenesis in Higher Animals

HANS–DIETER SÖLING AND JOCHEN KLEINEKE

I. INTRODUCTION

In all higher animals examined thus far, hepatic gluconeogenesis from pyruvate, L-lactate, or C_3-L-amino acids (L-alanine, L-serine) involves reactions in at least two cellular compartments: the mitochondrial matrix and the cytosol. The first step in gluconeogenesis from pyruvate—namely, the carboxylation of pyruvate to oxaloacetate by pyruvate carboxylase—is restricted to the mitochondrial matrix space. Reports of a "soluble" pyruvate carboxylase activity in liver (Seubert *et al.*, 1968) can be explained by mitochondrial breakage (Walter and Anabitarte, 1971; Weiss *et al.*, 1974). In addition, since mitochondrial matrix enzymes are synthesized extramitochondrially, a certain percentage of pyruvate carboxylase activity in the postmitochondrial supernatant may result from freshly synthesized pyruvate carboxylase that has not yet entered the mitochondria.

The second step in the sequence, from pyruvate to phosphoenolpyruvate, is catalyzed by phosphoenolpyruvate carboxykinase (PEP-CK). This enzyme has been demonstrated intra- as well as extramitochondrially in all species examined to date. These findings point to the possibility of species differences in the regulation of gluconeogenesis. In this chapter we present experimental data on species differences in the regulation of hepatic gluconeogenesis and try to relate them as far as possible to biochemical facts (e.g., differences in the enzyme pattern or the enzyme distribution, transport phenomena). Only differences in the "short-term" regulation of gluconeogenesis are considered.

II. GLUCONEOGENESIS *IN VITRO*

Homogenates from rat liver are unable to form glucose from L-lactate, whereas homogenates from pigeon liver have a rather high gluconeogenic capacity under the same conditions (Krebs, *et al.*, 1964; Gevers and Krebs, 1966). This is understandable because in pigeon liver, the PEP-CK activity is much higher than in rat liver and is nearly completely localized within the mitochondria (which stay more or less intact during the preparation of liver homogenates). In rat liver homogenates the comparatively low PEP-CK activity is further diluted by the homogenization medium, since it is localized in the cytosol.

A. Isolated Perfused Liver

Maximum rates for the net formation of glucose from a variety of precursors are summarized in Tables I and II. There are several striking differences among the species compared thus far. Some can be explained rather easily; for example, the lower capacity for glucose formation from fructose by guinea pig liver as compared to rat liver is due to lower activity of fructokinase in guinea pig liver, whereas the activity of sorbitol dehydro-

Table I. Rates of net formation of glucose by isolated perfused livers from different mammalian species[a]

Species	No exogenous precursor	Precursor (μmole glucose formed/g wet liver/min)								References
		L-Lactate	Pyruvate	L-Alanine	Glycerol	Fructose	Propionate	Oxalacetate	L-Malate	
Rat	0.15 ±0.02	0.97 ±0.11	0.54 ±0.02	—	—	—	0.54[b] ±0.05	0.62 ±0.10	—	Söling *et al.* (1970) Volkmann (1973)
		0.92	0.88	0.52	—	2.54	—	—	—	Haeckel and Haeckel (1971)
Guinea pig	0.18 ±0.02	1.51 ±0.06	1.26 ±0.09	0.62 ±0.07	0.70 ±0.08	1.34 ±0.19	—	0.54 ±0.02	0.29 ±0.09	Söling *et al.* (1970)
		1.16	0.83	0.56	—	1.06	—	—	—	Haeckel and Haeckel (1971)
Mouse (normal)	0.52 ±0.10	1.30 ±0.05	1.39[b] ±0.06	0.59[b] ±0.13	0.52[b] ±0.04	1.52[b] ±0.17	—	—	—	Loten *et al.* (1974)
Mouse (obese) ob/ob	0.80 ±0.07	1.46 ±0.11	—	—	—	—	—	—	—	Loten *et al.* (1974), Assimacopoulos-Jeannet *et al.* (1973)
Rabbit	0.15 ±0.002	0.70 ±0.05	—	—	—	—	—	—	—	Söling, Janson, and Kuhn, unpublished data
	0.12[c]	0.42[c]	—	—	—	—	—	—	—	Arinze *et al.* (1973)
Cat	—	0.52[c]	—	—	—	—	—	—	—	Arinze and Hanson (1973)
Pigeon	0.23 ±0.05	3.31 ±0.37	0.25 ±0.05	0.22 ±0.04	—	—	—	—	—	Söling *et al.* (1973)

[a] The animals were usually starved for 48 hours, pigeons for 72 hours. The precursor concentrations were 10 to 20 mM.
[b] Data were calculated from data of Table 1 of Assimacopoulos-Jeannet *et al.* (1973).
[c] Data were estimated from figures given in the reference.

Table II. Rates of net formation of glucose from various precursors in the perfused liver from female rats starved for 48 hours

Substrate added	Glucose formed (μmole/g liver/min)
Nil	0.14 ±0.03
L-Lactate	1.06 ±0.09
Pyruvate	1.02 ±0.08
Glycerol	0.48 ±0.03
Dihydroxyacetone	2.07 ±0.30
D-Fructose	2.68 ±0.25
D-Galactose	0.36 ±0.11
Sorbitol	1.79 ±0.06
Xylitol	0.67 ±0.11
L-Serine	0.98 ±0.04
L-Alanine	0.66 ±0.12
L-Proline	0.55 ±0.10
L-Ornithine	0.19 ±0.05
L-Arginine	0.27
L-Valine	0.12
L-Isoleucine	0.22
L-Threonine	0.40 ±0.04
β-Aminobutyrate	0.23 ±0.04
Sodium L-glutamate	0.31 ±0.04
Ammonium L-glutamate	0.29 ±0.04
L-Glutamine	0.45 ±0.08
L-Aspartate	0.23 ±0.04
L-Asparagine	0.51 ±0.11
Oxaloacetate	1.22 ±0.12
Succinate	0.24 ±0.03
α-Oxoglutarate	0.25 ±0.05
Fumarate	0.19 ±0.05
L-Malate	0.23 ±0.07

Source: Data from Ross, Hems, and Krebs, (1967).

genase is much higher than in rat liver (Table III). In all species tested, L-lactate is the best precursor for gluconeogenesis among the metabolites entering gluconeogenesis below PEP (Table I). However, the rate at which gluconeogenesis from L-lactate occurs differs widely among the species compared. The highest rates were observed in pigeon liver (Söling *et al.*, 1973). Gluconeogenesis from high concentrations of pyruvate also exhibit remarkable species differences: in rat liver gluconeogenesis from pyruvate is in the same range as that from L-lactate, in guinea pig liver it is slightly lower (Söling *et al.*, 1970), and in pigeon liver there is no measurable glucose formation from pyruvate (Söling *et al.*, 1973). Moreover, pigeon livers are also unable to form measurable amounts of glucose from L-alanine (Söling *et al.*, 1970). In contrast to kidney, the second gluconeogenic organ, rat and guinea pig livers are unable to form significant amounts of glucose from L-aspartate or citric acid cycle intermediates (Ross *et al.*, 1967a; Söling *et al.*, 1970). Propionate, on the other hand, is a good precursor for hepatic gluconeogenesis not only in ruminants but also in rats, guinea pigs, and pigeons (Fig. 1). Since the conversion of propionate to glucose involves steps of the Krebs cycle (from succinyl-CoA to L-malate), the low capacity of the liver to form glucose from citric acid cycle intermediates must result mainly from limited transport of the precursors into the liver cells.

B. Isolated Liver Cells

All preparations of isolated liver cells that have been used for the study of gluconeogenesis are modifications of the method reported by Berry and Friend (1969). Only a limited number of results comparing gluconeogenesis in different species are available. The comparison between rat and pigeon liver cells (Table IV) confirms the findings obtained with isolated, perfused livers from rats and pigeons (Söling *et al.*, 1973). Gluconeogenesis from L-lactate was much higher in cells from pigeon liver as compared to rat liver, and there were negligible rates of glucose formation from pyruvate by pigeon liver cells. The functional state of isolated guinea pig liver cells usually was impaired in comparison to isolated perfused guinea pig livers.

Table III. Activities of fructokinase (EC 2.7.1.4) and sorbitol dehydrogenase (EC 1.1.1.14) in livers from fed and starved rats and guinea pigs

Liver	Metabolic condition of animal	Rates (μmole/g liver/min) Fructokinase	Sorbitol-dehydrogenase
Rat	Fed	6.97 ±0.63	6.90 ±0.67
	Starved (48 hours)	5.21 ±0.36	14.44 ±1.99
Guinea pig	Fed	0.18 ±0.03	23.53 ±0.77
	Starved (48 hours)	0.72 ±0.09	45.20 ±2.68

Source: Data from Söling and Bernhard, unpublished results.

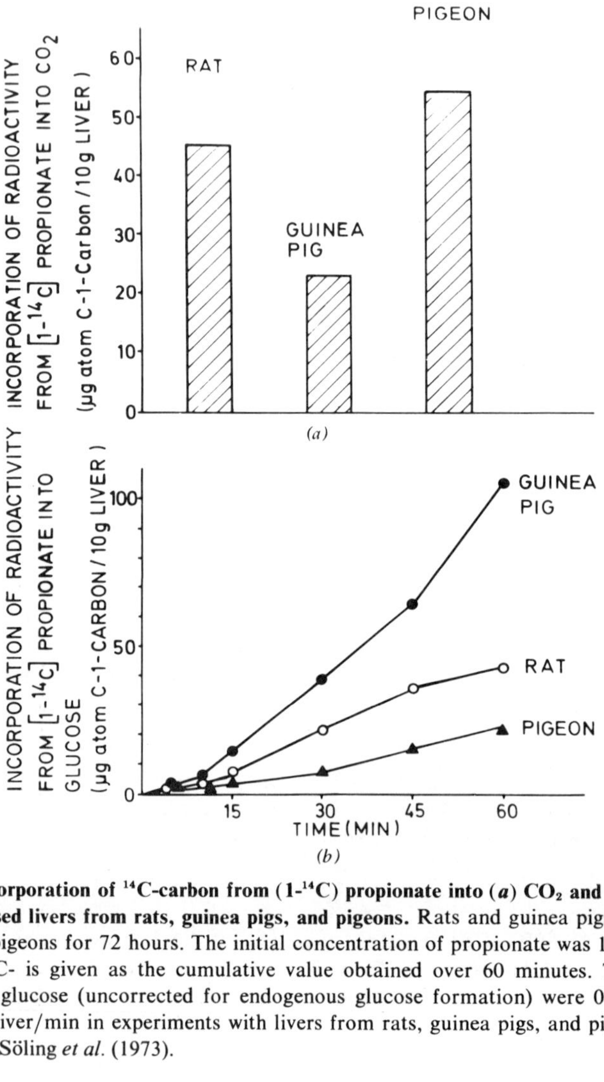

Figure 1. Incorporation of ^{14}C-carbon from (1-^{14}C) propionate into (*a*) CO_2 and (*b*) glucose by isolated, perfused livers from rats, guinea pigs, and pigeons. Rats and guinea pigs were starved for 48 hours, pigeons for 72 hours. The initial concentration of propionate was 10 mM. Incorporation of ^{14}C- is given as the cumulative value obtained over 60 minutes. The mean net formations of glucose (uncorrected for endogenous glucose formation) were 0.50, 0.87, and 0.46 μmole/g liver/min in experiments with livers from rats, guinea pigs, and pigeons, respectively. *Source:* Söling *et al.* (1973).

Therefore, the results obtained with these cells must be interpreted with great caution.

C. Estimation of Gluconeogenic Rates in Man

Because of the difficulties in discriminating between glucose production and gluconeogenesis on one hand, and the cumbersome techniques needed for estimating hepatic blood flow on the other, few data concerning gluconeogenesis in man are available. Some recent data from Dietze and his associates are summarized in Table V. The rate of gluconeogenesis during infusion of L-lactate is within the range observed with isolated, perfused livers from various vertebrate species (cf. Table I). However, more specific information is presented in chapters 14 and 15 of this volume.

III. INTRACELLULAR LOCALIZATION OF PEP-CK AND FORMATION OF PEP FOR GLUCONEOGENESIS IN DIFFERENT SPECIES

It was first proposed by Lardy *et al.*, (1965) that in rat liver the formation of PEP from L-lactate proceeds by a reversal of the Borst cycle. Borst

Table IV. Net formation of glucose from various precursors by isolated cells from rat, pigeon, and guinea pig[a]

Experimental condition	Rate for species (μmole/g dry wt/min)		
	Rat	Pigeon	Guinea pig
Control	0.17 ±0.01	0.03 ±0.01	0.41 ±0.03
L-Lactate	2.59 ±0.02	9.60 ±0.67	—
L-Lactate + hexanoate	—	11.17 ±0.55	—
L-Lactate + L-trytophane	—	9.88 ±0.39	—
L-Lactate + glucagon	—	9.91 ±0.29	—
L-Lactate + pyruvate (9:1; mole/mole)	1.67 ±0.12	7.40 ±0.14	1.38 ±0.10
Pyruvate	2.78 ±0.02	1.90 ±0.03	—
Pyruvate + glucagon	—	1.80 ±0.01	—
L-Alanine	1.53 ±0.05	—	—
Propionate	1.54 ±0.05	2.09 ±0.04	—
Propionate + glucagon	—	2.16 ±0.07	—
Glycerol	1.48 ±0.04	—	—

[a] The precursor concentration was always 10 mM at the beginning of the incubation. The incubation lasted for 60 minutes at 37°C. The concentrations of hexanoate, L-tryptophane, and glucagon were 1 mM, 2.45 mM, and 3×10^{-9} M, respectively.
Source: Data from Söling, Kehl, Heinz, Janson, and Kuhn, unpublished results.

Table V. Net formation of glucose and gluconeogenesis by human liver in the postabsorptive state and during starvation[a]

Starvation period (hours)	Total net formation of glucose	Gluconeogenesis (nmole glucose/g liver/min)				
		Total	From L-lactate[b]	From pyruvate	From L-alanine	From glycerol
15	459 ±18	174 ±9	93 ±0.5	6.0 ±0.1	26 ±2.5	32 ±2
60	230 ±34	218 ±26	138 ±3.0	7.0 ±0.2	42 ±7.5	46 ±11
120	279 ±22	176 ±8	109 ±0.3	6.5 ±0.2	20 ±2.5	56 ±9

[a] The rates of gluconeogenesis from L-lactate, pyruvate, L-alanine, and glycerol were calculated under the assumption that the metabolites taken up by the liver were completely used for gluconeogenesis and that energy for gluconeogenesis was derived from oxidation of other metabolites (e.g., fatty acids).

[b] During an infusion of 30 μmole/kg/min of L-lactate or 46 μmole/kg/min of fructose, gluconeogenesis increased from 175 nmole/g liver/min to 509 and 580 nmole/g liver/min, respectively.

Source: Data from Dietze *et al.* (1973a, 1973b) and Dietze (1974).

(1963) was the first to propose that the transport of reducing equivalents from the cytosol into the mitochondria not only involves a translocation of L-malate but might also be accompanied by an outward transport of L-aspartate and α-oxoglutarate and an inward transport of L-glutamate. This proposal was made in view of the finding that the K_m for the outward transport of oxaloacetate generated intramitochondrially by the oxidation of L-malate exceeded the physiological levels of oxaloacetate by several orders of magnitude. In the model proposed by Lardy *et al.* (1965) for the formation of PEP from L-lactate, the anions L-aspartate, α-oxoglutarate, and L-glutamate move in the same direction as in the Borst-model, the only difference being that during gluconeogenesis from L-lactate, oxaloacetate is generated intramitochondrially by way of carboxylation of pyruvate, which is subsequently converted to PEP in the cytosol. The reducing equivalents necessary for the formation of phosphoglyceraldehyde in the triosephosphate dehydrogenase reaction are generated mainly by the oxidation of L-lactate to pyruvate (Fig. 7).

According to the mechanism proposed by Lardy, *et al.* (1965), when pyruvate is the gluconeogenic precursor reducing equivalents for the triosephosphate dehydrogenase step must be generated intramitochondrially and must be transported out of the mitochondrial matrix space together with the carbon skeleton of oxaloacetate in the form of L-malate. In turn, L-malate is oxidized by cytosolic malate dehydrogenase to oxaloacetate, which is then converted to PEP. The situation is more complicated when L-alanine is the glucogenic precursor, because of the distribution of urea cycle steps between the cytosol and the mitochondrial matrix space. A theoretical model, proposed by J. R. Williamson *et al.*, (1968) suggests that the reducing equivalents necessary for the conversion of PEP to glucose are generated partially intramitochondrially (and are transported into the cytosol as L-malate) and partially by the oxidation of L-malate generated extramitochondrially from fumarate, which is produced in the cytosol by cleavage of argininosuccinate in the urea cycle.

The model described for the formation of PEP in rat liver is supported by several experimental findings, among which the strongest are the differential effects of aminooxyacetate (J. R. Williamson, 1970; Arinze *et al.*, 1973) and of glisoxepide (Söling and Seck, 1975) on gluconeogenesis from pyruvate as compared to L-lactate. Aminooxyacetate completely inhibits gluconeogenesis from L-lactate in the isolated perfused rat liver. Since aminooxyacetate is a strong inhibitor of glutamate-oxaloacetate aminotransferase, and since according to the model proposed by Lardy *et al.* (1965) only gluconeogenesis from L-lactate involves the transamination reaction, this finding is in line with the proposed reaction mechanism. Also supporting this model is the inhibition of gluconeogenesis from L-lactate by isolated rat liver cells by glisoxepide, an inhibitor of L-aspartate transport through the inner mitochondrial membrane (Söling and Seck, 1975). The whole model is based on the assumption that PEP for gluconeogenesis is formed almost exclusively in the cytosol. There is sufficient evidence that this is so in rat liver. Therefore the following questions arose.

(a) Does the proposed model hold for species in which a high percentage of total PEP-CK activity is located intramitochondrially?
(b) Can PEP formed intramitochondrially be used for gluconeogenesis, and if so to what extent and how is this regulated?

There are several indications that intramitochondrially formed PEP contributes to some extent to gluconeogenesis from pyruvate and L-lactate in guinea pig liver (Söling *et al.*, 1970). We found an incomplete inhibition of gluconeogenesis from L-alanine by quinolinate (an inhibitor of PEP-CK: Veneziale *et al.*, 1967) in isolated, perfused guinea pig livers at concentrations that completely inhibited gluconeogenesis in the isolated, perfused rat liver (Fig. 2). J. R. Williamson (1970) and Arinze *et al.* (1973) could not completely inhibit gluconeogenesis from L-lactate in isolated, perfused guinea pig livers by aminooxyacetate when it was completely blocked in isolated, perfused rat livers. We found that in rabbit liver gluconeogenesis from 20 mM L-lactate was virtually completely suppressed by aminooxyacetate (control 0.70 ± 0.05 μmole/g liver/min versus 0.13 ± 0.03 μmole/g liver/min glucose formed in the presence of 0.2 mM aminooxyacetate), even though livers from starved rabbits contain a high percentage of the total PEP-CK activity inside the mitochondria. This finding would be in line with the proposals made by Johnson *et al.* (1970) (see also Chapter 9, this volume). Quinolinate, on the other hand, failed to inhibit gluconeogenesis from L-lactate in the isolated, perfused rabbit liver at concentrations (2.5 mM) that were effective in rat liver (not shown). Measurements of the intracellular uptake of quinolinate by rabbit and rat liver revealed that there was a negligible uptake by rabbit liver in comparison to rat liver (Table VI). This may explain the discrepancy between the inhibition by aminooxyacetate and quinolinate in rabbit liver.

In pigeon liver gluconeogenesis from L-lactate is completely resistant to inhibition by aminooxyacetate or quinolinate (Figs. 3 and 2). On the other hand, mitochondrial and soluble glutamate-oxaloacetate aminotransferases in pigeon liver are as sensitive to inhibition by aminooxyacetate as the corresponding rat liver enzymes (Söling *et al.*, 1973). In line with these results is the finding that glisoxepide, an inhibitor of the L-aspartate/L-glutamate antiport system, did inhibit gluconeogenesis from L-lactate in isolated liver cells from rats but not from pigeons (Söling and Seck, 1975). Therefore in pigeon liver gluconeogenesis from L-lactate seems not to involve transamina-

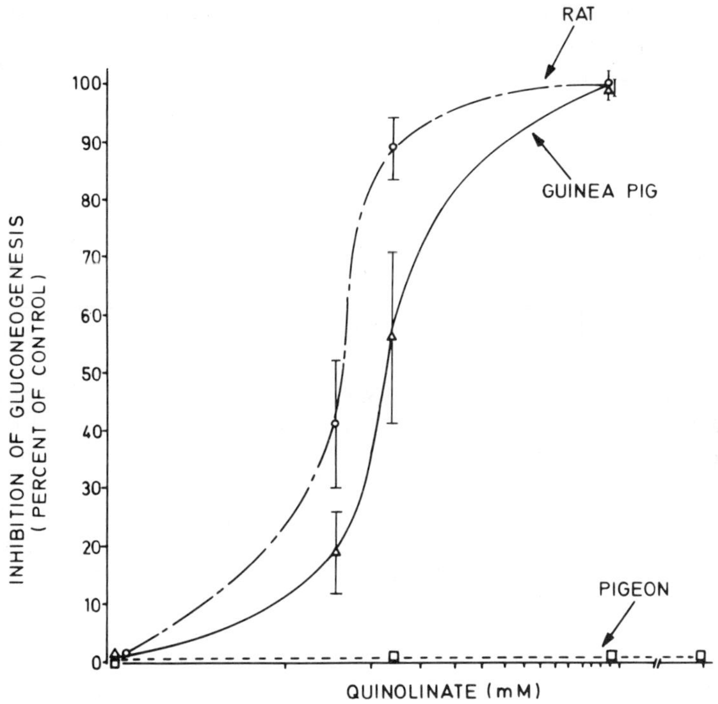

Figure 2. Effects of increasing concentrations of quinolinate on the net formation of glucose from L-alanine (10 mM) by isolated, perfused livers from rats and guinea pigs, and from 10 mM L-lactate by isolated, perfused pigeon livers. Rats and guinea pigs were starved for 48 hours, pigeons for 72 hours. *Source:* Söling *et al.* (1970).

tion or a translocation of L-aspartate, since the PEP used for gluconeogenesis is synthesized intramitochondrially. This requires the translocation of PEP from the matrix space to the cytosol, as we discuss later in the chapter. As can be deduced from these results, there are considerable species differences in the generation of cytosolic PEP for gluconeogenesis. Rat and pigeon are the extremes on the scale of species variations. It seems, however, that when L-lactate is the glucogenic precursor, all species have in common the requirement for a sufficiently high concentration of extramitochondrial L-malate. We will return to this point later.

We have already mentioned that the isolated, perfused pigeon liver is unable to form glucose from pyruvate. This results in part from an insufficient rate of production of reducing equivalents by pyruvate oxidation (Söling *et al.,* 1973). In the isolated, perfused rat and guinea pig liver, the ratio of pyruvate used to glucose formed is close to 4, which means that about the same amounts of pyruvate had been oxidized and converted into glucose.

Table VI. Intracellular uptake of quinolinate by isolated, perfused livers from rats and rabbits[a]

		Ratios		
		^3H radioactivity/^{14}C radioactivity		
Species	Time (min)	Medium	Liver tissue	Medium/liver tissue
Rabbit (I)	20	1.860	1.819	1.02
	40	—	—	—
	60	1.944	1.794	1.08
Rabbit (II)	20	1.778	1.642	1.08
	40	1.947	1.674	1.16
	60	1.858	1.620	1.15
Rat (I)	20	2.072	1.392	1.43
	40	2.167	1.452	1.49
	60	2.021	1.319	1.53
Rat (II)	20	1.965	1.504	1.31
	40	1.935	1.498	1.29
	60	2.944	1.929	1.53

[a] The animals were starved for 48 hours prior to operation. After an equilibration period in the perfusion chamber, L-lactate (final concentration 20 mM), quinolinate (final concentration 2.5 mM) with 7 μCi of (U-^{14}C) quinolinate, and 12 μCi of ^{3}H-inulin were added to the perfusion medium (total volume; 150 ml). Tissue aliquots and aliquots from the medium were taken at 20, 40, and 60 minutes for analysis of ^{3}H- and ^{14}C-radioactivity. The samples were lyophilized and oxidized in a Packard 305 sample oxidizer. Radioactivity was determined in a Packard 3380 liquid scintillation spectrometer equipped with an absolute activity analyzer. Additional controls were done by internal standardization. The ratio of the ratios (last column) indicates the degree of uptake of quinolinate by the liver. The higher the ratio, the more quinolinate has been taken up by the inulin-inaccessible space. The results show that isolated livers from rats accumulate significantly more quinolinate than those from rabbits.

However, it is not clear yet whether lack of reducing equivalents in the cytosol is the only reason for the lack of gluconeogenesis from pyruvate in pigeon liver, since ethanol led only to a slight stimulation of glucose formation from pyruvate. This is true despite the fact that the oxidation of ethanol proceeded at a rate similar to that found in rat liver (Söling *et al.*, 1973).

IV. FUNCTION OF INTRAMITOCHONDRIAL PEP-CK IN NONGLUCONEOGENIC REACTIONS

It is of interest to ask whether other reactions besides glucose formation might compete for PEP synthetized in the PEP-CK reaction. In fetal rat liver PEP-CK activity can be demonstrated intramitochondrially long before PEP-CK activity can be measured in the cytosol (Ballard and Hanson, 1967). This indication that intramitochondrially formed PEP must be used for purposes other than gluconeogenesis is supported by the fact that FDPase and G-6-Pase, the two other key gluconeogenic enzymes, exhibit also no or only very low activity up to the seventeenth day of gestation (Burch *et al.*, 1963; Ballard and Oliver, 1962).

Since in pigeon liver almost all PEP-CK activity is mitochondrial (see Table XVIII), it seemed possible that even in adult pigeon liver at least some of the intramitochondrially synthetized PEP could be used for reactions similar to those occurring in fetal rat liver mitochondria. For this pur-

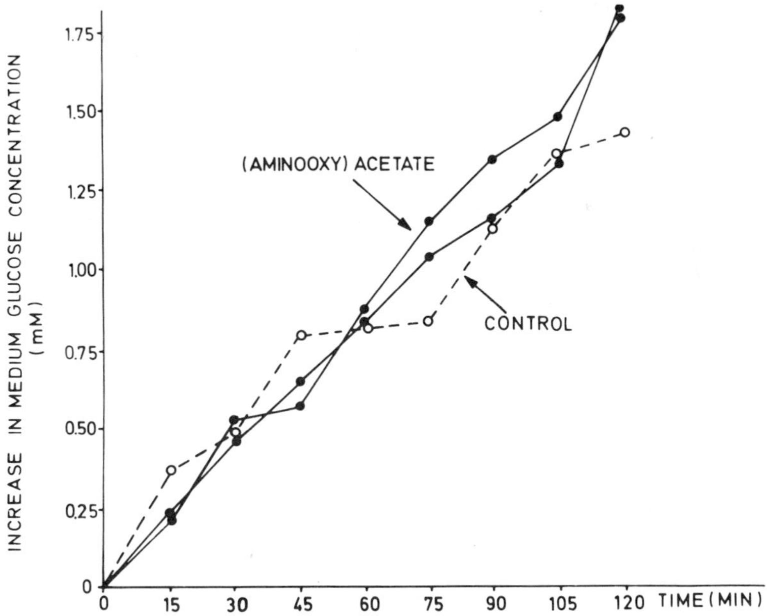

Figure 3. Absence of an effect of aminooxyacetate on net formation of glucose from L-lactate by isolated, perfused pigeon livers. The L-lactate was added at zero time to give a final concentration of 10 mM. At the same time an intraportal infusion of 35 μmoles of L-lactate/hr was initiated and continued over the whole experiment. After 60 minutes aminooxyacetate was added to the perfusion medium to give a final concentration of 0.2 mM. *Source:* Söling *et al.* (1973).

pose we have examined whether synthesis of *N*-acetylneuraminic acid (NANA) and its incorporation into mitochondrial proteins might occur at an appreciable rate in pigeon liver mitoplasts (Söling *et al.*, 1975). Mitoplasts were used to ensure complete separation of any possible contaminants from cytosol or other particulate fractions (lysosomes, Golgi apparatus, microsomes).

As revealed by the methods described in the legends to Figs. 4 and 5, it could be clearly shown that PEP formed from (U-^{14}C)-L-malate is used as precursor for the synthesis of *N*-acetylneuraminic acid (NANA) and that this NANA is incorporated into glycoproteins of the inner mitochondrial membrane.

In the meantime, we have demonstrated also the occurrence of all necessary enzymes in pigeon liver mitoplasts (NANA-9-P synthetase; NANA-9-P phosphatase; CMP-NANA synthase; sialic acid transferase). Interestingly, we find in mitochondria from fetal rat liver (seventeenth day of gestation) a 5 to 10 times higher rate of incorporation of carbon from L-malate into the sialic acid moiety of mitochondrial glycoproteins as compared to mitochondria from adult rat liver. This corresponds with a 5 to 10 times higher specific activity of PEP-CK found in fetal rat liver mitochondria as compared to mitochondria from the liver of adult rats. In pigeon liver mitochondria, NANA formation seems to be small compared to the total capacity for intramitochondrial formation of PEP. However, it points to the possibility that metabolic functions other than gluconeogenesis might exist for intramitochondrial PEP and that these functions could exhibit species differences in view of the dramatic species differences in the intramitochondrial activity of PEP-CK.

V. REGULATION OF GLUCONEOGENESIS IN RUMINANTS AND NONRUMINANTS

Under physiological circumstances, monogastric animals cover their need for carbohydrates mainly by normal food intake. Only a small portion is formed by gluconeogenesis. Ruminants, on the other hand, are very dependent on gluconeogenesis. Apart from amino acids, the main gluconeogenic precursor is propionate, formed in the rumen by microorganisms. Propionate is converted in the liver to glucose either by way of succinate-malate or after conversion in the gut wall to L-lactate. As can be expected from the different roles played by gluconeogenesis in ruminant and in nonruminant animals, several fundamental regulatory differences have been observed:

1. The activities of two of the glucogenic key enzymes—namely, pyruvate carboxylase and glucose-6-phosphatase—are already high in livers

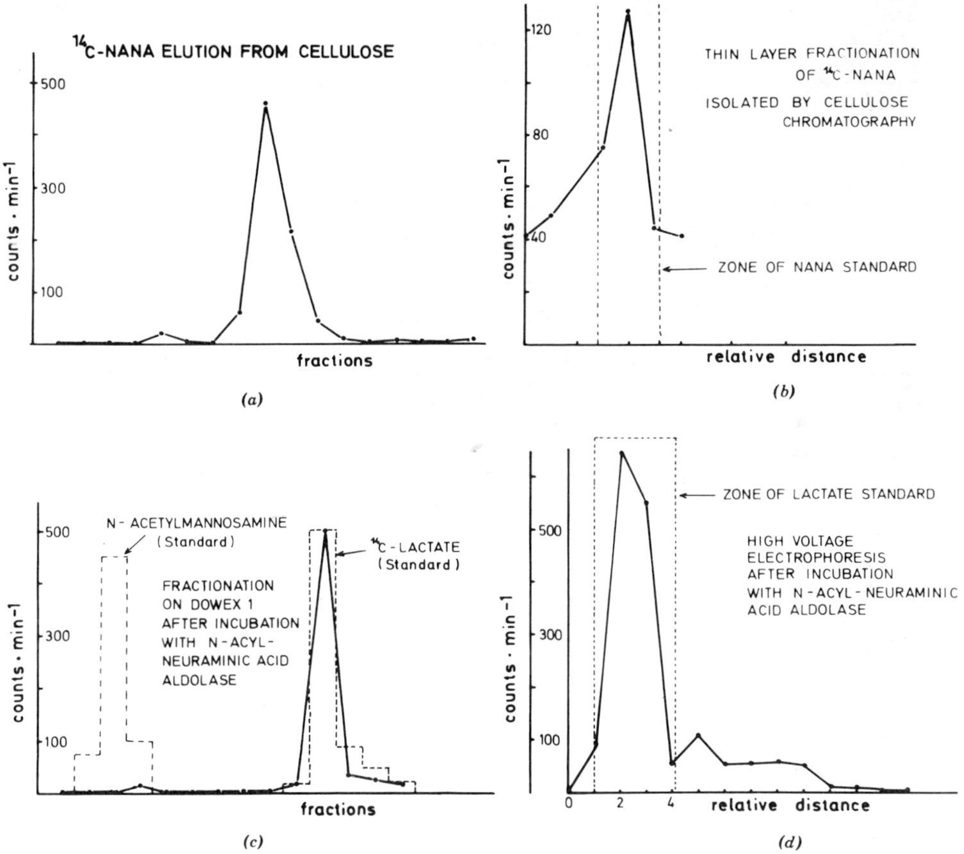

Figure 4. Elution of (^{14}C)-labeled *N*-acetylneuraminic acid (NANA) from the cellulose column (*a*) and detection of the radioactivity at the spot of authentic *N*-acetylneuraminic acid on a thin layer plate (*b*). After splittage of *N*-acetylneuraminic acid by *N*-acyl neuraminic acid aldolase (*c*), all the radioactivity can be recovered in the C_3 unit stemming from phosphoenolpyruvate (The pyruvate generated in the aldolase reaction had been converted to L-lactate in the presence of NADH by lactate dehydrogenase). Lactate had been separated by anion exchange chromatography. (*d*) Essentially the same as (*c*) except that in this case lactate was separated by high voltage electrophoresis. *Source*: From Söling *et al.* (1974).

from fetal ruminants but not from nonruminant fetuses (Ballard *et al.*, 1969).

2. The capacity for gluconeogenesis in ruminant fetal liver is nearly as high as in the adult liver, whereas it is more or less absent in fetal livers of nonruminant animals (Ballard *et al.*, 1969).

3. In accordance with the low amount of carbohydrate absorbed from the gut, glucokinase activity is absent in ruminant liver (Ballard *et al.*, 1969).

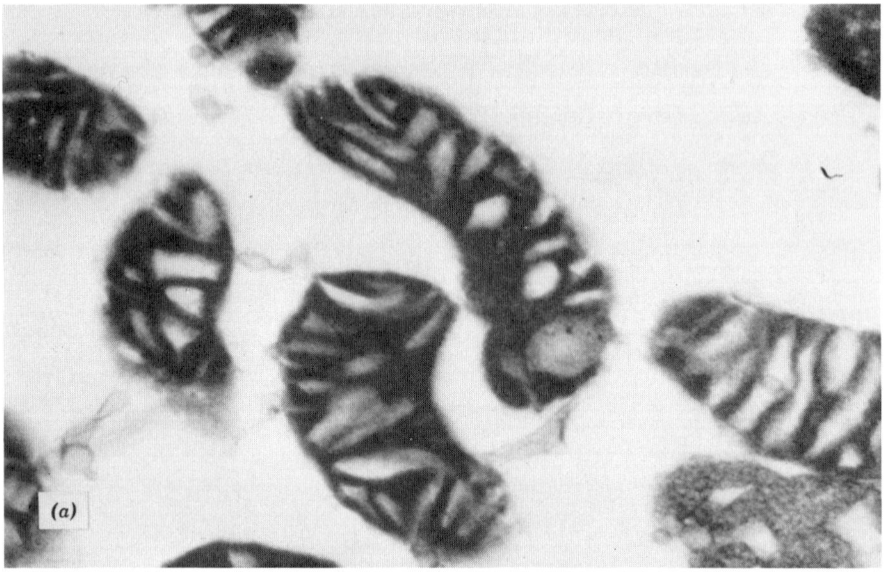

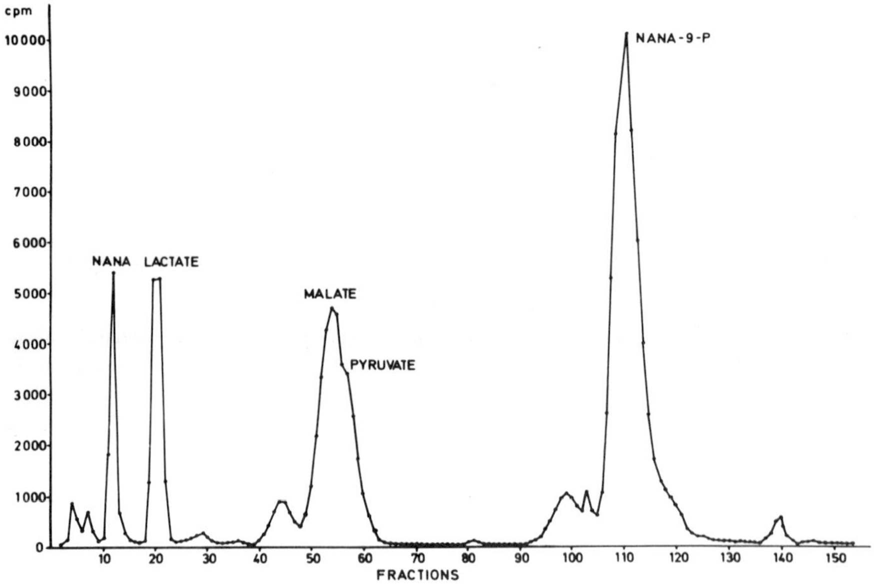

(b)

Figure 5. Synthesis of (^{14}C)-labeled *N*-acetylneuraminic acid (NANA) and its 9-phosphate (NANA-9-P) from (U-^{14}C)phosphoenolpyruvate by purified pigeon liver mitoplasts. (*a*) Purified pigeon liver mitoplasts (60,000-fold magnification). We thank Dr. Cora Creutzfeldt and Miss H. Dörler for their help in preparing the electron-micrographs. (*b*) Anion exchange chromatogram of the reaction products after incubation of the mitoplasts in the presence of 0.75 mM (1-^{14}C)phosphoenolpyruvate and 10 mM cold *N*-acetyl-mannosamine-6-phosphate. *Source*: From Söling *et al.* (1974).

4. The total glucose entry rate as a function of total body weight is similar in ruminants and nonruminants. However, the percentage of this entry rate resulting from gluconeogenesis is considerably higher in ruminants.

Recently Atwal and Sauer (1973) compared gluconeogenesis from various precursors using beef liver slices. At saturating substrate concentrations, the quantitative relationship between the maximal rates for glucose formation from different gluconeogenic precursors was about the same as that found in nonruminants. With gluconeogenesis from pyruvate considered to be 100%, they noted that glucose synthesis from propionate and glycerol were 71 and 78%, respectively. Moreover, acetylcarnitine as well as dibutyryl-3,5-cAMP stimulated gluconeogenesis from pyruvate. This similarity between gluconeogenic rates in various species under the conditions of substrate saturation is in sharp contrast to the relative importance of different precursors in various species under natural *in vivo* conditions, reemphasizing the basic importance of substrate availability for the regulation of gluconeogenesis. We shall restrict ourselves to these few remarks, but we recommend for further details the excellent review of Ballard, *et al.* (1969).

VI. SPECIES DEPENDENT ALTERATION IN THE MITOCHONDRIAL REDOX STATE

The redox states of the cytosolic and mitochondrial $NAD^+/NADH$ systems as calculated from the corresponding substrate couples in the livers from different species are summarized in Table VII. In comparison to rat liver (D. H. Williamson *et al.,* 1967), the cytoplasmic redox state is more reduced in guinea pig and rabbit liver, whereas the mitochondrial redox state is more oxidized in these species (Willms *et al.,* 1970; Garber and Hanson, 1971a, 1971b). Upon starvation, the cytosolic redox couple becomes more reduced in rat and rabbit liver but does not change or becomes even slightly oxidized in guinea pig liver. Under the same condition the mitochondrial redox state is shifted toward reduction in rat liver but toward oxidation in guinea pig and rabbit liver (D. H. Williamson *et al.,* 1967; Garber and Hanson, 1971a, 1971b). It has been argued that in species with intramitochondrial PEP-CK activity (e.g., rabbits and guinea pigs), an oxidation of the mitochondrial $NAD^+/NADH$ system favors mitochondrial PEP synthesis, since it would increase the steady-state concentration of oxaloacetate in the mitochondria. Garber and Hanson (1971a, 1971b) could indeed show with isolated mitochondria from guinea pig and rabbit liver that a shift toward oxidation favors mitochondrial PEP formation

Table VII. Redox states of the cytosolic and mitochondrial NAD^+/NADH systems in the liver of various mammalian species in vivo and vitro

	NAD^+/NADH ratio calculated from								
	Lactate dehydrogenase system (cytoplasm)				3-Hydroxybutyrate dehydrogenase system (mitochondria)				
		Animals starved (hours)		Animals fat fed (hours)		Animals starved (hours)		Animals fat fed (hours)	
Species	Animals well fed	48	96	72	Animals well fed	48	96	72	Reference
Rat	725	528	—	—	8	6	—	—	D. H. Williamson *et al.* (1970)
Guinea pig	293	344	330	—	12	33	41	—	Garber and Hanson (1971a)
	333	—	—	216	18	—	—	8	Willms *et al.* (1970)
Rabbit	482	—	342	—	9	—	21	—	Garber and Hanson (1971b)

Isolated, perfused guinea pig liver	3-Hydroxybutyrate dehydrogenase system (mitochondria)	
Control (11.1 mM glucose)	42	Willms *et al.* (1970)
11.1 mM glucose + 0.72 mmole/h hexanoate (30 min)	41	
Control (2 mM L-lactate)	42	Arinze *et al.* (1973)
2 mM L-lactate + 0.2 mM octanoate (6 min)	35	

over malate and aspartate synthesis. Similarly, the inhibition of gluconeogenesis by fatty acids in species with intramitochondrial PEP-CK (Söling *et al.*, 1970; Arinze *et al.*, 1973) has been explained by a more reduced mitochondrial redox state in these species and concomitant enhanced utilization of the remaining oxaloacetate by the citrate synthase reaction, due to increased availability of acetyl-CoA (Garber and Hanson, 1971a, 1971b; Arinze *et al.*, 1973). Moreover, Arinze *et al.*, (1973) and Arinze and Hanson (1973) have shown that an infusion of 3-hydroxybutyrate decreases gluconeogenesis in guinea pig and cat liver, and they discuss a similar mechanism for rabbit liver. Recently Jomain-Baum and Hanson (1973) demonstrated similarities between mitochondria from guinea pig and human liver, which led them to extend their explanation to the effect of fatty acids on gluconeogenesis in human liver. However, the slight reduction of the mitochondrial $NAD^+/NADH$ system observed in guinea pig liver *in vitro* following the administration of hexanoate (Willms *et al.*, 1970) or octanoate (Arinze *et al.*, 1973) (Table VII), does not sufficiently account for the observed inhibition of gluconeogenesis. A decrease in the concentration of pyruvate due to increased reduction of the cytoplasmic $NAD^+/NADH$ system offers a better explanation for the observed inhibition of gluconeogenesis by fatty acids under these conditions.

VII. SPECIES DEPENDENT REGULATION OF GLUCONEOGENESIS AND FATTY ACID OXIDATION

Krebs *et al.* (1964) noted that acetoacetate stimulates gluconeogenesis from rat kidney cortex slices. Subsequently others reported that in isolated, perfused rat liver (Söling, 1965; Struck *et al.*, 1965; Struck *et al.*, 1966; Herrera *et al.*, 1966; J. R. Williamson *et al.*, 1966, 1968; Ross *et al.*, 1967b; Teufel *et al.*, 1967; Söling *et al.*, 1968) as well as in isolated rat liver cells (Söling, Kehl and Heinz, unpublished results), gluconeogenesis from L-lactate and pyruvate is strongly stimulated by long-chain and medium-chain fatty acids. Several authors proposed that this stimulation resulted mainly from an enhancement of pyruvate carboxylation by an increased intramitochondrial level of acetyl-S-CoA (Krebs *et al.*, 1965; Söling *et al.*, 1964, 1968; J. R. Williamson, 1967; J. R. Williamson *et al.*, 1968; Ruderman *et al.*, 1971). Stimulation of gluconeogenesis by long-chain fatty acids would be favored by an inhibition of the adenine nucleotide translocator, resulting in an increased intramitochondrial ATP/ADP ratio. This would lead to deinhibition of pyruvate carboxylase. However it seems doubtful whether the inhibition of the adenine translocator by fatty acids plays an important physiological role. The intramitochondrial ATP/ADP ratio increases in the presence of medium-chain fatty acids, but this effect is due to an accumulation of AMP rather than to an inhibition of the adenine translocator (Stucki *et al.*, 1972). Under physiological conditions pyruvate carboxylase and pyruvate dehydrogenase most probably compete for intramitochondrial pyruvate. It has been shown recently (Wieland *et al.*, 1972a) that fatty acid oxidation favors the interconversion of pyruvate dehydrogenase into the inactive form. This by itself would favor pyruvate carboxylation. The effect of fatty acids is probably mediated by an increase in the intramitochondrial ATP/ADP ratio. Using the isolated, perfused rat liver, investigators have found that the formation of $^{14}CO_2$ from (1-^{14}C) pyruvate is inhibited in the presence of fatty acids (Teufel *et al.*, 1967; Söling *et al.*, 1968). Moreover, fatty acids could lead to an inhibition of active pyruvate dehydrogenase by increasing the intramitochondrial concentrations of acetyl-CoA and/or decreasing the intramitochondrial NAD^+/NADH ratio.

In livers from guinea pigs or pigeons, gluconeogenesis responds to the addition of fatty acids in a way completely different from that observed in rat liver: albumin-bound oleate at a concentration of 2 mM significantly inhibits gluconeogenesis from L-lactate as well as the net uptake of L-lactate by isolated, perfused guinea pig livers (Söling *et al.*, 1970, Arinze *et al.*, 1973). Under the same conditions gluconeogenesis or uptake of L-lactate remained completely unaffected in pigeon liver (Söling *et al.*, 1973). Loten *et al.* (1974) likewise failed to observe any effect of oleate on gluconeogenesis from L-lactate in isolated, perfused mouse liver. Hexanoate also inhibited gluconeogenesis from L-lactate and L-lactate uptake in the isolated, perfused guinea pig liver (Söling *et al.*, 1970), and was ineffective in pigeon liver (Söling *et al.*, 1973). (Table VIII). When pyruvate was the gluconeogenic precursor, hexanoate led to a small but already significant stimulation of gluconeogenesis in guinea pig liver (Söling *et al.*, 1970). In the isolated, perfused pigeon liver, where no glucose formation occurred above the endogenous rate in the presence of pyruvate, hexanoate or oleate failed to increase glucose formation (Söling *et al.*, 1973).

The change in the overall concentration of acetyl-CoA and CoA-SH during stimulated fatty acid oxidation was the same in the isolated, perfused guinea pig liver and rat liver (Söling *et al.*, 1970). Moreover, the lack of stimulation of gluconeogenesis from L-lactate by fatty acids in guinea pig liver cannot be explained by different kinetic properties of pyruvate carboxylase from rat and guinea pig liver (Söling *et al.*, 1970). The inhibition of gluconeogenesis from L-lactate in guinea pig liver by oleate or hexanoate can be explained best by a decrease in the concentration of pyruvate due to a reduction of the cytosolic NAD^+/NADH system.

The lactate/pyruvate ratio of *in vivo* freeze-clamped guinea pig liver is considerably higher than it is in rat liver (Willms *et al.*, 1970a) when

Table VIII. Net uptake of L-lactate and the formation of glucose by isolated, perfused livers from 48-hour-starved guinea pigs and rats under various conditions[a]

Experimental condition	Guinea pig				Rat			
	Lactate used	Glucose formed	Lactate/glucose	n	Lactate used	Glucose formed	Lactate/glucose	n
L-Lactate	3.05 ±0.14	1.51 ±0.06	2.02	17	2.50 ±0.12	0.97 ±0.11	2.58	11
L-Lactate + oleate	2.41 ±0.18	1.09 ±0.10	2.21	10	3.66 ±0.62	1.75 ±0.23	2.09	6
L-Lactate + oleate + L-carnitine	2.82 ±0.09	1.18 ±0.04	2.38	3	—	—	—	—
L-Lactate + hexanoate	2.14 ±0.19	1.16 ±0.07	1.84	6	3.24 ±0.16	1.66 ±0.01	1.95	7
L-Lactate + glucagon	2.91 ±0.12	1.57 ±0.05	1.85	11	3.45 ±0.28	1.64 ±0.16	2.09	4
L-Lactate + dibutyryl-cAMP	3.24 ±0.16	1.57 ±0.06	2.06	5	3.60 ±0.28	2.43 ±0.21	1.48	4

[a] Rates for both species expressed in micromoles per gram per minute. The concentration of L-lactate at the beginning of the experiment was always 20 mM. Oleate was added to the medium at a final concentration of 2 mM. The oleate was bound to albumin according to Ross *et al.* (1967b). L-Carnitine was added at a final concentration of 1.5 mM. Hexanoate was infused intraportally throughout the whole experiment at a rate of 0.72 mmole/hr. Glucagon (350 μg) was added at the beginning of the experiment followed by further additions of 150 μg after 15, 30, and 45 minutes. Dibutyryl-cAMP was added to the medium to give a final concentration of 0.1 mM. The volume of the perfusion medium was 150 ml in all experiments. Mean values ±SEM.
Source: Data from Söling *et al.* (1970).

measured in the absence of exogenous fatty acids. At the same time the redox state of the 3-hydroxybutyrate/acetoacetate system is more positive in guinea pig liver (Willms *et al.*, 1970a). In the presence of medium-or long-chain fatty acids, the lactate/pyruvate ratio in isolated perfused guinea pig livers increases without a significant change or even with a decrease of the 3-hydroxybutyrate/acetoacetate ratio (Willms *et al.*, 1970a). This can be observed in experiments with livers from fed guinea pigs at low medium concentrations of L-lactate as well as in experiments with livers from starved guinea pigs and high concentrations of external L-lactate (Willms *et al.*, 1970a). For a given concentration of L-lactate in the perfusion medium, the concentration of pyruvate is always less in experiments with guinea pig livers than it is in similar tests with rat livers.

According to McClure *et al.* (1971a), the physiological K_m of rat liver pyruvate carboxylase for pyruvate is in the order of 50 μM. If the K_m of pyruvate for pyruvate carboxylase from guinea pig liver were in the same range, this would mean that changes in the pyruvate concentration below 100 μM will affect directly the rate of pyruvate carboxylation. In the presence of 20 mM L-lactate, the concentrations of pyruvate in the perfusion medium were 200 μM in experiments with guinea pig livers (Söling, Kleineke, and Janson, unpublished data). In the presence of exogenous fatty acids the concentration of pyruvate in the medium was lower but still above 100 μM. However, we do not know the concentration of pyruvate at the site of pyruvate carboxylase under these conditions. That the concentration of pyruvate in the medium does not have to be identical with the intracellular concentration of pyruvate has recently been emphasized by Sies *et al.* (1972) on the basis of *in vivo* experiments. There are, therefore, reasons to assume that the intramitochondrial concentration of pyruvate is not identical with the overall pyruvate concentration in the cell.

The distribution of pyruvate across the inner mitochondrial membrane is probably governed by the gradients of other anions across this membrane. These gradients depend, among other factors, on the ΔpH between the matrix and the extramatrix space and on the redox states of the NAD^+/NADH systems in these two compartments. At present any estimation of the concentration of pyruvate at the site of pyruvate carboxylase is purely speculative. Nevertheless, a decrease of the pyruvate concentration due to a shift in the redox state would offer the most reasonable explanation for an inhibitory effect of fatty acids on gluconeogenesis from L-lactate in guinea pig liver. This idea is supported by the finding that in the presence of high concentrations of pyruvate, fatty acids no longer inhibit gluconeogenesis but stimulate it, in experiments with isolated, perfused guinea pig livers (Söling *et al.*, 1970). While the formation of $^{14}CO_2$ from (1-^{14}C)pyruvate by isolated perfused rat liver is inhibited by fatty acids (Söling *et al.*, 1968), no effect can be observed with guinea pig livers (Söling *et al.*, 1970). Apparently pyruvate oxidation is inhibited by fatty acid oxidation in rat but not in guinea pig liver. It has already been mentioned that the redox state of the intramitochondrial $NAD^+/NADH$ system as calculated from the 3-hydroxybutyrate/acetoacetate ratio becomes more reduced in rat liver, but remains unchanged or becomes even more oxidized in guinea pig liver during enhanced fatty acid oxidation. This may explain the lack of inhibition of pyruvate oxidation in guinea pig liver, since NADH is a strong inhibitor of active pyruvate dehydrogenase.

Different factors account for the effect of fatty acids on gluconeogenesis in pigeon liver. In the presence of 20 mM L-lactate, the concentrations of pyruvate in the perfusion medium were similar to those measured in experiments with guinea pig livers (200 μM). Addition of fatty acids have no effect on the rate of gluconeogenesis from or the uptake of L-lactate, or on the lactate/pyruvate ratio in the medium. In the presence of a continuous infusion of L-lactate aminooxyacetate (0.2 mM) significantly increased the lactate/pyruvate ratio in the medium (Fig. 6) but did not inhibit gluconeogenesis.

We have no evidence that isolated guinea pig liver mitochondria in the presence of fatty acids exhibit changes in the ATP/ADP ratio different from those noted in rat liver. If this is true, we are left with three possible mechanisms: (*a*) the effect of fatty acids in rat liver does not result from an effect on the ATP/ADP ratio in the mitochondrial matrix; (*b*) gluconeogenesis from pyruvate and L-lactate is limited by pyruvate carboxylase in rat but not in guinea pig and pigeon liver; (*c*) the effect of fatty acids on the intramitochondrial ATP/ADP ratio is basically the same in both species but is overcompensated in guinea pig liver by a fall in the intramitochondrial concentration of pyruvate. In summary, one can state that fatty acids stimulate gluconeogenesis in rat liver, but this type of regulation cannot be generalized for other species.

The question remains whether fatty acids may still play a *permissive* role in those species where they do not have a *regulatory* function. When the oxidation of endogenous fatty acids is blocked by α-bromomyristic acid, gluconeogenesis from L-lactate is strongly inhibited in isolated, perfused guinea pig livers (Söling, Willms, and Janson, unpublished data). Similar observations have been made with isolated, perfused rat livers by McGarry *et al.* (1973) when fatty acid oxidation was inhibited by decanoyl-D(+)-carnitine. However, this inhibition could be seen only when the concentration of the glucogenic precursor L-lactate was high, not when the precursor concentrations were set to more physiological levels. We were unable to confirm the findings of McGarry, *et al.* (1973) in experiments with isolated liver cells from rats starved for 48 hours (Table IX). As can be seen, deca-

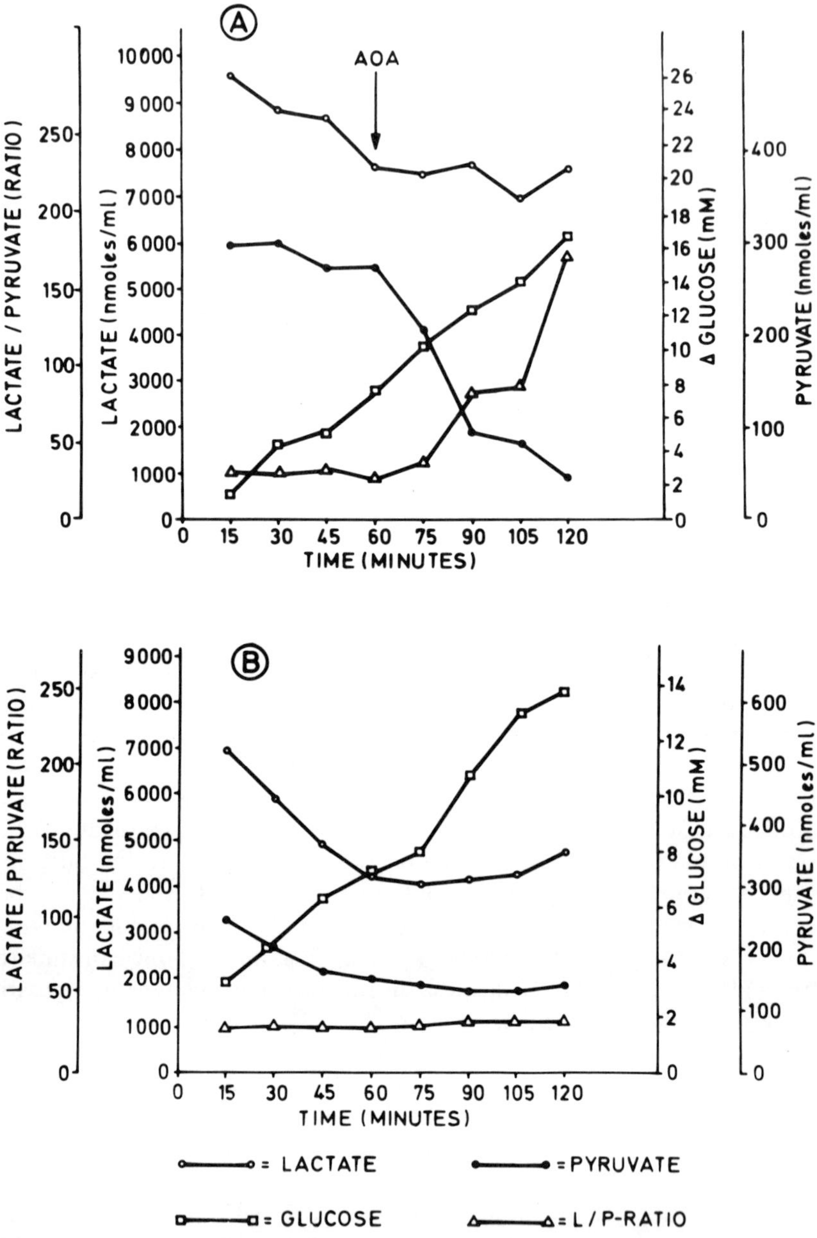

Figure 6. Effects of aminooxyacetate on the net production of glucose and the lactate/pyruvate ratio in the medium in experiments with isolated, perfused pigeon livers. The experiments were performed as described in the legend to Fig. 3. After addition of aminooxy acetate (*a*), the lactate/pyruvate ratio shows a dramatic increase, mainly because of a drop of the concentration of pyruvate. The rate of glucose production remains unaltered. The control experiment (*b*) shows a constant lactate/pyruvate ratio over the duration of the whole experiment. *Source:* H. D. Söling and G. Janson, unpublished experiments.

Table IX. Effect of decanoyl-D (+) carnitine (2 mM) on gluconeogenesis and ketogenesis by isolated liver cells from 48-hour-starved rats and guinea pigs and 72-hour-starved pigeons[a]

		Net formation of glucose (nmole/g dry wt/min)			Net formation of ketone bodies (nmole/g dry wt/min)		
Concentration of L-lactate + pyruvate	Decanoyl-D(+) carnitine	Rat	Guinea pig	Pigeon	Rat	Guinea pig	Pigeon
∅	∅	340 ±11	410 ±16	490 ±11	650 ±30	270 ±12	1560 ±85
	+	260 ±14	380 ±29	370 ±6	270 ±38	280 ±20	1730 ±32
1 mM	∅	450 ±18	640 ±33	3560 ±52	500 ±57	270 ±42	1220 ±53
	+	300 ±25	640 ±25	3470 ±202	340 ±28	290 ±12	1310 ±106
2 mM	∅	570 ±7	820 ±45	5650 ±32	470 ±53	300 ±29	970 ±75
	+	350 ±10	770 ±33	5190 ±309	310 ±21	260 ±28	1020 ±128
3 mM	∅	740 ±70	880 ±37	6570 ±106	490 ±28	330 ±8	790 ±64
	+	420 ±46	820 ±12	6420 ±21	270 ±49	320 ±25	670 ±425
4 mM	∅	710 ±57	980 ±45	7070 ±53	380 ±42	340 ±33	730 ±148
	+	470 ±14	850 ±29	6850 ±96	340 ±67	320 ±21	520 ±32
10 mM	∅	1670 ±70	1390 ±53	7330 ±85	620 ±14	400 ±45	590 ±64
	+	1190 ±21	1350 ±45	6860 ±85	440 ±28	440 ±20	610 ±32
2 mM + 2 mM oleate	∅	440 ±24			1950 ±92		
	+	430 ±38			670 ±39		
10 mM + 2 mM oleate	∅	1690 ±134			2210 ±53		
	+	1190 ±22			900 ±11		

[a] Lactate + pyruvate was added to the medium at the initial total concentrations given; the L-lactate to pyruvate ratio was always 9/1. Except when the addition of exogenous albumin bound oleate is indicated, ketogenesis resulted from the oxidation of endogenous precursors.

Source: Söling, Janson, Kuhn, and Heinz, unpublished data.

noyl-D(+)-carnitine effectively inhibits gluconeogenesis from L-lactate plus pyruvate even at concentrations of L-lactate + pyruvate as low as 1 mM when no exogenous fatty acids were supplied. This supports the idea that fatty acid oxidation is essential in rat liver for maximal rates of gluconeogenesis from physiological levels of precursors. In isolated liver cells from guinea pigs or pigeons, decanoyl-D-(+)-carnitine failed to affect gluconeogenesis from L-lactate plus pyruvate (Table IX), which is in accordance with the finding that exogenous fatty acids failed to enhance gluconeogenesis in isolated, perfused livers from these species. Under *in vivo* conditions an elevation of the plasma level of long-chain fatty acids leads in dogs to a decline in the rate of glucose production due to stimulation of release of insulin from the β-cells of the pancreas (Seyffert and Madison, 1967; Crespin *et al.*, 1973). Stimulation of insulin release by fatty acids has also been observed in the human (Balasse and Ooms, 1973). However, glucose production and gluconeogenesis are not identical. In the work of Seyffert and Madison (1967), the dogs were maintained on a fat-poor, carbohydrate-rich diet until 15 hours before the experiment. Therefore, it is likely that most of the glucose that was released prior to lipid infusion came from breakdown of liver glycogen. The release of insulin by fatty acids could of course inhibit glucose production resulting from glycogenolysis but not necessarily that from gluconeogenesis. Therefore a direct regulatory role of hepatic fatty acid oxidation for gluconeogenesis under physiological conditions is unlikely. A permissive role is possible but must be substantiated by further experiments. The effects of fatty acids observed in some species (e.g., rats) most probably can be obtained only under unphysiological conditions.

VIII. SPECIES DIFFERENCES IN THE RESPONSE TO SHORT-TERM HORMONAL EFFECTS

Glucose release from livers of fed animals is enhanced by glucagon (Sokal *et al.*, 1964; Glinsman and Mortimore, 1968; Jefferson *et al.*, 1968; Exton *et al.*, 1971a, 1972), epinephrine (Sokal *et al.*, 1964), 3′,5′-cAMP, and dibutyryl-3′,5′-cAMP (Glinsman and Mortimore, 1968; Jefferson *et al.*, 1968, Exton *et al.*, 1971a, 1972; Williams *et al.*, 1971) in all vertebrate species tested. This effect is attributed to a stimulation of glycogenolysis (For a review see Söling and Seufert, 1974). Increased glycogenolysis is not necessarily accompanied by a stimulation of gluconeogenesis.

Glucagon, epinephrine, and 3′,5′-cAMP stimulate gluconeogenesis from a variety of precursors in the isolated, perfused rat liver (Sokal, 1966; Exton and Park, 1968; Lewis *et al.*, 1970; Exton *et al.*, 1971a). Glucagon is active at near-physiological levels (Exton *et al.*, 1972). According to Exton *et al.* (1971b), the sensitivity of the isolated, perfused rat liver to glucagon decreases in the order: glycogenolysis > gluconeogenesis > ureogenesis > ketogenesis. Weinstein *et al.* (1973) found under similar conditions the following order: glycogenolysis > ureogenesis > ketogenesis > triglyceride formation and secretion. The effect of glucagon on gluconeogenesis from L-lactate, pyruvate, and L-alanine has been explained by positing a stimulation of fatty acid oxidation following increased lipolysis of liver triglycerides (J. R. Williamson, 1967). But other investigators (Fröhlich and Wieland, 1971) have noted that the effect of glucagon could be separated from the effect of fatty acids. Whether all effects of glucagon on liver metabolism can be simply explained on the basis of an increased phosphorylation of cellular membrane proteins, as proposed by Zahlten *et al.* (1972), remains an open question.

The mechanism of action of glucagon on gluconeogenesis from pyruvate, L-lactate and L-alanine has been explained by Exton *et al.* (1971b) as due to stimulation of cytosolic PEP-CK activity. In favor of this view is the finding that glucagon stimulates gluconeogenesis from propionate (Blair *et al.*, 1973b), a compound whose metabolism does not involve pyruvate carboxylase. Since glucagon acts within a few minutes after addition to the perfusion or incubation medium, its effect is not due to an increase in the concentration of PEP-CK protein. On the other hand, there is no evidence that glucagon or 3′,5′-cAMP exerts a direct stimulatory effect on this enzyme. Therefore, it seems likely that changes in the metabolite pattern owing to the action of glucagon are caused by effects on other metabolic reactions. This in turn would lead to increased gluconeogenesis by either allosteric regulatory mechanisms or by changes in substrate concentrations. Recently Blair *et al.* (1973a) and Zahlten *et al.* (1973) reported that glucagon also stimulates gluconeogenesis from fructose, dihydroxyacetone, glycerol, and xylitol in experiments with isolated, perfused rat livers and isolated rat liver cells. The reactions most likely to be affected by glucagon under these conditions would be the fructose-1,6-diphosphatase-phosphofructokinase step. Also pyruvate oxidation could be inhibited, which would leave more C_3-units for pyruvate carboxylation (Zahlten *et al.*, 1973). Possibly the stimulatory effect of glucagon or 3′,5′-cAMP on gluconeogenesis from precursors entering the gluconeogenic pathway below PEP are mediated by the same mechanisms by which gluconeogenesis from fructose, glycerol, dihydroxyacetone, and xylitol is stimulated. According to recent results of Kneer *et al.* (1974), epinephrine stimulates gluconeogenesis in isolated rat hepatocytes at the PFK-FDPase step in a manner independent of 3′,5′-cAMP.

When the effect of glucagon, epinephrine, or dibutyryl-3′,5′-cAMP was examined in isolated, perfused livers from fed guinea pigs, we observed strong stimulation of glycogenolysis (Söling *et al.*, 1970). However, in isolated, perfused livers from guinea pigs starved for 48 hours, glucagon as well as 3′,5′-cAMP and dibutyryl-3′,5′-cAMP failed completely to affect gluconeogenesis from, and utilization of, L-lactate under conditions revealing a significant stimulation in isolated, perfused rat livers (Table VIII). Ureogenesis and ketogenesis, on the other hand, were stimulated in livers from both species by glucagon and 3′,5′-cAMP (Söling *et al.*, 1970). Loten *et al.* (1974) also failed to observe stimulation of gluconeogenesis from L-lactate by glucagon or 3′,5′-cAMP in isolated, perfused mouse livers, although both agents stimulated glycogenolysis in the same system.

Several authors (Glinsman and Mortimore, 1968; Mackrell and Sokal, 1969; Menahan and Wieland, 1969; Johnson *et al.*, 1972) have described inhibition of glucagon-stimulated gluconeogenesis by insulin in experiments with isolated, perfused rat livers or isolated rat liver cells. However, Zahlten *et al.* (1973) and Söling, Kehl, and Heinz (Table X) did not observe an inhibition of glucagon-stimulated gluconeogenesis from various precursors in isolated rat liver cells. In the absence of stimulation of gluconeogenesis by glucagon or 3′,5′-cAMP, significant *in vitro* effects of insulin on gluconeogenesis have generally not been noted. However, Rudorf *et al.* (1970, 1971) have described a significant inhibition of gluconeogenesis from L-alanine by insulin *in vitro* in experiments with isolated, perfused livers from mice and rats. *In vivo* measurements of "glucose production" do not

Table X. Effects of glucagon, insulin, and glucagon + insulin on glucose formation by isolated liver cells from 48-hour-starved rats in the presence of various precursors[a]

Experimental condition	Net formation of glucose (nmole/g/min)				
	No precursor	L-Lactate	Propionate	Glycerol	L-Alanine
Control	63 ±5	649 ±51	399 ±23	629 ±52	354 ±13
Glucagon	—	866 ±56	481 ±44	660 ±34	380 ±9
Insulin	—	700 ±62	482 ±18	773 ±50	336 ±31
Glucagon + insulin	—	838 ±60	662 ±35	816 ±49	424 ±21

[a] The initial precursor concentration was always 10 mM. The initial concentration of glucagon was 2.7×10^{-7} M and that of insulin 1 mU/ml. The incubation was performed for 60 minutes at 37°C.

Source: Data from Söling and Seufert (1974).

determine the rate of gluconeogenesis. Therefore any data concerning short-term *in vitro* effects of insulin on gluconeogenesis have to be considered with great caution. The effect of insulin on gluconeogenesis in the isolated, perfused guinea pig liver has not yet been examined. Long-term treatment with glucagon or insulin will of course influence gluconeogenesis also by an increase or decrease of the activity of regulatory enzymes. However, it is generally accepted that these effects need at least 2 to 4 hours to produce significant changes. Taunton *et al.* (1972) and Greene *et al.* (1974) reported recently that in rat and human liver, respectively, glucagon and insulin injection causes dramatic changes in the activities of phosphofructokinase, pyruvate kinase, and fructose-1,6-diphosphatase as early as 4 minutes after administration. However, in an extensive study (Söling, Kuhn, and Janson, in preparation) of their effect we were unable to confirm the data of Taunton *et al.* (1972).

IX. ANION TRANSLOCATION THROUGH THE INNER MITOCHONDRIAL MEMBRANE AND ITS POSSIBLE ROLE IN SPECIES DEPENDENT REGULATION OF GLUCONEOGENESIS

We restrict the discussion of anion movements to anionic species involved in exchange reactions related to gluconeogenesis, although concomitant cation and/or proton movements cannot be ignored. Since the initial reports of Chappell and co-workers (Chappell, 1968; Chappell and Haarhoff, 1967), several more carrier systems have been reported to exist in the inner mitochondrial membrane (for a more detail discussion of this subject see Chapter 5, this volume). Most of these systems have been characterized by substrate selectivity and sensitivity to specific inhibitors (for review see Chappell, 1968; Meijer, 1971; Klingenberg, 1970; Harris, 1972).

As far as it is known, mitochondrial membranes from vertebrates have similar if not identical carrier systems. Qualitative differences reported thus far are restricted to those between different tissues—for example, the activity of the tricarboxylate carrier is either low (England and Robinson, 1969) or completely absent in rat heart mitochondria but has a high capacity in rat liver mitochondria (Scott *et al.*, 1972; Sluse *et al.*, 1971). The few studies done with liver mitochondria from animals other than the rat refer to the transport of phosphoenolpyruvate in mitochondria from ox liver (Robinson, 1971) and the transport of oxaloacetate in mitochondria from guinea pig, rabbit, and pigeon liver (Gimpel *et al.*, 1973). The transport of oxaloacetate is of considerable interest, since it has been shown that oxaloacetate, at least at physiological concentrations (1–4 nmole/g liver), cannot penetrate rat liver mitochondrial membranes (Harris and Berent, 1970; Haslam and Griffiths, 1968). The maximal rate of penetration found by these authors at saturating concentrations of oxaloacetate was 15 nmole/mg protein/min with an apparent K_m value of 40 μM, when following the rate of oxidation of intramitochondrial NAD(P)H by added oxaloacetate. It was suggested that this compound is transported, like L-malate, by way of the dicarboxylate carrier system. This possibility has been recently reinvestigated using rapid reaction spectrophotometry (Gimpel *et al.,* 1973). The apparent K_m and V_{max} determined under this condition were somewhat higher—namely 80 to 130 μM and 80 to 270 nmole/mg protein/min, respectively. In addition, these authors could demonstrate that both the oxidation of intramitochondrial NAD(P)H in the presence of oxaloacetate and the efflux of oxaloacetate generated intramitochondrially were inhibited by 2-*n*-butylmalonate. Moreover, when testing mitochondria from different species, they found that in comparison to rat liver mitochondria, mitochondria from rabbit and pigeon liver were relatively impermeable (0.1–1 nmole/mg protein/min) to oxaloacetate even at concentrations up to 10 mM. The permeability for this compound in guinea pig liver mitochondria was about half the rate found in rat liver mitochondria.

Gimpel *et al.* (1973) discuss the physiological significance of these differences and note that there is some relationship between the capacity of the oxaloacetate translocating system and the intracellular distribution of PEP-CK: in guinea pig or pigeon liver, where PEP for gluconeogenesis is partially or exclusively synthesized intramitochondrially, a translocation of oxaloacetate would be of minor or no importance for gluconeogenesis, whereas in rat liver, where the synthesis of PEP for gluconeogenesis proceeds mainly in the cytosol, a high rate of translocation of oxaloacetate would be expected. In rabbit liver PEP formation from oxaloacetate takes place in both the mitochondria and the cytosol (Johnson *et al.,* 1970; Garber and Hanson, 1971; see Chapter 9, this volume). However, studies with isolated livers from starved rabbits perfused with L-lactate in the presence of aminooxyacetate show that 60 to 80% of carbon utilized for glucose formation is brought about by outward transport of intramitochondrial aspartate (Arinze *et al.,* 1973).

It should be also mentioned briefly that according to recent findings of Bradford and McGivan (1973) and McGivan *et al.* (1973) the rate of glutamate deamination by rat liver mitochondria is limited by the transport of glutamate into the mitochondria. Since the rate of glutamate uptake was estimated to be relatively slow in comparison to other mitochondrial carrier systems (5 nmole/mg protein/min vs. 50–100 nmole/mg protein/min at 10°C), this particular transport, which is indirectly linked to the outward transport of aspartate, may become rate limiting. That anion translocation by way of the mitochondrial membrane could become rate limiting under certain experimental conditions was first shown in the isolated, perfused rat liver (J. R. Williamson *et al.*, 1970) using 2-*n*-butylmalonate, a competitive inhibitor of the malate/phosphate exchange (Robinson and Chappell, 1967). The 2-*n*-butylmalonate inhibited gluconeogenesis from pyruvate or lactate between 50 and 80%. The possible mechanisms for this effect on gluconeogenesis in rat liver on one hand, and pigeon liver on the other, are illustrated in Figs. 7 and 8. Interestingly, inhibition of gluconeogenesis by 2-*n*-butylmalonate was accompanied by a proportional inhibition of oxygen uptake in isolated, perfused pigeon livers (Söling *et al.*, 1973), whereas the oxygen uptake by isolated, perfused rat livers remained unaffected by 2-*n*-butylmalonate (J. R. Williamson *et al.*, 1970; Söling *et al.*, 1973), although the inhibition of gluconeogenesis was similar in both species. In both species a sufficient cycling of L-malate between the intra- and the extramatrix space seems to be a precondition for gluconeogenesis from L-lactate, irrespective of whether PEP is formed intra- or extramitochondrially. A direct competitive inhibition of the transport of PEP by 2-*n*-butylmalonate in pigeon liver seems to be unlikely, since, at least in rat liver mitochondria,

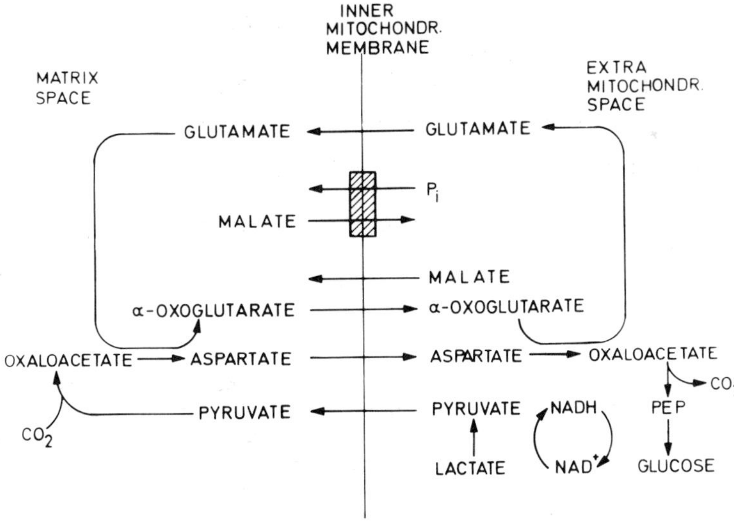

Figure 7. Anion movements across the inner mitochondrial membrane in rat liver during gluconeogenesis from L-lactate. The transamination cycle involves a continuous outflow of α-oxoglutarate in exchange for extramatrix L-malate. Since L-malate is formed mainly in the matrix space, it first must be translocated into the extramatrix space, involving an exchange against extramatrix inorganic phosphate. Therefore, an inhibition of the latter exchange mechanism by *n*-butylmalonate (indicated by the hatched bar) could block gluconeogenesis from L-lactate by inhibiting the transamination cycle. *Source*: Söling *et al.* (1973).

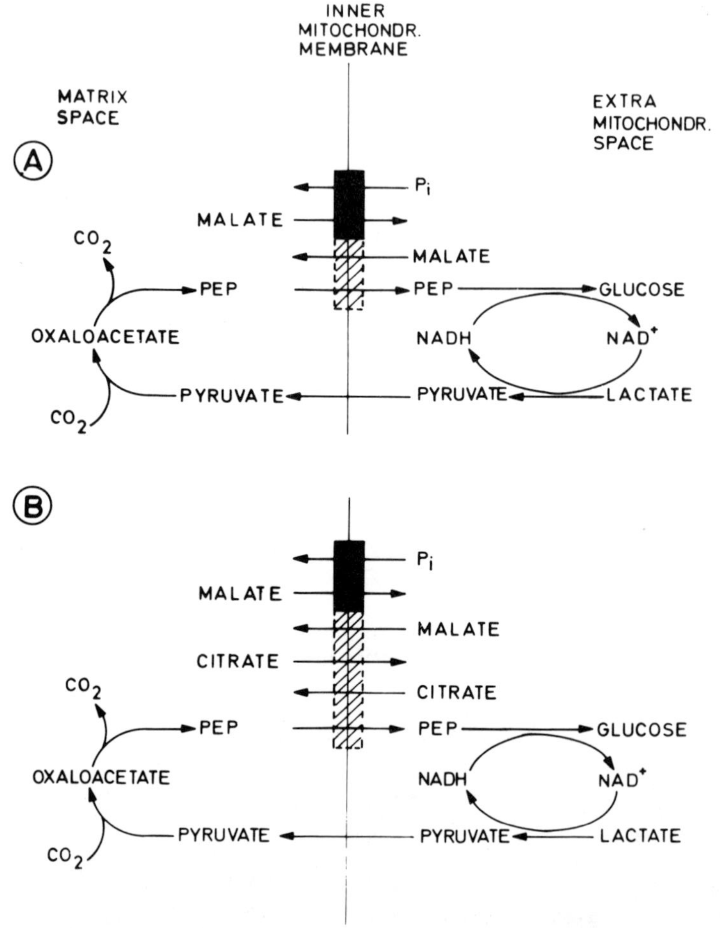

Figure 8. Anion movements across the inner mitochondrial membrane in pigeon liver during gluconeogenesis from L-lactate. Phosphoenolpyruvate is formed intramitochondrially and is translocated into the cytosol in exchange for L-malate (*a*) or for citrate (*b*). The outward transport of phosphoenolpyruvate is inhibited by *n*-butylmalonate either directly (hatched bar) or indirectly (solid bar) by the inhibitory effect of *n*-butylmalonate on the L-malate/phosphate exchange. *Source*: Söling *et al.* (1973).

a K_i of 3.5 mM (Palmieri *et al.*, 1972) was estimated for the tricarboxylate carrier system. This value is about 10 times higher than the K_i determined for the dicarboxylate carrier (Robinson and Williams, 1970).

Direct inhibition of gluconeogenesis by benzene-1,2,3-tricarboxylate, a competitive inhibitor of the tricarboxylate carrier (Robinson *et al.*, 1971b), could not be demonstrated in isolated, perfused rat or pigeon liver, even at concentrations as high as 50 mM (Söling *et al.*, 1973). However, the transport of PEP itself was sensitive to this inhibitor when tested with isolated mitochondria from both species (Kleineke, unpublished data). It is probable that this lack of inhibition noted in perfused rat and pigeon liver is due to a slow uptake of benzene-1,2,3-tricarboxylate by the liver.

Glisoxepide, a sulfonylurea compound, inhibits gluconeogenesis from L-lactate and L-alanine but not from pyruvate in isolated rat liver cells (Table XI*a*) (Söling and Seck, 1975) but did not inhibit gluconeogenesis from L-lactate in isolated pigeon liver cells (Table XI*b*). However glisoxepide specifically inhibited the translocation of L-aspartate through the inner membrane of rat liver mitochondria (Söling and Seck, 1975). This confirms the concept that in rat liver, gluconeogenesis from L-lactate, but not from pyruvate, involves the translocation of L-aspartate, whereas in pigeon liver,

Table XI*a*. Effects of glisoxepide on gluconeogenesis from L-alanine, L-lactate, and pyruvate by isolated rat liver cells[a]

	Rates (μmole/g dry wt/min)			
Experimental condition	Concentration of glisoxepide (M)	Glucose production	Formation of L-lactate + pyruvate	Lactate/pyruvate ratio
Control, no substrate	∅	0.37 ±0.01	0.31 ±0.03	3.2 ±0.5
L-Alanine, control	∅	1.54 ±0.05	1.91 ±0.21	3.3 ±0.2
L-Alanine	5×10^{-4}	0.85 ±0.10	2.28 ±0.22	5.4 ±0.4
L-Alanine	1×10^{-4}	0.88 ±0.03	2.07 ±0.08	5.1 ±0.4
L-Alanine	5×10^{-5}	1.20 ±0.01	1.86 ±0.01	4.5 ±0.1
L-Lactate, control	∅	2.53 ±0.02	—	10.6 ±0.9
L-Lactate	5×10^{-4}	1.23 ±0.02	—	20.8 ±1.1
L-Lactate	1×10^{-4}	1.77 ±0.10	—	15.6 ±0.3
Pyruvate, control	∅	2.79 ±0.02	—	1.0 ±0.2
Pyruvate	5×10^{-4}	2.56 ±0.01	—	1.0 ±0.2

[a] The initial precursor concentration was always 10 mM. Cells were isolated as described by Söling and Seck (1975). Cells corresponding to 10 to 15 mg dry weight were incubated in a total volume of 2 ml and shaken in a waterbath at 37°C and 120 strokes/min for 60 minutes. The incubation was stopped by addition of 1 ml of 1 N $HClO_4$ (mean values ±SEM) ($n = 4$).
Source: Data from Söling and Seck (1975).

Table XI*b*. Effects of glisoxepide on gluconeogenesis from L-lactate (initial concentration 10 mM) by isolated pigeon liver cells[a]

Experimental condition	Concentration of glisoxepide (M)	Glucose formation (μmole/g dry wt/min)	Lactate/pyruvate
Control, no substrate	Ø	0.48 ±0.02	—
L-lactate	Ø	9.60 ±0.06	25.4 ±0.8
L-lactate	1×10^{-3}	7.59 ±0.05	35.6 ±3.2
L-lactate	5×10^{-4}	8.33 ±0.27	35.5 ±1.2
L-lactate	1×10^{-4}	9.43 ±0.04	33.4 ±0.1

[a] Isolation and incubation was performed as described in the footnote to Table XI*a*. Each incubation vessel contained cells corresponding to 3–5 mg dry weight (mean values ± SEM) (n = 4).
Source: Data from Söling and Seck (1975).

gluconeogenesis from L-lactate is independent of the transport of L-aspartate (Söling *et al.*, 1973).

Inhibition of the adenine nucleotide translocator by long-chain fatty acids has been observed in liver mitochondria from rats, guinea pigs, ground squirrels, and rhesus monkeys (Wojtczak *et al.*, 1969; Shug *et al.*, 1971; Vaartjes *et al.*, 1973). Metabolic conditions known to alter tissue levels of acyl-CoA (starvation, diabetes, hibernation) caused corresponding changes in the rate of adenine nucleotide translocation (Lerner *et al.*, 1972). The tricarboxylate carrier is also inhibited by acyl-CoA (Halperin *et al.*, 1972; Shug and Shrago, 1973). However, the overall long-chain acyl-CoA levels measured *in vivo* are 3 to 5 times higher (McLean *et al.*, 1971), than those that cause a 90% inhibition of the adenine nucleotide translocation *in vivo* (Lerner *et al.*, 1972; Shug *et al.*, 1971) and 20 to 30 times higher than those that block the tricarboxylate carrier (Shug and Shrago, 1973). This casts some doubt on the importance of long-chain acyl-CoA in the regulation of adenine nucleotide translocation, although it is possible that most of the long-chain acyl-CoA present in the liver is bound to proteins. Moreover, inhibition of the tricarboxylate carrier system in pigeon liver should decrease gluconeogenesis from L-lactate, but even high concentrations of long-chain fatty acids had no effect (Söling *et al.*, 1973).

Other inhibitors of single carrier systems (e.g., the sulfhydryl group blocking reagents *N*-ethylmaleimide—NEM—or mersalyl) are too unspecific to permit a distinguishing between a direct effect at the site of the carrier or an unspecific inhibition at some other locus. This is particularly true in complex systems like isolated cells, perfused organs, or *in vivo* experiments. As already mentioned, liver mitochondria from various vertebrates do not have qualitatively different carrier systems. From studies of the tricarboxylate carrier system in rat liver mitochondria, it is known that the specificity of this carrier is broad (Kleineke *et al.*, 1973). However no reports concerning the specificity of the same carrier system in different species are available.

The apparent K_m values reported for the tricarboxylate carrier system in rat liver are between 0.1 and 0.4 mM for most substrates, reflecting approximately the physiological concentration range of these substrates in the cytosol. Since these data are derived from rat liver mitochondria, the kinetics of the transport of phosphoenolpyruvate were determined in liver mitochondria from rats, guinea pigs, and pigeons. The transport of phosphoenolpyruvate in exchange for mitochondrial citrate by way of the tricarboxylate carrier system is a first-order process in all species tested. There is a decrease in the apparent K_m in the order rat > guinea pig > pigeon (Table XII). The V_{max} values decrease in the same order. Both K_m and V_{max} of the inward transport of PEP by way of the tricarboxylate carrier decrease with increasing activity of intramitochondrial PEP-CK. The

Table XII. Kinetic constants of phosphoenolpyruvate uptake by way of the tricarboxylate carrier system in mitochondria from different species[a]

Species	K_m (mM)	V_{max} at 10°C (nmole/min/mg)
Rat	1.0	62
Guinea pig	0.35	55
Pigeon	0.048	13

[a] The mitochondria were preloaded with citrate as described by Kleineke *et al.* (1973) and incubated at 10°C in the presence of different concentrations of phosphoenolpyruvate. The reaction was terminated by addition of benzene-1,2,3-tricarboxylate (50 mM final concentration). The mitochondria were separated by rapid centrifugation in a modified Eppendorf centrifuge with swing-out rotor and cooling device. Analysis was performed as described by Kleineke *et al.* (1973). The incubation medium contained (final concentrations): mannitol 200 mM; triethanolamine-Cl (pH 7.5) 25 mM; *n*-butylmalonate 5 mM; rotenone 2 μg/ml; antimycin A 0.05 μg/ml, and about 10 mg of mitochondrial protein.

Source: Data from Kleineke, Tager, Sauer, and Söling, in preparation.

maximal velocities reported here are higher than those determined for rat (Palmieri *et al.*, 1972; Robinson, 1973) and ox (Robinson, 1971) liver mitochondria. The rate noted for pigeon liver, which was the slowest under our conditions, can still account for a gluconeogenic rate of more than 3 μmole/g liver/min, since the rate of exchange is doubled every 5°C (Robinson *et al.*, 1971a) and the liver contains about 70 to 90 mg of mitochondrial protein per gram of liver.

A new approach to determine the kinetic constants of the inward and the outward transport was made by Sluse *et al.* (1972, 1973). These authors reported that K_m values for the outward transport of dicarboxylate anions by way of the oxoglutarate carrier system in rat heart mitochondria were 1 to 3 orders of magnitude higher than the respective inward K_m values. The model proposed by Sluse *et al.* (1972, 1973) for the translocation reaction is a "rapid equilibrium random bi-bi" mechanism (Cleland's nomenclature, 1963a–1963c), which implies that the anions transported bind independently in a nonordered random manner at the different sites of the membrane carrier and that the Michaelis constants determined represent the true dissociation constants of the respective substrate-translocator complex. The techniques used by Sluse *et al.* (1972, 1973) for heart mitochondria have not been applied to liver mitochondria or to other translocator systems.

Large differences in the binding constants for the same anion at the inner and the outer surface of the inner mitochondrial membrane would favor the existence of an extra-/intramitochondrial concentration gradient that would affect the direction of metabolic flux. We have studied the kinetic constants of the PEP-citrate exchange in isolated pigeon liver mitochondria using Sluse's technique (Söling, 1974, Kleineke and Söling, 1974). The external anion (Fig. 9) was PEP, the internal counteranion (Fig. 10) citrate. The kinetic data calculated from the respective replots of slopes and intercepts are summarized in Table XIII. The apparent K_m values for the inward (K_A) and the outward (K_B) transport differ three orders of magnitude from each other and are identical with the dissociation constants of the carrier-substrate complex K_{iA} and K_{iB}, respectively. (Table XIII). In these experiments the calculation of the internal substrate concentration (and by this the outward K_m) was derived from measurement of the different distribution spaces of mitochondria for (^{14}C)-sucrose and tritiated water, respectively; this is the general method of determining the matrix water space (for review, see Tedeschi 1971). This determination is based on the assumptions implied by the "two space model": (1) the extramatrix, osmotically inactive "substrate space" is identical with the "sucrose space," and (2) each mitochondrion has two separate "spaces"—one permeable and the other relatively impermeable to sucrose. Assuming a one-space model where each individual mitochondrion is made up of a single internal space and the difference in sucrose permeability is mimicked by a size distribution of the mitochondrial

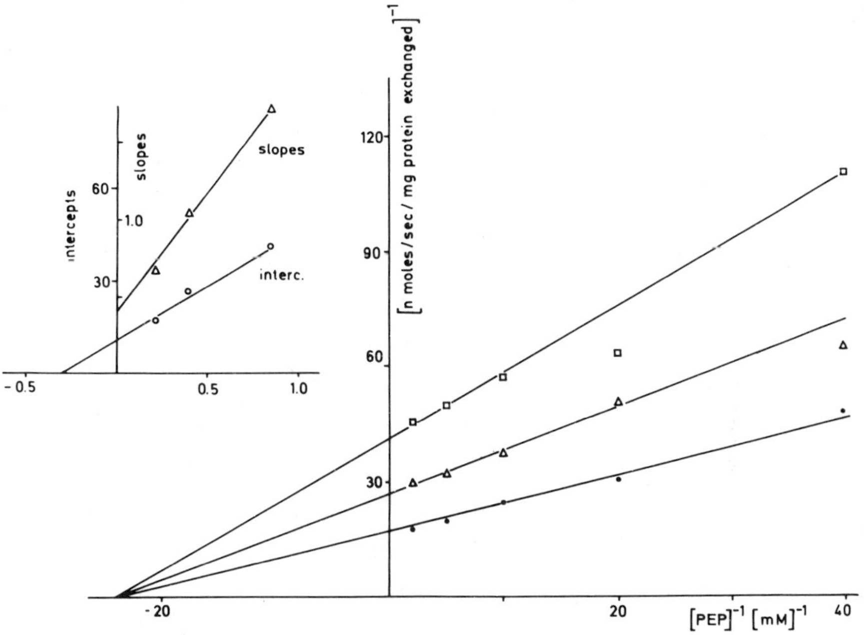

Figure 9. Double reciprocal plots of exchange of external phosphoenolpyruvate with internal citrate at three different concentrations of internal citrate. The experiments were performed with isolated pigeon liver mitochondria, as described in footnote *a*, Table XII. The insert gives the replots of the slopes and intercepts versus the internal concentration of citrate. The internal citrate concentration was varied according to Sluse *et al.* (1972). *Source*: Kleineke and Söling, in preparation.

population, the internal volume would increase about two to five times and consequently the outward K_m values would decrease. Nevertheless these results show that large differences between the binding constants for inward and outward transport exist in liver mitochondria and can be found also for the tricarboxylate carrier system. In view of the species dependent differences of the apparent $K_{m\text{PEP}}$ values for the inward transport of PEP, one would also expect species dependent differences for the outward transport. Further work on this problem is in progress.

Harris *et al.* (Harris, 1970, Harris and Bangham, 1972; Harris and Berent, 1970) have pointed out that the Gibbs-Donnan relation governs the out/in distribution of anions in isolated mitochondria from rat liver according to the equation:

$$\frac{[A^-_{\text{out}}]}{[A^-_{\text{in}}]} = \left[\frac{[B^{2-}_{\text{out}}]}{[B^{2-}_{\text{in}}]}\right]^{1/2} = \left[\frac{[C^{3-}_{\text{out}}]}{[C^{3-}_{\text{in}}]}\right]^{1/3}$$

Therefore the out/in concentration gradient of a given anion species de-

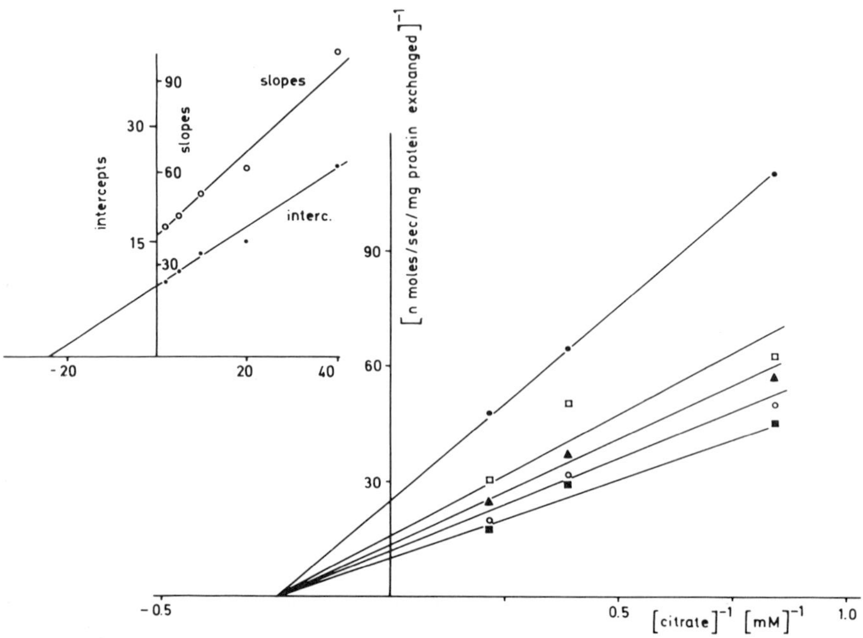

Figure 10. Double reciprocal plots of exchange of internal citrate with external phosphoenolpyruvate at five different concentrations of external phosphoenolpyruvate. (Same experiments as those of Fig. 9.) The insert gives the replots of the slopes and intercepts versus the external concentration of phosphoenolpyruvate. *Source*: Kleineke and Söling, in preparation.

pends strongly on the out/in proton gradient (Δ pH). It means further that the concentration change of a multiply charged anion will affect the intramitochondrial concentration of a singly charged anion. For example, a decrease of the intramitochondrial concentration of citrate due to either increased metabolic utilization or increased intramitochondrial pH, will drop the internal pyruvate concentration, influencing the activity of pyruvate utilizing enzymes. Multiply charged anions will compete strongly with less charged anions for the limited internal positive charge. Thus the intramitochondrial accumulation of citrate leads to an extrusion of pyruvate out of the mitochondria (Harris and Berent, 1970). There is no reason to assume that the Gibbs-Donnan relation holds only for rat liver mitochondria. However the following questions remain unanswered.

1. Are there species dependent differences in the physiological Δ pH between the intra- und extramatrix space?
2. Is the translocation of anions for gluconeogenesis affected by the

known species differences in the

$$\frac{NAD^+_{out}/NADH_{out}}{NAD^+_{in}/NADH_{in}} \text{ ratio?}$$

3. Is the anion distribution according to the Gibbs-Donnan relation affected by limitations in the specific translocator system, and if so, does this interaction differ between species?

Some of the results obtained with isolated mitochondria may be artifacts due to changes in membrane configuration during mitochondrial isolation. This is in part supported by the finding (Harris *et al.*, 1971) that the addition of external colloids can diminish the permeability of the inner membrane of isolated rat liver mitochondria toward several anions. Similarly, some of the species differences could reflect a different "quality" of mitochondria separated from specific cells.

Can metabolic fluxes be limited at the step of anion translocation? Berry (1971) observed, in experiments with isolated rat liver cells, that the reduction of pyruvate was limited by the transfer of reducing equivalents from the mitochondria into the cytosol. Scholz *et al.* (personal communication) found that in the isolated perfused rat liver the oxidation of ethanol is not limited by the transfer of reducing equivalents into the cytosol, the contrary was found by Berry and Kun (1972) with isolated rat liver cells. Also, the transfer of hydrogen through the inner mitochondrial membrane can limit hepatic gluconeogenesis in the rat—for example, in the presence of a mixture of sorbitol, glycerol, and pyruvate (Werner and Berry, 1974) or xylitol (J. R. Williamson *et al.*, 1971a).

Table XIII. Constants of the transport of anions in both directions by way of the tricarboxylate carrier system in pigeon liver mitochondria

Direction of transport	Values[a]		
	K_m (mM)	K_i (mM)	V_{app} (nmole/sec/mg protein)
Out → in	0.038	0.041	0.093
In → out	3.76	3.99	0.111

[a] Abbreviations: K_m = Michaelis-Menten constant, K_i = true binding constant, V_{app} = apparent maximum velocity at + 4°C.

Very recently Müllhofer *et al.* (1974) have proposed that in the isolated, perfused rat liver the rate of gluconeogenesis from 5 mM L-lactate, a concentration easily reached during severe muscle work, is limited by the rate of hydrogen transport into the mitochondria. Gluconeogenesis was stimulated under these circumstances by dibutyryl-cAMP, probably by increasing hydrogen transfer. The main supports for this finding were that small concentrations of pyruvate stimulated gluconeogenesis from L-lactate and that the stimulatory effect of dibutyryl-cAMP disappeared when the capacity for the transfer of hydrogen equivalents had been enhanced by prior treatment with triiodothyronine, an inducer of mitochondrial glycerol-3-phosphate oxidase. Since gluconeogenesis from L-lactate is not stimulated by glucagon or dibutyryl-cAMP in livers from guinea pigs and mice, it is possible that under certain conditions gluconeogenesis is limited in a species dependent manner by anion translocation.

X. SPECIES DEPENDENT REGULATION OF SPECIFIC ENZYMATIC STEPS

A. Relative Activities of Gluconeogenic and Glycolytic Enzymes in Livers from Different Species

Among the four "classical" gluconeogenic enzymes—pyruvate carboxylase, PEP-CK, fructose-1,6-diphosphatase, and glucose-6-phosphatase—PEP-CK activity exhibits by far the greatest species differences. The activities of the other three enzymes vary only to a minor degree from species to species (Tables XV to XXII and XXIV and XXV). Among the glycolytic enzymes, glucokinase shows the greatest species differences. The enzyme ratios (gluconeogenic enzyme activity/glycolytic enzyme activity) differ greatly (cf. glycolytic and glucogenic enzyme activities in Tables XV to XXII and XXIV and XXV and see Table XIV. However, enzyme activity determinations are carried out under V_{max} conditions that certainly do not occur *in vivo*. Moreover, at least some of these enzymes (pyruvate carboxylase, pyruvate kinase, fructose-1,6-diphosphatase, phosphofructokinase) are controlled by homotropic and/or heterotropic allosteric mechanisms. The species differences in enzyme activities, enzyme ratios, and intracellular distribution of enzymes point already to different capacities for gluconeogenesis and different types of regulation in these species.

B. Regulation of Pyruvate Carboxylase in Different Species

The activity of pyruvate carboxylase varies only slightly among species compared to the large differences observed for PEP-CK (Table XV). The

Table XIV. Enzyme activity ratios of glucogenic and glycolytic enzymes in livers from fed and starved rats, guinea pigs, and pigeons

	Ratios			
Species[a]	Glucose-6-phosphatase/ hexokinase + glucokinase	Fructose-1,6-diphosphatase/ phosphofructokinase	Pyruvate carboxylase/ pyruvate kinase	PEP-CK/ pyruvate kinase
Rat (fed)	1.37	8.89	0.123	0.088
Rat (starved)	4.12	18.88	0.308	0.253
Guinea pig (fed)	5.28	2.51	0.784	2.133
Guinea pig (starved)	17.53	4.05	0.917	3.352
Pigeon (fed)	—	—	0.280	2.538
Pigeon (starved)	—	—	0.385	2.117

[a] Rats and guinea pigs were starved for 48 hours, pigeons for 72 hours.
Source: Data from Söling *et al.* (1970, 1973).

Table XV. Activity of pyruvate carboxylase in livers from various vertebrate species

Species	Metabolic state	Temperature (°C)	Enzyme activity (μmole/g liver/min)	Reference
Rat	Fed	30°	4.43 ±0.22	Weiss *et al.* (1972)
Rat	Diabetic (acute)	30°	5.20 ±0.28	Weiss *et al.* (1972)
Rat	Diabetic (chronic)	30°	9.31 ±0.39	Weiss *et al.* (1972)
Rat	Fed	30°	4.64 ±0.22	Söling *et al.* (1973)
Rat	Starved (48 hours)	30°	11.67 ±1.77	Söling *et al.* (1973)
Guinea pig	Fed	30°	5.14 ±0.39	Söling *et al.* (1973)
Guinea pig	Starved (48 hours)	30°	7.26 ±0.40	Söling *et al.* (1973)
Sheep	Fed	30°	2.52	Taylor *et al.* (1971)
Sheep	Starved (6 days)	30°	7.64	Taylor *et al.* (1971)
Sheep	Diabetic	30°	24.59	Taylor *et al.* (1971)
Dog	Fed	Not given	0.90	Garthoff *et al.* (1972)
Pigeon	Fed	30°	7.60 ±1.31	Söling *et al.* (1973)
Pigeon	Starved (48 hours)	30°	11.41 ±1.28	Söling *et al.* (1973)
Man	Fasted overnight	30°	1.1 ±0.7	Böttger *et al.* (1969)

kinetics of pyruvate carboxylase and its regulation have been extensively studied (for review see Utter and Scrutton, 1969). However a systematic evaluation of species differences is limited to a few parameters or species. Acetyl-CoA is the most important allosteric activator of pyruvate carboxylase (Utter and Keech, 1960). As listed in Table XVI the K_a value for acetyl-CoA varies over a rather narrow range. Significantly lower values have only been reported for the enzyme from rabbit and chicken liver (Utter and Fung, 1971). However, Scrutton and White, (1974) found a significantly lower K_a value for acetyl-CoA with human pyruvate carboxylase (61 μM) as compared with the rat enzyme (130 μM) when testing at physiological pyruvate concentrations (0.1 mM) and at pH 7.2, instead of using optimal test conditions, where this value for both enzymes was very similar (Table XVI). Whereas the chicken liver enzyme is completely inactive in the absence of acetyl-CoA, the enzymes from rat and human liver retain about 20% of their maximal activity under the same condition (for review see Scrutton and White, 1974).

Pyruvate carboxylase from chicken liver is also activated by other acyl-CoA derivatives (Scrutton and Utter, 1967). The K_a for propionyl-CoA at pH 7.8 was 110 μM. We found a K_a value of 57 μM for propionyl-CoA with purified rat liver pyruvate carboxylase at pH 7.7 (Seufert, Volkmann, and Söling, unpublished data).

Different inhibitor constants for the allosteric inhibitor acetoacetyl-CoA have been described by Utter and Fung (1971). Their K_i values for the liver

Table XVI. Activator constants (K_a) for acetyl-CoA with liver pyruvate carboxylase from different species

Species	K_a, Acetyl-CoA (μM)	pH	References
Rat	19	7.8	Utter and Fung (1971)
Guinea pig	24	7.9	Söling *et al.* (1970)
Rabbit	9	7.8	Utter and Fung (1971)
Pigeon	20	7.7	Dugal and Louis (1974)
Chicken	2.2	7.8	Utter and Fung (1971)
Pig	23[a]	8.0	Warren and Tipton (1974)
Human	15	7.8	Scrutton and White (1974)

[a] After purification of acetyl-CoA, a value of 5.7 was found.

enzyme from chicken were 3.2×10^{-4} M, from rabbit 9.7×10^{-4} M, and from rat 6.6×10^{-4} M, respectively. Since neither Seufert *et al.* (1971) nor McClure and Lardy (1971) were able to find an inhibitory effect of acetoacetyl-CoA on rat liver pyruvate carboxylase, however, the inhibitory effect of acetoacetyl-CoA seems questionable. The same holds for the inhibition of pyruvate carboxylase by malonyl-CoA, which had been described for pyruvate carboxylase from chicken (Scrutton and Utter, 1967) and rat (Wimhurst and Manchester, 1970a) liver, but could not be confirmed independently by Seufert *et al.* (1971) and McClure and Lardy (1971) with rat liver pyruvate carboxylase. According to Scrutton and White (1974), the chicken liver enzyme is activated by tris(hydroxymethyl)aminomethane$^+$ and inhibited by α-oxoglutarate, whereas the enzymes from rat and human liver are not affected by these two compounds. Interestingly, Scrutton *et al.* (1965) found that of the two partial reactions catalyzed by pyruvate carboxylase from chicken liver, only the first partial reaction (the formation of the ternary enzyme-biotin-CO_2 complex) was stimulated by acetyl-CoA. However McClure *et al.* (1971c) and Seufert *et al.* (1971), working with rat liver pyruvate carboxylase, found activation of both partial reactions by acetyl-CoA. This variation probably reflects technical differences in the method of approach rather than a real species difference. We have found (Söling *et al.*, 1970) a lower pH optimum for pyruvate carboxylase from guinea pig liver as compared with the rat liver enzyme, when tested at constant concentrations of either ATP-Mg^{2-} or ATP-Mn^{2-}. Also the apparent K_m values for ATP and pyruvate were similar for the two enzymes. However, the K_m values for pyruvate have to be reconsidered in view of the "low" and "high" K_m values for pyruvate found by McClure *et al.* (1971a) and Scrutton and White (1974) with rat and human liver pyruvate carboxylase, respectively.

Scrutton *et al.* (1972) have recently reported that calf liver pyruvate carboxylase contains bound Mn^{2+} plus bound Mg^{2+} (combined stoichiometrically with biotin), whereas pyruvate carboxylase from chicken and turkey liver contains only bound Mn^{2+}. The situation for rat liver pyruvate carboxylase (McClure *et al.*, 1971a) seems to be similar to that the of calf liver enzyme. When Mn^{2+} was replaced by Mg^{2+} in chicken liver pyruvate carboxylase, the kinetic constants did not change significantly, but the binding of Mg^{2+} to the avian enzyme seems to be more labile than the binding to the calf liver enzyme (Scrutton *et al.*, 1972). The total pyruvate carboxylase activity in rat liver depends on the state of nutrition of the animal. Its activity increases in starvation (see Table XV) and after injection of cortisol (Henning *et al.*, 1964) and decreases after administration of insulin (Prinz and Seubert, 1964) and in states of high carbohydrate feeding. We found that the increase in total pyruvate carboxylase activity in livers from starved guinea pigs and pigeons was far less than in livers from starved rats (Table XV). Adaptation of the amount of pyruvate carboxylase protein in liver occurs too slowly to be responsible for short-term changes of the rate of pyruvate carboxylation.

Although a variety of known parameters could directly affect pyruvate carboxylation (pyruvate, acetyl-CoA, ADP, ATP/ADP ratio), it is not clear which factor is most important under physiological conditions. The most important factor certainly is the concentration of pyruvate, which is regulated mainly by the supply of substrate (L-lactate, pyruvate, L-alanine), the redox state, and the rate of pyruvate oxidation. But assuming a fixed concentration of pyruvate, what is the physiological importance of the other factors? Ruderman *et al.* (1971) have shown that the rate of gluconeogenesis from L-alanine in the isolated, perfused rat liver was more or less proportional to the overall concentration of acetyl-CoA. Another factor for which a regulatory role has been claimed is the ATP-Mg^{2-}/ADP-Mg^{1-} ratio. In 1963 Keech and Utter (1963) described an inhibition of chicken liver pyruvate carboxylase by ADP which was competitive with ATP. The same inhibition was noted by Walter and Stucki (1970) and by McClure and Lardy (1971) for the rat liver enzyme. Walter and Stucki (1970) and Stucki *et al.* (1972) have extensively studied this type of regulation in isolated rat liver mitochondria. Since the regulation of pyruvate carboxylase activity by the ATP-Mg^{2-}/ADP Mg^{1-} ratio has also been seen with pyruvate carboxylase from yeast (Miller and Atkinson, 1972) and *Penicillum camemberti* (Stan, 1972), it may be a fundamental type of regulation and may exist in the livers of a variety animal species including those not yet studied.

In evaluating the relative importance of acetyl-CoA on one hand and the ATP-Mg^{2-}/ADP-Mg^{1-} ratio on the other as physiological regulators for pyruvate carboxylase, it is important to distinguish clearly between a permissive and a regulatory role of a metabolic factor (or factors). Acetyl-CoA is needed for a sufficient rate of pyruvate carboxylation and is therefore permissive. However if the intramitochondrial level of acetyl-CoA is well above the concentration needed for maximum stimulation and/or if the physiological variations of the acetyl-CoA concentrations are small, it is not likely to play a regulatory role. The same holds for the ATP-Mg^{2-}/ADP-Mg^{1-} ratio. J. R. Williamson (1967) has presented some data on the intramitochondrial concentration of acetyl-CoA in rat liver, calculated as suggested by Garland *et al.* (1965) and based on measurements in isolated rat liver mitochondria. These values range between 75 and 250 pmole/μl matrix water. Stucki *et al.* (1972) give values for isolated rat liver mitochondria of about 250 and 500 pmole/μl matrix water in the absence and in the presence, respectively, of an ADP-regenerating system. We found very similar values in our own experiments (Kleineke and Söling, unpublished results). The K_a

value for acetyl-CoA increases with decreasing pH as shown for pyruvate carboxylase from chicken (Scrutton and Utter, 1967), rat (McClure and Lardy, 1971; Söling *et al.*, 1970), and guinea pig (Söling *et al.*, 1970) liver. However, the matrix space is more alkaline than the cytosol, and even if the intramitochondrial pH is approximately 7.0, the actual acetyl-CoA concentration would be saturating.

It seems reasonable, therefore, to assume that the intramitochondrial concentration of acetyl-CoA probably does not regulate pyruvate carboxylase under physiological conditions. Changes in the intramitochondrial ATP-Mg^{2-}/ADP-Mg^{1-} ratio, which are difficult to detect by the usual measurements of the ATP/ADP ratio in the whole organ, are more likely to regulate pyruvate carboxylation under physiological circumstances. This contrasts to the overall ATP/ADP ratios, which are between 3 and 4 in the livers of all species tested. The mitochondrial matrix ATP/ADP ratio is probably much lower. Because of the much higher affinity of the adenine nucleotide translocator for the inward transport of ADP as compared to ATP, a low intramitochondrial ATP/ADP ratio can be maintained within the mitochondrial matrix even in the presence of a high external ATP/ADP ratio (Klingenberg *et al.*, 1969). The ATP/ADP ratio in the matrix of freshly isolated rat liver mitochondria is between 0.5 and 1.0 (Klingenberg *et al.*, 1969; Heldt *et al.*, 1972; Stucki *et al.*, 1972). A measurable inhibition of pyruvate carboxylation begins at ratios below 1.0 (McClure and Lardy, 1971). When the ATP/ADP ratio in the matrix space of isolated rat liver mitochondria is varied by an ADP regenerating system or by the addition of fatty acids, the rate of carboxylation changes in a parallel manner (Stucki *et al.*, 1972). A "pulse" of external ADP leads to a rapid decrease in the rate of pyruvate carboxylation, followed by a rapid recovery. The change in the rate of pyruvate carboxylation follows directly the change in the ATP/ADP ratio in the matrix water space (Fig. 11). Under physiological conditions, where the concentration of pyruvate is low, changes of the intramatrix ATP-Mg^{2-}/ADP-Mg^{1-} ratio could favor pyruvate carboxylation by affecting the rate of pyruvate oxidation. The foregoing discussion is based largely on experiments with isolated mitochondria and must be substantiated using newer techniques that allow the determination of intra- and extramitochondrial concentrations of metabolites in the whole organ. First reports obtained by centrifugation of freeze-dried liver in organic solvents (Elbers *et al.*, 1974) or with the technique reported recently by Zuurendonk and Tager (1974) are promising.

C. Species Differences in the Intracellular Localization and Regulation of PEP-CK

The total activity of PEP-CK exhibits much greater interspecies differences than pyruvate carboxylase (Tables XVII and XV). Even more striking are

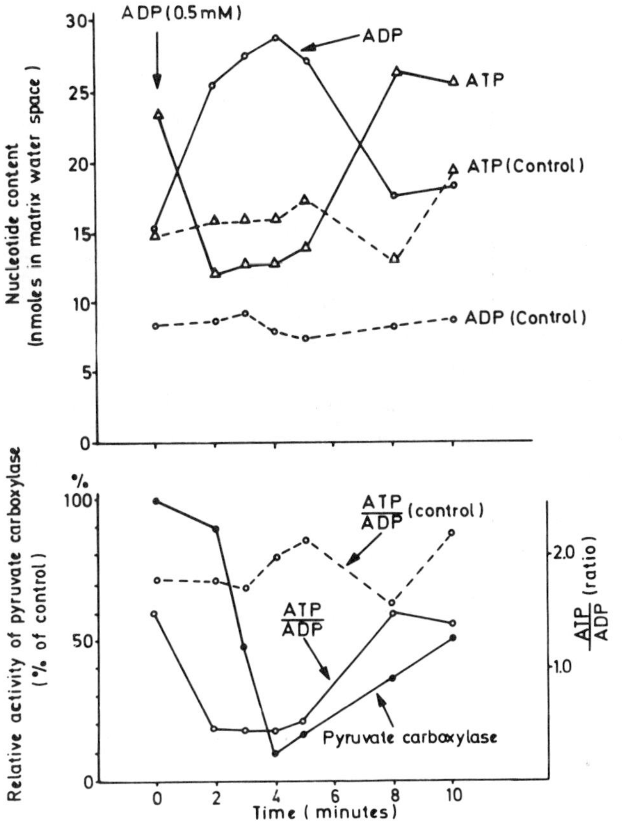

Figure 11. Effect of a pulse of ADP on intramitochondrial levels of adenosine diphosphate and ATP and the activity of pyruvate carboxylase in an experiment with isolated rat liver mitochondria. Isolated rat liver mitochondria were incubated in the following medium (final concentrations): pyruvate, 1 mM; K-HCO_3, 10 mM; K-phosphate (pH 7.4), 6.6 mM; triethanolamine-Cl (pH 7.4), 6.6 mM; $MgSO_4$, 10 mM; sucrose to make the mixture isotonic. The reactions were started by the addition of mitochondria (equivalent to about 4 mg protein). In the experiments designed for the determination of intramatrix concentrations of adenine nucleotides, the incubation mixture contained trace amount of ^3H_2O and ^{14}C-sucrose and was layered over silicon oil, which was layered on top of $HClO_4$. In the experiments designed for the determination of pyruvate carboxylation, the assay mixture contained trace amounts of $[^{14}C]HCO_3^-$ and was not layered over silicon oil. In the control experiments, the incubation mixture did not contain ADP. In the other experiments ADP was added at zero time to give a final concentration of 0.5 mM. The incubations were carried out at 10°C and were terminated either by spinning the mitochondria through the silicon layer into $HClO_4$ in a modified Eppendorf centrifuge or, when pyruvate carboxylase activity was determined, by addition of cold $HClO_4$. The matrix content of ATP and ADP was calculated as the difference of the nucleotide content in the total mitochondrial water space and the nucleotide content in the mitochondrial sucrose space. *Source*: Kleineke and Söling, unpublished experiments.

Table XVII. Total activity of phosphoenolpyruvate carboxykinase in livers from various vertebrates

Species	Metabolic state	Direction of measurement	Temperature (°C)	Enzyme activity (μmole/g liver/min)	Reference
Rat	Standard fed	PEP → OAA	37°	2.1	Reshef and Hanson (1972)
Rat	Standard fed	OAA → PEP	30°	3.3	Söling *et al.* (1973)
	Starved, 48 hours	OAA → PEP	30°	9.7	Söling *et al.* (1973)
Rat	Standard fed	Not given	30°	2.3–2.7	Böttger *et al.* (1969)
Rat	Standard fed	OAA → PEP	30°	9.0–10.0[a]	Nordlie and Lardy (1963)
Guinea pig	Standard fed	OAA → PEP	30°	14.0	Söling *et al.* (1970)
Guinea pig	Starved, 48 hours	OAA → PEP	30°	26.6	Söling *et al.* (1970)
Guinea pig	Standard fed	PEP → OAA	30°	14.0	Holten and Nordlie (1965)
Guinea pig	Standard fed	Not given	30°	8.8	Böttger *et al.* (1969)
Guinea pig	Standard fed	OAA → PEP	30°	17.5–33.3[a]	Nordlie and Lardy (1963)

Mouse	Standard fed	OAA → PEP	30°	3.3	Nordlie and Lardy (1963)
Hamster	Standard fed	OAA → PEP	30°	2.1	Nordlie and Lardy (1963)
Rabbit	Standard fed	OAA → PEP	30°	5.9	Nordlie and Lardy (1963)
Rabbit	Standard fed	OAA → PEP	30°	4.4	Usatenko (1970)
Pig	Fed	PEP → OAA	37°	9.4	Swiatek *et al.* (1970)
Dog	Fed	Not given	Not given	1.8	Garthoff *et al.* (1972)
Man	Overnight fasted	OAA → PEP	30°	11.2	Diesterhaft *et al.* (1971)
Man	Overnight fasted	Not given	30°	10.2	Böttger *et al.* (1969)
Turtle	Fed	PEP → OAA	Not given	4.0	Penney and Kornecki (1973)
Bullfrog	Fed	PEP → OAA	37°	21.8	Nagano *et al.* (1973)
Pigeon	Fed	PEP → OAA	25°	8.5[b]	Gevers (1967)
Pigeon	Starved, 48 hours	PEP → OAA	25°	8.7[b]	Gevers (1967)
Pigeon	Fed	OAA → PEP	30°	68.8	Söling *et al.* (1973)
Pigeon	Starved, 72 hours	OAA → PEP	30°	62.7	Söling *et al.* (1973)

[a] Values calculated from data given in the references.
[b] Values calculated from dry weight data by assuming a wet weight/dry weight ratio of 3.8.

the dramatic differences between the distribution of PEP-CK activity between the cytosol and the mitochondria (Table XVIII) as first described by Nordlie and Lardy (1963). The extremes are rat, mouse, and hamster on one hand and the pigeon on the other. In rat liver the total PEP-CK activity is only about 10% of the total activity in pigeon liver, but more than 90% of this activity is localized in the cytosol (Nordlie and Lardy, 1963; Usatenko, 1972). In pigeon liver, on the other hand, most of the PEP-CK activity is found in the mitochondrial matrix space (Söling *et al.*, 1973). The hepatic cytosol activity measured in this species is mainly due to mitochondrial breakage (Söling *et al.*, 1973). In livers from fed rabbits the distribution of PEP-CK is similar to that in pigeon liver, although the activity in the cytosol is slightly higher than could be explained by mitochondrial breakage (Johnson *et al.*, 1970). During starvation or acute diabetes, the cytosol activity of PEP-CK in rabbit liver increases significantly. In livers from guinea pigs (Nordlie and Lardy, 1963; Söling *et al.*, 1970), sheep (Taylor *et al.*, 1971; Garthoff *et al.*, 1972), pig (Swiatek *et al.*, 1970), and man (Diesterhaft *et al.*, 1971), considerable activities of PEP-CK are found, as well in the cytosol as in the mitochondria (Table XVIII). However, under conditions of increased gluconeogenesis, such as in starvation or acute diabetes, only the activity in the cytosol increases, whereas the mitochondrial activity remains unchanged. This has been shown for the following species: rat (Shrago *et al.*, 1963), guinea pig (Shrago *et al.*, 1963; Söling *et al.*, 1970), and rabbit (Johnson *et al.*, 1970). In sheep (Taylor *et al.*, 1971) and pig (Swiatek *et al.*, 1970) liver, diabetes or starvation leads to an increase of cytosotic and mitochondrial PEP-CK activities. In pigeon liver, where nearly all activity is localized intramitochondrially, no increase in total or cytosolic PEP-CK activity occurs after 72 hours of starvation (Söling *et al.*, 1973). In cows, starvation, ketosis, or treatment with steroids leads to a decrease of cytoplasmic but not of mitochondrial PEP-CK activity (Heitzman *et al.*, 1972). However Ballard *et al.* (1968) did not observe any effect of starvation on the cytosol enzyme in cow liver. As mentioned earlier, the very small activity of PEP-CK in the cytosol in pigeon liver can be explained by the occurrence of mitochondrial breakage during the preparation of the postmitochondrial supernatant. On the other hand, the low intramitochondrial PEP-CK activity in rat liver cannot be explained by a slight contamination of the mitochondrial preparation with cytosolic PEP-CK. Ballard and Hanson (1969) reported an immunochemical difference between the enzyme from mitochondria and cytosol. In addition, we could show that in intact rat liver mitochondria the oxaloacetate-CO_2 exchange reaction that is mediated by PEP-CK and is completely dependent on exogenous ATP (Janssens, Kleineke and Söling, unpublished data) can be blocked by carboxyatractyloside, a strong inhibitor of the adenine nucleotide translocator in the inner mitochondrial membrane. From the relatively low activity of mitochondrial PEP-CK it is evident that the synthesis of PEP for gluconeogenesis in rat liver proceeds nearly exclusively in the cytosol. Ballard and Hanson (1967) reported that in fetal rat liver during development, mitochondrial PEP-CK activity appears before the cytosol enzyme. Therefore we still do not know whether the mitochondrial PEP-CK activity found in livers from adult rats is a metabolic atavism or has some physiological significance.

The mechanism of the PEP-CK reaction has been studied in a variety of species. Unfortunately, all the most thorough studies of the reaction mechanism have been carried out with enzymes from mitochondria, and even these results contain many discrepancies. Miller and Lane (1968), working with pig liver mitochondrial PEP-CK, have mainly studied the carboxylation reaction; they propose a mechanism in which a ternary enzyme–Mn^{2+}–PEP complex is formed, followed by a random order binding of IDP and CO_2. In a concerted reaction, the central complex is then converted into a new quarternary enzyme–Mn^{2+}–oxaloacetate–ITP complex followed by the release of the reaction products. Barns *et al.* (1972) have studied sheep kidney mitochondrial PEP-CK. They describe another mechanism for the carboxylation reaction according to which a ternary enzyme–Mn^{2+}–IDP complex is formed, followed by the binding of PEP and CO_2 in a random sequential manner. From the quarternary central complex, oxaloacetate and ATP are released in a random sequential fashion (Barns *et al.*, 1972). Felicioli *et al.* (1970) used chicken liver mitochondrial PEP-CK to investigate the direction of decarboxylation (PEP-synthesis) and concluded that the reaction proceeded according to a sequentially ordered bi-ter mechanism in which ITP binding precedes binding of oxaloacetate. The reaction products are released in the order CO_2, PEP, and IDP. Further differences are found with respect to the $^{14}CO_2$-oxaloacetate exchange reaction. The reaction is stimulated only by nucleosidetriphosphate (ITP) with the pig liver mitochondrial enzyme (Chang and Lane, 1966) whereas with the sheep kidney mitochondrial enzyme the exchange reaction is stimulated by nucleoside triphosphate and diphosphate (IDP) Barns *et al.* (1972). In our own experiments with rat liver mitochondria the $^{14}CO_2$-oxaloacetate exchange reaction was stimulated equally by ATP and by ADP, which supports the results of Barns *et al.* (1972).

Whether the different reaction mechanisms just mentioned reflect true species differences is doubtful. More likely, they can be attributed to variation in technical approach and interpretation of results. The study of the reaction mechanism is difficult because even if there is a preferred sequence of binding and release of substrates and products, the enzyme can form ternary enzyme–Mn^{2+}–substrate (product) complexes with ITP or IDP or

Table XVIII. Distribution of phosphoenolpyruvate carboxykinase activity between cytosol and mitochondria in livers from various vertebrates

Species	Metabolic state	Enzyme activity (units/g wet liver)		Soluble activity/ mitochondrial activity (ratio)	Reference
		Soluble	Mitochondrial		
Rat	Fed	1.39	0.53	2.62	Heitzman *et al.* (1972)
Rat	Fed	3.00	0.16	18.75	Usatenko (1970)
Rat	Starved	6.20	0.19	32.63	Usatenko (1970)
Rat	Fed	7.28–8.42	0.25–0.83	29.12–10.14	Nordlie and Lardy (1963)
Mouse	Fed	3.29	0	∞	Nordlie and Lardy (1963)
Hamster	Fed	2.06	0	∞	Nordlie and Lardy (1963)
Rabbit	Fed	1.60	1.70	0.94	Usatenko (1970)
Rabbit	Starved	3.40	2.10	1.62	Usatenko (1970)
Rabbit	Fed	—	3.41	0	Nordlie and Lardy (1963)
Rabbit	Fed	0.60	1.80	0.33	Johnson *et al.* (1970)
Rabbit	Starved	3.05	2.08	1.47	Johnson *et al.* (1970)
Rabbit	Diabetic	2.68	1.74	1.54	Johnson *et al.* (1970)

Guinea pig	Fed	3.20	10.86	0.29	Holten and Nordlie (1965)
Guinea pig	Fed	3.35	10.64	0.31	Söling *et al.* (1970)
Guinea pig	Fasted	12.75	13.48	0.95	Söling *et al.* (1970)
Pig	Fed	6.34	2.60	2.43	Swiatek *et al.* (1970)
Pig	Fasted	18.06	4.30	4.20	Swiatek *et al.* (1970)
Sheep	Fed	3.24	1.30	2.49	Taylor *et al.* (1971)
Sheep	Fasted	2.95	0.92	3.21	Taylor *et al.* (1971)
Sheep	Diabetic	7.06	2.24	3.15	Taylor *et al.* (1971)
Dog	?	0.56	1.05	0.53	Garthoff *et al.* (1972)
Cow	Fasted	6.12	7.04	0.87	Heitzman *et al.* (1972)
Bullfrog	Fed	17.6	0.40	44.0	Nagano *et al.* (1973)
Turtle	Fed	2.00	0.80	2.50	Penney and Kornecki (1973)
Man	Starved overnight	4.30	3.90	1.10	Diesterhaft *et al.* (1971)
Pigeon	Fed	2.08[a]	6.70[a]	0.31	Gevers (1967)
Pigeon	Fed	7.56	58.00	0.13	Söling *et al.* (1973)

[a] Values were calculated from data given in Gevers (1967). Phosphoenolpyruvate carboxykinase activity is not completely restricted to these two compartments but can also be found in other fractions, especially the nuclear compartment.

PEP or oxaloacetate. Differences between mitochondrial and cytosolic PEP-CK have been studied in rat liver by Ballard (1970), in guinea pig by Holten and Nordlie (1965), in chicken by Jo *et al.* (1974), and in human liver by Diesterhaft *et al.* (1971). Ballard and Hanson (1969) found that antibodies against PEP-CK from rat liver cytosol did not cross-react with rat liver mitochondrial PEP-CK. Diesterhaft *et al.* (1971) could separate mitochondrial and cytosolic PEP-CK from human liver by DEAE ion exchange chromatography at pH 7.0, where the mitochondrial and cytosolic enzymes appeared as cationic and anionic proteins, respectively. Jo *et al.* (1974) found four forms of PEP-CK in embryonic chicken liver by DEAE-cellulose chromatography. Two forms were cytosolic and two mitochondrial. However, in livers from adult chickens only one cytosolic and one mitochondrial form of PEP-CK were found. They did not differ with respect to apparent K_m values for oxaloacetate, GTP, PEP, and GDP. Holten and Nordlie (1965) reported that the mitochondrial PEP-CK from guinea pig liver was inhibited by AMP and was more active with Mg^{2+} than with Mn^{2+} at pH 8.0, whereas the cytosolic enzyme was not affected by the nucleotide and was more active in the presence of Mn^{2+}. Otherwise, the two were kinetically indistinguishable. For most species, K_m values between 1.4×10^{-4} M and 4×10^{-3} M for oxaloacetate have been reported. However, Ballard (1970) has pointed out that much more reliable data can be had from an assay system in which oxaloacetate is not added as a substrate but is generated from L-malate by the malate dehydrogenase reaction. He found apparent K_m values for oxaloacetate with the cytosolic or mitochondrial form of PEP-CK from rat or sheep liver to be in the range of 2.5×10^{-5} M. In the presence of Mn^{2+} the K_m for oxaloacetate of mitochondrial PEP-CK decreased to about 9×10^{-6} M.

The molecular weights of PEP-CK from mitochondria or cytosol from various species are all in the same range, between 71,000 and 75,000 (Chang and Lane, 1966; Ballard and Hanson, 1969; Barns and Keech, 1972). Only Jo *et al.* (1974) describe an apparent molecular weight of 24,000 based on gel filtration measurements. The similarities of kinetic and molecular parameters lead to the conclusion that the properties of PEP-CK in the cytosol and in the mitochondria are rather similar, which is in sharp contrast to the characteristics of other isoenzymes found in these two compartments (e.g., glutamate-oxaloacetate aminotransferase).

D. Species Dependent Regulation of Pyruvate Oxidation

When the concentration of pyruvate is low, competition for pyruvate between pyruvate oxidation and pyruvate carboxylation might play a role in the regulation of gluconeogenesis. Pyruvate dehydrogenase activity is regulated in two ways.

1. Pyruvate dehydrogenase is converted from an active to an inactive form by the phosphorylation of a serine group in the enzyme molecule by ATP and a specific protein kinase. The enzyme can be reconverted to the active form by dephosphorylation catalyzed by a specific phosphatase (Linn *et al.*, 1969a; Wieland and von Jagow-Westermann, 1969).

2. Active pyruvate dehydrogenase is inhibited by acetyl-CoA (competitively with CoA-SH) and by NADH (competitively with NAD^+) (Garland and Randle, 1964; Bremer, 1969). The interconversion of pyruvate dehydrogenase has been demonstrated in pork (Linn *et al.*, 1969b) and rat (Söling and Bernhard, 1971; Wieland *et al.*, 1972; Patzelt *et al.*, 1973) liver. Since the interconversion reactions have been found in a variety of other organs from rats and cows (Linn *et al.*, 1969b, Wieland and von Jagow-Westermann, 1969; Weiss *et al.*, 1971; Taylor *et al.*, 1973; Siess *et al.*, 1971) and even in *Neurospora crassa* (Wieland *et al.*, 1972), it is reasonable to assume that this is a common mechanism for regulation of hepatic pyruvate dehydrogenase in all higher species.

The interconversion of pyruvate dehydrogenase seems to be regulated mainly by the ATP/ADP ratio in the mitochondrial matrix space. Factors that lead to a decrease in this ratio [e.g., high concentrations of fructose (Söling and Bernhard, 1971)] result in an increase in the active form of the enzyme and enhance the rate of pyruvate oxidation, whereas factors [e.g., α-oxoglutarate in experiments with isolated rat liver mitochondria (Portenhauser and Wieland, 1972)] that increase the ATP/ADP ratio, lower the amount of active pyruvate dehydrogenase (Walajtys *et al.*, 1974). The conversion of pyruvate dehydrogenase to the active form is stimulated by pyruvate (Patzelt *et al.*, 1973). Correlation between the percentage of active pyruvate dehydrogenase and pyruvate concentration in the medium during perfusion of isolated rat liver indicates a half-maximum stimulation at a pyruvate concentration of about 1×10^{-3} M (Patzelt *et al.*, 1972). No significant effects were seen at concentrations around and below 1×10^{-4} M, a concentration closer to the level of pyruvate in rat liver. Without knowing the pyruvate concentration in the mitochondrial matrix, however, it seems impossible to predict whether changes in the levels of liver pyruvate noted physiologically have a regulatory significance. Increased fatty acid oxidation favors the conversion of active pyruvate dehydrogenase to the inactive form in rat liver (Patzelt *et al.*, 1973) and inhibits pyruvate oxidation in the isolated, perfused rat liver (Teufel *et al.*, 1967; Söling *et al.*, 1968). Interestingly, fatty acids fail to inhibit pyruvate oxidation in the isolated, perfused guinea pig liver (Söling *et al.*, 1970). No data are available concerning species differences in the inhibition constants for the inhibition of active pyruvate dehydrogenase by acetyl-CoA or NADH. It seems reasonable to assume that species differences in the redox state of the mitochondrial NAD^+/NADH system can result in different rates of pyruvate oxidation.

E. Pyruvate Kinase and the Species Dependent Regulation of Gluconeogenesis

The V_{max} activity of hepatic pyruvate kinase shows considerable species differences (see Table XIX). In most species (except birds), total pyruvate kinase is higher than the activities of pyruvate carboxylase or PEP-CK. In rat liver pyruvate kinase activity is considerably higher than the activities of these two gluconeogenic enzymes. Therefore, a net formation of glucose can occur only if pyruvate kinase is partially or completely inhibited or inactive, or if there are separate compartments for PEP in the cytosol of the hepatocyte. The latter would, of course, demand at least two compartmentalized enolases, one equilibrated with the PEP generated in the PEP-CK reaction, another responsible for the glycolytic formation of PEP situated close to pyruvate kinase. Up to now, the discussion has been restricted principally to the first possibility—namely, allosteric control of pyruvate kinase.

In rat liver, four isoenzymes of pyruvate kinase have been described (Tanaka *et al.*, 1965; Osterman *et al.*, 1973); PK 1 = L-type pyruvate kinase, PK 2, PK 3 = M_1-type pyruvate kinase, and PK 4 = M_2-type pyruvate kinase; PK 2, PK 3, and PK 4 cross-react immunologically (Osterman *et al.*, 1973). The regulation of PK 1-type pyruvate kinase has been reviewed recently by Seubert and Schoner (1971). The PK 1-type pyruvate kinase exhibits strong homotropic cooperativity with respect to PEP (Susor and Rutter, 1968; Rozengurt *et al.*, 1969; Llorente *et al.*, 1970; Schoner *et al.*, 1970; Carbonell *et al.*, 1973). Fructose-1,6-diphosphate is a strong activator; L-alanine and ATP-Mg^{2-} are potent inhibitors. The effect of fructose-1,6-diphosphate is enhanced, while that of L-alanine and ATP-Mg^{2-} is inhibited by increasing the proton concentration (Rozengurt *et al.*, 1969; Schoner *et al.*, 1970). Glucose-1,6-diphosphate has an effect similar to that of fructose-1,6-diphosphate (Koster *et al.*, 1972). The affinity of the enzyme for fructose-1,6-diphosphate and PEP can be reduced by agents that oxidize some or all of the SH-groups of the enzyme (van Berkel *et al.*, 1973).

The distribution of the total hepatic pyruvate kinase activity between PK 1-type pyruvate kinase and PK 4-type pyruvate kinase is species specific (Osterman and Fritz, 1973): in adult rat liver, 95% of the total activity is PK 1. In some species (hamster, guinea pig, monkey) up to 30% of total pyruvate kinase activity is of the PK 4 type. In cat and human liver, pyruvate kinase activity was equally distributed between these two types. The PK 1 type of pyruvate kinase has also been found in avian livers [chicken (Leveille, 1969), pigeon (Gabrielli and Baldi, 1972)], but nothing has been reported on the relative percentage of the PK 4 type of pyruvate kinase in these species.

Recently van Berkel *et al.* (1972) have reported that PK 1 in rat liver is located only in the parenchymal liver cells, whereas the other types of pyruvate kinase are mainly in the Kupffer cells. It is not clear whether a similar type of distribution exists in other species. When tested in the absence of fructose-1,6-diphosphate, the apparent K_m of the enzyme for PEP exhibits considerable species differences [rat, mouse, pig, chicken (Leveille, 1969)]. These differences between species are not found when the enzyme is assayed in the presence of fructose-1,6-diphosphate. However, these experiments were performed with whole liver homogenates, and the differences may be due to interspecific dissimilarities in the distribution of total pyruvate kinase activity between the PK 1 and PK 4 isoenzymes.

Diets rich in carbohydrate increase the total pyruvate kinase activity in livers from rats, have a marginal effect in the mouse liver, and have virtually no effect in avian livers (pigeon, chicken—see Table XIX). Insulin deficiency has the same effect in rats, spiny mice, and man (Table XIX). However, these changes in total pyruvate kinase activity are probably due to *de novo* synthesis of the enzyme and cannot explain how pyruvate kinase is regulated on a short-term basis. Under physiological conditions the concentration of ATP in the hepatic cytosol is 2 to 3 mM and alanine is 1 to 1.5 mM. This would be sufficient to inhibit completely PK 1 activity, as long as the concentration of fructose-1,6-diphosphate is low. Therefore, fluctuation in the concentration of fructose-1,6-diphosphate probably regulates the activity of PK 1-type pyruvate kinase. The effect of fructose-1,6-diphosphate may be supported by a similar effect of glucose-1,6-diphosphate (Koster *et al.*, 1972). J. R. Williamson *et al.* (1969a) have shown that there is no uniform response in the level of fructose-1,6-diphosphate and no clear correlation between the rate of gluconeogenesis and the level of fructose-1,6-diphosphate. Llorente *et al.* (1970) have pointed out that since fructose-1,6-diphosphate binds to enzymes like aldolase and fructose-1,6-diphosphatase, the concentration of free fructose-1,6-diphosphate may be considerably smaller than the observed value. This makes it difficult to measure the free fructose-1,6-diphosphate concentration in any tissue.

The PK 4 type of pyruvate kinase is not inhibited by ATP and exhibits only very low cooperativity with PEP (Carbonell *et al.*, 1973). It is inhibited by L-alanine and various other L-amino acids (Carbonell *et al.*, 1973), and this inhibition can be overcome by fructose-1,6-diphosphate. In the absence of inhibitory concentrations of L-amino acids, the effect of fructose-1,6-diphosphate is minimal. The inhibition by concentrations of L-alanine as

Table XIX. Activity of pyruvate kinase in the liver from various vertebrate species

Species	Metabolic condition	Temperature (°C)	Enzyme activity (μmole/g liver/min)	Reference
Rat[a] (Wistar strain 1)	Fed	25°	27.2 ±2.6	Eggleston and Krebs (1969)
Rat[a] (Wistar strain 1)	Starved (48 hours)	25°	9.9 ±0.7	Eggleston and Krebs (1969)
Rat[a] (Wistar strain 2)	Fed	25°	60.8 ±7.5	Eggleston and Krebs (1969)
Rat[a] (Wistar strain 2)	Starved (48 hours)	25°	28.8 ±1.9	Eggleston and Krebs (1969)
Rat (Wistar)	Fed	30°	21.76 ±0.75	Willms *et al.* (1970b)
Rat (Wistar)	Diabetic	30°	12.00 ±0.26	Willms *et al.* (1970b)
Acomys cahirinus (spiny mouse)	Nondiabetic	30°	27.95 ±2.17	Willms *et al.* (1970b)
Acomys cahirinus	Diabetic (nonketotic)	30°	28.45 ±4.00	Willms *et al.* (1970b)
Obese-hyperglycemic mouse	Lean, fed	30°	32.51 ±2.03	Willms *et al.* (1970b)
Obese-hyperglycemic mouse	Obese, fed	30°	56.48 ±4.79	Willms *et al.* (1970b)
New Zealand obese mouse	Lean, fed	30°	24.39 ±1.09	Willms *et al.* (1970b)
New Zealand obese mouse	Obese, fed	30°	31.27 ±1.83	Willms *et al.* (1970b)
Guinea pig	Fed	30°	6.56 ±0.26	Söling *et al.* (1973)
Guinea pig	Starved (48 hours)	30°	7.92 ±0.40	Söling *et al.* (1973)
Hamster[b]	Fed	25°	16.0	Osterman and Fritz (1973)
Rabbit[b]	Fed	25°	16.5	Osterman and Fritz (1973)

Cow	Nonlactating (fed)	37°	12.40 ±2.22	Baird and Heitzman (1970)
Cow	Lactating (fed)	37°	15.34 ±3.21	Baird and Heitzman (1970)
Cat[b]	Fed	25°	12	Osterman and Fritz (1973)
Dog[b]	Fed	25°	37	Osterman and Fritz (1973)
Monkey[b]	Fed	25°	14	Osterman and Fritz (1973)
Pigeon	Fed	30°	27.11 ±4.25	Söling *et al.* (1973)
Pigeon	Starved (48 hours)	30°	29.60 ±5.13	Söling *et al.* (1973)
Man	Nondiabetic, starved overnight	25°	6.2	Schmidt and Schmidt (1963)
Man	Untreated diabetics	25°	11.8	Schmidt and Schmidt (1963)
Man	Nondiabetic, starved overnight	25°	7.9	Wallenfels *et al.* (1958)
Man	Untreated diabetics	25°	11.8	Wallenfels *et al.* (1958)
Man[c]	Starved overnight	30°	29.0 ±2	Willms *et al.* (1970b)
Man	Diabetic (insulin deficiency type)	30°	19.5 ±6	Willms *et al.* (1970b)
Man[d]	—[d]	25°	8.0	Osterman and Fritz (1973)

[a] It should be noted that when similar but different strains are examined in the same species with identical analytical methods, large activity differences can be obtained.
[b] Data estimated from a graphical presentation.
[c] Data calculated under the assumption that 1 g wet liver is equivalent to 100 mg of protein in an undiluted 100,000 *g* supernatant.
[d] Tissue was taken several hours after death.

high as 2 mM is nearly completely overcome by only 10 μM fructose-1,6-diphosphate (Carbonell *et al.*, 1973). In rat liver ATP-Mg^{2-} and the cooperative effect of PEP may control pyruvate kinase in addition to an inhibition of the enzyme by L-alanine. Friedrichs and Schoner (1974) did not observe a unidirectional shift in the concentration of L-alanine during stimulation of gluconeogenesis by glucagon using perfused rat liver. In the absence of an exogenous precursor, glucagon doubled the intrahepatic concentration of L-alanine but decreased the levels of this amino acid in the presence of 10 mM L-lactate. In both cases, glucagon stimulated gluconeogenesis. Therefore, the importance of changes in the intrahepatic concentration of L-alanine for the regulation of hepatic pyruvate kinase remains to be established. Moreover, the phosphorylation of liver pyruvate kinase by a 3′,5′-cAMP dependent protein kinase has recently been described (Ljungström *et al.*, 1974; Helmquist *et al.*, 1974). this phosphorylation was accompanied by an inactivation of the enzyme. Similar observations have been made by Schoner (personal communication). Since glucagon leads to increased activity of 3′,5′-cAMP dependent proteine kinases, this effect would more readily explain the decreased pyruvate kinase activity during glucagon-stimulated gluconeogenesis than would regulation of the enzyme's activity by changes in the levels of L-alanine.

The importance of the various metabolic factors just discussed depends also on the relative contributions of PK 1 and PK 4 to total hepatic pyruvate kinase activity. Osterman *et al.* (1973) have reported that only 3 to 10% of total pyruvate kinase activity in rat kidney cortex is PK 1 and about 90% is the PK 4 type, which is in accordance with similar findings of Imamura and Tanaka (1972), Imamura *et al.* (1972), Susor and Rutter (1968), and Carbonell *et al.* (1972). Nevertheless, regulation of gluconeogenesis in kidney cortex is very similar to that in rat liver. Species differences in the PK 1/PK 4 ratio seem to be of minor importance. The fact that the extra energy cost for hepatic gluconeogenesis from C_3 precursors is very low in different species [rat, guinea pig, pigeon (Söling *et al.*, 1973)] supports this statement and indicates the tight control of pyruvate kinase during gluconeogenesis.

F. Nucleoside Diphosphate Kinase and PEP Synthesis

The formation of PEP from oxaloacetate involves the splitting of one γ-phosphate bond of GTP (ITP) per mole of PEP formed. Since GTP formed by substrate-level phosphorylation in the matrix space does not penetrate the inner mitochondrial membrane (or does so only at a negligible rate), GTP must be formed mainly by transphorylation from ATP via nucleosidediphosphate kinase. In species in which PEP is formed also intramitochondrially, intramitochondrial GTP can be regenerated either by substrate-level phosphorylation or by the nucleosidediphosphate kinase reaction. The activity of mitochondrial NDP kinase exhibits large species differences, although the absolute activities reported by several authors differ considerably because the different methods used for estimation. Mitochondria from guinea pig and rabbit liver (Ishihara and Kikuchi, 1968; Garber and Hanson, 1971) have considerably lower NDP kinase activity than those from pigeon liver (Ishihara and Kikuchi, 1968; Gevers, 1967), while the activity in mitochondria from rat and human liver lies in between these two groups (Table XX). In agreement with this are recent data of Bryla *et al.* (1973), who showed that in isolated guinea pig liver mitochondria the formation of GTP by substrate-level phosphorylation is essential for maximum rates of PEP formation. In these experiments, which were based mainly on the use of an uncoupler of oxidative phosphorylation, the subsequent activation of ATPase by the uncoupler was avoided by the additional use of oligomycin. An earlier report (Garber and Ballard, 1970) had suggested that the major part of the GTP necessary for PEP formation in guinea pig liver mitochondria was generated by the NDP kinase reaction. However, the uncoupler concentration used (40 μM 2,4-dinitrophenol) probably activated ATPase, removing GTP for the regeneration of ATP by NDP kinase from these studies. We conclude that in all species the GTP formed *extra*mitochondrially is generated from ATP by the NDP kinase reaction. In species such as the guinea pig and rabbit, where the *intra*mitochondrial activity of NDP kinase is low, *intra*mitochondrially formed PEP may contribute to gluconeogenesis. GTP is mainly formed by substrate level phosphorylation and only partially by NDP kinase. In pigeon liver (and perhaps in livers from other birds), where PEP for gluconeogenesis is nearly exclusively formed intramitochondrially and the intramitochondrial activity of NDP kinase is high (Gevers, 1967; Ishihara and Kikuchi, 1968), GTP used for PEP formation is probably derived completely from the NDP kinase reaction. The situation in human liver is unclear. The distribution of PEP-CK activity between mitochondria and cytosol is similar to that in guinea pig liver, but mitochondrial activity of NDP kinase is higher than it is in the livers of guinea pigs or rabbits. Total NDP kinase, on the other hand, is considerably lower in human liver. Therefore, it is likely that in human liver substrate-level phosphorylation contributes to gluconeogenesis.

G. Species Differences at the Fructose-1,6-Diphosphatase (FDPase)–Phosphofructokinase (PFK) Step

Comparative studies of the V_{max} activities of FDPase and PFK and of the FDPase/PFK ratio in various species are limited. Some data are collected

Table XX. Activity of nucleotide diphosphate kinase in mitochondria and cytosol from livers of various mammalian species[a]

	Mitochondria		Cytosol	
Species	nmole/mg protein/min	μmole/g liver/min[b]	nmole/mg protein/min	μmole/g liver/min
Rat	59.3 ±3.1	3.71 ±0.36	791 ±203	70.4 ±16.1
Guinea pig	5.7 ±1.4	0.34 ±0.09	1108 ±152	76.1 ±11.6
Pigeon	340.6 ±28.7	19.70 ±1.90	172 ±47	10.4 ±2.0
Man	50.0 ±8.4	3.20 ±0.50	214 ±23	8.0 ±1.5

[a] The test was performed in direction of ATP formation under the following conditions (final concentrations): *N*-2-hydroxyethylpiperazine-*N*-2-ethane sulfonate 50 mM, pH 7.6; GTP 1 mM, ADP 0.08 mM; NADP 0.5 mM; glucose 60 mM; $MgCl_2$ 1.7 mM; KCl 10 mM; hexokinase 10 μg/ml; glucose-6-phosphate dehydrogenase 5.6 μg/ml. The temperature was 30°C. Measurements were performed in a split beam spectrophotometer. The results were corrected for NADPH formation in the presence of ADP but absence of GTP, and in the presence of GTP but absence of ADP.

[b] Values in column 3 (micromoles) calculated under the simplifying assumption that in all four species 1 g of wet liver contains 60 mg of mitochondrial protein.

Source: Data from Söling and Bernhard, unpublished results.

in Tables XIV, XXI, and XXII. There are large difference of the FDPase/PFK ratio in livers from rats and guinea pigs (Söling *et al.*, 1970), but in all species examined so far, the activity of FDPase was always significantly higher than that of PFK (cf. Tables XXI and XXII). Since both enzymes are under strong allosteric control, a comparison of the V_{max} activities is of limited value. Studies of the kinetic properties of FDPase are mainly restricted to FDPase from rabbit liver (for a review, see Pontremoli and Horecker, 1970). Olson and Marquard (1972) have reported that antibodies against chicken liver FDPase cross-reacted with other avian hepatic FDPases (turkey and budgerigar) but not with FDPase from rabbit or rat liver. Unfortunately these authors have not yet discovered whether the immunological differences were accompanied by kinetic differences. In our hands rat liver FDPase did not behave like rabbit liver FDPase. The rat enzyme had a different electrophoretic mobility on polyacrylamide gels and sedimentation properties on sucrose gradients (Fig. 12). These differences correspond to well-defined aggregational states of the enzyme (Fig. 13). Rabbit FDPase migrated as a single band (Fig. 14), which in polyacrylamide gradient gel electrophoresis corresponded to the trimer of the rat liver enzyme. This was observed with purified enzymes and with crude liver extracts as well.

Upon incubation of rat liver FDPase with SH-agents (mercaptoethanol, dithiothreitol, CoA-SH), the oligomeric forms of the enzyme were converted to the monomeric form (Fig. 14), whereas incubation at different pH values or in the presence of either EDTA or 5′-AMP did not produce this effect. The molecular weights of FDPase from the livers of different species show only slight differences [rabbit 143,000 (Traniello *et al.*, 1971), cow 127,000 (Byrne *et al.*, 1971), rat 135,000 (Söling and Bernhard, 1974)].

If appropriate conditions are used during purification of FDPase, the pH optimum of FDPase from livers of a variety of species is below 8 [cow (Geller *et al.*, 1971), rabbit (Traniello *et al.*, 1971), rat (Fig. 13), carp (Villañueva and Marcus, 1974)]. Hepatic FDPase from different species is inhibited by 5′-AMP (Taketa and Pogell, 1963) and fructose-1,6-diphosphate (Taketa and Pogell, 1965; Nakashima *et al.*, 1970). Since the K_i value for 5′-AMP inhibition depends strongly on the experimental conditions [e.g., the concentration of K^+ and Mg^{2+} (Gonzales *et al.*, 1972)], a comparison of species differences with respect to 5′-AMP inhibition is difficult. Interestingly, the AMP inhibition of FDPase seems to be temperature regulated in poicilotherms (Gonzales *et al.*, 1972). In carp liver the lower sensitivity toward AMP inhibition observed at higher temperature can be compensated for by a decrease in pH in such a way that the regulation of FDPase by AMP is rather stable over the temperature range of 5 to 25°C (Villañueva and Marcus, 1974).

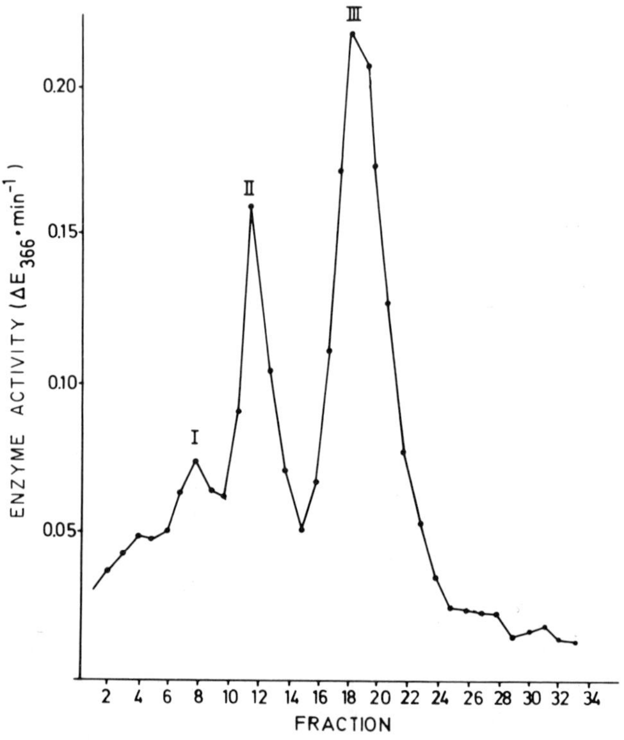

Figure 12. Gradient centrifugation of purified rat liver fructose-1,6-diphosphatase. Centrifugation was carried out according to Martin and Ames (1961). A linear gradient from 5 to 20% sucrose was 50 mM with respect to tris-Cl (pH 8.8). Centrifugation was carried out with a SW 65 rotor for 9 hours at 43,500 rpm in an L-2 65-B Spinco ultracentrifuge. Centrifugations at pH 6.0 and pH 7.5 revealed a similar pattern. *Source*: Söling and Bernhard, in preparation.

Pontremoli (for a review, see Pontremoli and Horecker, 1970) has found that purified FDPase from rabbit liver can be severalfold activated by incubation with cysteamine (Pontremolli *et al.*, 1967), CoA-SH (Nakashima *et al.*, 1969) and especially with homocystine (Nakashima *et al.*, 1970). We could not confirm these effects with purified rat liver FDPase (Fig. 15). In contrast to our purified rat liver enzyme, which had a pH optimum below pH 8, the enzyme used by Pontremoli *et al.* (1967) and by Nakashima *et al.* (1969, 1970) had a pH optimum around 9. However Traniello *et al.* (1971) showed that the alkaline pH optimum of purified rabbit liver FDPase is an artifact resulting from the proteolytic cleavage of a small peptide from the enzyme molecule during purification. We suspect therefore that the activating effect of the SH-agents is noted only with the "alkaline" enzyme. Traniello *et al.* (1971) mention very briefly that "neutral" FDPase from

Table XXI. Activity of fructose-1,6-diphosphatase in livers from various vertebrates[a]

Species	Metabolic state	Enzyme activity (μmole/g liver/min)	Reference
Rat (Buffalo strain)	Fed	7.00	Sato and Tsuiki (1969)
Rat (Wistar)	Fed	12.96 ±0.60	Söling *et al.* (1970)
Rat (Wistar)	Starved (48 hours)	12.35 ±0.5	Söling *et al.* (1970)
Rat (Wistar)	Diabetic	10.51 ±0.39	Willms *et al.* (1970b)
Mouse	Fed	11.70	Sato and Tsuiki (1969)
Obese-hyperglycemic mouse, (C 57 B1 73/63-06, Bar-Harbor)	Fed	29.32 ±2.28	Willms *et al.* (1970b)
Acomys cahirinus	Fed, (non diabetic)	12.98 ±0.62	Willms *et al.* (1970b)
Acomys cahirinus	Fed, (diabetic)	12.52 ±0.85	Willms *et al.* (1970b)
New Zealand obese mouse	Fed (lean)	13.54 ±0.84	Willms *et al.* (1970b)
New Zealand obese mouse	Fed (obese)	10.31 ±0.75	Willms *et al.* (1970b)
Guinea pig	Fed	5.14 ±0.26	Söling *et al.* (1970b)
Guinea pig	Starved (48 hours)	4.40 ±0.28	Söling *et al.* (1970b)
Rabbit	Fed	4.10 ±0.20	Mokrasch *et al.* (1956)
Sheep	Fed	11.95 ±0.93	Wallace and Newsholme (1967)
Domestic fowl	Fed	19.15 ±2.01	Wallace and Newsholme (1967)
Man[b]	Normal (overnight fast)	11.0 ±1.5	Willms *et al.* (1970b)
Man[b]	Diabetic (insulin deficiency-type)	18.3 ±1.5	Willms *et al.* (1970b)
Man[b]	Diabetic (maturity onset type)	19.5 ±3.0	Willms *et al.* (1970b)

[a] For better interpretation of the data presented, it is recommended to check the original literature for precise experimental conditions.
[b] Values calculated from the data in Willms *et al.* (1970b) under the assumption of a protein concentration of 100 mg protein/ml undiluted 100,000 *g* supernatant.

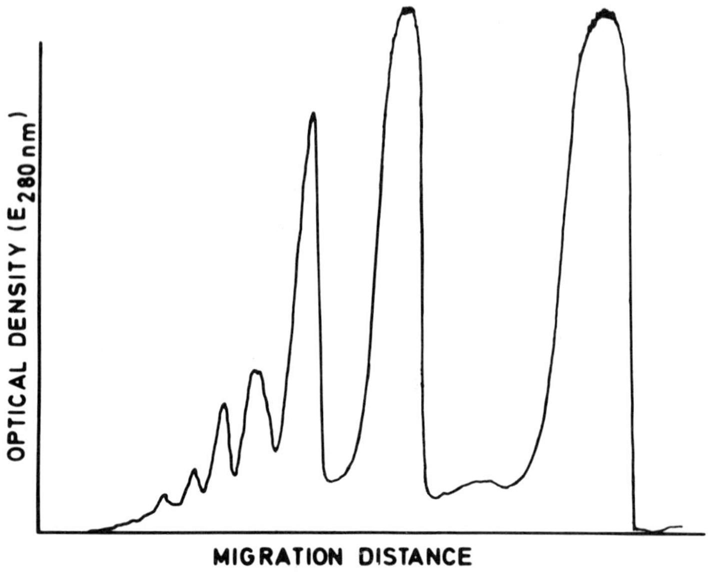

(a)

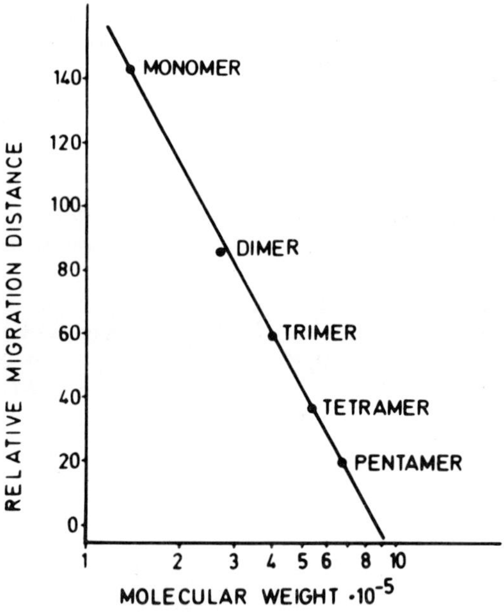

(b)

Figure 13. (*a*) Electrophoresis of purified rat liver fructose-1,6-diphosphatase in a continuous polyacrylamide gradient system according to Rüchel *et al.* (1973). Scan of the stained gel. (*b*) Relationship between relative migration distance and molecular weight of the different aggregational states of rat liver fructose-1,6-diphosphatase as calculated from gradient polyacrylamide gel electrophoresis and calibration with bovine serum albumin. All aggrega-

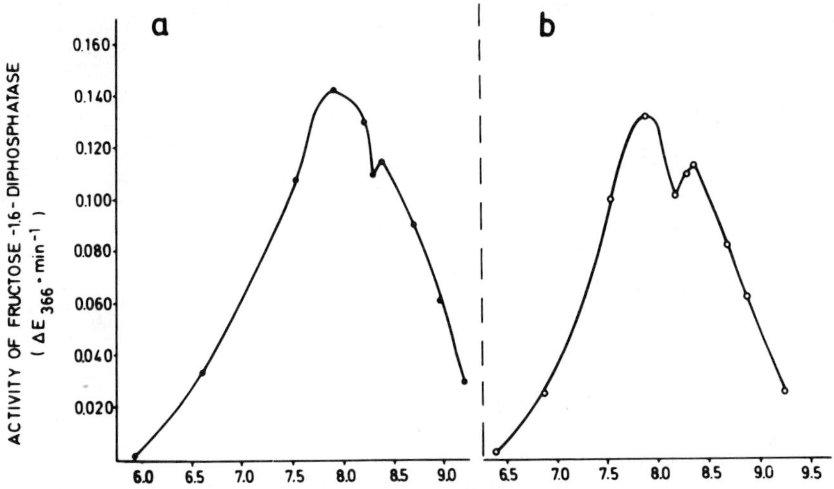

(c)

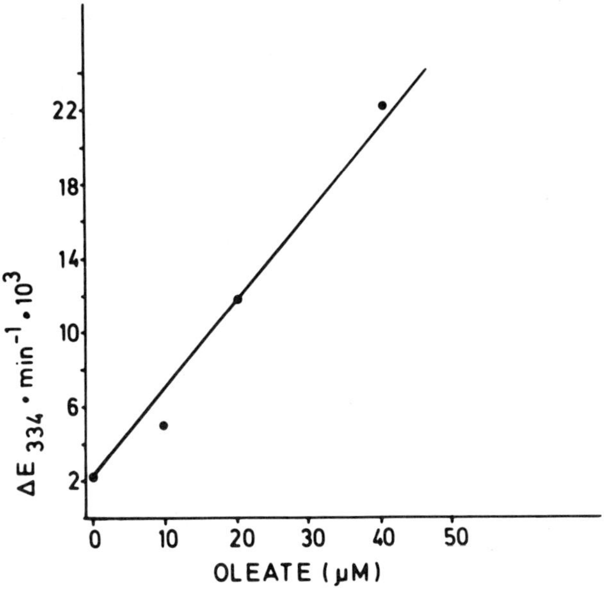

(d)

tional states were enzymically active. *Source*: Söling and Bernhard, in preparation. (*c*) Determination of pH dependence of purified rat liver fructose-1,6-diphosphatase in the presence (left) and absence (right) of 1mM EDTA. The divalent cation was Mg^{2+} (10 mM). *Source*: Söling and Bernhard, in preparation. (*d*) Stimulation of activity of diluted purified rat liver fructose-1,6-diphosphatase by oleate. The assay system was essentially the same as described by Carlson *et al.* (1973). The enzyme concentration was 1.85 μg/ml. The assay was performed in the presence of 30 μM $MnCl_2$. *Source*: Söling and Bernhard, unpublished results.

Table XXII. Activity of phosphofructokinase in livers from various vertebrate species

Species	Metabolic state	Temperature (°C)	Enzyme activity (μmole/g liver/min)	Reference
Rat (Wistar)	Fed	30°	1.46 ±0.11	Söling *et al.* (1970)
Rat	Starved (48 hours)	30°	0.65 ±0.12	Söling *et al.* (1970)
Rat (Wistar)	Diabetic	30°	0.41 ±0.02	Willms *et al.* (1970b)
Rat	Fed	25°	0.48 ±0.03	Wallace and Newsholme (1967)
Guinea pig	Fed	30°	2.05 ±0.21	Söling *et al.* (1970)
Guinea pig	Starved (48 hours)	30°	1.09 ±0.08	Söling *et al.* (1970)
Acomys cahirinus	Nondiabetic	30°	1.61 ±0.06	Willms *et al.* (1970b)
Acomys cahirinus	Diabetic	30°	1.78 ±0.14	Willms *et al.* (1970b)
Acomys cahirinus	Diabetic, ketotic	30°	0.46	Willms *et al.* (1970b)
New Zealand obese mouse	Lean	30°	1.11 ±0.04	Willms *et al.* (1970b)
New Zealand obese mouse	Obese	30°	1.05 ±0.05	Willms *et al.* (1970b)
Sheep	Fed	25°	0.52 ±0.12	Wallace and Newsholme (1967)
Domestic fowl	Fed	25°	1.35 ±0.35	Wallace and Newsholme (1967)
Man[a]	Fasted overnight	30°	0.94 ±0.25	Willms *et al.* (1970b)
Man[a]	Diabetic (insulin-deficiency type)	30°	0.45 ±0.11	Willms *et al.* (1970b)
Man[a]	Diabetic (maturity onset type)	30°	0.60 ±0.19	Willms *et al.* (1970b)

[a] Values calculated from the data of Willms *et al.* (1970b) under the assumption that 1 g wet liver is equivalent to 100 mg of protein in the undiluted 100,000 *g* supernatant fraction.

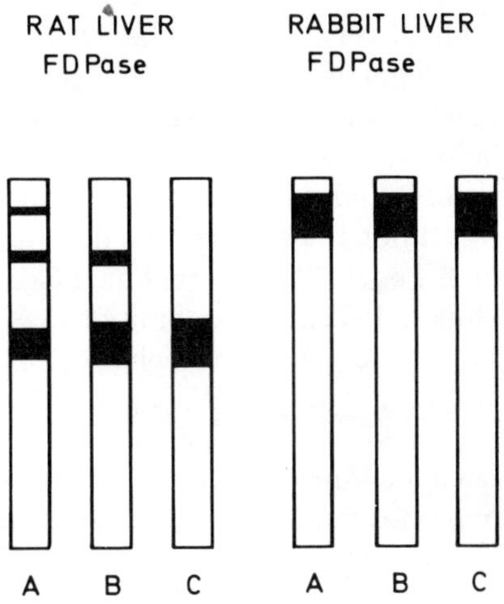

Figure 14. Different behaviors of purified fructose-1,6-diphosphatase from rat and rabbit liver during electrophoresis in 7.5% polyacrylamide gel. The electrode buffer system was tris-borate 42 mM, pH 8.8. Electrophoresis was run for 160 minutes at 4 mA per gel. Prior to electrophoresis the enzymes were preincubated for 1 hour at 37°C in 20 mM tris-Cl pH 8.0 without (*a*) or with addition of either 4 mM CoA-SH (*b*) or 10 mM CoA-SH (*c*). *Source*: Söling and Bernhard, in preparation.

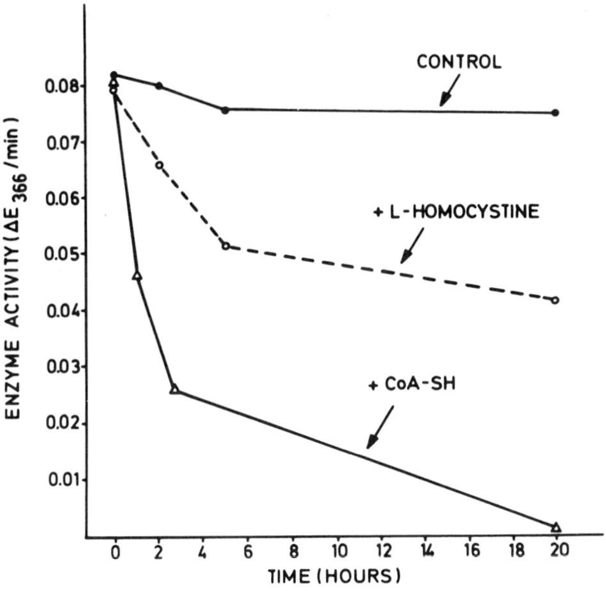

Figure 15. Effect of incubation of purified rat liver fructose-1,6-diphosphatase with L-homocystine or CoA-SH. Here 1.4 nmole of fructose-1,6-diphosphatase protein was incubated with 200 nmole of L-homocystine or with 2000 nmole of CoA-SH in 100 mM tris-Cl, pH 8.5 at 25°C, for the times indicated in the abscissa. Aliquots of the incubation mixture were taken at the times indicated for the measurement of fructose-1,6-diphosphatase activity. Enzyme activity was measured under the following conditions (final concentrations given): triethanolamine-Cl (pH 7.5), 45 mM; L-histidine, 1 mM; $NADP^+$, 0.15 mM; $MnCl_2$, 1 mM; phosphohexose isomerase, 2 μg/ml; glucose-6-phosphate dehydrogenase, 4.6 μg/ml, and appropriate amounts of fructose-1,6-diphosphatase. *Source*: Söling and Bernhard, in preparation.

rabbit liver could also be activated by SH-agents, but no data concerning this point were presented. Speculations about an interrelation between the redox state of the $NADP^+/NADPH$ or the glutathion system and the regulation of FDPase activity seem to be premature at this time.

ATP and ADP inactivate purified or crude FDPase, provided the enzyme was diluted. This has been described for FDPase from livers of rats (Taketa and Pogell, 1963), rabbits (Taketa and Pogell, 1963), and guinea pigs (Jones, 1972). This inhibition was not accompanied by a change in the K_m for fructose-1,6-diphosphate (Taketa *et al.*, 1971). The inhibition of the enzymes from rabbit (Pogell *et al.*, 1971) or guinea pig (Jones, 1972) liver is abolished by 3-phosphoglycerate. A variety of other compounds of unrelated chemical structure [methylene diphosphonate, 6-phosphogluconate, pyrophosphate, phosphoribosyl-pyrophosphate, oxaloacetate, and citrate (Taketa *et al.*, 1971)] likewise inhibited, completely or partially, the inactivation by ATP or ADP. This lack of specificity together with the requirement of a high degree of dilution of the enzyme makes it rather doubtful that inactivation by ATP or ADP is regulatory under physiological conditions. However, an enhancement of the AMP inhibition by citrate may play a regulatory role, particularly in guinea pig liver (Jones, 1972).

It has been proposed by Pogell *et al.* (1968) that rabbit liver FDPase is specifically activated by phosphofructokinase (PFK). Therefore, it would be of interest to know whether species differences exist for this type of metabolic regulation. We have examined this problem with FDPase from rat liver. Using the same assay system as Pogell *et al.* (1968), we found that rabbit muscle PFK did activate purified as well as crude rat liver FDPase (Table XXIII). However, albumin from various sources at similar concentrations activated to the same extent as PFK (Table XXIII). Untreated human serum had by far the strongest activating effect. Moreover, if a PFK/FDPase interaction were to play a significant role, homologous liver PFK should be more effective than the muscle enzyme. We have measured the effect of highly purified, homogeneous rat liver PFK on rat liver FDPase (Table XXIII). Under these conditions PFK was less effective than bovine serum albumin. Furthermore the unspecific activation of FDPase by proteins is the same regardless of whether the enzyme is inhibited by 5′-AMP. Rat liver FDPase, which had been partially desensitized against 5′-AMP inhibition by pyridoxal 5′-phosphate (Marcus and Hubert, 1968), was also activated to about the same extent as the native enzyme (Söling and Bernhardt, unpublished experiments). Therefore an enzyme-enzyme interaction between PFK and FDPase is probably not of specific physiological significance for control of the activity of FDPase. The activating effect of protein fraction from rat liver homogenate obtained by acid precipitation at pH 6.0 (Pogell *et al.*, 1968) also was not specific in our

Table XXIII. Unspecific stimulation of fructose-1,6-diphosphatase activity by various proteins[a]

Protein factor added	Final activity (% of control without protein factor)
Phosphofructokinase (rat liver)	140
Phosphofructokinase (rabbit muscle)	410
Lactate dehydrogenase (pig heart)	203
Pyruvate kinase (rabbit muscle)	135
Bovine serum albumin (Behring-Werke)	292
Bovine serum albumin (Serva)	380

[a] A 4 (w/v) rat liver homogenate was prepared with 0.154 M KCl and centrifuged for 1 hour at 36,000 *g*. The supernatant was filtered through glasswool and dialyzed against quartz-distilled water. This was used as a source for FDPase. FDPase was incubated at 30°C for 5 minutes in the following incubation mixture (final concentrations): Tris-HCl, pH 7.5, 30 mM; $NADP^+$ 0.25 mM; $MnCl_2$ 0.05 mM; phosphoglucose isomerase 5 μg/ml; glucose-6-phosphate dehydrogenase 2 μg/ml, and "activating" proteins equivalent to a final concentration of 0.3 mg/ml. After 5 minutes the reaction was initiated by the addition of fructose-1,6-P_2 (0.08 mM final concentration).

experiments with rat liver FDPase. When corrected for intrinsic FDPase activity, the activating effect of the pH 6.0 fraction paralleled its protein concentration but was no more effective than bovine serum albumin at similar concentrations (Söling and Bernhard, unpublished results) and certainly less active than native human serum. On the basis of experiments with rabbit liver FDPase, it was claimed that the factor in the pH 6 sediment is identical with long-chain fatty acids (Baxter *et al.*, 1972). Baxter *et al.* (1972) and Carlson *et al.* (1973) demonstrated an activation of rabbit liver FDPase by oleate. We have confirmed this finding for rat liver FDPase (Fig. 13) and have shown that the effect is independent of the concentration of this enzyme. This contrasts with the activating effect of proteins on FDPase, which decreases with increasing concentrations of FDPase. However, one must bear in mind that free (unbound) long-chain fatty acids are not present in the cell at concentrations necessary to activate FDPase. Allen and Blair (1972) have described an activation of rabbit liver FDPase by phospholipids, a finding we confirmed with rat liver FDPase.

Phosphofructokinase from rat liver (Underwood and Newsholme 1965;

Brand 1974) exhibits homotropic cooperativity with respect to fructose-6-phosphate. ATP and citrate are negative, whereas 5′-AMP, 3′,5′-cAMP, and ADP are positive allosteric regulators [rat liver enzyme (Underwood and Newsholme, 1965; Brand, 1974), rabbit liver enzyme (Kemp, 1971), sheep liver enzyme (Passoneau and Lowry, 1964; Brock, 1969)]. The liver enzyme behaves like the muscle enzyme (for a review see Mansour, 1972). The allosteric regulation of liver PFK by ATP [first described for the muscle enzyme by Lardy and Parks (1956)] is strongly dependent on the enzyme's concentration. The higher the pH value and enzyme concentration, the less sensitive the enzyme is to ATP inhibition. Studies comparing the kinetic properties of liver phosphofructokinase from different species are lacking.

Since glucagon stimulates gluconeogenesis from a variety of substrates including dihydroxyacetone, glycerol, xylitol, and fructose, it is conceivable that this hormone interacts at the fructose-1,6-diphosphatase/phosphofructokinase step. However, a glucagon mediated increase in 3′,5′-cAMP would inhibit gluconeogenesis by stimulating phosphofructokinase. An increase in citrate and/or a decrease of 5′-AMP after glucagon administration has not been described. Since glucagon or 3′,5′-AMP fails to stimulate gluconeogenesis in isolated, perfused livers from guinea pigs or mice, a different type of regulation at this site in these species seems to be possible, although proof for such a difference is lacking.

For the regulation of gluconeogenic or glycolytic flux rates through the FDPase/PFK step, metabolites that exert adverse effects on either of the two enzymes are of primary interest. The metabolites most likely to play a physiological regulatory role are 5′-AMP and fructose-1,6-diphosphate. The importance of direct allosteric control by fructose-1,6-diphosphate of FDPase and PFK in liver is not completely understood. Unlike the situation in muscle PFK, fructose-1,6-diphosphate has virtually no effect on the cooperative kinetic of the rat liver enzyme with fructose-6-phosphate (Brand, 1974). The physiological importance of the inhibition of liver FDPase by fructose-1,6-diphosphate is uncertain, since the inhibition begins at concentrations of fructose-1,6-diphosphate above 30 μM but remains incomplete up to 1 mM and higher (Taketa and Pogell, 1965; Nakashima *et al.*, 1970). The role of citrate is more complicated. The citrate mediated protection of fructose-1,6-diphosphatase against inactivation by ATP or ADP is very unspecific and is probably of minor physiological importance. However, Jones (1972) noted that concentrations of citrate 0.4 mM and higher enhance the inhibition of guinea pig liver fructose-1,6-diphosphatase by 5′-AMP. Therefore, *in vivo* changes of citrate levels below 0.4 mM should be of minor physiological importance, since the citrate concentrations found in rat liver under a variety of metabolic or nutritional conditions were always less than 0.3 mM (for review, see Greenbaum *et al.*, 1971). The overall concentration of citrate in the isolated, perfused guinea pig liver is between 0.4 and 0.9 mM in various experimental conditions (Haeckel and Haeckel, 1971). Whether these higher levels of citrate can affect already the regulation of FDPase in the whole organ has be be proven.

The allosteric effect of citrate on rat liver PFK is rather complex and needs further elucidation until its physiological importance can be finally assessed. Citrate enhances the inhibitory effect of ATP on PFK of the livers of sheep (Passoneau and Lowry, 1964) and rats (Underwood and Newsholme, 1965; Brand, 1974). The concentrations of citrate necessary to achieve an inhibitory effect is inversely proportional to the concentration of ATP. The increased sensitivity toward inhibition by ATP in the presence of citrate is overcome by AMP and fructose-6-phosphate [sheep liver enzyme (Passoneau and Lowry, 1964), rat liver enzyme (Underwood and Newsholme, 1965; Brand, 1974)]. Under physiological conditions the situation is further complicated by the inhibitory effect of physiological concentrations of ATP, which can be counteracted effectively by ADP (Brand, 1974). Kemp (1971) found that at neutral pH rabbit liver PFK was much less sensitive to the inhibitory effect of citrate and the activating affect of AMP and ADP as compared to the muscle enzyme. Under similar experimental conditions, rat liver PFK was much more sensitive against citrate inhibition (Brand, 1974) than was the rabbit liver enzyme. Whether this represents true species differences or differences in the experimental approach remains to be clarified.

Hepatic PFK [rat liver (Underwood and Newsholme, 1965; Brand, 1974), rabbit liver (Kemp, 1971)] is strongly activated by NH_4^+ ions. However, this does not mean that NH_4^+ ions play a regulatory role for the FDPase/PFK step, especially in view of the finding of Greenbaum *et al.* (1971) that the overall level of NH_4^+ ions in rat liver changes only marginally under very different metabolic conditions. Clark *et al.* (1974) suggest that the PFK-FDPase "futile" cycle in rat liver *in vivo* is regulated mainly by the adenine nucleotides. However, these findings have been questioned by Hue and Hers (1974a).

We have recently discovered (Brand and Söling, 1975) that rat liver phosphofructokinase belongs to the group of enzymes that can be chemically modified by phosphorylation and dephosphorylation (Fig. 16). A cyclic nucleotide independent PFK-kinase and a PFK-phosphatase have been identified in rat liver. The "kinase" activity decreases rapidly during starvation and reappears within 6 hours after refeeding (Fig. 17). The dephosphorylated PFK dissociates immediately into subunits. The nonphosphorylated, inactive subunits reassociate upon phosphorylation to form a stable active monomer containing four phosphates (one phosphate per protomer). In the presence of high concentrations of FDP, the inactive nonphosphorylated protomers can also reassociate to an active nonphos-

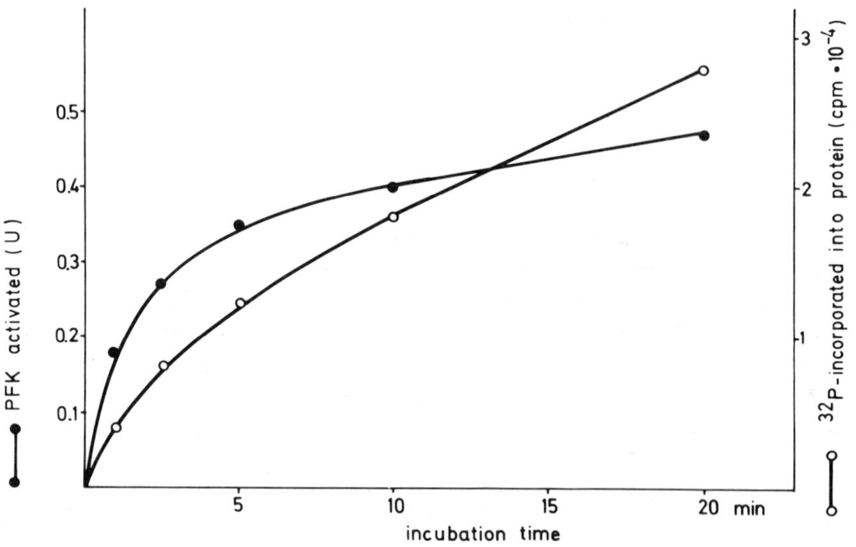

(a)

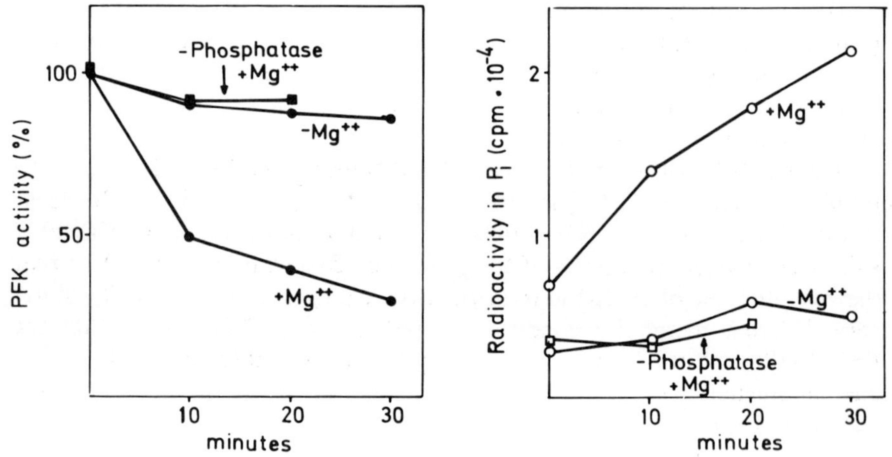

(b)

Figure 16. (*a*) Incorporation of ^{32}P radioactivity from ATP(^{32}P) into rat liver phosphofructokinase (○——○) and simultaneous increase in phosphofructokinase activity (●——●). Rat liver phosphofructokinase had been inactivated prior to the activation incubation. Radioactive phosphofructokinase was separated from the incubation by a specific double antibody technique. *Source:* Brand and Söling (1975). (*b*) Inactivation of enzymically active, ^{32}P-labeled rat liver phosphofructokinase by enzymic dephosphorylation. The left-hand graph plots the decrease in enzymatic activity, which occurred only in presence of the phosphatase + Mg^{2+} but not with either of the two factors alone. The right-hand graph plots the release of inorganic phosphate, which again occurred only in the presence of phosphatase + Mg^{2+} *Source:* Brand and Söling (1975).

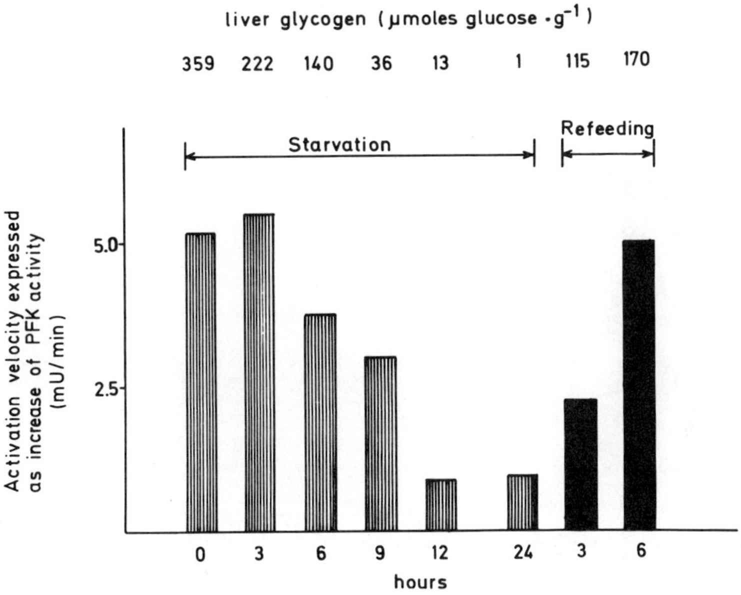

Figure 17. Changes of phosphofructokinase activating activity in livers from starved and from starved refed rats. The numbers in the scale (upper part of figure) represent the corresponding glycogen concentrations. *Source:* Brand and Söling (1975).

phorylated monomer. However, this monomer is unstable and dissociates immediately into the active protomers when the FDP is removed. The phosphorylation of the unstable, nonphosphorylated, active monomer formed in the presence of FDP seems to proceed much more rapidly than the phosphorylation and subsequent aggregation of the inactive protomers. A tentative model of this type of regulation is depicted in Fig. 18.

A direct correlation between the degree of PFK phosphorylation and the rate of "futile cycling" at the FDPase-PFK step remains to be established. However, the rapid adaptation of the interconversion to starvation and refeeding favors regulatory importance for this process (Brand and Söling, 1975). Recently we demonstrated the PFK-kinase-phosphatase system in pig liver, also.

Uyeda and Luby (1974) have reported a specific inhibition of phosphofructokinase by FDPase. FDPase from chicken liver enhanced the inhibition of PFK from rabbit muscle and chicken liver by ATP or 3-phosphoglycerate. Other proteins tested were without effect. The authors believe they have excluded the possibility that the interaction of FDPase with the PFK reaction product fructose-1,6-diphosphate was responsible for the effect. We were able to confirm the stimulation of ATP inhibition of liver

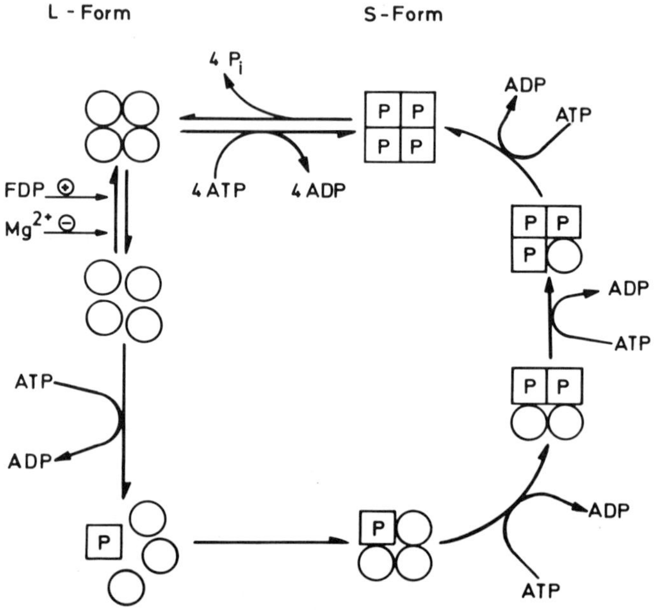

Figure 18. Interconversion of different forms of rat liver phosphofructokinase. The enzyme can exist in two enzymatically active tetrameric states in which the L-form is unphosphorylated and labile, dissociating upon removal of fructose-1, 6-diphosphate into inactive protomers. The phosphorylated S-form remains stable and active even in the absence of fructose-1, 6-diphosphate. The inactive, nonphosphorylated protomers (open circles) can be phosphorylated by a kinase in the presence of ATP-Mg^{2-}. The phosphorylation of one protomer initiates the formation of a hybrid tetramer. This facilitates the phosphorylation of a second protomer in this tetramer, which in turn leads to further facilitation of the next phosphorylation step. This is in accordance with the cooperative-type activation curve when PFK activation is plotted against ATP-Mg^{2-} concentration. The model is supported further by the findings that the phosphorylation reaction follows hyperbolic kinetics in the presence of fructose-1, 6-diphosphate and that the phosphatase-catalyzed dephosphorylation of the phosphorylated tetramer leads to inactivation of the enzyme in the absence but not in the presence of fructose-1, 6-diphosphate. *Source:* Brand and Söling (1975).

PFK by liver FDPase for the two enzymes from rat liver (Fig. 19). However, in recent experiments we were able to prove definitely that the action of FDPase is mediated exclusively by its effect on the concentration of fructose-1,6-diphosphate (Söling and Bernhard, unpublished results).

H. Species Differences in the Regulation at the Glucose-6-Phosphatase–Hexokinase-Glucokinase Step

The formation of glucose-6-phosphate and the cleavage of this substance are catalyzed by reactions that are irreversible under physiological conditions. The hepatic ATP-glucose-6-phosphate phosphotransferase can be separated into three species having a low K_m for glucose ("hexokinases," K_m glucose $= \sim 0.01$ mM) and low substrate specificity, and one species having a high K_m for glucose ("glucokinase," K_m glucose $= \sim 10$ mM) and high substrate specificity (Viñuela *et al.*, 1963). The hexokinase activity is allosterically inhibited by glucose-6-phosphate. According to Sapag-Hagar *et al.* (1969), glucokinase is exclusively located in hepatocytes, whereas hexokinase activity is in nonparenchymal cells. However, Werner *et al.* (1972) found that liver parenchymal cells do contain appreciable amounts of hexokinases. In all species tested thus far, glucokinase activity decreases

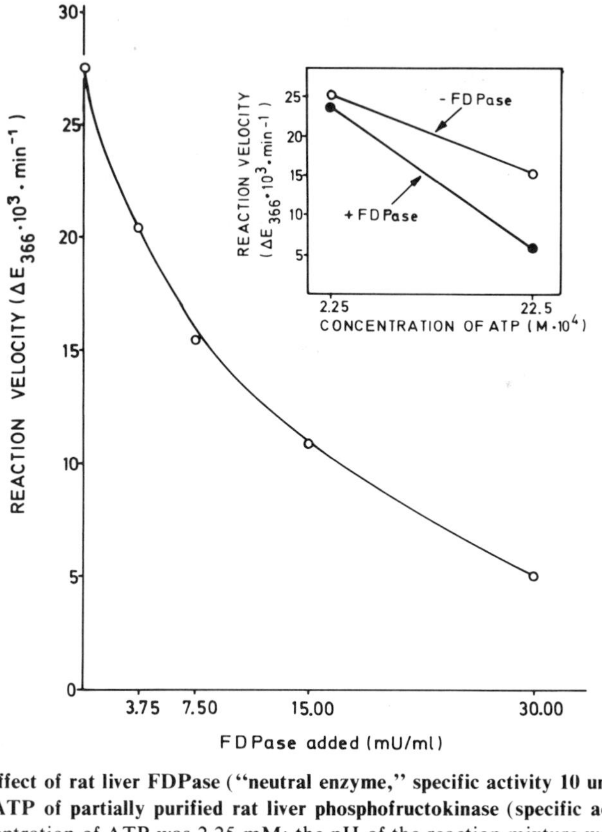

Figure 19. Effect of rat liver FDPase ("neutral enzyme," specific activity 10 units/mg) on the inhibition by ATP of partially purified rat liver phosphofructokinase (specific activity 5 units/mg. The concentration of ATP was 2.25 mM; the pH of the reaction mixture was 8.0 at 30°C. The test mixture contained 41 mU/ml of phosphofructokinase (This activity of phosphofructokinase was measured under optimal conditions. Under the test conditions employed here, the activity was only 7.5 mU/ml.) The insert shows that FDPase had no inhibiting effect on phosphofructokinase in the presence of a low ATP concentration. The concentration of FDPase in the insert experiment was 15 mU/ml. The assay system was test system A (via pyruvate kinase-lactate dehydrogenase) of Uyeda and Luby (1974).

Table XXIV. Activities of hexokinase and glucokinase in livers from various vertebrate species

Nutritional classification	Species	Rate (μmole/g liver/min)	
		Hexokinase	Glucokinase
Omnivorous	Rat	0.665 ±0.015	3.83 ±0.15
	Mouse	0.698 ±0.033	3.08 ±0.18
	Possum	0.610 ±0.066	6.57 ±0.28
	Pig	0.635 ±0.035	3.17 ±0.15
	Man (normal)	0.852 ±0.055[a]	0.764 ±0.073[a]
	Man (diabetic, juvenile type)	0.382 ±0.073[a]	0.323 ±0.117[a]
	Man (Diabetic, maturity, onset type)	0.794 ±0.059[a]	1.294 ±0.250[a]
Carnivorous	Dog	0.500 ±0.060	1.83 ±0.13
	Cat	0.410 ±0.033	<0.033
Herbivorous	Guinea pig	0.387 ±0.050	1.25 ±0.18
	Rabbit	0.278 ±0.015	1.32 ±0.03
	Sheep	0.128 ±0.021	<0.03
	Cow	0.387 ±0.070	<0.03
	Quokka	0.392 ±0.038	<0.03
	Tammar	0.205	<0.03

[a] Data from Ballard (1965) except for the values for man, which are from Willms *et al.* (1970b). They were recalculated from the original data under the assumption that the undiluted cytosolic fraction of 1 g liver contains 100 mg of protein.

under conditions of starvation, diabetes, and restriction of carbohydrates (Sols *et al.*, 1964). The activity of this enzyme reveals extreme species variability, whereas the activity of glucose-6-phosphatase shows only minor variations (Tables XXIV and XXV). Allosteric effectors of G-6-Pase that act under physiological conditions are not known yet. All vertebrate species tested exhibit a similar increase in G-6-Pase activity in response to endocrine or nutritional changes that diminish glucokinase activity.

In this discussion of species differences in the regulation of gluconeogenesis, we focus mainly on glucokinase. Interestingly, glucokinase activity is high in species in which PEP-CK activity is mainly in the cytosol (e.g., rat) is lower in species in which PEP-CK is partly intramitochondrial and partly in the cytosol (e.g., guinea pig, man). It is virtually absent in birds, among which PEP-CK is almost totally mitochondrial. When the net

Table XXV. Activity of glucose-6-phosphatase in livers from various vertebrate species

Species	Metabolic state	Temperature (°C)	Enzyme activity (μmole/g liver/min)	Reference
Rat	Fed	37°	7.23 ±0.26	Söling *et al.* (1970)
Rat	Starved (48 hours)	37°	13.87 ±0.75	Söling *et al.* (1970)
Rat	Fed	30°	8.70 ±0.56	Wimhurst and Manchester (1970b)
Rat	Starved	30°	24.75 ±0.98	Wimhurst and Manchester (1970b)
Rat	Diabetic	30°	30.15 ±1.85	Wimhurst and Manchester (1970b)
Acomys cahirinus	Fed (nondiabetic)	37°	12.79 ±0.68	Willms *et al.* (1970b)
Guinea pig	Fed	37°	7.40 ±0.15	Söling *et al.* (1970)
Guinea pig	Starved (48 hours)	37°	13.13 ±0.54	Söling *et al.* (1970)
Sheep[a]	Fed	37°	12.8	Filsell *et al.* (1969)
Sheep[a]	Fasted (6 days)	37°	22.0	Filsell *et al.* (1969)
Goat	Fed	37°	19.6	Purser and Bergen (1969)
Cow	Nonlactating (fed)	37°	21.93 ±4.83	Baird and Heitzmann (1970)
Cow	Lactating (fed)	37°	25.65 ±2.51	Baird and Heitzmann (1970)
Man	Overnight fasted	37°	18.1 ±2.41	Öckerman (1967)
	Overnight fasted	37°	16.2 ±1.86	Ricketts (1963)

[a] Calculated from the data by assuming a protein concentration of 200 mg/g wet liver.

formation of glucose by isolated, perfused livers or isolated liver cells is examined, the liver is normally from starved animals, whose glucokinase activity is significantly diminished. But even then large species differences persist.

The species differences observed for the maximum rates of glucose production from L-lactate in experiments with isolated, perfused livers or isolated liver cells cannot be explained by citing the recycling of carbon due to large differences in hepatic glucokinase activity. The intracellular and the medium concentrations of glucose under these conditions are far below the K_m glucose for glucokinase. Also the rate of glucose formation from fructose in rat liver is in the same range as that from L-lactate in pigeon liver, which would not be expected if gluconeogenesis from L-lactate in rat liver is limited by a higher activity of glucokinase. However, after ingestion of a meal high in carbohydrate, the concentration of glucose in the portal blood is in the range of the enzyme's $K_{m\ \mathrm{glucose}}$. Under these conditions, differences in the activity of glucokinase may contribute to a different mode of regulation of the hepatic glucose balance. It was shown several decades ago that the net glucose balance across the liver of dogs (Soskin *et al.*, 1938) depended on the concentration of glucose ("Soskin effect"). Similar observations have been made by McCraw *et al.* (1968), Glinsman *et al.* (1969), and Walli *et al.* (1974) in experiments with isolated, perfused rat livers. At concentrations higher than 11 mM there is a net uptake of glucose, at concentrations around 11 mM there is no significant net change, and at glucose concentrations below 10 mM there is a net production of glucose. As Fig. 20 indicates, the "Soskin effect" is virtually absent in guinea pig and chicken liver, where the activity of glucokinase is low or absent. The isolated, perfused rat liver changes from a net uptake of L-lactate at the lower glucose concentration to a net production at the higher glucose concentration, but the livers of the two other species continue to take up L-lactate at the higher glucose concentration (not shown). Therefore, as long as gluconeogenesis is studied *in vitro* with livers or liver cells from starved animals in the absence of exogenous glucose, the glucokinase/glucose-6-phosphatase step plays no rate-limiting or regulatory role for gluconeogenesis. *In vivo,* however, especially in the absorptive state, the response of gluconeogenesis to changes of the portal glucose concentration may be controlled to a high degree at this step in the species that possess sufficiently high activities of glucokinase. The mechanism for such a control at the glucokinase-G-6-Pase step has been worked out by Hue and Hers (1974b). According to these authors, the net flux of glucose and the rate of recycling is regulated mainly by the concentration of glucose itself. An increase of glucose concentration within the physiological range pulls glucose into glycogen by a direct stimulation of the glycogen synthetase reaction.

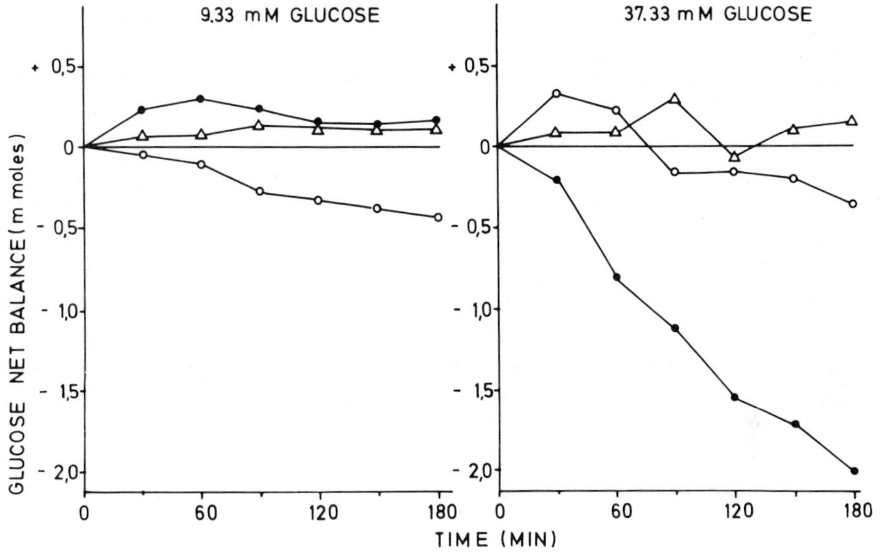

Figure 20. Changes of the net glucose balance in experiments with isolated, perfused livers from fed rats (solid points), guinea pigs (open points),and chickens (triangles) after increasing the glucose concentration in the medium from 9.33 to 33.33 mM. Only rat livers respond with a net uptake a glucose. *Source:* Söling and Seufert (1974).

This is accompanied by a decrease in glucose-6-phosphate, which in turn diminishes the rate of hydrolysis of glucose-6-phosphate by G-6-Pase. This type of regulation, of course, cannot be operative in species lacking glucokinase (e.g., pigeons) or in animals in which glucokinase activity is decreased to very low values because of lack of insulin.

XI. "FUTILE CYCLING", ENERGY COST, AND SPECIES DIFFERENCES IN THE REGULATION OF GLUCONEOGENESIS

The net rate of gluconeogenesis is affected by the extent to which unidirectionally catalyzed gluconeogenic reactions are counteracted by the corresponding glycolytic enzymes. Each mole of PEP, fructose-1,6-diphosphate, or glucose that is reconverted to pyruvate, fructose-6-phosphate, or glucose-6-phosphate, accounts for the additional split of one γ-phosphate bond from ATP (or GTP), thus for additional consumption of oxygen. Without this "futile cycling" of carbon, 6 moles of ATP (GTP) are used per mole of glucose formed from L-lactate or pyruvate. Assuming a tightly coupled system, this would correspond to an oxygen consumption of 2 g-atoms per mole of glucose formed. Any "futile cycling" should result in oxygen/glucose ratios higher than 2.

We have compared the oxygen/glucose ratio in isolated, perfused livers from rats, pigeons, and guinea pigs (Table XXVI), and have found very similar values in the three species—all below 2.5. Similar oxygen/glucose ratios have been measured in experiments with isolated, perfused rat livers by Scholz (personal communication). These values are surprisingly close to the theoretically expected value of 2. They do not, however, exclude a considerable rate of "futile cycling" (see Söling *et al.*, 1973). If 50% of the fructose-6-phosphate formed in the fructose-1,6-diphosphatase reaction is reconverted to fructose-1,6-diphosphate by phosphofructokinase, this would lead to the consumption of 7 instead of 6 moles of ATP per mole of glucose synthesized from L-lactate or pyruvate. The result would be an increase of the oxygen/glucose ratio from 2.0 to 2.33, which is well within the measured range. Similarly, if 1 out of 3 moles of PEP formed from L-lactate or pyruvate is reconverted to pyruvate in the pyruvate kinase reaction, 7 instead of 6 moles ATP (GTP) per mole of glucose formed would be consumed.

The existence of "futile cycling" at the step of glucose-6-phosphatase/hexokinase-glucokinase and fructose-1,6-diphosphatase-phosphofructokinase has been claimed recently by Rognstad *et al.* (1973) and Clark *et al.* (1973) using isolated rat liver cells. Clark *et al.* (1974) estimated that the rate of recycling at the glucose-6-phosphatase/hexokinase-glucokinase step in livers of 2-day-old rats is as low as 0.13 μmole/g liver/min. In livers from fed mice the loss of tritium from (2-^{3}H) glucose in comparison to the tritium loss from (6-^{3}H) glucose gave a rate of recycling at the glucose-6-phosphatase/hexokinase-glucokinase step of about 1 μmole/g liver/min (Hue and Hers, 1974a). The rate of recycling at the glucose-6-phosphatase/hexokinase-glucokinase step certainly is much lower under gluconeogenic conditions (e.g., starvation) or in species with low or absent acitivity of glucokinase. Clark *et al.* (1974) have estimated the rate of recycling at the fructose-1,6-diphosphatase/phosphofructokinase step by rat liver *in vivo* by measuring the loss of tritium from (2-^{3}H)- and (5-^{3}H)-glucose. The highest rate of recycling (obtained in the presence of the hypoglycemic agent phenethylbiguanide) was 0.20 μmole/g wet liver/min. The recycling rate under "physiological" conditions was only about 0.06 μmole/g wet liver/min. Hue and Hers (1974a) suggested that in liver, the loss of tritium from (2-^{3}H)- or (5-^{3}H)-glucose cannot be taken as a measure of the rate of recycling at the fructose-1,6-diphosphatase/phosphofructokinase step. They found that hydrogen exchange in position 5 of fructose-6-phosphate can occur in mouse liver without conversion to fructose-1,6-diphosphate by phosphofructokinase. The observed loss of tritium results from the combined ac-

Table XXVI. Quantitative relationship between glucose formation and oxygen uptake by isolated, perfused livers from starved pigeons, rats, and guinea pigs[a]

Species	Number of experiments	Net formation of glucose (μmole/g liver/min)	Oxygen uptake (μg-atom/g liver/min) in relation to addition of L-lactate + pyruvate: Before	Oxygen uptake (μg-atom/g liver/min) in relation to addition of L-lactate + pyruvate: After	Δ Oxygen uptake (μg-atom/g liver/min)	Oxygen uptake/glucose formation (ratio)
Pigeon	5	1.09 ±0.11	4.40 ±0.57	7.12 ±0.80	2.72 ±0.42	2.47 ±0.39
Rat	5	0.69 ±0.09	6.26 ±0.15	7.97 ±0.31	1.71 ±0.35	2.42 ±0.29
Guinea pig	1	0.84	3.75	5.75	2.00	2.38

[a] Gluconeogenesis was induced by addition of 195 μmole L-lactate + pyruvate (10:1) to the medium and start of an intraportal infusion of 815 μmole/hr of L-lactate + pyruvate (10:1). The initial perfusion volume was 200 ml.

Source: Data from Söling *et al.* (1973).

tions of *trans* aldolase and triosephosphate isomerase. This is substantiated by the finding that in mouse liver there is a negligible loss of tritium from the C-3 position of glucose relative to the (^{14}C) activity of (U-^{14}C)-glucose. The data of Clark *et al.* (1974), therefore, do not definitively establish recycling between fructose-6-phosphate und fructose-1,6-diphosphate. The loss of tritium from (3-^{3}H)-glucose in mice *in vivo* (Hue and Hers, 1974a) corresponds well to the activity of the Cori cycle as judged from the similar loss of tritium from (6-^{3}H)-glucose. On the other hand, Rognstad *et al.* (1973) measured the appearance of (^{14}C)-radioactivity from (1-^{14}C)-galactose in the C-1 and C-6 positions of glucose, in experiments with isolated rat liver cells. They found a measurable incorporation into the C-6 position of glucose even under conditions of maximal net gluconeogenesis. This incorporation was not affected by increasing the L-lactate pool. This result can be explained only by the operation of the phosphofructokinase reaction even under conditions of maximal gluconeogenesis. Thus the available data indicate that the rate of "futile cycling" at the fructose-1,6-diphosphatase/phosphofructokinase step must be very low under physiological conditions, definitely far less than recycling at the glucose-6-phosphatase/hexokinase-glucokinase step. Furthermore, the net flux of gluconeogenesis from L-lactate or pyruvate certainly is not limited by the operation of "futile cycling" at these two steps. This is best illustrated by noting that when L-lactate or pyruvate at saturating concentrations was metabolized together with 10 mM glycerol, by isolated, perfused rat livers, the rate of glucose formation was close to the theoretical sum of the glucogenic rates with either precursor alone (Ross *et al.*, 1967b).

The rate of recycling at the pyruvate carboxylase–PEP-CK/pyruvate kinase step has been estimated by Friedmann *et al.* (1971) in rats. The rate depended on the nutritional state of the animal but did not exceed 0.35 μmole/g liver/min. In our experiments (Table XXVI) we assumed that recycling occurs mainly at the pyruvate carboxylase–PEP-CK/pyruvate kinase step so that the oxygen/glucose ratios would correspond to maximum recycling rates of 30 to 40%. In view of the different techniques applied, these results fit relatively well with the recycling rates measured by Friedmann *et al.* (1971), who used isotope techniques in livers from starved rats.

Finally these data do not support the calculations of J. R. Williamson (1971), which suggest that the energy cost of gluconeogenesis is as high as 18 ATP equivalents per mole of glucose formed. It is, on the other hand, surprising that pigeon and rat liver exhibit the same oxygen/glucose ratios in spite of the considerable differences in the location and absolute activities of PEP-CK and in the ratios of pyruvate carboxylase/pyruvate kinase and PEP-CK/pyruvate kinase. The different capacities of livers from different vertebrate species cannot be explained on the basis of different rates of "futile cycling" at the three steps discussed.

XII. CONCLUDING REMARKS

The picture presented in this chapter is oversimplified because it does not represent all factors involved in the short-term regulation of gluconeogenesis. The problem of sex differences, for example, may vary to a large extent from species to species. Another fact not considered at all is that the hepatocytes are not uniform cells but differ from each other depending mostly on their position within the liver lobule. It is already known that centrolobular hepatocytes differ electronmicroscopically from hepatocytes situated in the periphery of the lobules (e.g., Loud, 1968). Only recently Katz and Jungermann (1975) have clearly demonstrated that the isolated, perfused rat liver can produce lactate from fructose and at the same time utilize lactate for glucose formation. They suggested the existence of a hepatocyte population situated in the periphery of the lobules that synthesizes glycogen, glucose, and urea, which can oxidize fatty acids and form ketone bodies. A second, centrolobular-located population, can carry out glycolysis and fatty acid synthesis. Histochemical evidence was also presented that glycogen was stored nearly exclusively in the peripheral hepatocytes, which are rich in glucose-6-phophatase activity, whereas glucose-6-phosphatase activity was virtually absent in the centrolobular hepatocytes (Sasse *et al.*, 1975).

It is most likely that this differentiation between hepatocytes results from variations in oxygen tension, cencentration gradients of one or several metabolites and/or hormones, or hemodynamic gradients along the small portal veins to the central vein of the liver. Since light microscopy has indicated differences in the morphological features of liver lobules from different vertebrate species, hemodynamic differences and different "gradients" of oxygen tension, metabolites, and hormones probably also exist. Further research is needed to evaluate the degree to which these factors affect the occurrence of different types of hepatocytes, thus contributing to species differences in the regulation of gluconeogenesis.

REFERENCES

Allen, M. B. and Blair, J. McD. (1972). *Biochem. J.* **130,** 1167–1169.

Arinze, I. J. and Hanson, R. W. (1973). *FEBS Lett.* **31,** 280–282.

Arinze, I. J., Garber, A. J., and Hanson, R. W. (1973). *J. Biol. Chem.* **248,** 2266–2274.

Assimacopoulos-Jeannet, F., Exton, J. H., and Jeanrenaud, B. (1973). *Am. J. Physiol.* **225,** 25–32.

Atwal, A. S. and Sauer, F. (1973). *Int. J. Biochem.* **4,** 283–292.

Baird, G. D., and Heitzman, R. J. (1970). *Biochem. J.* **116,** 865–874.

Balasse, E. O. and Coms, H. A. (1973). *Diabetologia* **9,** 145.

Ballard, F. J. (1965). *Comp. Biochem. Physiol.* **14,** 437–443.

Ballard, F. J. (1970). *Biochem. J.* **120,** 809–814.

Ballard, F. J. and Hanson, R. W. (1967). *Biochem. J.,* **104,** 866–871.

Ballard, F. J., Hanson, R. W., Kronfeld, D. S., and Raggi, F. (1968). *J. Nutr.* **95,** 160–173.

Ballard, F. J. and Hanson, R. W. (1969). *J. Biol. Chem.* **244,** 5625–5630.

Ballard, F. J., Hanson, R. W., and Kronfeld, D. S. (1969). *Fed. Proc.* **28,** 218–231.

Ballard, F. J. and Oliver, J. T. (1962). *Nature, (Lond.)* **195,** 498–499.

Barns, R. J. and Keech, D. B. (1972). *Biochim. Biophys. Acta* **276,** 284–296.

Barns, R. J., Keech, D. B., and O'Sullivan, W. J. (1972). *Biochim. Biophys. Acta* **289,** 212–224.

Baxter, R. C., Carlson, C. W., and Pogell, B. M. (1972). *J. Biol. Chem.* **247,** 2969–2971.

Berry, M. N. (1971). *Biochem. Biophys. Res. Commun.* **44,** 1449–1456.

Berry, M. N. and Friend, D. S. (1969). *J. Cell Biol.* **43,** 506–520.

Berry, M. N. and Kun, E. (1972). *Eur. J. Biochem.* **27,** 395–400.

Blair, J. B., Cooke, D. E., and Lardy, H. A. (1973a). *J. Biol. Chem.* **248,** 3601–3607.

Blair, J. B., Cooke, D. E., and Lardy, H. A. (1973b). *J. Biol. Chem.* **248,** 3608–3613.

Borst, P. (1963). (P. Karlson, ed.), *Funktionelle und morphologische Organisation der Zelle* Springer-Verlag, Berlin, pp. 113–119.

Böttger, I., Brdiczka, D., Pette, D., and Wieland, O. (1969). *Eur. J. Biochem.* **8,** 113–119.

Bradford, N. M. and McGivan, J. D. (1973). *Biochem. J.* **134,** 1023–1029.

Brand, I. (1974). Ph. D. thesis, University of Göttingen, Göttingen, Germany.

Brand, I. and Söling, H. D. (1975). *FEBS Lett.* **57,** 163–168.

Bremer, J. (1969). *Eur. J. Biochem.* **8,** 535–540.

Brock, D. J. H. (1969). *Biochem. J.* **113,** 235–242.

Bryla, J., Smith, C. M., and Williamson, J. R. (1973). *J. Biol. Chem.* **248,** 4003–4007.

Burch, H. B., Lowry, O. H., Kuhlmann, A. M., Skerjance, J., Diamant, E. J., Lowry, St. R., and v. Dippe, P. (1963). *J. Biol. Chem.* **238,** 2267–2273.

Byrne, W. L., Rajagopalan, G. T., Griffin, L. D., Ellis, E. H., Harris, T. M., Hochachka, P., Reid, L., and Geller, A. M. (1971). *Arch. Biochem. Biophys.* **146,** 118–133.

Carbonell, J., Feliu, J. E., Marco, R., and Sols, A. (1973). *Eur. J. Biochem.* **37,** 148–156.

Carlson, C. W., Baxter, R. C., Ulm, E. H., and Pogell, B. M. (1973). *J. Biol. Chem.* **248,** 5555–5561.

Chang, H. C. and Lane, M. D. (1966). *J. Biol. Chem.* **241,** 2413–2420.

Chappell, J. B. (1968). *Brit. Med. Bull.* **24,** 150–157.

Chappell, J. B. and Haarhoff, K. N. (1967). In *Biochemistry of Mitochondria* (E. C. Slater, Z. Kaniuga, and L. Wojtczak, eds.), Academic Press and Polish Scientific Publishers, London and Warsaw, pp. 75–91.

Clark, D. G., Rognstad, R., and Katz, J. (1973). *Biochem. Biophys. Res. Commun.* **54,** 1141–1148.

Clark, M. G., Bloxham, D. P., Holland, P. C., and Lardy, H. A. (1974). *J. Biol. Chem.* **249,** 279–290.

Cleland, W. W. (1963). *Biochim. Biophys. Acta* **67,** 104–137.

Cleland, W. W. (1963). *Biochim. Biophys. Acta* **67,** 173–187.

Cleland, W. W. (1963). *Biochim. Biophys. Acta* **67,** 188–196.

Crespin, St. R., Greenough, W. B., and Steinberg, D. (1972). *Diabetes* **21,** 1179–1184.

Diesterhaft, M., Shrago, E., and Sallach, H. J. (1971). *Biochem. Med.* **5,** 297–303.

Dietze, G. (1974). *Habilitationsschrift,* University of Munich.

Dietze, G., Wicklmayr, M., Grunst, J., Hepp, K. D., Brauns, S., and Mehnert, H. (1973). *Verh. Dtsch. Gesellsch. Inn. Med.* **79,** 811–914.

Dietze, G., Wicklmayr, M., Hepp, K. D., Dames, H., and Mehnert, H. (1973). In *Diabetes Mellitus,* Vol. 3 (A. Behringer, ed.) International Donau Symposium, Wilhelm Maudrich Verlag, Vienna, Munich, Bern, pp. 619–624.

Dugal, B. S. and Louis, B. (1974). *Acta Biol. Med. Germ.* **32,** 1–16.

Eggleston, L. V. and Krebs, H. A. (1969). *Biochem. J.* **114,** 877–879.

Elbers, R., Heldt, H. W., Schmucker, P., Soboll, S., and Wiese, H. (1974). *Hoppe-Seyler's Z. Physiol. Chem.* **355,** 378–393.

England, P. J. and Robinson, B. H. (1969). *Biochem. J.* **112,** 8.

Exton, J. H. and Park, C. R. (1968). *J. Biol. Chem.* **243,** 4189–4196.

Exton, J. H., Robinson, G. A., Sutherland, E. A., and Park, C. R. (1971a). *J. Biol. Chem.* **246,** 6166–6177.

Exton, J. H. Ui, M., Lewis, S. B., and Park, C. R. (1971b). In *Regulation of Gluconeogenesis* (H. D. Söling and B. Willms, eds.), Academic Press, New York, pp. 160–178.

Exton, J. H., Lewis, S. B., Ho, R. J., and Park, C. R. (1972). *Adv. Cycl. Nucleot. Res.* **1,** 91–101.

Felicioli, R. A., Barsacchi, R., and Ipata, P. L. (1970). *Eur. J. Biochem.* **13,** 403–409.

Filsell, O. H., Jarrett, I. G., Taylor, P. H., and Keech, D. B. (1969). *Biochim. Biophys. Acta* **184,** 54–63.

Friedmann, B., Goodman, E. H., Jr., Saunders, H. L., Kostos, V., and Weinhouse, S. (1971). *Arch. Biochem. Biophys.* **143,** 566–578.

Friedrichs, D. and Schoner, W. (1974). *Biochim. Biophys. Acta* **343,** 341–355.

Fröhlich, J. and Wieland, O. (1971). *Eur. J. Biochem.* **19,** 557–562.

Gabrielli, F. and Baldi, S. (1972). *Eur. J. Biochem.* **31,** 209–214.

Garber, A. J. and Ballard, F. J. (1970). *J. Biol. Chem.* **245,** 2229–2240.

Garber, A. J. and Hanson, R. W. (1971a). *J. Biol. Chem.* **246,** 589–598.

Garber, A. J. and Hanson, R. W. (1971b). *J. Biol. Chem.* **246,** 5555–5562.

Garland, P. B. and Randle, P. J. (1964). *Biochem. J.* **91,** 6c–7c.

Garland, P. B., Shepherd, D., and Yates, D. W. (1965). *Biochem. J.* **97,** 587–594.

Garthoff, L., Wolf, G., and Mehlman, M. A. (1972). *Proc. Soc. Exp. Biol. Med.* **141,** 532–535.

Geller, A. M., Rajagopolan, G. T., Ellis, E. H., and Byrne, W. L. (1971). *Arch. Biochem. Biophys.* **146,** 134–143.

Gevers, W. (1967). *Biochem. J.* **103,** 141–152.

Gevers, W. and Krebs, H. A. (1966). *Biochem. J.* **98,** 720–735.

Gimpel, J. A., DeHaan, E. J., and Tager, J. M. (1973). *Biochim. Biophys. Acta,* **292,** 582–591.

Glinsman, W. H., and Mortimore, G. E. (1968). *Am. J. Physiol.* **215,** 553–559.

Glinsman, W. H., Hern, E. P., and Lynch, A. (1969). *Am. J. Physiol.* **216,** 698–703.

Gonzales, A. M., Gonzales, F., Golf, S., and Marcus, F. (1972). *J. Biol. Chem.* **247,** 6067–6070.

Greenbaum, A. L., Gumaa, K. A., and McLean, P. (1971). *Arch. Biochem. Biophys.* **143,** 617–663.

Greene, H. L., Taunton, O. D., Stifel, F. B., and Herman, R. H. (1974). *J. Clin. Invest.* **53,** 44–51.

Haeckel, R. and Haeckel, H. (1971). In *Regulation of gluconeogenesis* (H.-D. Söling and B. Willms, eds.), Academic Press, New York, pp. 127–139.

Halperin, M. L., Robinson, B. H., and Fritz, J. B. (1972). *Proc. Nat. Acad. Sci. U.S.* **69,** 1003–1007.

Harris, E. J. (1970). *FEBS Lett.* **11,** 225–228.

Harris, E. J. (1972). In *Transport and Accumulation in Biological Systems* (E. J. Harris, ed.), Butterworths, London, and University Park Press, Baltimore, Md., pp. 289–345.

Harris, E. J. and Berent, C. (1970). *FEBS Lett.* **10,** 6–12.

Harris, E. J. and Bangham, J. A. (1972). *J. Membrane Biol.* **9,** 141–154.

Harris, E. J., Tate, C., Manger, J. R., and Bangham, J. A. (1971). *Bioenergetics* **2,** 221–232.

Haslam, J. M. and Griffiths, D. E. (1968). *Biochem. J.* **109,** 921–928.

Heitzman, R. J., Herriman, L. D., and Mallinson, C. B. (1972). *FEBS Lett.* **20,** 19–21.

Heldt, H. W., Klingenberg, M., and Milovancev, M. (1972). *Eur. J. Biochem.* **30,** 434–440.

Henning, H. V., Huth, W., and Seubert, W. (1964). *Biochem. Biophys. Res. Commun.* **17,** 496–501.

Herrera, G. M., Kamm, D., Ruderman, N., and Cahill, G. F. (1966). *Adv. Enzyme Regul.* **4,** 225–235.

Hjelmquist, G., Andersson, J., Edlung, B., and Engström, L. (1974). *Biochem. Biophys. Res. Commun.,* **61,** 559–563.

Holten, D. D. and Nordlie, R. C. (1965). *Biochemistry* **4,** 723–731.

Hue, L. and Hers, H. G. (1974a). *Biochem. Biophys. Res. Commun.* **58,** 532–539.

Hue, L. and Hers, H. G. (1974b). *Biochem. Biophys. Res. Commun.* **58,** 540–548.

Imamura, K. and Tanaka, T. (1972). *J. Biochem.* **71,** 1043–1051.

Imamura, K., Taniuchi, K., and Tanaka, T. (1972). *J. Biochem.* **72,** 1001–1015.

Ishihara, N. and Kikuchi, G. (1968). *Biochim. Biophys. Acta* **153,** 733–748.

Jefferson, L. S., Exton, J. H., Sutherland, E. W., and Park, C. R. (1968). *J. Biol. Chem.* **243,** 1031–1038.

Jo, J. S., Ishihara, N., and Kikuchi, G. (1974). *Arch. Biochem. Biophys.* **160,** 246–254.

Johnson, D. C., Ebert, K. A., and Ray, P. D. (1970). *Biochem. Biophys. Res. Commun.* **39,** 750–755.

Johnson, M. E. M., Das, N. M., Butcher, F. R., and Fain, J. N. (1972). *J. Biol. Chem.* **247,** 3229–3235.

Jomain-Baum, M. and Hanson, R. W. (1973). *FEBS Lett.* **29,** 145–148.

Jones, C. T. (1972). *Biochem. J.* **130,** 23.

Katz, N., and Jungermann, K. (1975). *Hoppe-Seyler's Z. Physiol. Chem.* **356,** 244.

Keech, D. B., and Utter, M. F. (1963). *J. Biol. Chem.* **238,** 2609–2614.

Kemp, R. G. (1971). *J. Biol. Chem.* **246,** 245–252.

Kleineke, J. and Söling, H. D. In preparation.

Kleineke, J., Sauer, H., and Söling, H. D. (1973). *FEBS Lett.* **29,** 82–86.

Klingenberg, M. (1970). *FEBS Lett.* **6,** 145–154.

Klingenberg, M., Wulf, R., Heldt, H. W., and Pfaff, E. (1969). In *Mitochondria, Structure and Function* (L. Ernster and Z. Drahota, eds.) Academic Press, New York, pp. 59–77.

Kneer, N. M., Clark, M. G., Bosch, A. L., and Lardy, H. A. (1974). *Diabetes* **23,** Suppl. 1, 347.

Krebs, H. A. and Eggleston, L. V. (1965). *Biochem. J.* **94,** 3c–4c.

Krebs, H. A., Dierks, C., and Gascoyne, T. (1964). *Biochem. J.* **93,** 112–121.

Krebs, H. A., Speake, R. N., and Hems, R. (1965). *Biochem. J.* **94,** 712–720.

Koster, J. F., Slee, R. G., Staal, G. E. J., and Berkel, T. J. C. van (1972). *Biochim. Biophys. Acta* **258,** 763–768.

Lardy, H. A. and Parks, R. E., Jr. (1956). In *Enzymes: Units of Biological Structure and Function* (O. H. Gaebler, ed.), Academic Press, New York, p. 584.

Lardy, H. A., Paetkau, V., and Walter, P. (1965). *Proc. Nat. Acad. Sci. U.S.* **53,** 1410–1415.

Lerner, E., Shug, A. L., Elson, C., and Shrago, E. (1972). *J. Biol. Chem.* **247,** 1513–1519.

Leveille, G. A. (1969). *Comp. Biochem. Physiol.* **28,** 733–740.

Lewis, S. B., Exton, J. H., Ho, R. J., and Park, C. R. (1970). *Fed. Proc.* **29,** 379.

Linn, T. C., Pettit, F. H., and Reed, L. J. (1969a). *Proc. Nat. Acad. Sci. (N.Y.)* **62,** 234–241.

Linn, T. C., Pettit, F. H., Hucho, F., and Reed, L. D. (1969b). *Proc. Nat. Acad. Sci. (N.Y.)* **64,** 227–234.

Ljungström, O., Hjelmquist, G., and Engström, L. (1974). *Biochim. Biophys. Acta* **358,** 289–298.

Llorente, P., Marco, M., and Sols, A. (1970). *Eur. J. Biochem.* **13,** 45–54.

Loten, E. G., Assimacopoulos-Jeannet, F., Le Marchand, Y., Singh, A., and Jeanrenaud, B. (1974). *Adv. Enzyme Regul.* **12,** 45–71.

Loud, A. V. (1968). *J. Cell. Biol.* **37,** 27–35.

Mackrell, D. J. and Sokal, J. E. (1969). *Diabetes* **18,** 724–732.

Mansour, T. E. (1972). In *Current Topics in Cellular Regulation,* Vol. 5 (B. L. Horecker and E. R. Stadtman, eds.), Academic Press, New York, pp. 1–46.

Marcus, F. and Hubert, E. (1968). *J. Biol. Chem.* **243,** 4923–4925.

Martin, R. G. and Ames, B. N. (1961). *J. Biol. Chem.* **236,** 1372–1379.

McClure, W. R. and Lardy, H. A. (1971). *J. Biol. Chem.* **246,** 3591–3596.

McClure, W. R., Lardy, H. A., and Kneifel, H. P. (1971a). *J. Biol. Chem.* **246,** 3569–3578.

McClure, W. R., Lardy, H. A., Wagner, M., and Cleland, W. W. (1971b). *J. Biol. Chem.* **246,** 3579–3583.

McClure, W. R., Lardy, H. A., and Cleland, W. W. (1971c). *J. Biol. Chem.* **246,** 3584–3590.

McCraw, E. F., Peterson, M. J., Yarnell, G., and Ashmore, J. (1968). *Adv. Enzyme Regul.* **6,** 57–65.

McGarry, J. D., Meier, J. M., and Foster, D. W. (1973). *J. Biol. Chem.* **248,** 270–278.

McGivan, J. D., Bradford, N. M., Crompton, M., and Chappell, J. B. (1973). *Biochem. J.* **134,** 209–215.

McLean, P., Gumaa, K. A., and Greenbaum, A. L. (1971). *FEBS Lett.* **17,** 345–350.

Meijer, A. J. (1971). Thesis, University of Amsterdam.

Miller, A. L. and Atkinson, D. E. (1972). *Arch. Biochem. Biophys.* **152,** 531–538.

Miller, R. S. and Lane, M. D. (1968). *J. Biol. Chem.* **243,** 6041–6049.

Mokrasch, L. C., Davidson, W. D., and McGilvery, R. W. (1956). *J. Biol. Chem.* **222,** 179–184.

Müllhofer, G., Loy, E. L., Wollenberg, P., and Krämer, R. (1974). *Hoppe-Seyler's Z. Physiol. Chem.* **355,** 239–254.

Nagano, H., Itoh, H., and Shukuya, R. (1973). *FEBS Lett.* **33,** 125–127.

Nakashima, K., Pontremoli, S., and Horecker, B. L. (1969). *Proc. Nat. Acad. Sci. U.S.* **64,** 947–951.

Nakashima, K., Horecker, B. L., and Pontremoli, S. (1970). *Arch. Biochem. Biophys.* **141,** 578–587.

Nordlie, R. C. and Lardy, H. A. *J. Biol. Chem.* **238,** 2259–2263.

Öckerman, P. A. (1967). *Clin. Chim. Acta* **17,** 201–206.

Olson, J. P. and Marquardt, R. R. (1972). *Biochim. Biophys. Acta* **268,** 453–467.

Osterman, J. and Fritz, P. J. (1973). *Comp. Biochem. Physiol.* **44B,** 1077–1085.

Osterman, J., Fritz, P. J., and Wuntch, T. (1973). *J. Biol. Chem.* **248,** 1011–1018.

Palmieri, F., Stipani, I., and Quagliariello, E. (1972). *Eur. J. Biochem.,* **26,** 587–594.

Passoneau, J. V. and Lowry, O. H. (1964). *Adv. Enzyme Regul.* **2,** 265–276.

Patzelt, C., Löffler, G., and Wieland, O. H. (1973). *Eur. J. Biochem.* **33,** 117–122.

Penney, D. G. and Kornecki, E. H. (1973). *Comp. Biochem. Physiol.,* **46B,** 405–415.

Pogell, B. M., Tanaka, A., and Siddons, R. C. (1968). *J. Biol. Chem.* **243,** 1356–1367.

Pogell, B. M., Taketa, K., and Sarngadharan, M. G. (1971). *J. Biol. Chem.* **246,** 1947–1948.

Pontremoli, S. and Horecker, B. L. (1970). In *Current Topics in Cellular Regulation,* Vol. 2 (B. L. Horecker and E. A. Stadtman, eds.), Academic Press, New York, pp. 174–200.

Pontremoli, S., Traniello, S., Enser, M., Shapiro, S., and Horecker, B. L. (1967). *Proc. Nat. Acad. Sci. U.S.* **58,** 286–293.

Portenhauser, R. and Wieland, O. (1972). *Eur. J. Biochem.* **31,** 308–314.

Prinz, W. and Seubert, W. (1964). *Biochem. Biophys. Res. Commun.* **16,** 582–585.

Purser, D. B. and Bergen, W. G. (1969). *J. Diary Sci.* **52,** 790–795.

Reshef, L. and Hanson, R. W. (1972). *Biochem. J.* **127,** 809–818.

Ricketts, R. R. (1963). *Clin. Chim. Acta* **14,** 160–162.

Robinson, B. H. (1971a). *FEBS Lett.* **14,** 309–312.

Robinson, B. H. (1971b). *FEBS Lett.* **16,** 267–271.

Robinson, B. H. (1973). In *Symposia of the Society for Experimental Biology* Cambridge University Press, Cambridge, pp. 195–214.

Robinson, B. H. and Chappell, J. B. (1967). *Biochem. Biophys. Res. Commun.* **28,** 249–255.

Robinson, B. H. and Williams, G. R. (1970). *Biochim. Biophys. Acta* **216,** 63–70.

Robinson, B. H., Williams, G. R., Halperin, M. L., and Leznoff, C. C. (1971a). *J. Biol. Chem.* **246,** 5280–5286.

Robinson, B. H., Williams, G. R., Halperin, M. L., and Leznoff, C. C. (1971b). *Eur. J. Biochem.* **20,** 65–71.

Rognstad, R., Clark, D. G., and Katz, J. (1973). *Biochem. Biophys. Res. Commun.* **54,** 1149–1156.

Ross, B. D., Hems, R., and Krebs, H. A. (1967a). *Biochem. J.* **102,** 942–951.

Ross, B. D., Hems, R., Freedland, R. A., and Krebs, H. A. (1967b). *Biochem. J.* **105,** 869–875.

Rozengurt, E., Jiménez de Asúa, L., and Carminatti, H. (1969). *J. Biol. Chem.* **244,** 3142–3147.

Rüchel, R., Mesecke, S., Wolfrum, D. I., and Neuhoff, V. (1973). *Hoppe-Seyler's Z. Physiol. Chem.* **354,** 1351–1368.

Ruderman, N. B., Toewes, C. J., and Shafrir, E. (1971). In *Regulation of Gluconeogenesis* (H.-D. Söling and B. Willms, eds.), Academic Press, New York, pp. 194–206.

Rudorff, K. H., Huchtermeyer, H., Windeck, R., and Staib, W. (1970). *Eur. J. Biochem.* **16,** 481–486.

Rudorff, K. H., Windeck, R., and Staib, W. (1971). In *Regulation of Gluconeogenesis* (H.-D. Söling and B. Willms, eds.), Academic Press, New York, pp. 237–244.

Sapag-Hagar, M., Marco, R., and Sols, A. (1969). *FEBS Lett.* **3,** 68–71.

Sasse, D., Katz, N., and Jungermann, K. (1975). *FEBS Lett.* **57,** 83–88.

Sato, K. and Tsuiki, S. (1969). *Arch. Biochem. Biophys.* **129,** 173–180.

Schnaitman, C. and Greenawalt, J. W. (1968). *J. Cell. Biol.* **38,** 158–175.

Schmidt, E. and Schmidt, F. W. (1963). *Klin. Wochenschr.* **41,** 637–643.

Schoner, W., Haag, U., and Seubert, W. (1970). *Hoppe-Seyler's Z. Physiol. Chem.* **351,** 1071–1088.

Scott, K. M., Jurkowitz, M., and Brierley, G. P. (1972). *Arch. Biochem. Biophys.* **153,** 682–694.

Scrutton, M. C. and Utter, M. F. (1967). *J. Biol. Chem.* **242,** 1723–1735.

Scrutton, M. C. and White, M. D. (1974). *Biochem. Med.* **9,** 271–292.

Scrutton, M. C., Keech, D. B., and Utter, M. F. (1965). *J. Biol. Chem.* **240,** 574–581.

Scrutton, M. C., Griminger, P., and Wallace, J. C. (1972). *J. Biol. Chem.* **247,** 3305–3313.

Seubert, W. and Schoner, W. (1971). In *Current Topics in Cellular Regulation,* Vol. 3 (B. L. Horecker and E. R. Stadtman, eds.), Academic Press, New York, pp. 237–268.

Seubert, W., Henning, H. V., Schoner, R. W., and l'Age, M. (1968). *Adv. Enzyme Regul.* **6,** 153–180.

Seufert, D., Herlemann, E. M., Albrecht, E., and Seubert, W. (1971). *Hoppe-Seyler's Z. Physiol. Chem.* **352,** 459–478.

Seyffert, W. A. and Madison, L. L. (1967). *Diabetes* **16,** 765–776.

Shrago, E., Lardy, H. A., Nordlie, R. C., and Foster, D. O. (1963). *J. Biol. Chem.* **238,** 3188–3192.

Shug, A. L. and Shrago, E. (1973). *Biochem. Biophys. Res. Commun.* **53,** 659–665.

Shug, A. L., Lerner, E., Elson, C., and Shrago, E. (1971). *Biochem. Biophys. Res. Commun.* **43,** 557–563.

Siess, E., Wittmann, J., and Wieland, O. (1971). *Hoppe-Seyler's Z. Physiol. Chem.* **352,** 447–452.

Sies, H., Tjinta, S., Brauser, B., and Bücher, T. (1972). *Adv. Enzyme Regul.* **10,** 309–322.

Sluse, F. E., Meijer, A. J., and Tager, J. M. (1971). *FEBS Lett.* **18,** 149–153.

Sluse, F. E., Ranson, M., and Liébecq, C. (1972). *Eur. J. Biochem.* **25,** 207–217.

Sluse, F. E., Goffart, G., and Liébecq, C. (1973). *Eur. J. Biochem.* **32,** 283–291.

Sokal, J. E. (1966). *Endocrinology* **78,** 538–548.

Sokal, J. E., Sarcione, E. J., and Henderson, A. M. (1964). *Endocrinology* **74,** 930–938.

Söling, H. D. (1965). *Klin. Wochenschr.* **43,** 914.

Söling, H. D. (1974). In *Regulation of Hepatic Metabolism,* Vol. 6 *Alfred Benzon Symposium* (F. Lundquist and N. Tygstrup, eds.), Munksgaard, Copenhagen (1973) pp. 48–61.

Söling, H. D. and Bernhard, G. (1971). *FEBS Lett.* **13,** 201–203.

Söling, H. D. and Bernhard, G. In preparation.

Söling, H. D. and Seufert, C. D. (1975) In *Handbook of Experimental Pharmacology* Vol. 32/2 (A. Hasselblatt and F. von Bruchhausen, eds.), Springer-Verlag, Berlin, Heidelberg, New York, pp. 413–434.

Söling, H. D. and Seck, A. (1975). *FEBS Lett.* **51,** 52–59.

Söling, H. D., Kattermann, R., and Schmidt, H. (1964). *Abstracts, Fifth Congress of the International Diabetes Federation, Toronto 1964, Excerpta medica* (Int. Congr. Ser.) **74,** 55.

Söling, H. D., Willms, B., Friedrichs, D., and Kleineke, J. (1968). *Eur. J. Biochem.* **4,** 364–373.

Söling, H. D., Willms, B., Kleineke, J., and Gehlhoff, M. (1970). *Eur. J. Biochem.,* **16,** 289–302.

Söling, H. D., Kleineke, J., Willms, B., Janson, G., and Kuhn, A. (1973). *Eur. J. Biochem.* **37,** 233–243.

Söling, H.-D., Luthe, H., and Lück, H. J. Unpublished observations.

Sols, A., Salas, M., and Viñuela, E. (1964). *Adv. Enzyme Regul.* **2,** 177–188.

Soskin, S., Essex, H. E., Herrick, J. F., and Mann, F. C. (1938). *Am. J. Physiol.* **124,** 558–567.

Stan, H. J. (1972). *Int. J. Biochem.* **3,** 573–582.

Struck, E., Ashmore, J., and Wieland, O. (1965). *Biochem. Z.* **343,** 107–110.

Struck, E., Ashmore, J., and Wieland, O. (1966). *Adv. Enzyme Regul.* **4,** 219–224.

Stucki, J. W., Brawand, F., and Walter, P. (1972). *Eur. J. Biochem.* **27,** 181–191.

Susor, W. A. and Rutter, W. J. (1968). *Biochem. Biophys. Res. Commun.* **30,** 14–20.

Swiatek, K. R., Chao, K. L., Chao, H. L., Cornblath, M., and Tildon, T. (1970). *Biochim. Biophys. Acta* **206,** 310–321.

Taketa, K. and Pogell, B. M. (1963). *Biochem. Biophys. Res. Commun.* **12,** 229–235.

Taketa, K. and Pogell, B. M. (1965). *J. Biol. Chem.* **240,** 651–662.

Taketa, K., Sarngadharan, M. G., Watanabe, A., Aoe, H., and Pogell, B. M. (1971). *J. Biol. Chem.* **236,** 5676–5683.

Tanaka, T., Harano, Y., Morimura, H., and Mori, R. (1965). *Biochem. Biophys. Res. Commun.* **21,** 55–60.

Taunton, O. D., Stifel, F. B., Greene, H. L., and Herman, R. H. (1972). *Biochem. Biophys. Res. Commun.* **48,** 1663–1669.

Taylor, P. H., Wallach, J. C., and Keech, D. B. (1971). *Biochim. Biophys. Acta* **237,** 179–191.

Taylor, S. I., Mikherjee, C., and Jungas, R. L. (1973). *J. Biol. Chem.* **248,** 73–81.

Tedeschi, H. (1971). In *Current Topics in Membranes and Transport,* Vol. 2 (F. Bronner and A. Kleinzeller, eds.), Academic Press, New York, London, pp. 207–231.

Teufel, H., Menahan, L. A., Shipp, J. C., Böning, S., and Wieland, O. (1967). *Eur. J. Biochem.* **2,** 182–186.

Traniello, S., Pontremoli, S., Tashima, Y., and Horecker, B. L. (1971). *Arch. Biochem. Biophys.* **146,** 161–166.

Underwood, A. H. and Newsholme, E. A. (1965). *Biochem. J.* **95,** 868–875.

Usatenko, M. S. (1970). *Biochem. Med.* **3,** 298–310.

Utter, M. F. and Keech, D. B. (1960). *J. Biol. Chem.* **235,** PC17–18.

Utter, M. F. Scrutton, M. C. (1969). In *Current Topics in Cellular Regulation,* Vol. 1 (B. L. Horecker and E. R. Stadtman, eds.), Academic Press, New York, pp. 253–296.

Utter, M. F. and Fung, C. H. (1971). In *Regulation of Gluconeogenesis* (H.-D. Söling and B. Willms, eds.), Academic Press, New York, pp. 1–10.

Uyeda, K. and Luby, L. J. (1974). *J. Biol. Chem.* **14,** 4562–4570.

Vaartjes, W. J., Kemp, A., Jr., Souverijn, J. H. M., and van den Bergh, S. G. (1973). *FEBS Lett.* **23,** 303–308.

Van Berkel, T. J. C., Koster, J. F., and Hülsmann, W. C. (1972). *Biochim. Biophys. Acta* **276,** 425–429.

Van Berkel, T. J. C., Koster, J. F., and Hülsmann, W. C. (1973). *Biochim. Biophys. Acta* **293,** 118–124.

Veneziale, C. M., Walter, P., Kneer, N., and Lardy, H. A. (1967). *Biochemistry* **6,** 2129–2138.

Villañueva, J. and Marcus, F. (1974). *J. Biol. Chem.* **249,** 745–749.

Viñuela, E., Salas, M., and Sols, A. (1963). *J. Biol. Chem.* **238,** PC1175–1177.

Volkmann, B. (1973). Thesis, University of Göttingen, Göttingen, Germany.

Walajtys, E. D., Gottesman, D. P., and Williamson, J. R. (1974). *J. Biol. Chem.* **249,** 1857–1865.

Wallace, J. C. and Newsholme, E. A. (1967). *Biochem. J.* **104,** 378–384.

Wallenfels, K., Creutzfeldt, W., and Summ, H. D. (1958). In *Diabetes Mellitus,* G. Thieme-Verlag, Stuttgart, pp. 330–336.

Walli, A. K., Siebler, G., and Schimassek, H. (1974). *Hoppe-Seyler's Z. Physiol. Chem.* **355,** 353–362.

Walter, P. and Stucki, J. W. (1970). *Eur. J. Biochem.* **12,** 508–519.

Walter, P. and Anabitarte, M. (1971). *FEBS Lett.* **12,** 289–292.

Weinstein, I., Klausner, H. A., and Heimberg, M. (1973). *Biochim. Biophys. Acta* **296,** 300–309.

Weiss, G., Müller-Ohly, B., and Seubert, W. (1972). *Biochem. Biophys. Res. Commun.* **46,** 225–230.

Weiss, G., Lamartiniere, C. A., Müller-Ohly, B., and Seubert, W. (1974). *Eur. J. Biochem.* **43,** 391–403.

Weiss, L., Löffler, G., Schirmann, A., and Wieland, O. (1971). *FEBS Lett.* **15,** 229–231.

Werner, H. V., Bartley, J. C., and Berry, M. N. (1972). *Biochem. J.* **130,** 1153–1155.

Werner, H. V. and Berry, M. N. (1974). *Eur. J. Biochem.* **42,** 315–324.

Wieland, O. and von Jagow-Westermann, B. (1969). *FEBS Lett.* **3,** 271–274.

Wieland, O. H., Patzelt, C., and Löffler, G. (1972a). *Eur. J. Biochem.* **26,** 426–433.

Wieland, O., Hartmann, H. M., and Siess, E. A. (1972b). *FEBS Lett.* **27,** 240–244.

Williams, T. F., Exton, J. H., Friedmann, N., and Park, C. R. (1971). *Am. J. Physiol.* **221,** 1645–1651.

Williamson, D. H., Lund, P., and Krebs, H. A. (1967). *Biochem. J.* **103,** 514–527.

Williamson, J. R. (1967). *Adv. Enzyme Regul.* **5,** 229–255.

Williamson, J. R. (1970). Presented at the Eighth International Congress of Biochemistry, Lucerne, Switzerland.

Williamson, J. R., Kreisberg, R. A., and Felts, W. P. (1966). *Proc. Nat. Acad. Sci.* (*N.Y.*) **56,** 247–254.

Williamson, J. R., Browning, E. T., and Olson, M. S. (1968). *Adv. Enzyme Regul.* **6,** 67–100.

Williamson, J. R., Browning, E. T., and Scholz, R. (1969). *J. Biol. Chem.* **244,** 4607–4616.

Williamson, J. R., Anderson, J., and Browning, E. T. (1970). *J. Biol. Chem.* **245,** 1717–1726.

Williamson, J. R., Jakob, A., and Scholz, R. (1971b). *Metabolism* **20,** 13–26.

Willms, B., Kleineke, J., and Söling, H.-D. (1970a). *Biochim. Biophys. Acta* **215,** 438–448.

Willms, B., Ben-Ami, P., and Söling, H.-D. (1970b). *Horm. Metab. Res.* **2,** 135–141.

Wimhurst, J. M. and Manchester, K. L. (1970a). *Biochem. J.* **120,** 79–93.

Wimhurst, J. M. and Manchester, K. L. (1970b). *Biochem. J.* **120,** 95–103.

Wojtczak, L., Bogucka, K., Sarzala, M. G., and Zaluska, H. (1969). *FEBS Symp.* **17,** 79–92.

Zahlten, R. N., Hochberg, A. A., Stratman, F. W., and Lardy, H. A. (1972). *Proc. Nat. Acad. Sci. U.S.* **69,** 800–804.

Zahlten, R., Stratman, F. W., and Lardy, H. A. (1973). *Proc. Nat. Acad. Sci. U.S.* **70,** 3213–3218.

Zuurendonk, P. F., and Tager, J. M. (1974). *Biochim. Biophys. Acta* **333,** 393–399.

CHAPTER TWELVE

Influence of Glucagon on the Metabolism of Fructose, D-Glyceraldehyde, and Dihydroxyacetone in Rat Liver

CARLO M. VENEZIALE

INTRODUCTION

The dedication of a book on gluconeogenesis to Henry A. Lardy is more than appropriate because he and his co-workers have made multiple interesting and important contributions to that field. However, students of biochemistry and medicine should appreciate that Professor Lardy also has made significant contributions to other fields, including other areas of carbohydrate metabolism, basic enzymology, oxidative phosphorylation, and reproduction. It would be wrong to identify him with gluconeogenesis only. The dedication to him of a book on each of those topics would not be an exaggerated tribute at all.

Glucagon has been reported to stimulate gluconeogenesis in isolated rat livers from a variety of precursors including bicarbonate (Garcia *et al.*,

1966), lactate and pyruvate (Schimassek and Mitzkat, 1963; Garcia *et al.*, 1966), and alanine (Garcia *et al.*, 1966; Mallette *et al.*, 1969). More recently it has been shown that glucagon also stimulates gluconeogenesis from fructose (Veneziale, 1971) and from D-glyceraldehyde and dihydroxyacetone (Veneziale, 1972a). The latter findings have been independently confirmed by others (Blair *et al.*, 1973; Garrison and Haynes, 1973; Tolbert and Fain, 1974).

Many aspects of fructose metabolism and its stimulation by glucagon are of interest to both biochemistry and medicine. People in the western hemisphere ingest large amounts of fructose daily, either in the form of the free monosaccharide or in the form of sucrose. In the United States, the average daily intake of fructose derived from sucrose alone is approximately 65 g (Yudkin, 1969a), and this excludes that derived from foods that are rich in fructose as the free monosaccharide (Hardinge *et al.*, 1965). It is relatively common to find persons whose sucrose intake is great enough to provide 125 to 150 g of fructose daily from that source alone (Yudkin, 1969b). With the widespread occurrence of such large amounts of fructose in the diet, it is not surprising that the results of fructose administration have been investigated. For example, dietary change from glucose to fructose over a short period and fructose feeding over a prolonged period have been shown to cause increases in liver weight in rats (Szepesi and Freedland, 1968; Baron *et al.*, 1971). The intravenous or oral administration of fructose to children has been found to result in increased uric acid concentrations in both urine and serum, indicating increased uric acid production (Perheentupa and Raivio, 1967). In addition, fructose loading causes a depletion of liver ATP and inorganic phosphate (Mäenpää *et al.*, 1968; Raivio *et al.*, 1969).

Medical interest in fructose metabolism also results because there are two inherited diseases of fructose metabolism. One is a harmless disturbance known as essential fructosuria; it is rare and is marked by the presence of fructose in the urine after the ingestion of foods containing fructose. The primary enzyme disorder is a lack of hepatic fructokinase, which catalyzes the phosphorylation of fructose to fructose-1-phosphate and initiates the major metabolic pathway of hepatic fructose metabolism to dihydroxyacetone phosphate and D-glyceraldehyde (for list of references, see Heinz *et al.*, 1968). Another disorder of fructose metabolism is hereditary fructose intolerance. Administration of fructose in these cases results in severe hypoglycemia. The enzyme defect in this disease is a deficiency of aldolase B, which splits fructose-1-phosphate into two three-carbon fragments (Froesch, 1966).

It has been known for some time that patients with diabetes mellitus convert administered fructose to glucose more rapidly than do normal persons (Miller *et al.*, 1956–1957; Felber *et al.*, 1959). It has since been discovered that both maturity onset diabetics and juvenile diabetics secrete glucagon inappropriately (Unger *et al.*, 1970). Thus augmented gluconeogenesis secondary to inappropriate secretion of glucagon could make a significant contribution to hyperglycemia in some patients. The stimulatory effect of glucagon on gluconeogenesis from D-glyceraldehyde (Veneziale, 1972a) might be medically significant. For example, if one assumes a similar response in the human (average adult liver weight is 1500 g), the rate of glucose production augmented by glucagon—as under conditions of inappropriate glucagon secretion (Unger *et al.*, 1970)—could amount to 117 g/24 hr. This estimate is based on a steady supply of substrate, a condition that could be met by the diets of many, as noted previously. Thus it becomes of potential medical importance to know the site of action of glucagon in its stimulation of gluconeogenesis from fructose and other precursors.

In the process of gluconeogenesis from pyruvate, the initial enzyme-catalyzed reactions involve both mitochondria and cytoplasm. The enzymes are believed to be pyruvate carboxylase, malate dehydrogenase, glutamate-oxalacetate transaminase, and phosphoenolpyruvate carboxykinase (see review by Exton *et al.*, 1970). These enzymes catalyze the conversion of pyruvate to the glycolytic-gluconeogenic intermediate, phosphoenolpyruvate. The site of action of glucagon in its stimulation of gluconeogenesis from pyruvate has been postulated to be some reaction in the conversion of pyruvate to phosphoenolpyruvate (Ross *et al.*, 1967; Exton *et al.*, 1970).

In contrast, gluconeogenesis from fructose requires none of these enzymes. As shown in Fig. 1, fructose carbon enters the glycolytic-gluconeogenic sequence as dihydroxyacetone phosphate and D-glyceraldehyde-3-phosphate. The required enzymes are fructokinase, aldolase B, and D-triokinase. Both D-glyceraldehyde and dihydroxyacetone enter the scheme as a result of their phosphorylation under the influence of D-triokinase. The segment of gluconeogenesis between dihydroxyacetone phosphate or glyceraldehyde-3-phosphate and glucose serves the process of gluconeogenesis, whether the carbon source for newly formed glucose is pyruvate or fructose.

Because of such pathway differences, the use of pyruvate and fructose (also dihydroxyacetone and D-glyceraldehyde) in gluconeogenic studies, in the absence and in the presence of glucagon, has provided the opportunity to evaluate which metabolic segment(s) is actually influenced by glucagon or its mediator, adenosine cyclic 3′,5′-monophosphate (cAMP). An action of glucagon in the early segment of pyruvate metabolism should have no influence on the rate of gluconeogenesis from fructose. An action of glucagon on the initial segment of fructose metabolism should have no influence on

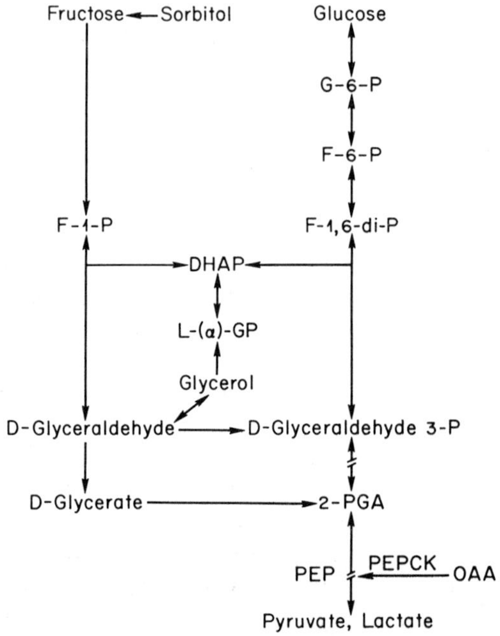

Figure 1. Metabolism of fructose in liver.

gluconeogenesis from pyruvate. Theoretically, an action of glucagon on that segment of gluconeogenesis common to both precursors should influence gluconeogenesis from both. Thus fructose, dihydroxyacetone, and D-glyceraldehyde have been useful substrates in localizing the actions of glucagon. Before presenting data derived from the use of this experimental approach and before exploring possible implications, it is appropriate to discuss individual enzyme-catalyzed reactions that are involved in the metabolism of fructose, D-glyceraldehyde, and dihydroxyacetone.

II. INTERMEDIARY METABOLISM OF FRUCTOSE, D-GLYCERALDEHYDE, AND DIHYDROXYACETONE

For the most part and unless otherwise stated, the details reviewed here are drawn from rat liver experimentation.

A. Metabolism of Fructose

The conversion of fructose to glucose is believed to occur in the cytoplasm of the hepatic parenchymal cell in accordance with the following sequence of reactions (Cori *et al.*, 1951; Hers and Kusaka, 1953; Hers, 1955; Hue and Hers, 1972):[1]

$$\text{fructose} + \text{ATP} \xrightarrow{\text{fructokinase}} \text{fructose-1-P} + \text{ADP} \quad (1)$$

$$\text{fructose-1-P} \xrightarrow{\text{aldolase } B} \text{dihydroxyacetone-P} + \text{D-glyceraldehyde} \quad (2)$$

$$\text{D-glyceraldehyde} + \text{ATP} \xrightarrow{\text{D-triokinase}} \text{glyceraldehyde-3-P} + \text{ADP} \quad (3)$$

$$\text{glyceraldehyde-3-P} + \text{dihydroxyacetone-P} \xrightarrow{\text{aldolase A}} \text{fructose-1,6-di-P} \quad (4)$$

$$\text{fructose-1,6-di-P} \xrightarrow{\text{fructose-1,6-di-P-ase}} \text{fructose-6-P} + P_i \quad (5)$$

$$\text{fructose-6-P} \xrightarrow{\text{phosphohexose isomerase}} \text{glucose-6-P} \quad (6)$$

$$\text{glucose-6-P} \xrightarrow{\text{glucose-6-P-ase}} \text{glucose} + P_i \quad (7)$$

$$\text{fructose} + 2\text{ATP} \rightarrow \text{glucose} + 2\text{ADP} + 2P_i$$

B. Metabolism of D-Glyceraldehyde

In contrast to fructose, D-glyceraldehyde is not a foodstuff constituent. It is important as a natural product mainly because it is formed enzymatically from fructose. At intracellular concentrations greater than those usually generated during the course of fructose metabolism, D-glyceraldehyde can be metabolized by reactions 8 and 9 (Wolf and Leuthardt, 1953; Leuthardt and Wolf, 1954; Moore, 1959):

$$\text{D-glyceraldehyde} + \text{NADH} + H^+ \xrightarrow[\text{dehydrogenase}]{\text{NAD-linked glycerol}} \text{glycerol} + \text{NAD}^+ \quad (8)$$

A nomenclature problem arises here because this enzyme activity has been ascribed (Holzer and Schneider, 1955) to alcohol dehydrogenase (EC 1.1.1.1). During a fiftyfold purification of alcohol dehydrogenase from horse liver, the ratio of NADH-dependent reducing activity with acetaldehyde as substrate to that with glyceraldehyde as substrate remained constant. This can be taken as evidence that one enzyme was responsible for both reductions. It has not been established that only one enzyme is in-

[1] For discussion of aldolases A and B from rabbit liver see Rajkumar *et al.* (1966).

volved in rat liver. At this time, therefore, it is appropriate to use the term "NAD-linked glycerol dehydrogenase" in discussing glycerol formation from glyceraldehyde in rat liver.

$$\text{D-glyceraldehyde} + \text{NADPH} + \text{H}^+ \xrightarrow[\text{dehydrogenase}]{\text{NADP—linked glycerol}} \text{glycerol} + \text{NADP}^+ \tag{9}$$

Although the name "aldose reductase (EC 1.1.1.21)" might be used to describe this activity, the more descriptive term "NADP-linked glycerol dehydrogenase" is preferable for even another reason. Aldose reductase from seminal vesicle, placenta (Hers, 1962a), and lens (Barman, 1969) reduces both D-glyceraldehyde and glucose; liver extracts do not catalyze the reduction of glucose (Hers, 1960, 1962a). Therefore, it is not certain that liver tissue contains the aldose reductase activity synonymous with that found so far in extrahepatic tissue.

The K_m for glyceraldehyde in reaction 8 has been reported to be 11 mM (D isomer) (Leuthardt and Wolf, 1954) or 30 mM (DL racemic mixture) (Holzer and Schneider, 1955); these values are manyfold greater than the intracellular concentrations found even in livers perfused with 4.0 mM D-glyceraldehyde (Veneziale, 1972a). Taking this into account, together with other kinetic considerations recently elaborated (including the low K_m value for D-glyceraldehyde in the triokinase reaction) (Sillero *et al.*, 1969; Frandsen and Grunnet, 1971; Veneziale, 1972b), it appears that NAD-linked glycerol dehydrogenase activity could become important only when the intracellular substrate concentration is abnormally high.

Similar considerations suggest that in the metabolism of D-glyceraldehyde under physiologic conditions and perhaps even in the D-glyceraldehyde-liver perfusion experiments cited (Veneziale, 1972a), D-triokinase is also more important than NADP-linked glycerol dehydrogenase. For example, in the case of the dehydrogenase, V_{max} was less than the corresponding value reported for D-triokinase, while K_m for substrate was greater. Nevertheless, it is possible that there are dietary and hormonal conditions in which the NADP-linked glycerol dehydrogenase assumes a more important role. This must be determined, as has already been done for the NAD-linked dehydrogenase (Veneziale, 1972b).

The results of experiments that utilized intact animals but were designed to study hepatic metabolism might be difficult to interpret (Landau and Merlevede, 1963). The metabolism of administered substrate by extrahepatic tissues would preclude the direct liver study that had been intended. For example, in guinea pig seminal vesicle it has been shown that D-glyceraldehyde is rapidly converted to glycerol (Prendergast and Veneziale, 1975); newly formed glycerol could then be returned to the general circulation, from which it could be cleared and metabolized by liver.

Another enzymatic pathway for D-glyceraldehyde is the oxidative one leading to glycerate formation (Kattermann *et al.*, 1961; Rauschenbach and Lamprecht, 1964; Heinz *et al.*, 1968):

$$\text{D-glyceraldehyde} + \text{NAD}^+ \xrightarrow[\text{dehydrogenase}]{\text{aldehyde}} \text{glycerate} + \text{NADH} + \text{H}^+ \quad (10)$$

Several lines of evidence have been presented in support of the hypothesis that this is a major route in rat liver, but the work has not been considered conclusive (Sillero *et al.*, 1969). In isolated liver experiments involving perfusion with [U-^{14}C]-fructose, the specific radioactivity of intracellular 3-phosphoglycerate was not as great as might have been predicted if the predominant fate of D-glyceraldehyde is oxidation to glycerate (Veneziale, 1971). However, the importance of reaction 10 must not be minimized on the basis of present knowledge because the contributions made by aldehyde dehydrogenase under varied conditions also remain to be determined. Perfusion of pig liver with D-glyceraldehyde resulted in some accumulation of glycerate in the perfusate. When fructose was substrate, only minimal amounts of glycerate accumulated (Sestoft *et al.*, 1972).

C. Dihydroxyacetone

The phosphorylation of dihydroxyacetone, as well as that of D-glyceraldehyde, is catalyzed by D-triokinase (Hers, 1962b), as follows:

$$\text{dihydroxyacetone} + \text{ATP} \xrightarrow{\text{D-triokinase}} \text{dihydroxyacetone-P} + \text{ADP} \quad (11)$$

A partially purified preparation of D-triokinase from rat liver was found to phosphorylate dihydroxyacetone with a maximal activity 2.5 times greater than the maximal activity of D-glyceraldehyde phosphorylation (Frandsen and Grunnet, 1971). The K_m for dihydroxyacetone in reaction 11 was 0.006 mM; that for D-glyceraldehyde was 0.019 mM.

Dihydroxyacetone can also be metabolized by liver glycerokinase as follows:

$$\text{dihydroxyacetone} + \begin{matrix}\text{ATP}\\ \text{or}\\ \text{UTP}\end{matrix} \xrightarrow{\text{glycerokinase}} \text{dihydroxyacetone-P} + \begin{matrix}\text{ADP}\\ \text{or}\\ \text{UDP}\end{matrix} \quad (12)$$

In two publications (Burch *et al.*, 1969; Exton and Park, 1969) it was stated that the first reaction of dihydroxyacetone "presumably" involves catalysis by glycerokinase. This statement might be misleading because it fails to reckon with catalysis by D-triokinase.

The purified enzyme from rat liver was shown to catalyze the phosphorylation of dihydroxyacetone twice as fast as that of glycerol, but it did not phosphorylate D-glyceraldehyde (Bublitz and Kennedy, 1954). Because dihydroxyacetone is not known to be a foodstuff constituent, the physiologic function of glycerokinase probably is related to glycerol metabolism.

Dihydroxyacetone can also be reduced to glycerol by glycerol dehydrogenases in rat liver preparations with either NADH (Wolf and Leuthardt, 1953) or NADPH (Moore, 1959) as co-factor; however, rates of reduction were much less than those observed when glyceraldehyde was substrate.

III. INFLUENCE OF FRUCTOSE ON HEPATIC ADENINE NUCLEOTIDES

Fructose loading causes severe disturbances in liver metabolism. Kjerulf-Jensen (1942) was the first to demonstrate that fructose-1-phosphate accumulates in liver after a fructose load (Kjerulf-Jensen, 1942; Heinz, 1968; Burch *et al.*, 1969). The accumulation is attributed to the inhibition of aldolase B by the inosine monophosphate (IMP) that arises from the degradation of adenine nucleotides (Ronca-Testoni *et al.*, 1970; Woods *et al.*, 1970). Normally, ATP and P_i inhibit the breakdown of AMP to IMP by inhibiting AMP deaminase. ATP also inhibits the breakdown of AMP to adenosine and P_i by inhibiting 5-nucleotidase. Fructose loading causes a depletion of liver ATP and P_i (Mäenpää *et al.*, 1968; Raivio *et al.*, 1969), thereby removing their inhibition of adenine nucleotide degradation and allowing the subsequent accumulation of enough IMP to inhibit aldolase B. In the fructose experiments to be described, the concentration used was not great enough to cause the fructose loading phenomenon (Veneziale, 1971). When administered intraperitoneally to the intact animal in large amounts, glycerol, but not dihydroxyacetone, also has been shown to result in hepatic ATP depletion (Burch *et al.*, 1970).

IV. EFFECT OF GLUCAGON IN ISOLATED LIVER FROM FASTED RAT

A. Fructose Metabolism

Rates of glucose release during perfusion with fructose are given in Table I. The use of quinolinate afforded the opportunity to observe the effects of glucagon at a time when flux through the phosphoenolpyruvate carboxykinase reaction was minimized. In most cases the rate of glucose release was approximately doubled as a result of the addition of glucagon. Hepatic glycogen content resulting from perfusion with fructose was also de-

Table I. Rates of hepatic glucose release during perfusion with and without fructose[a]

Interval (min)	No glucagon	Glucagon (5 μg)
	With fructose	
0–5	27 ±9 (4)	51 ±16 (4)
0–10	48 ±12 (7)	101 ±15 (7)
0–15	69 ±25 (4)	128 ±15 (5)
	Without fructose	
0–10	12 ±4 (5)	25 ±8 (6)
0–15	19 ±4 (5)	34 ±11 (6)

[a] Quinolinate (2.4 mM) was added after 30 minutes of perfusion. Fructose (2.0 mM) and glucagon were added 30 minutes later. Intervals are after time of fructose addition to perfusate. Data are given as rate, in micromoles formed during the interval, corrected to 10 g wet weight of liver (mean ±SD). Total perfusate volume, 100 ml. See Veneziale (1971) for other details.

termined. At 60 minutes, just before fructose was added, the mean (±SD) glycogen content was 11 μmole glucose/10 g wet wt (±5; $N = 7$). At 70 minutes, 10 minutes after fructose addition, it was 18 μmole/10 g (±7; $N = 6$). Therefore, the glucagon-induced increment in rate of glucose release in the presence of fructose cannot be accounted for by glycogenolysis.

The data on [^{14}C]glucose formation from [U-^{14}C]fructose (Table II) indicate that glucagon did stimulate gluconeogenesis from fructose. The concentrations of hepatic adenine nucleotides after perfusion with quinolinate and fructose are given in Table III. ATP depletion did not occur.

The rate of uptake of D-fructose by isolated rat livers was also studied (Veneziale, 1971). After 10 minutes of perfusion, approximately 40% of the added fructose had been cleared from the perfusate, both in the absence and in the presence of glucagon. Therefore, the action of glucagon must be on an intracellular aspect of the process.

B. D-Glyceraldehyde, Dihydroxyacetone, and Glycerol Metabolism

The rates of glycogenesis and glycogenolysis appear in Table IV; the rates of hepatic glucose release are in Table V. Considering the glyceraldehyde

Table II. Effect of glucagon on [^{14}C] glucose formation from [U-^{14}C] fructose in isolated liver[a]

[U-^{14}C]Fructose added (dpm/μmole)	Perfusion (min)	Total dpm in perfusate glucose	
		No glucagon	Glucagon (5 μg)
2.2×10^5	7	4.7×10^6	6.8×10^6
2.1×10^5	10	6.6×10^6	11×10^6
2.4×10^5	10	7.9×10^6	14×10^6
4.5×10^5	10	14×10^6	23×10^6

[a] Quinolinate was added after 30 minutes of perfusion; [U-^{14}C]fructose and glucagon were added 30 minutes later. Liver samples were taken at 7 or 10 minutes after addition of [U-^{14}C]fructose and extracted with perchloric acid. See Veneziale (1971) for details.

data, the mean (±SD) total carbohydrate formed—represented by that released and that converted to glycogen—was 92 ± 18 μmole/10 g/15 min. In the presence of glucagon, the rate of formation of new carbohydrate was that released minus that due to glycogenolysis: 167 – 12 μmole/10 g/15 min. Therefore, the effect of glucagon was to stimulate gluconeogenesis by 45 μmole/10 g/15 min ($P < 0.005$). For the dihydroxyacetone data, the augmentation was 35 μmole/10 g/15 min ($P < 0.025$). Similar calculations

Table III. Hepatic adenine nucleotides after perfusion with fructose[a]

Nucleotide	Rate (μmole/g wet wt)	
	No glucagon	Glucagon (5 μg)
ATP	2.54 ±0.22 (4)	2.81 ±0.31 (5)
ADP	0.81 ±0.11 (5)	0.80 ±0.12 (5)
AMP	0.17 ±0.05 (4)	0.18 ±0.04 (5)

[a] Quinolinate was added after 30 minutes of perfusion; fructose and glucagon were added 30 minutes later, and liver samples were taken 15 minutes after that and extracted with perchloric acid. Means ±SD. For details and metabolite data, see Veneziale (1971).

Table IV. Rates of glycogen formation[a]

Substrate[b]	N^c	Formation (μmole glucose/10 g/15 min)[d]
D-Glyceraldehyde	11	18 ±12
Dihydroxyacetone	8	14 ±10
Glycerol	6	18 ±11
D-Glyceraldehyde plus glucagon	5	−12 ±5

[a] See Table V for description of experimental conditions.
[b] Concentration in perfusate, 4.0 mM.
[c] Number of liver perfusions.
[d] Means (±SD) of increments found in individual livers 15 minutes after addition of substrate to perfusate; the mean 60 minute hepatic glycogen value was 17 ±8 μmole/10 g (N = 17).

with the glycerol data reveal that gluconeogenesis from glycerol was not stimulated by glucagon. For example, the total carbohydrate formed was 134 + 18 μmole/10 g/15 min, which is only 10 μmole less than the amount released in the presence of glucagon; the difference can be attributed to glycogen breakdown.

Table VI gives ^{14}C tracer data. The increase in ^{14}C incorporation in the presence of glucagon paralleled the increase in amount of glucose, demonstrating that the chemical measurement data of Table V do reflect

Table V. Effect of glucagon on rates of hepatic glucose release[a]

	Glucose release (μmole/10 g/15 min)	
Substrate	No glucagon	Glucagon
None	25 ±6 (5)	40 ±12 (5)
D-Glyceraldehyde	92 ±16 (20)	167 ±24 (14)
Dihydroxyacetone	114 ±19 (13)	175 ±23 (6)
Glycerol	134 ±27 (12)	162 ±16 (5)

[a] Quinolinate (2.4 mM) was added to the perfusate after 30 minutes of perfusion; substrates (all at 4.0 mM in perfusate) and glucagon were added 30 minutes later. Means ±SD. For details see Veneziale (1972a).

Table VI. [^{14}C]Glucose formation from D-[U-^{14}C]glyceraldehyde[a]

No glucagon		Glucagon (5 μg)	
μmole/10 g	dpm	μmole/10 g	dpm
90	6.6×10^6	150	10.7×10^6
91	6.0×10^6	165	9.5×10^6

[a] Quinolinate (2.4 mM) was added after 30 minutes of perfusion; substrate (4.0 mM, 9 μCi) and glucagon were added 30 minutes later; samples were taken 15 minutes after that. Values for dpm represent total circulating in perfusate glucose.

the metabolism of the added glyceraldehyde. The tracer data support the view, proposed here, that glucagon stimulates the conversion of glyceraldehyde to glucose.

Extensive intracellular glucogenic metabolite measurements have been made in livers perfused with fructose and D-glyceraldehyde. Although the results have been of some interest, they have failed to pinpoint a specific enzyme site of hormone action. During gluconeogenesis from D-glyceraldehyde, glucagon caused a decrease in the concentrations of lactate and

Table VII. Hepatic adenine nucleotides during perfusion with D-glyceraldehyde[a]

	Rate (μmole/g wet wt)			
	No glucagon		Glucagon (5 μg)	
Nucleotide	4 min	8 min	4 min	8 min
ATP	3.43 ±0.51 (4)	3.10 ±0.38 (4)	3.06 ±0.43 (4)	2.96 ±0.44 (4)
ADP	1.20 ±0.47 (4)	1.02 ±0.19 (4)	1.22 ±0.11 (4)	1.07 ±0.09 (4)
AMP	0.34 ±0.13 (4)	0.29 ±0.03 (4)	0.35 ±0.06 (4)	0.34 ±0.08 (4)

[a] Quinolinate (2.4 mM) was added after 30 minutes of perfusion; substrate (4.0 mM) and glucagon were added 30 minutes later; samples were taken at 4 and 8 minutes after that. For details, see Veneziale (1972a).

pyruvate and a slightly higher glucose-6-phosphate concentration relative to that of fructose-6-phosphate. The reader is referred to the original publications for detailed data and discussion (Veneziale, 1971 and 1972a). The concentrations of hepatic adenine nucleotides are given in Table VII; ATP depletion did not occur.

V. DISCUSSION

Recently it was reported (Taunton *et al.*, 1972) that minutes after the intravenous administration of large amounts of glucagon (150–300 μg) to anesthetized rats, the assayable activity of fructose diphosphatase increased. Fructose diphosphatase must still remain an attractive possibility as a glucagon-sensitive enzyme in the gluconeogenic process from even primitive precursors such as pyruvate. Stimulation of this enzyme would explain the augmented gluconeogenesis caused by glucagon from a variety of precursors including bicarbonate and alanine (Garcia *et al.*, 1966), pyruvate and lactate (Schimassek and Mitzkat, 1963), glycine (Kalant, 1956), fructose, D-glyceraldehyde, and dihydroxyacetone (Veneziale, 1971, 1972a), and, more recently, xylitol (Blair *et al.*, 1973). The finding by Schimassek and Mitzkat (1963) that in the rat liver perfused with lactate glucagon caused an increase of the ratio of fructose-6-phosphate to fructose diphosphate lends support to the hypothesis that fructose diphosphatase can be hormonally activated. The forward crossover was subsequently found by Williamson *et al.* (1969) also.

On the other hand, there are some arguments against glucagon stimulation of fructose diphosphatase.

1. The assayable activity of fructose diphosphatase in rat liver is one of the highest among all the enzymes required for fructose metabolism, suggesting that the enzyme does not catalyze a rate-limiting step even in the absence of hormone (Heinz *et al.*, 1968). The usefulness of the enzymologic approach has been questioned (Exton and Park, 1969), but counterarguments can be presented. For example, our present state of knowledge does not rule out the possibility that fructose diphosphatase is operating at or near optimal conditions. Therefore, data derived from enzyme assays conducted under first-order conditions with respect to enzyme concentration could reflect the *in vivo* situation.
2. Others have failed to demonstrate, by metabolite measurements, circumstances in which changes in the activity of fructose diphosphatase play a regulatory role in gluconeogenesis (Exton and Park, 1969).
3. We have failed to observe a stimulation of gluconeogenesis from glycerol in the presence of glucagon (Veneziale, 1972a).

4. Although the demonstration was for the enzyme in kidney, cAMP—the presumed mediator of glucagon-stimulated gluconeogenesis from all precursors—actually inactivated fructose diphosphatase (Holzer and Duntze, 1972). The effect of intravenous glucagon on fructose diphosphatase as reported by Taunton *et al.* (1972) might have been secondary to other hormonally caused events.

It may be shortsighted to maintain that the stimulation of fructose metabolism by glucagon must involve the molecular alteration of a particular enzyme, such as that demonstrated for skeletal muscle phosphorylase (Walsh *et al.*, 1972). An alternative speculation is that the hormone increases the availability of a metabolite to a particular cell compartment, which is analogous to the postulate that glucagon stimulates mitochondrial pyruvate uptake (Adam and Haynes, 1969). Such an effect could result from morphologic alterations due to hormones, and this has already been suggested (Garcia *et al.*, 1966). A second alternative speculation for glucagon activation of fructose metabolism could involve changes in a cell compartment (pH, ionic strength, ion concentration, etc.), thereby facilitating reactions occurring within that compartment.

But these speculations are not easily subjected to testing by experimentation. It might be more fruitful to continue to investigate the possibility that glucagon activates an enzyme of fructose (and also D-glyceraldehyde and dihydroxyacetone) metabolism or inactivates a glycolytic enzyme (a counterenzyme) by using metabolic techniques that include isolated intact hepatic parenchymal cells (Veneziale and Lohmar, 1973) and cell-free preparations (McDermott and Veneziale, 1972; Veneziale, 1972b) in addition to the isolated, perfused liver system. To this end, the latter area has already provided highly suggestive data. For example glucagon is not likely to influence fructokinase or aldolase B (Fig. 1) because it stimulates the metabolism of D-glyceraldehyde as well as that of fructose. The cell does not require D-triokinase for metabolism of glycerol, but this enzyme is necessary for metabolism of D-glyceraldehyde. Recalling that the enzyme phosphorylates both D-glyceraldehyde and dihydroxyacetone and that glucagon stimulated gluconeogenesis from both substrates but not from glycerol, it becomes reasonable to suspect that D-triokinase activity is subject to glucagon regulation. This remains a possibility because it has been demonstrated (Veneziale, 1972b) that this enzyme is subject to regulation by diet and cortisone and is not a constitutive enzyme, as previously thought.

An alternative possibility, which would not exclude any of those already mentioned, is that glucagon (cAMP) inhibits hepatic phosphofructokinase activity. This would diminish futile cycling of substrate between the phosphofructokinase and fructose diphosphatase reactions (Newsholme and

Gevers, 1967), thus leading to an increase in flux from fructose diphosphate to glucose. Indeed, Taunton *et al.* (1972) have already demonstrated that glucagon administered intravenously to intact rats decreases the assayable activity of hepatic phosphofructokinase. Blair *et al.*, (1973) showed that there was a decrease in the concentration of fructose diphosphate in rat livers perfused with dihydroxyacetone (or xylitol) and glucagon. From estimation of the fructose diphosphatase-phosphofructokinase substrate cycle by isotopic techniques, Clark *et al.* (1974a) concluded that regulation of hepatic glucose metabolism could occur by regulation of the cycle. Both findings from Dr. Lardy's laboratory would be consistent with involvement of phosphofructokinase in the regulation of hepatic gluconeogenesis. Subsequent work by this group with isolated hepatocytes (Clark *et al.*, 1974b) has provided convincing evidence that both glucagon and cAMP act to inhibit the flux through phosphofructokinase while increasing it through fructose diphosphatase.

Using isolated hepatic parenchymal cells incubated with dihydroxyacetone as substrate, we have shown that monobutyryl cAMP increases the rate of gluconeogenesis (Veneziale and Lohmar, 1973), decreases the concentration of fructose diphosphate (Veneziale, 1974), and decreases the assayable activity of phosphofructokinase (Veneziale and Swenson, 1974). Our data, which tended to single out phosophofructokinase as a controlling enzyme of gluconeogenesis, have the advantage of being derived from cells primarily responsible for glucose formation (Veneziale and Swenson, 1975). Inhibition of phosphofructokinase by cAMP could explain how glucagon stimulates the conversion of fructose, D-glyceraldehyde, and dihydroxyacetone (perhaps pyruvate also) to glucose in isolated, perfused livers. The foregoing work indicates a need for detailed direct investigation of phosphofructokinase and fructose diphosphatase of mammalian liver with regard to possible regulation by cAMP.

ADDENDUM

Since the submission of this chapter considerable progress has been made in the author's laboratory on studies of the regulation of carbohydrate synthesis in isolated rat hepatocytes from 24-hour fasted rats. As shown by electron microscopy, our hepatocyte preparation comprised a homogeneous population of intact parenchymal cells that formed glucose at rates as great as those of intact liver. Cells were suspended in an ionic buffer solution containing 1.5% gelatin and incubated with substrate in the presence and absence of hormones. After 10 minutes of incubation, homogenates were prepared and centrifuged at $27{,}500 \times g$ for 30 minutes, and the supernatant fractions were quickly assayed for the activities of multiple glycolytic-gluconeogenic enzymes.

Glucagon and monobutyryl cAMP inhibited the assayable activities of phosphofructokinase and pyruvate kinase to 20 to 25% of control values. Epinephrine did not influence the assayable activities of either enzyme, although it stimulated gluconeogenesis as markedly as did glucagon and monobutyryl cAMP. The activity of fructose diphosphatase was not influenced by glucagon, monobutyryl cAMP, epinephrine, and insulin.

When liver cell homogenates were subjected to centrifugation at higher forces (37,400 $\times$ *g* for 60 minutes or greater), the assayable activity of phosphofructokinase in supernatant fractions began to diminish. Only 17% of the control activity—that is, activity of the supernate after centrifugation at 27,500 $\times$ *g* for 30 minutes—was found in the supernate after centrifugation at 114,000 $\times$ *g* for 60 minutes. Additional loss of phosphofructokinase activity was observed in the high-speed supernates prepared from cells that had been incubated with epinephrine. However, in these fractions, pyruvate kinase activity did not differ from control values.

The foregoing results demonstrated (1) a behavior of phosphofructokinase which was not predictable on the basis of its known solubility properties and (2) differential effects of glucagon and epinephrine on the activity of the enzyme. These observations suggest that separate mechanisms are operative in stimulation of gluconeogenesis by glucagon and epinephrine.

REFERENCES

Adam, P. A. J. and Haynes, R. C., Jr. (1969). *J. Biol. Chem.* **244,** 6444.

Barman, T. E. (1969). *Enzyme Handbook,* Vol. 1, Springer, New York, p. 41.

Baron, P., Griffaton, G., and Lowy, R. (1971). *Enzymes* **12,** 481.

Blair, J. B., Cook, D. E., and Lardy, H. A. (1973). *J. Biol. Chem.* **248,** 3601.

Bublitz, C. and Kennedy, E. P. (1954). *J. Biol. Chem.* **211,** 951.

Burch, H. B., Lowry, O. H., Meinhardt, L., Max, P., Jr., and Chyu, K. (1970). *J. Biol. Chem.* **245,** 2092.

Burch, H. B., Max, P., Jr., Chyu, K., and Lowry, O. H. (1969). *Biochem. Biophys. Res. Commun.* **34,** 619.

Clark, M. G., Bloxham, D. P., Holland, P. C., and Lardy, H. A. (1974a). *J. Biol. Chem.* **249,** 279.

Clark, M. G., Kneer, N. M., Bosch, A. L., and Lardy, H. A. (1974b). *J. Biol. Chem.* **249,** 5695.

Cori, G. T., Ochoa, S., Slein, M. W., and Cori, C. F. (1951). *Biochim. Biophys. Acta* **7,** 304.

Exton, J. H., Mallette, L. E., Jefferson, L. S., Wong, E. H. A., Friedmann, N., Miller, T. B., Jr., and Park, C. R. (1970). *Rec. Progr. Horm. Res.* **26,** 411.

Exton, J. H. and Park, C. R. (1969). *J. Biol. Chem.* **244,** 1424.

Felber, J.-P., Renold, A. E., and Zahnd, G. R. (1959). *Bibl. Paediatr.* **70,** 467.

Frandsen, E. K. and Grunnet, N. (1971). *Eur. J. Biochem.* **23,** 588.

Froesch, E. R. (1966). In *The Metabolic Basis of Inherited Disease,* 2nd ed. (J. B. Stanbury, J. B. Wyngaarden, and D. S. Fredrickson, eds.), McGraw-Hill, New York.

Garcia, A., Williamson, J. R., and Cahill, G. F., Jr. (1966). *Diabetes* **15,** 188.

Garrison, J. C. and Haynes, R. C., Jr. (1973). *J. Biol. Chem.* **248,** 5333.

Hardinge, M. G., Swarner, J. B., and Crooks, H. (1965). *J. Am. Diet. Ass.* **46,** 197.

Heinz, F. (1968). *Hoppe-Seyler's Z. Physiol. Chem.* **349,** 859.

Heinz, F., Lamprecht, W., and Kirsch, J. (1968). *J. Clin. Invest.* **47,** 1826.

Hers, H. G. (1955). *J. Biol. Chem.* **214,** 373.

Hers, H. G. (1960). *Biochim. Biophys. Acta* **37,** 120.

Hers, H. G. (1962a). In *Methods in Enzymology,* Vol. 5 (S. P. Colowick and N. O. Kaplan, eds.), Academic Press, New York, pp. 333–335.

Hers, H. G. (1962b). In *The Enzymes,* Vol. 6, 2nd ed. (P. D. Boyer, H. Lardy and K. Myrbäck, eds.), Academic Press, New York, pp. 75–83.

Hers, H. G. and Kusaka, T. (1953). *Biochim. Biophys. Acta* **11,** 427.

Holzer, H. and Duntze, W. (1972). In *Biochemical Regulatory Mechanisms in Eukaryotic Cells* (E. Kun and S. Grisolia, eds.), Wiley, New York, pp. 115–136.

Holzer, H. and Schneider, S. (1955). *Klin. Wochenschr.* **33,** 1006.

Hue, L. and Hers, H. G. (1972). *Eur. J. Biochem.* **29,** 268.

Kalant, N. (1956). *Arch. Biochem. Biophys.* **65,** 469.

Kattermann, R., Dold, U., and Holzer, H. (1961). *Biochem. Z.* **334,** 218.

Kjerulf-Jensen, K. (1942). *Acta Physiol. Scand.* **4,** 249.

Landau, B. R. and Merlevede, W. (1963). *J. Biol. Chem.* **238,** 861.

Leuthardt, F. and Wolf, H. P. (1954). *Helv. Chim. Acta* **37,** 1732.

Mäenpää, P. H., Raivio, K. O., and Kekomäki, M. P. (1968). *Science* **161,** 1253.

Mallette, L. E., Exton, J. H., and Park, C. R. (1969). *J. Biol. Chem.* **244,** 5724.

McDermott, K. M. and Veneziale, C. M. (1972). *Arch. Biochem. Biophys.* **150,** 733.

Miller, M., Craig, J. W., Drucker, W. R., and Woodward, H., Jr. (1956–1957). *Yale J. Biol. Med.* **29,** 335.

Moore, B. W. (1959). *J. Am. Chem. Soc.* **81,** 5837.

Newsholme, E. A. and Gevers, W. (1967). *Vitam. Horm.* (*N.Y.*) **25,** 1.

Perheentupa, J. and Raivio, K. (1967). *Lancet.* **2,** 528.

Prendergast, F. G. and Veneziale, C. M. (1975). *J. Biol. Chem.* **250,** 1282.

Raivio, K. O., Kekomäki, M. P., and Mäenpää, P. H. (1969). *Biochem. Pharmacol.* **18,** 2615.

Rajkumar, T. V., Woodfin, B. M., and Rutter, W. J. (1966). In *Methods in Enzymology,* Vol. 9 (S. P. Colowick and N. O. Kaplan, eds.), Academic Press, New York, pp. 491–498.

Rauschenbach, P. and Lamprecht, W. (1964). *Hoppe-Seyler's Z. Physiol. Chem.* **339,** 277.

Ronca-Testoni, S., Raggi, A., and Ronca, G. (1970). *Biochim. Biophys. Acta* **198,** 101.

Ross, B. D., Hems, R., and Krebs, H. A. (1967). *Biochem. J.* **102,** 942.

Schimassek, H. and Mitzkat, H. J. (1963). *Biochem. Z.* **337,** 510.

Sestoft, L., Tønnesen, K., Hansen, F. V., and Damgaard, S. (1972). *Eur. J. Biochem.* **30,** 542.

Sillero, M. A. G., Sillero, A., and Sols, A. (1969). *Eur. J. Biochem.* **10,** 345.

Szepesi, B. and Freedland, R. A. (1968). *Can. J. Biochem.* **46,** 1459.

Taunton, O. D., Stifel, F. B., Greene, H. L., and Herman, R. H. (1972). *Biochem. Biophys. Res. Commun.* **48,** 1663.

Tolbert, M. E. M. and Fain, J. N. (1974). *J. Biol. Chem.* **249,** 1162.

Unger, R. H., Aguilar-Parada, E., Müller, W. A., and Eisentraut, A. M. (1970). *J. Clin. Invest.* **49,** 837.

Veneziale, C. M. (1971). *Biochemistry.* **10,** 3443.

Veneziale, C. M. (1972a). *Biochemistry.* **11,** 3286.

Veneziale, C. M. (1972b). *Eur. J. Biochem.* **31,** 59.

Veneziale, C. M. (1974). *Fed. Proc.* **33,** 1456.

Veneziale, C. M. and Lohmar, P. H. (1973). *J. Biol. Chem.* **248,** 7786.

Veneziale C. M. and Swenson, R. P. (1975). *Mayo Clin. Proc.* **50,** 271.

Walsh, D. A., Brostrom, C. O., Brostrom, M. A., Chen, L., Corbin, J. D., Reimann, E., Soderling, T. R., and Krebs, E. G. (1972). *Adv. Cycl. Nucleot. Res.* **1,** 33.

Williamson, J. R., Browning, E. T., Thurman, R. G., and Scholz, R. (1969). *J. Biol. Chem.* **244,** 5055.

Wolf, H. P. and Leuthardt, F. (1953). *Helv. Chim. Acta* **36,** 1463.

Woods, H. F., Eggleston, L. V., and Krebs, H. A. (1970). *Biochem. J.* **119,** 501.

Yudkin, J. (1969a). *Nutr. News* **32,** 9.

Yudkin, J. (1969b). *Nutr. Today* **4,** 16.

CHAPTER THIRTEEN

A Possible Role for Pyrophosphate in the Control of Hepatic Glucose Metabolism

J. W. R. LAWSON, R. W. GUYNN, NEAL CORNELL, AND RICHARD L. VEECH

I. INTRODUCTION[1]

Enzyme pairs such as phosphofructokinase and fructose diphosphatase can catalyze substrate cycles in which the net reaction is hydrolysis of ATP to ADP + P_i. Recent work in Professor Lardy's laboratory and elsewhere has

[1] Abbreviations used in this chapter: Ald, aldolase; FDPase, fructose diphosphatase; G-6-Pase, glucose-6-phosphatase; Myk, myokinase; PFK, phosphofructokinase; PP_iase, pyrophosphatase; TPI, triosephosphate isomerase; UDPG-PP_iase, uridine-diphosphoglucose pyrophosphorylase; DHAP, dihydroxyacetone phosphate; EDTA, ethylenediamine tetraacetic acid; FDP, fructose-1,6-diphosphate; F-6-P, fructose-6 phosphate; G-1-P, glucose-1-phosphate; G-6-P, glucose-6-phosphate; PEP, phosphoenolpyruvate; 2-PG, 2-phosphoglyceric acid; 3-PG, 3-phosphoglyceric acid; P_i, orthophosphate; PP_i, pyrophosphate; UDPG, uridine-diphosphoglucose; UTP, uridine triphosphate. Square brackets [] designate concentration; Σ represents the sum of all species present. Ionic strength (I) has the dimensions of moles/liter and is calculated from the formula: $I = \frac{1}{2} \Sigma C_i z_i^2$.

attempted to quantitate the rates of some rat liver substrate cycles (Clark *et al.*, 1974; Rognstad *et al.*, 1973). It had earlier been suggested (Newsholme and Gevers, 1967) that substrate cycling might account for the observation that the ATP requirement for gluconeogenesis from lactate exceeds the theoretical value of 6 (Krebs *et al.*, 1964). Since a decrease in citrate concentration theoretically releases the inhibition exerted by ATP on PFK (Passonneau and Lowry, 1964; Paetkau and Lardy, 1967; Bloxham and Lardy, 1973), it was further proposed that substrate cycles constitute an important control mechanism for gluconeogenesis and glycolysis (Newsholme and Gevers, 1967; Scrutton and Utter, 1968) that becomes operative when the level of cytoplasmic citrate is decreased (J. R. Williamson *et al.*, 1969).

The widely accepted view that PFK is one control point of glycolysis—perhaps the critical point—is based largely on the correlation between the catalytic properties of the enzyme *in vitro* and changes in the levels of its substrates and effectors under different conditions of flux through the glycolytic pathway (for reviews see Krebs, 1972; Mansour, 1972, Ramaiah, 1974). Inhibition of glucose phosphorylation is also assumed to result from decreased activity of PFK, since slower rates through PFK lead to increases in [F-6-P] and its isomerase partner [G-6-P], an inhibitor of hexokinase (Weil-Malherbe and Bone, 1951). Thus phosphorylation of F-6-P and dephosphorylation of FDP are thought to have a central role in the control of both glycolysis and gluconeogenesis.

It is clear from Table I that the capacities of glucose-6-phosphatase and fructose-1,6-diphosphatase to dephosphorylate their respective substrates exceed by a factor of 4 the capacities of the synthetic enzymes to phosphorylate these compounds. Without effective coordinate control of both phosphorylation and dephosphorylation, substrate cycling through these reactions would be expected, since PFK and FDPase catalyze thermodynamically favorable reactions. The regulatory properties associated with the PFK and FDPase reactions are extremely complex, and mutations affecting those regulatory properties would be expected to result in a number of genetic "diseases." No such diseases are known; therefore it seems worthwhile to reconsider the phosphorylation-dephosphorylation processes from the point of view of the most rugged and simplest form of control—that is, the near-equilibrium control, where the ratio of [substrate]/[product] is determined largely by the laws of mass action.

Obviously, the [glucose]/[G-6-P] and [F-6-P]/[FDP] ratios will depend on the simultaneous rates of a number of reactions. Although it has been argued (see Chapter 4, this volume; Guuma *et al.*, 1971) that existence of intracellular microenvironments or compartmentation complicates interpretation of metabolite measurements, a large body of evidence collected over many years favors the view that near-equilibrium reactions occur both in

Table I. Activity of enzymes capable of forming or phosphorylating glucose or fructose-6-phosphate in rat liver

Enzyme	Activity[a] (μmole/min/g fresh wt)	K_m (M)	Reference to K_m
Hexokinase	0.8	10^{-5}	Sols (1968)
Glucokinase	2.5	10^{-2}	Sols (1968)
Glucose-6-phosphatase	16.7	4×10^{-4}	Nordlie *et al.* (1966)
Phosphofructokinase	2.5	—	
Fructose-1-6-diphosphatase	10.0	10^{-6}	Taketa *et al.* (1965)
Alkaline phosphate	3.6	—	

[a] Activities are from Weber *et al.* (1967) and represent the activities of enzymes measured in the same rat livers at pH 7.4 and 37°C. Alkaline phosphatase activity was measured in the author's laboratory at pH 9.0, 38°C, using para-nitrophenyl phosphate as substrate.

mitochondria and cytoplasm *in vivo*. The measured tissue content of a number of metabolites fits the expected values for K_{eq} of many pyridine nucleotide-linked dehydrogenases, kinases, isomerases, and transaminases (Krebs and Veech, 1969). In this chapter we assume that metabolite content measured in whole tissue is closely enough related to the concentrations at reaction sites to permit comparison of the concentrations of a substrate couple with those of co-factors with which it could be reacting, and to judge whether the reaction is near-equilibrium. Thus the ratio [glucose]/[G-6-P] may be compared with:

$$\frac{[\text{glucose}]}{[\text{G-6-P}]} = \frac{1}{K_{\text{hexokinase}}} \times \frac{[\text{ADP}]}{[\text{ATP}]}$$

$$\frac{[\text{glucose}]}{[\text{G-6-P}]} = \frac{K_{\text{glucose-6-phosphatase}}}{[\text{P}_i]}$$

$$\frac{[\text{glucose}]}{[\text{G-6-P}]} = \frac{1}{K_{\text{glucose transphosphorylase}}} \times \frac{[\text{P}_i]}{[\text{PP}_i]}$$

A similar set of equations can be written for the [fructose-6-P]/[fructose diphosphate] ratio.

A recently developed method (Guynn *et al.*, 1974) makes it possible to estimate the free cytoplasmic [PP_i] and to evaluate the relationships (above) involving the [glucose]/[G-6-P] and [F-6-P]/[FDP] ratios. Redetermination of several equilibrium constants under conditions approximating those *in vivo* and measurement of the tissue content of relevant metabolites and cofactors leads us to the conclusion that *in vivo* the glucose-6-phosphatase transphosphorylation reaction described by Nordlie (1964) apparently achieves near-equilibrium. The same is true for a fructose-6-phosphate transphosphorylation reaction catalyzed by an enzymatic pathway that is not known at present. If it is true that transphosphorylation reactions involving glucose and F-6-P achieve near-equilibrium, it follows that substrate cycling between glucose-G-6-P and F-6-P-FDP can occur without energy loss and could account for the pattern of isotopic labeling found by Lardy and his coworkers. It further follows that the major effector of the [F-6-P]/[FDP] ratio or the [glucose]/[G-6-P] ratio is the [PP_i]/[P_i] ratio.

II. METHODS

A. Rats

Male Sprague-Dawley rats (Carworth, Kensington, New Jersey) weighing 200 to 300 g were used. All animals were meal-fed (8:30 A.M.–11:30 A.M.) the NIH-07 stock diet composed of 51.3% carbohydrate, 24.4% protein, 4.7% fat, and all received water *ad libitum*. Starved rats received no food for 48 hours before sacrifice.

B. Diabetic Rats

Rats were made diabetic by injection of 150 mg streptozotocin/kg body weight into the tail vein. This dose is sufficient to eliminate more than 97% of the pancreatic β-cell function within 24 hours (Junod *et al.*, 1969). The streptozotocin was prepared immediately prior to injection in 0.025 M citrate buffer pH 4.5. The control group received the buffer solution only; both groups received 5% glucose solution *ad lib* until sacrifice. The livers were freeze clamped 27 hours after injection (5 hours postprandial), stored at −60°C, and processed the next morning. All streptozotocin animals were severely diabetic, as indicated by urine glucose concentrations of greater than 0.5% and very high concentrations of ketone bodies (Keto-Diastic, Ames Co., Elkhart, Indiana).

C. Diphenyleneiodonium-Treated Rats

Rats were injected intraperitoneally with a 0.5% aqueous solution of bis(diphenyleneiodonium) sulfate, pH 7.0, 10 mg/kg body weight, 2 hours after the beginning of the feeding period. One hour after injection the livers were freeze clamped and processed. This combination of schedule and dose has been shown to produce glycolysis, inhibit gluconeogenesis and to stimulate lactate production in 2-day-old suckling rats; in suckling animals whole body and blood glucose concentrations were lowered by diphenyleneiodonium, and similar results were obtained for blood glucose with adult rats (Holland *et al.*, 1973).

D. Treatment of Tissue

After cervical dislocation, livers were rapidly removed and freeze clamped in aluminum tongs precooled in liquid nitrogen. Perchloric acid extracts of the liver powder were prepared as described previously (Veech *et al.*, 1972).

E. Metabolite Measurements

We measured G-1-P, UTP, and UDPG and used them to calculate the free cytoplasmic PP_i concentration as described by Guynn *et al.* (1974). For the bicarbonate measurement 2 g of the frozen liver was suspended in an ice cold solution containing 15 mM K_2HPO_4, 5mM EDTA, 2.5 mM NaF, homogenized at $-20°C$ using an Ultrasonic Sonifier Cell Disruptor (Plainview, New York) for 3 minutes at 50 W, and analyzed by the method of Van Slyke and Neill (1924). The EDTA and NaF were added to inhibit glycolysis and prevent loss of CO_2 by lactic acid production. Measurements made immediately or 4 hours after EDTA and NaF addition gave identical values for a given sample. Hepatic CO_2 concentrations were assumed to be equal to the hepatic venous CO_2 concentrations. Samples of hepatic venous blood were obtained from animals under light ether anesthesia by passing a catheter through the jugular vein and heart into the inferior vena cava until the hepatic vein was reached. Blood was drawn, the pCO_2 measured, and the $[CO_2]$ calculated. Using the measured $[HCO_3^{1-}]$ and the calculated $[CO_2]$, the intracellular pH was estimated by the Henderson-Hasselbach equation with $pK_0 = 6.11$. The $[P_i]$ was measured as described by Guynn *et al.* (1972) except that the glyceraldehyde-3-phosphate dehydrogenase and 3-phosphoglycerate kinase were centrifuged and the $(NH_4)_2SO_4$ removed, to ensure that the reaction was complete in 20 minutes. All other metabolites were measured as previously described by Veloso *et al.* (1972) and Veech *et *al.* (1972). The level of cytoplasmic oxaloacetate was calculated according to J. R. Williamson *et al.* (1967). Isocitrate was calculated from the measured citrate using the value of 18 for the [citrate]/[isocitrate] ratio (Veloso *et al.*, 1973). Because measured FDP in liver includes much FDP that is enzyme bound *in vivo* (Section IV) free [FDP] was calculated from the measured [DHAP] using the relationship:

$$[\mathrm{FDP}] = \frac{[\mathrm{DHAP}]^2}{(K_{\mathrm{TPI}} \cdot K_{\mathrm{Ald}})}$$

where $K_{\mathrm{Ald}} = 0.99 \times 10^{-4}$ M and $K_{\mathrm{TPI}} = 9$ in rat liver *in vivo* (Veech *et al.*, 1969a). Calculation of pyridine nucleotide ratios and phosphorylation states was performed as previously described (Veech *et al.*, 1969; Krebs and Veech, 1969a), except that the [ATP]/[ADP] × [P_i] ratio was calculated using the total measured phosphate concentration.

F. Reagents

The reagents—FDPase, alkaline phosphatase, α-D-G-1-P, UTP, UDPG, D-F-6-P (Grade I), and D-FDP (Grade I)—were obtained from Sigma Chemical Co. (St. Louis, Missouri), D-G-6-P, D-G-1,6-diphosphate, and all other enzymes were purchased from Boehringer-Mannheim Corp., New York; D-glucose and D-fructose were products of Baker Chemical Co., Phillipsburg, New Jersey. Streptozotocin and bis(diphenyleneiodonum) sulfate were gifts from the Upjohn Co., Kalamazoo, Michigan, and Dr. M. G. Clark of the Institute for Enzyme Research, Madison, Wisconsin, respectively. Other chemicals were reagent grade and were products of Mallinckrodt Chemical Works, St. Louis.

G. Instruments

Metabolite assays were performed on a Zeiss PMQ II spectrophotometer modified to hold 13 cuvettes. The total magnesium concentrations were analyzed with a Varian Techtron Model AA-5 atomic absorption spectrometer, pH and $P_{\mathrm{CO_2}}$ measurements were made at 38°C using a Radiometer Copenhagen pH meter (type 27) equipped with a gas monitor (type PHA927A) and the E5021 pH microelectrode unit.

H. Observed Equilibrium Constants of Glucose-6-Phosphatase, FDPase, and Myokinase

The observed equilibrium constants of G-6-Pase and FDPase at 38°C, pH 6.99, free [Mg^{2+}] = 0.9 to 1.0 mM, and $I = 0.25$ were measured in both directions using varying amounts of substrates. The detailed reaction mixtures and procedures are given in Tables II and III. The ionic strength of the potassium phosphate buffer was 0.22, and other components of the reaction mixtures increased the ionic strength to 0.25. Reactions were considered to be at equilibrium when the concentration ratios were identical after approaching equilibrium from both directions.

The observed equilibrium constant of myokinase was measured (in both directions) at varying pH and free [Mg^{2+}]. The variation of K_{obs} with free

Table II. Determination of the equilibrium constant of the glucose-6-phosphatase reaction[a]

	Equilibrium concentrations (M)			
Experiment	[Glucose] $\times 10^{+1}$	[P_i] $\times 10^{+2}$	[G-6-P] $\times 10^{+5}$	$K_{obs} \times 10^{-2}$
1 (*r*)	1.05	9.96	9.47	1.10
	1.06	9.42	9.30	1.07
2 (*r*)	0.51	9.69	4.49	1.10
	0.52	9.83	4.65	1.11
3 (*f*)	1.08	9.69	9.14	1.14
	1.08	9.69	9.64	1.09
4 (*f*)	1.09	9.83	9.80	1.09
	1.09	9.42	9.64	1.06

[a] Initial reaction mixtures contained: 106 mM glucose (Experiment 1); 52 mM glucose (Experiment 2); 106 mM glucose and 0.24 mM G-6-P (Experiment 3); 106 mM glucose and 0.48 mM G-6-P (Experiment 4). All reaction mixtures contained 6.04 mM $MgCl_2$ and 98 mM potassium phosphate buffer pH 6.99 and were done in duplicate. The reactions were initiated with 18 units alkaline phosphotase per milliliter of final reaction mixture and incubated for 120 minutes at 38°C. After the pH was measured, the reaction was stopped with perchloric acid (final concentration 0.84 M) and neutralized with 1.5 M $KHCO_3$ in 2.3 M KOH. After centrifugation, the extracts were analyzed for glucose, G-6-P and P_i by standard enzymatic assays. The forward reaction (*f*) is in the direction of G-6-P utilization; the reverse reaction (*r*) is in the direction of G-6-P production. All values are at $I = 0.25$, pH 6.99 ± 0.01, and free [Mg^{2+}] $= 0.97 \pm 0.01$ mM.

$$K_{obs} = \frac{[\sum P_i]\,[\text{glucose}]}{[\sum \text{G-6-P}]}$$

[Mg^{2+}] at a constant pH was also measured in the direction of ADP formation. The details of the initial reaction mixtures and of the procedures are given in Tables IV and V.

For all the equilibrium measurements, the enzymes were centrifuged at 31,000 *g* for 15 minutes to remove the excess $(NH_4)_2SO_4$. Immediately before use, they were resuspended in an appropriate volume of water containing 1 μM EDTA.

Table III. Determination of the equilibrium constant of the fructose-1, 6-diphosphatase reaction[a]

	Equilibrium concentrations (M)			
Experiment	[F-6-P] $\times 10^{+3}$	[P_i] $\times 10^{+1}$	[FDP] $\times 10^{+6}$	$K_{obs} \times 10^{-2}$
1 (*r*)	5.25	1.02	3.48	1.54
	5.25	1.05	3.19	1.73
2 (*r*)	20.8	1.02	10.5	2.03
	21.2	1.02	10.8	2.01
3 (*f*)	10.8	1.01	6.48	1.66
	10.8	1.05	6.16	1.84
4 (*f*)	11.0	1.02	7.34	1.54
	10.8	1.05	7.10	1.60

[a] Initial reaction mixtures contained: 5.1 mM F-6-P (Experiment 1); 21 mM F-6-P (Experiment 2); 11 mM F-6-P and 0.05 mM FDP (Experiment 3); 11 mM F-6-P and 0.2 mM FDP (Experiment 4). All reaction mixtures contained 6.05 mM $MgCl_2$ and 104 mM potassium phosphate buffer pH 6.99 and were done in duplicate. The reactions were initiated with 0.4 unit fructose-1,6-diphosphatase per milliliter of final reaction mixture and were incubated for 120 minutes at 38°C. After the pH was measured, the reaction was stopped with perchloric acid (final concentration 0.84 M) and neutralized with 1.5 M $KHCO_3$ in 2.3 M KOH. After centrifugation the extracts were analyzed for F-6-P, P_i, and FDP by standard enzymatic assays. The forward reaction (*f*) is in the direction of FDP utilization; the reverse reaction (*r*) is in the direction of FDP production. All values are at $I = 0.25$, pH $= 6.99 \pm 0.01$, and free [Mg^{2+}] $= 0.89 \pm 0.01$ mM.

$$K_{obs} = \frac{[\sum P_i]\,[\sum \text{F-6-P}]}{[\sum \text{FDP}]}$$

Table IV. Determination of the equilibrium constant for the ionic reaction: $2ADP^{3-} \rightleftarrows ATP^{4-} + AMP^{2-}$

		Equilibrium concentrations (mM)				K	
Experiment[a]	pH	ATP	AMP	ADP	Free $[Mg^{2+}]$	K_{obs}	free $[Mg^{2+}] = 0$
1 (*f*)	6.91	0.633	0.733	0.687	1.08	0.983	0.401
	6.89	0.627	0.726	0.687	1.10	0.964	0.398
2 (*r*)	6.91	0.633	0.759	0.700	1.08	0.980	0.400
	6.90	0.640	0.752	0.700	1.08	0.982	0.403
3 (*f*)	6.97	0.633	0.759	0.700	0.38	0.981	0.462
	6.97	0.633	0.746	0.713	0.38	0.929	0.436
4 (*r*)	6.98	0.646	0.772	0.726	0.37	0.946	0.444
	6.98	0.659	0.722	0.739	0.36	0.871	0.409

[a] Initial reaction mixtures contained 0.94 mM ATP, 1.04 mM AMP, 0.03 mM ADP (Experiments 1 and 3); 0.04 mM ATP, 0.14 mM AMP, 1.92 mM ADP (Experiments 2 and 4); 2.5 mM $MgCl_2$ (Experiments 1 and 2); 1.21 mM $MgCl_2$ (Experiments 3 and 4). All reaction mixtures contained 9.7 mM potassium phosphate buffer pH 6.99 and were adjusted to $I = 0.25$ with KCl. The reactions were run in duplicate in both forward and reverse directions and were initiated with 0.2 unit myokinase per milliliter of reaction mixture and were incubated 30 minutes at 38°C. After measuring pH, the reactions were stopped by addition of perchloric acid (final concentration 0.84 M) and the samples centrifuged for 20 minutes at 35,000 *g*. Aliquots of the supernatants were neutralized with 1.5 M $KHCO_3$ in 2.3 M KOH, centrifuged for 20 minutes at 35,000 *g*, and analyzed for ATP, ADP, and AMP by standard enzymatic assays.

Table V. Variation of K_{obs} of the myokinase reaction with free $[Mg^{2+}]$ at pH 6.99

	Equilibrium concentrations (mM)				K	
Trial[a]	ATP	AMP	ADP	Free $[Mg^{2+}]$	K_{obs}	Free $[Mg^{2+}] = 0$
1	0.388	0.417	0.478	0.10	0.708	0.484
2	0.400	0.433	0.439	0.25	0.899	0.470
3	0.381	0.413	0.394	0.44	1.014	0.451
4	0.413	0.446	0.420	0.66	1.044	0.429
5	0.413	0.452	0.420	0.83	1.060	0.424
6	0.407	0.436	0.459	1.95	0.842	0.360
7	0.368	0.404	0.510	4.45	0.572	0.324
8	0.342	0.375	0.581	9.98	0.380	0.316

[a] Initial reaction mixtures contained 0.49 mM ATP, 0.27 mM ADP, 0.54 mM AMP, 9.5 mM potassium phosphate buffer, pH 6.99, and 0.36 to 14.4 mM total $MgCl_2$. Ionic strength was adjusted to 0.25 with KCl. The reactions, occurring in the direction of ADP formation, were initiated with 0.2 unit of myokinase per milliliter of reaction mixture and were incubated 20 minutes at 38°C. The pH was measured, adjusted to pH 6.99 if necessary, and incubated for an additional 10 minutes. The latter steps were repeated until the pH remained at 6.99. Perchloric acid was then added, and the samples were treated as described in Table IV. The free magnesium concentration was calculated as described in Section II.

III. CALCULATIONS

A. Equilibrium Constants of the Inorganic Pyrophosphatase, Hexokinase, Phosphofructokinase, and the Glucose-PP_i and F-6-P-PP_i Phosphotransferase Reactions

The equilibrium position of the following pyrophosphatase reaction

$$\Sigma PP_i \rightleftarrows 2\Sigma P_i \tag{1}$$

lies very far to the right; thus only traces of PP_i would remain at equilibrium. Therefore the equilibrium constant $K_{obs(PP_i ase)}$ is difficult to measure directly because of the analytical problems of measuring PP_i especially in the presence of large amounts of P_i. It can be determined indirectly, however, from two or more reactions whose sum is Eq. 1. One such set of reactions is

$$2\Sigma\text{ATP} \rightleftarrows 2\Sigma\text{ADP} + 2\Sigma\text{P}_i \tag{2}$$

$$\Sigma\text{AMP} + \Sigma\text{PP}_i \rightleftarrows \Sigma\text{ATP} \tag{3}$$

$$2\Sigma\text{ADP} \rightleftarrows \Sigma\text{ATP} + \Sigma\text{AMP} \tag{4}$$

$$\Sigma\text{PP}_i \rightleftarrows 2\Sigma\text{P}_i \tag{1}$$

and

$$K_{\text{obs(PP}_i\text{ase)}} = \frac{[\text{P}_i]^2}{[\text{PP}_i]} = \frac{[K_{\text{obs(2)}}]^2[K_{\text{obs(4)}}]}{[K_{\text{obs(3)}}]}$$

where $K_{\text{obs(2)}}$ = observed equilibrium constant for the hydrolysis of ATP to ADP + P_i

$K_{\text{obs(3)}}$ = observed equilibrium constant for the hydrolysis of ATP to AMP + PP_i

$K_{\text{obs(4)}}$ = observed equilibrium constant for the myokinase reaction

Since the observed equilibrium constants for reactions 2 and 3 have been determined by Guynn and Veech (1973) and Guynn *et al.* (1974), respectively, under physiological conditions, the $K_{\text{obs(PP}_i\text{ase)}}$ for the hydrolysis of inorganic pyrophosphate can be calculated if the appropriate values for Eq. 4 are known.

Once the K_{obs} for pyrophosphate hydrolysis is known, the observed equilibrium constants at physiological conditions of the following reactions can be calculated in a similar manner by combination with the observed equilibrium constants of G-6-Pase and FDPase, respectively:

$$\text{glucose} + \Sigma\text{PP}_i \leftrightarrows \Sigma\text{G-6-P} + \text{P}_i \tag{5}$$

$$\Sigma\text{F-6-P} + \Sigma\text{PP}_i \leftrightarrows \Sigma\text{FDP} + \Sigma\text{P}_i \tag{6}$$

where

$$K_{\text{obs(glucose-PP}_i)} = \frac{[\Sigma\text{G-6-P}][\text{P}_i]}{[\text{glucose}][\text{PP}_i]} = \frac{K_{\text{obs(PP}_i\text{ase)}}}{K_{\text{obs(G-6-Pase)}}} \tag{7}$$

$$K_{\text{obs(F-6-P-PP}_i)} = \frac{[\Sigma\text{FDP}][\text{P}_i]}{[\Sigma\text{F-6-P}][\text{PP}_i]} = \frac{K_{\text{obs(PP}_i\text{ase)}}}{K_{\text{obs(FDPase)}}} \tag{8}$$

By combining the observed equilibrium constants of G-6-Pase and FDPase with the corresponding value of Eq. 2, the observed equilibrium constants and ΔG_{obs} of hexokinase and phosphofructokinase can also be calculated for physiological conditions:

$$K_{obs(HK)} = \frac{[\Sigma\text{G-6-P}][\Sigma\text{ADP}]}{[\text{glucose}][\Sigma\text{ATP}]} = \frac{K_{obs(ATPase)}}{K_{obs(G-6-Pase)}} \tag{9}$$

$$K_{obs(PFK)} = \frac{[\Sigma\text{FDP}][\Sigma\text{ADP}]}{[\Sigma\text{F-6-P}][\Sigma\text{ATP}]} = \frac{K_{obs(ATPase)}}{K_{obs(G-6-Pase)}} \tag{10}$$

B. Myokinase Equilibrium

The equilibrium constant of the ionic equation for the myokinase reaction is defined as follows:

$$K = \frac{[\text{ATP}^{4-}][\text{AMP}^{2-}]}{[\text{ADP}^{3-}]^2} \tag{11}$$

This constant cannot be measured directly, since under physiological conditions there exist several acid forms and magnesium complexes that are analytically indistinguishable. Only the sum of all the species present can be determined experimentally. These sum concentrations can then be used to define an observed constant:

$$K_{obs} = \frac{[\Sigma\text{ATP}][\Sigma\text{AMP}]}{[\Sigma\text{ADP}]^2} \tag{12}$$

However, K and K_{obs} are related by the acid dissociation and magnesium binding constants of the components of the reactions and the identity equation of all species, for example,

$$[\Sigma\text{ATP}] = [\text{ATP}^{4-}] + [\text{HATP}^{3-}] + [\text{MgATP}^{2-}] + [\text{MgHATP}^{1-}] \tag{13}$$

By substitution of the definitions of the acid dissociation ($K_{a\ldots}$) and magnesium binding constant ($K_{b\ldots}$) for ATP, Eq. 13 becomes

$$= [\text{ATP}^{4-}]\left(1 + \frac{[\text{H}^+]}{K_{a(\text{ATP})}} + K_{b(\text{ATP}^{4-})}[\text{Mg}^{2+}] + \frac{K_{b(\text{HATP})}[\text{H}^+][\text{Mg}^{2+}]}{K_{a(\text{ATP})}}\right)$$

or

$$[\Sigma\text{ATP}] = [\text{ATP}^{4-}]f_{\text{ATP}} \tag{14}$$

where f_{ATP} equals the sum of the terms in parentheses. Similar equations can be derived for ADP (Eqs. 15 and 16) and AMP (Eqs. 17 and 18).

$$[\Sigma ADP] = [ADP^{3-}] + [HADP^{2-}] + [MgADP^{1-}] + [MgHADP^{0}] \tag{15}$$

$$[\Sigma ADP] = [ADP^{3-}]f_{ADP} \tag{16}$$

$$\text{where} \quad f_{ADP} = 1 + \frac{[H^+]}{K_{a(ADP)}} + \frac{K_{b(HADP)}[H^+][Mg^{2+}]}{K_{a(ADP)}}$$

$$[\Sigma AMP] = [AMP^{2-}] + [HAMP^{1-}] + [MgAMP^{0}] \tag{17}$$

$$[\Sigma AMP] = [AMP^{2-}]f_{AMP} \tag{18}$$

$$\text{where} \quad f_{AMP} = 1 + \frac{[H^+]}{K_{a(AMP)}} + K_{b(AMP)}[Mg^{2+}]$$

The expression of the relationship between K and K_{obs} can now be obtained by combining Eqs. 1, 2, 4, 6, and 8 to yield

$$K = K_{obs} \times \left(\frac{f_{ADP}^2}{f_{ATP}\, f_{AMP}}\right) \tag{19}$$

Equation 19 can be used to calculate K when K_{obs} is measured at a known pH and free Mg^{2+}. Once K is known, Eq. 19 can be used to calculate K_{obs} under any conditions of pH and free Mg^{2+}.

C. Calculation of Thermodynamic Data

The ΔG_{obs} values were calculated from the relationship:

$$\Delta G_{obs} = -RT \ln K_{obs}$$

where the gas constant R is 1.987 cal/kg-mole. For concentration ratios of substrates and products other than equilibrium values, ΔG^0_{obs} values were calculated from the relationship

$$\Delta G_{obs} = \Delta G^0_{obs} + RT \ln \frac{[\text{products}]}{[\text{reactants}]}$$

For all calculations, unit activity has been used for the standard state of liquid water.

D. Calculation of Free Mg^{+2}

With the aid of a computer program, the free [Mg^{2+}] was calculated by solving simultaneous equations containing the composite binding constants and total substrate concentrations (Veloso *et al.*, 1973). Guynn and Veech (1973) have listed the acid dissociation and magnesium binding constants that best approximate the conditions under which the equilibrium constants were measured in the present studies. No values for magnesium-G-6-P and magnesium-F-6-P complexes are available, and the stability constants for those complexes were taken as equal to the stability constant of magnesium-G-1-P (Treveleyan *et al.*, 1952). Under the conditions used here, the concentration of magnesium-FDP was insignificantly low, and no stability constant was required for that complex.

IV. RESULTS

A. Equilibrium Constants of the Glucose-6-Phosphatase and Fructose-1,6-Diphosphatase Reactions

The equilibrium constant for the glucose-6-phosphatase reaction was measured at $I = 0.25$ and 38°C using alkaline phosphatase as the catalyst. Consistent results were obtained with different initial substrate concentrations and by approaching equilibrium from both sides (Table II). The mean value (±SEM) obtained for the constant defined as

$$K_{\text{obs(G-6-Pase)}} = \frac{[\text{glucose}][\Sigma P_i]}{[\Sigma \text{G-6-P}]} \qquad (20)$$

was $(1.10 + 0.01) \times 10^{+2}$ M at pH 6.99 ± 0.01 and free [Mg^{2+}] = $(0.97 \pm 0.01) \times 10^{-3}$ M. This value corresponds to a standard free energy ΔG^0_{obs} of −2.92 ± 0.01 kcal/mole for the hydrolysis of G-6-P. Values reported in the literature for the free energy of hydrolysis differ by 0.7 kcal/mole. Ginodman (1954) calculated a value of −2.6 kcal/mole at pH 7 and 25°C, whereas for the same conditions Atkinson *et al.* (1961) found −3.3 kcal/mole for free [Mg^{2+}] from 0 to 0.3 mM. Meyerhof and Green (1949) calculated −3.0 kcal/mole at 38°C, pH 8.5, [H_2O] = 55.5 M, and unspecified ionic strength and magnesium concentration. These reported values were obtained under higher concentrations of glucose and of inorganic phosphate than were used in this study (2.5–20 times) and consequently represent values at very high ionic strength. The value reported here may be considered more applicable to intracellular conditions, since it was measured under the appropriate physiological conditions of temperature, ionic strength, and free magnesium.

Table III gives the results of the determination of the observed equilibrium constant for the hydrolysis of FDP by FDPase. Approaching equilibrium from both directions using different initial concentrations produced similar values. The mean value (±SEM) for the constant defined as

$$K_{obs(FDPase)} = \frac{[\Sigma F\text{-}6\text{-}P][\Sigma P_i]}{[\Sigma FDP]} \tag{21}$$

was $(1.74 \pm 0.07) \times 10^{+2}$ at pH 6.99 ± 0.01, free $[Mg^{2+}] = (0.89 \pm 0.01) \times 10^{-3}$ M, $I = 0.25$, and 38°C. This corresponds to a standard free energy change ΔG^0_{obs} of −3.18 kcal/mole for the hydrolysis of FDP.

B. Equilibrium Constant of the Myokinase Reaction

Table IV presents the results of the determination of the equilibrium constant for the ionic reaction

$$2ADP^{3-} \rightleftarrows ATP^{4-} + AMP^{2-}$$

The average value (±SEM) as calculated from Eq. 19 was

$$K = \frac{[ATP^{4-}][AMP^{2-}]}{[ADP^{3-}]^2} = 0.419 \pm 0.0009 \tag{22}$$

This value is in reasonable agreement with values of 0.325 found by Blair (1970) and 0.375 found by Bowin and Kewin (1956). The theoretical variation of $K_{obs(Myk)}$ with free $[Mg^{2+}]$ at pH 6.99 was then calculated using this value of K in Eq. 19. The resultant curve appears in Fig. 1. Table V lists the results of the experimental determination of the variation of the observed equilibrium constant with free $[Mg^{2+}]$ at pH 6.99. There is reasonably good agreement between the theoretically predicted points and the experimental points (Fig. 1). The displacement of the experimental curve toward lower free magnesium concentrations is most likely due to one or more errors in the values of the binding constants used. At the physiological free $[Mg^{2+}]$ of 0.8 to 1.-mM, however, the predicated value of 1.05 agrees well with the measured value of 1.06. Therefore, the value that will be used for subsequent calculations at 38°, $I = 0.25$, pH 6.99, and free $[Mg^{2+}] = 0.9$–1.0 mM is

$$K_{obs(Myk)} = \frac{[\Sigma ATP][\Sigma AMP]}{[\Sigma ADP]^2} = 1.05$$

which corresponds to a ΔG^0_{obs} value of −0.03 kcal/mole. The value of

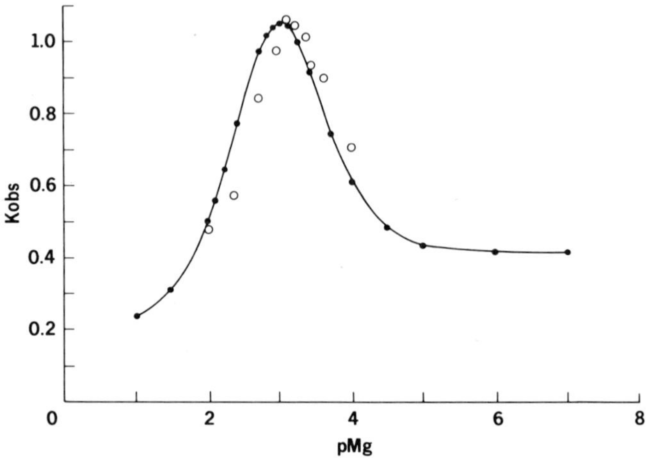

Figure 1. Theoretical and experimental dependence on free $[Mg^{2+}]$ of the K_{obs} for myokinase. The solid curve represents the theoretical relationship, and experimental points are indicated. Computational and experimental methods are described in the text.

$K_{obs(Myk)}$ at free $[Mg^{2+}] = 0$ at pH 6.99 is calculated from Eq. 19 to be 0.391 ($\Delta G^0_{obs} = 0.58$ kcal/mole).

C. Observed Equilibrium Constant for Inorganic Pyrophosphate Hydrolysis

The observed equilibrium constants for the hydrolysis of ATP to ADP and P_i (reaction 2) under the same physiological conditions as the myokinase measurements have been previously determined to be 9.86×10^5 M at $[Mg^{2+}] = 0$ and 2.19×10^5 M at free $[Mg^{2+}] = 10^{-3}$ M (Guynn and Veech, 1973). Values for the hydrolysis of ATP to AMP and PP_i (reaction 3) under identical conditions were found to be 10.8×10^6M ($[Mg^{2+}] = 0$) and 9.98×10^6 M (free $[Mg^{2+}] = 10^{-3}$ M) (Guynn *et al.*, 1974a). Using these values with the myokinase value, the observed equilibrium constants for the hydrolysis of inorganic pyrophosphate under physiological conditions are calculated (Section III) to be $3.52 \times 10^{+4}$ M at $[Mg^{2+}] = 0$ and $5.05 \times 10^{+3}$ M at free $[Mg^{2+}] = 10^{-3}$ M. The corresponding values of ΔG^0_{obs} for this reaction are -6.51 kcal/mole ($[Mg^{2+}] = 0$) and -5.3 kcal/mole (free $[Mg^{2+}] = 10^{-3}$ M). Flodgaard and Fleron (1974) calculated values of ΔG^0_{obs} from equilibrium measurements to be -5.77 kcal/mole (free $[Mg^{2+}]$ $= 0$) and -4.33 kcal/mole (free $[Mg^{2+}] = 3.1 \times 10^{-3}$ M) at $I = 0.20$, free $[K^+] = 167$ mM, and pH 7.4. At zero ionic strength and metal ion concentration, $\Delta G^{0'}$ was calculated to be (-4.98 ± 0.05) kcal/mole. Wood *et al.* (1966) calculated $\Delta G^{0'}_{obs}$ values to be -8.0 ± 0.3 kcal/mole $[Mg^{2+}] = 0)$, -6.4 ± 0.3 kcal/mole (free $[Mg^{2+}] = 0.5 \times 10^{-3}$ M) and -4.5 ± 0.3 kcal/mole (free $[Mg^{2+}] = 5 \times 10^{-3}$ M) at $I = 0.1$, free $[K^{+}] = 10^{-1}$ M, pH 7.0, and 25°C.

D. Observed Equilibrium Constants for the Hexokinase and Phosphofructokinase Reactions at pH 7, free $[Mg^{2+}] = 10^{-3}$ M, $I = 0.25$, and 38°C

Using the value 2.19×10^{5} M for $K_{obs(2)}$ of reaction 2 [Guynn *et al.* (1973)] the observed equilibrium constant for the hexokinase reaction:

$$\text{glucose} + \Sigma\text{ATP} \rightleftarrows \text{ADP} + \Sigma\text{G-6-P}$$

can be calculated for physiological conditions as given by Eq. 9 to be $1.99 \times 10^{+3}$ with $\Delta G^{0'}_{obs}$ of -4.69 kcal/mole. From the total equilibrium concentrations obtained by Robbins and Boyer (1957) for this reaction, the values of $K_{obs(HK)}$ and $\Delta G^{0'}_{obs(HK)}$ are calculated to be $2.62 \times 10^{+3}$ and -4.76 kcal/mole at free $[Mg^{2+}] = 10^{-3}$ M, pH 7.0, 30°C, and unspecified ionic strength. These values are in excellent agreement with the values calculated here.

Similarly, the observed equilibrium constant for the phosphofructokinase reaction:

$$\Sigma\text{F-6-P} + \Sigma\text{ATP} \rightleftarrows \Sigma\text{ADP} + \Sigma\text{FDP}$$

can be calculated for physiological conditions as shown by Eq. 10 to be $1.26 \times 10^{+3}$ and $\Delta G_{obs(PFK)} = -4.42$ kcal/mole. Hanson *et al.* (1973) measured the observed equilibrium constant at pH 8, 30°C, free $[Mg^{2+}] = 4 \times 10^{-3}$ M, and unspecified ionic strength to be $(2.3 \pm 0.3) \times 10^{3}$ ($\Delta G^{0'}_{obs} = -4.66$ kcal/mole).

E. Observed Equilibrium Constants of the Glucose-Inorganic Pyrophosphate and F-6-P-Inorganic Pyrophosphate Phosphotransferase Reactions

Using the calculated value of 5.05×10^{3} M for the K_{obs} of the hydrolysis of inorganic pyrophosphate in combination with the measured constants of $1.10 \times 10^{+2}$ M for the G-6-Pase reaction and $1.74 \times 10^{+2}$ M for the FDPase reaction, the observed equilibrium constants for reactions 5 and 6 as calculated by Eq. 7 and 8 can be determined for physiological conditions:

$$K_{obs}(\text{glucose} - PP_i) = \frac{[\Sigma\text{G-6-P}][\Sigma P_i]}{[\text{glucose}][\Sigma PP_i]} = 45.9$$

$$K_{obs}(\text{F-6-P} - PP_i) = \frac{[\Sigma\text{FDP}][\Sigma P_i]}{[\Sigma\text{F-6-P}][\Sigma PP_i]} = 29.0$$

The corresponding standard free energies are −2.38 kcal/mole for the phosphorylation of glucose and −2.08 kcal/mole for the phosphorylation of F-6-P by inorganic pyrophosphate. No values for these constants have been found in the literature. Table VI is a summary of the observed equilibrium constants and standard free energies measured and calculated for reactions used in this chapter.

F. Effects of Starvation, Streptozotocin, and Diphenyleneiodonium on Rat Liver Metabolites

Table VII gives the results of metabolite determinations in livers of rats subjected to different hormonal and dietary states. As expected, streptozotocin-induced diabetes and starvation yielded similar changes in metabolite levels, since both conditions cause increased gluconeogenesis. Evidence for the increased rate of gluconeogenesis *in vivo* in both conditions is given by the large decrease in lactate (300%). Significant decreases (30–70%) also occurred in the other metabolites of the glycolytic pathway: G-6-P, F-6-P, FDP, DHAP, 3-PG, 2-PG, and PEP; the largest decrease was a tenfold decline in pyruvate. In the diabetic animals there was almost a threefold increase in glucose, whereas in starvation there was a 30% decrease. There were no changes in [citrate], [isocitrate], or [NH_4^+] (not measured in starvation), but there was a significant decrease in [oxaloacetate] (70–80%). As expected there were very large increases in the ketone bodies: [acetoacetate] (7- to 25-fold) and [β-hydroxybutyrate] (15- to 40-fold). However, there was almost complete compensation for the metabolic acidosis as indicated by the return of the intracellular pH value to almost normal in the diabetic group. Although bicarbonate and CO_2 concentrations were not measured in the starved group, one would expect the changes to be similar. As anticipated from the high ATP requirement of gluconeogenesis, the concentration of ATP declined and the concentrations of ADP, AMP, and P_i increased. There were no significant changes in the concentrations of UTP and UDPG, but the calculated concentration of PP_i decreased by 30% in both conditions.

Table VI. Definitions and values of equilibrium constants used in this chapter[a]

Reaction	Definition of K_{obs}	Value of K_{obs}	ΔG^0 (kcal/mole)
Glucose + ΣATP ⇄ ΣG-6-P + ΣADP	$\frac{[\Sigma G\text{-}6\text{-}P][\Sigma ADP]}{[glucose][\Sigma ATP]}$	1990	−4.69
ΣG-6-P ⇄ glucose + ΣP_i	$\frac{[glucose][\Sigma P_i]}{[\Sigma G\text{-}6\text{-}P]}$	110 ± 1 M	−2.90
Glucose + ΣPP_i ⇄ ΣG-6-P + ΣP_i	$\frac{[\Sigma G\text{-}6\text{-}P][\Sigma P_i]}{[glucose][\Sigma PP_i]}$	45.9	−2.37
ΣF-6-P + ΣATP ⇄ ΣFDP + ΣADP	$\frac{[\Sigma FDP][\Sigma ADP]}{[\Sigma F\text{-}6\text{-}P][\Sigma ATP]}$	1260	−4.41
ΣFDP ⇄ ΣF-6-P + ΣP_i	$\frac{[\Sigma F\text{-}6\text{-}P][\Sigma P_i]}{[\Sigma FDP]}$	174 ±7 M	−3.19
ΣF-6-P + ΣPP_i ⇄ ΣP_i + ΣFDP	$\frac{[\Sigma FDP][\Sigma P_i]}{[\Sigma F\text{-}6\text{-}P][\Sigma PP_i]}$	29.0	−2.08
2 ΣADP ⇄ ΣATP + ΣAMP	$\frac{[\Sigma ATP][\Sigma AMP]}{[\Sigma ADP]^2}$	1.05	−0.03
ΣATP ⇄ ΣAMP + ΣPP_i	$\frac{[\Sigma AMP][\Sigma PP_i]}{[\Sigma ATP]}$	$9.98 \times 10^{+6}$ M	−9.96
ΣATP ⇄ ΣADP + ΣP_i	$\frac{[\Sigma ADP][\Sigma P_i]}{[\Sigma ATP]}$	2.19×10^5 M	−7.60
ΣPP_i ⇄ $2\Sigma P_i$	$\frac{[\Sigma P_i]^2}{[\Sigma PP_i]}$	5.05×10^3 M	−5.27

[a] Values for K_{obs} and ΔG^0 are for 0.25 ionic strength, approximately 1 mM free $[Mg^{2+}]$, pH 7.0, and 38°C. The state of liquid H_2O is taken as 1.

Table VII. Metabolite content of freeze-clamped liver[a]

Metabolite	Control (9)		Streptozotocin (7)		Diphenyleneiodonium (4)		Starved 48 hours (6)	
Glucose	6850	±70	18800	±400	11200	±900	4730	±150
Glucose-6-P	158	±5	108	±7	580	±71	74	±10
Glucose-1-P	12.1	±0.1	11.8	±0.7	32	±3	7.4	±0.8
Fructose-6-P	47	±1	26	±2	128	±11	20	±4
Fructose diphosphate (calc.)	1.4	±0.2	0.48	±0.03	1.0	±0.1	0.76	±0.06
Dihydroxyacetone phosphate	34	±2	21	±1	30	±1	26	±2
α-Glycerophosphate	166	±10	456	±51	529	±28	230	±20
3-Phosphoglycerate	309	±9	124	±11	331	±17	140	±20
2-Phosphoglycerate	42	±1	17	±1	59	±5		
Phosphoenolpyruvate	132	±7	57	±6	201	±18	72	±7
Pyruvate	189	±14	15	±1	113	±6	19	±1
Lactate	1090	±40	312	±30	2810	±130	270	±20
Citrate	402	±33	340	±24	123	±8	430	±20
Isocitrate (calc.)	23	±2	19	±1	6.8	±0.5	28	±2
α-Ketoglutarate	296	±10	55	±2	23	±1	120	±7
NH_4^+	577	±24	589	±35	1810	±260		
Malate	549	±56	375	±24	662	±32	360	±30
Oxaloacetate (calc.)	25	±4	4.6	±0.4	6.8	±0.6	7.2	±0.8
HCO_3^-	18200	±500	13700	±500	12600	±800		
CO_2	1140	±30	912	±65	890	±89		
pH (calc.)	7.31		7.29		7.26			
Acetoacetate	92	±6	2310	±380	44	±2	650	±40
3-Hydroxybutyrate	123	±8	4920	±730	142	±7	1820	±130

Table VII. (Continued)

Metabolite	Control (9)		Streptozotocin (7)		Diphenyl-eneiodonium (4)		Starved 48 hours (6)	
ATP	2700	±60	2290	±60	2300	±130	2490	±120
ADP	1060	±40	1740	±30	1340	±70	1380	±80
AMP	235	±21	522	±20	293	±22	343	±24
UTP	429	±24	348	±12	727	±110	360	±12
UDPG	378	±28	416	±25	385	±43	356	±24
P_i	3810	±130	4580	±110	4620	±27	4840	±170
PP_i	3.3	±0.5	2.2	±0.2	13.5	±2.2	2.2	±0.2

[a] Values are reported as nanomoles per gram of fresh weight ±SEM. The number of animals is given in parentheses. The values for fructose diphosphate, pH, isocitrate, oxaloacetate, and pyrophosphate were calculated as described in the text.

Also shown in Table VII are the metabolite changes after treatment with diphenyleneiodonium, a known inhibitor of gluconeogenesis (Hoiland *et al.* 1973). Evidence that this decreased gluconeogenesis and increased glycolysis *in vivo* is the threefold rise in lactate. In general, the changes in the other glycolytic intermediates under conditions of glycolysis were opposite to those described for conditions of gluconeogenesis. These were increases in the concentrations of glucose (85%), G-6-P (400%), G-1-P (260%), F-6-P (27%), 2PG (40%), PEP (50%), no change in FDP, DHAP, 3-PG, and a decrease in pyruvate (60%). [Citrate] and [isocitrate] decreased 33%, [α-ketoglutarate] decreased thirteenfold, [NH_4^+] increased threefold, and [oxaloacetate] decreased 75%; the [malate] was unchanged. Although there were decreases in the bicarbonate and CO_2 concentrations (45 and 30%), the lactic acidosis was almost completely compensated for as indicated by the return of the pH to normal. The [acetoacetate] decreased significantly; the [α-hydroxybutyrate] remained unchanged. The total ATP concentration decreased with a simultaneous increase in [UTP], [ADP], [P_i], and [PP_i] but there was no change in [AMP] or [UDPG].

It should be noted that the calculated concentration of free FDP is much lower than the measured total concentrations of 15 μM [control meal-fed rats, (unpublished observations)]. The concentration of free FDP was calculated because of the possibility of significant enzyme binding of FDP

(Veech *et al.* 1969a), making the general assumption that aldolase is present in excess relative to PFK and FDPase (Clark *et al.*, 1974). Approximate estimates of the concentrations of aldolase, FDPase, and PFK, the three FDP-utilizing enzymes, can be made from the maximum activities determined by assays and the turnover numbers of the enzymes. By assuming $K_m = K_s$ (= dissociation constant of enzyme-substrate complex) and using this in conjunction with the total concentration of FDP and the concentration of each enzyme, the free [FDP] can be calculated in a manner analogous to the calculation of the free [Mg^{2+}] (Guynn *et al.*, 1973). The results of such calculations (Table VIII) indicate that enzyme binding could account for a minimum of 75% of the total FDP concentration. Therefore, the concentration of free FDP has been calculated from the measured [DHAP] that approximates the free [DHAP] (Veech *et al.*, 1969a).

G. Effects of Starvation, Streptozotocin, and Diphenyleneiodonium on Rat Liver Redox and Phosphorylation States

Starvation, streptozotocin, and diphenyleneiodonium decreased the cytoplasmic [NAD^+]/[NADH] ratio two- to fourfold and the cytoplasmic [$NADP^+$]/[NADPH] ratio four- to fivefold (Table IX). In the mitochondria the ratio [NAD^+]/[NADH] was also decreased 1.6- to 2.5-fold. The ratio [ΣATP]/[ΣADP] × [ΣP_i] decreased approximately 50% under all conditions, whereas the [ΣATP]/[ΣAMP] [ΣPP_i] ratio decreased 1.5- to 2.5-fold upon starvation and streptozotocin and reached a minimum of 7.5-fold lower after diphenyleneiodonium treatment. The [ΣPP_i]/[ΣP_i]^2 ratio decreased by 2.5-fold in the starvation and streptozotocin-treated groups; however, in contrast to the other groups it increased almost threefold in the diphenyleneiodonium-treated group. It is important to note that the [ΣPP_i]/[ΣP_i]^2 ratio changes in opposite directions relative to the control value, depending on whether glycolysis or gluconeogenesis was occurring.

H. Effects of Starvation, Streptozotocin, and Diphenyleneiodonium on the Ratios Measured *in vivo* of the HK, G-6-Pase, PFK, FDPase, and the Glucose-PP_i and F-6-P-PP_i Transphosphorylase Reactions in Rat Liver

Table X compares ratios calculated from measurements of metabolites *in vivo* with the equilibrium constants measured *in vitro*. Those data show that under all conditions the ratios of the hexokinase, glucose-6-phosphates, phosphofructokinase, and fructose-1,6-diphosphatase reactions are 2 to 6

Table VIII. Estimation of enzyme-bound FDP in tissue

Tissue	Enzyme	Enzyme activity (μmole/min/g wet wt of tissue)	Turnover number (μmole substrate/min/μmole enzyme)	Calculated enzyme concentration (μM)	K_m (μm)	Bound [FDP] (μM)	Reference[a]
Rat liver	Aldolase	5.5					(1)
Rabbit liver			460				(2)
Rat liver				12			(1)
Rabbit liver					2.0		(3)
Rat liver	FDPase	6.9	1560			7.8	(4)
				4.4			(5)
					1.0		(6)
						3.5	
Rat liver	PFK	2.5					(7)
Rabbit muscle			3700				(8)
Rat liver				0.7			
Rabbit muscle					420		(8)
						0.0	

[a] References are as follows: (1) Veech *et al.* (1969a); (2) Morse and Horecker (1968); (3) Rutter *et al.* (1963); (4) Wallace and Newsholme (1967); (5) Pogell and McGilvery (1954); (6) Taketa *et al.* (1965); (7) Ashmore and Weber (1968); (8) Hanson *et al.* (1973).

Table IX. Calculation of redox and phosphorylation states and myokinase ratio in rat liver[a]

Ratio	Control (9)	Streptozotocin (7)	Diphenyleneiodonium (4)	Starved 48 hours (6)
Cytoplasmic [NAD]/[NADH]	1560 ±110	450 ±40	360 ±20	720 ±60
Mitochondrial [NAD]/[NADH]	15.5 ±1.2	9.5 ±0.4	6.4 ±0.6	7.7 ±0.4
Cytoplasmic [NADP]/[NADPH]	0.0138 ±0.0014	0.0029 ±0.0002	0.0033 ±0.0002	0.0042 ±0.0004
[ΣATP]/[ΣAMP] [ΣPP$_i$] M^{-1}	5,100,000 ±1.2 × 10^6	2,100,000 ±0.17 × 10^6	690,000 ±0.2 × 10^6	3,500,000 ±0.5 × 10^6
[ΣATP]/[ΣADP] [ΣP$_i$] M^{-1}	695 ±56	291 ±15	389 ±58	366 ±32
[PP$_i$]/[P$_i$]^2 M^{-1}	0.224 ±0.032	0.105 ±0.008	0.598 ±0.077	0.093 ±0.011
[ΣATP] [ΣAMP]/[ΣADP]^2	0.560 ±0.021	0.396 ±0.008	0.380 ±0.032	0.448 ±0.030

[a] Values are means ± SEM for the number of animals listed in parentheses. Methods of calculations are described in the text. Where indicated, units are liters per mole; other values are dimensionless.

Table X. Metabolite ratios determined in freeze-clamped liver compared with value of K_{eq}

Ratio[a]	Value of K_{eq} *in vitro*	Control (9)	Streptozotocin (7)	Diphenyleneiodonium (4)	Starved 48 hours (6)
$\frac{[\Sigma G\text{-}6\text{-}P]\,[\Sigma ADP]}{[glucose]\,[\Sigma ATP]}$	1990	0.0092 ±0.0007	0.0043 ±0.0003	0.030 ±0.003	0.0087 ±0.0005
$\frac{[glucose]\,[\Sigma P_i]}{[\Sigma G\text{-}6\text{-}P]}$ M	110	0.161 ±0.012	0.817 ±0.047	0.094 ±0.003	0.334 ±0.0005
$\frac{[\Sigma G\text{-}6\text{-}P]\,[\Sigma P_i]}{[glucose]\,[\Sigma PP_i]}$	45.9	33.7 ±5.9	12.1 ±0.8	18.8 ±2.7	32.9 ±3.1
$\frac{[\Sigma FDP]\,[\Sigma ADP]}{[\Sigma F\text{-}6\text{-}P]\,[\Sigma ATP]}$	1260	0.011 ±0.001	0.014 ±0.001	0.0048 ±0.0004	0.021 ±0.003
$\frac{[\Sigma F\text{-}6\text{-}P]\,[\Sigma P_i]}{[\Sigma FDP]}$ M	174	0.149 ±0.017	0.251 ±0.019	0.576 ±0.074	0.127 ±0.015
$\frac{[\Sigma FDP]\,[\Sigma P_i]}{[\Sigma F\text{-}6\text{-}P]\,[\Sigma PP_i]}$	29.0	37.3 ±3.8	41.4 ±5.4	3.2 ±0.8	83.6 ±9.0

[a] Valued given are means ±SEM for the number of animals listed in parentheses. Where indicated, units are moles per liter; otherwise the values are dimensionless.

orders of magnitude less than the corresponding equilibrium constants. On the other hand, the glucose-PP_i and F-6-P-PP_i transphosphorylase metabolite ratios in all conditions are in good agreement with the equilibrium constants *in vitro*.

V. DISCUSSION

As we see from data in Table X, the ratios of [substrates]/[products] for the hexokinase, glucose-6-phosphatase, phosphofructokinase, and fructose diphosphatase reactions are 250 to 10^6 times lower than the values predicted from the equilibrium constants determined *in vitro*. For those enzymes, wide displacement from the equilibrium values might be the expected result. However, it is surprising, indeed, that the observed ratios for

$$[\Sigma\text{G-6-P}][\Sigma P_i]/[\text{glucose}][\Sigma PP_i]$$

and

$$[\Sigma\text{FDP}][\Sigma P_i]/[\Sigma\text{F-6-P}][\Sigma PP_i]$$

agree closely with the calculated equilibrium ratios (Table X). In four cases the observed and calculated ratios differ by only a factor of 0.4 or less; in three other instances the factor is 2 to 4, and in a single case ([ΣFDP] [ΣP_i]/[ΣF-6-P] [ΣPP_i] for diphenyleneiodonium-treated animals) the factor is 9. Thus in seven out of eight comparisons, the agreement between observed and calculated ratios is remarkably good. In preliminary experiments similar results have been obtained with normal and ischemic muscle and normal brain.

Although transphosphorylation reactions involving PP_i and glucose or F-6-P are apparently in states of near-equilibrium *in vivo*, these results must be interpreted with caution. The appearance of near-equilibrium in these experiments could simply be fortuitous. That seems unlikely, however, in view of the number of metabolite measurements involved in making the final comparisons, and the variety of physiological conditions employed. The greatest problem in interpreting these results concerns the enzymatic mechanisms by which the near-equilibria are achieved.

Near-equilibrium in the [ΣG-6-P] [ΣP_i]/[glucose] [ΣPP_i] ratio is unlikely to result from the glucose-6-phosphatase transphosphorylation activity (Nordlie and Arion, 1964). The major reservations concerning the significance of that activity *in vivo* are: (*a*) the K_m for PP_i is about three orders of magnitude higher than the [PP_i] *in vivo*; (*b*) the pH profile for the transphosphorylase activity shows an optimum in the range of pH 4 to 5, and the predicted activity at physiological pH's would be insignificantly low. Nordlie (1974) has recently discussed these and other reservations in detail. Our conclusion in this regard is that the observed near-equilibrium value of the [ΣG-6-P] [ΣP_i]/[glucose] [ΣPP_i] ratio *in vivo* indicates the operation of an efficient transphosphorylase enzyme or enzyme system, but further work is required to decide which enzyme or series of enzymes acts as the catalyst.

In mammalian tissues there is no known catalyst for the reaction: F-6-P + $PP_i \rightleftarrows$ FDP + P_i. However, Reeves *et al.* (1974) have recently reported that *Entamoeba histolytica* contains an enzyme for that reaction. On the basis of size and stability properties, the PP_i-utilizing enzyme is distinct from the ATP-utilizing phosphofructokinase, but it is not yet certain (R. E. Reeves, personal communication) whether it also catalyzes the FDPase reaction (FDP $\rightleftarrows$ F-6-P + P_i). At pH 7.0 the K_m for PP_i is 14 μM, which is of the same order as rat liver [PP_i] (Table VII). Experiments are under way to see whether a similar enzyme can be detected in rat liver.

In has been proposed (Atkinson, 1968) that energy charge ([ATP] + [0.5 ADP]/[ATP] + [ADP] + [AMP]) affects specific enzymes and determines the rates and directions of flux in pathways such as glycolysis and gluconeogenesis. Since calculation of *Keq* for the glucose- and fructose-6-phosphate-transphosphorylases (Eq. 7 and 18) requires the use of Eq. 2, 3, and 4, which contain the parameters of energy charge, it might be asked whether the near-equilibria in the transphosphorylase processes can be accounted for by energy charge. On inspection of the data in Table VII, however, it is clear that energy charge does not correlate with the levels of glycolytic-gluconeogenic intermediates. In particular, there is no consistent relationship between energy charge (0.81, 0.69, 0.76, 0.75) and the [F-6-P]/[FDP] (33.6, 54.2, 128, 26.4) or [glucose]/[G-6-P] (43.4, 174, 19.3, 63.9) ratios.

Although the experimental results reported here show that near-equilibrium is maintained *in vivo* for the [ΣG-6-P][ΣP_i]/[glucose] [ΣPP_i] and [ΣFDP][ΣP_i]/[ΣF-6-P][ΣPP_i] ratios, we view these results primarily as suggestions for additional experiments. Yet if, as indicated by these data, those near-equilibria are characteristic of mammalian metabolism, several significant consequences follow.

First, substrate cycling as described by Newsholme and Gevers (1967) and studied by Clark *et al.* (1974) and Rognstad *et al.* (1973) is "futile" in that cycling involves net hydrolysis of ATP to ADP + P_i with a resultant energy loss of approximately 11 kcal/mole in the cell. The existence of efficient enzymes for the transphosphorylase process could account for the isotopic labeling data found by Clark *et al.* and Rognstad. Such cycling would occur without energy loss. Second,changes in the steady-state ratio of [F-6-P]/[FDP] would be a simple result of changes in the $[PP_i]/[P_i]$ ratio rather than a consequence of complex, coordinate allosteric controls acting on phosphofructokinase and fructose diphosphatase (Exton, 1972). Similarly, coordination between the [glucose]/[G-6-P] and [F-6-P]/[FDP] couples for glycolysis and gluconeogenesis would result from the common interaction of those couples with the $[PP_i]/[P_i]$ couple. Finally, perhaps the most far-reaching consequence of these results concerns the activity of the liver in maintaining blood glucose levels. Since glucose freely permeates the liver cell membrane, the existence of near-equilibrium in the hepatic $[\Sigma\text{G-6-P}][\Sigma P_i]/[\text{glucose}][\Sigma PP_i]$ ratio would suggest that uptake of blood glucose by the liver and hepatic release of glucose into the blood would ultimately be determined by only three factors: glucose-6-phosphate, PP_i, and P_i concentrations within the liver.

VI. SUMMARY

1. The equilibrium constants of the following reactions have been determined at 0.25 ionic strength, approximately 1 mM free Mg^{2+} and 38°C:

$$2\Sigma\text{ADP} \rightleftarrows \Sigma\text{ATP} + \Sigma\text{AMP}$$

$$\Sigma\text{fructose diphosphate} \rightleftarrows \text{fructose-6-phosphate} + \Sigma P_i$$

$$\Sigma\text{glucose-6-phosphate} \rightleftarrows \text{glucose} + \Sigma P_i$$

2. Using the equilibrium constants for the reactions listed in paragraph 1, and the previously determined ΔG^0 values for $\Sigma\text{ATP} \rightleftarrows \Sigma\text{ADP} + \Sigma P_i$ and $\Sigma\text{ATP}\ \Sigma\text{AMP} + \Sigma PP_i$, the following equilibrium constants were calculated:

$$\Sigma PP_i \rightleftarrows 2\Sigma P_i$$

$$\Sigma\text{fructose-6-phosphate} + \Sigma\text{ATP} \rightleftarrows \Sigma\text{fructose diphosphate} + \Sigma\text{ADP}$$

$$\Sigma\text{glucose} + \Sigma\text{ATP} \rightleftarrows \Sigma\text{glucose-6-phosphate} + \Sigma\text{ADP}$$

$$\Sigma\text{fructose-6-phosphate} + \Sigma PP_i \rightleftarrows \Sigma\text{fructose diphosphate} + \Sigma P_i$$

$$\Sigma\text{glucose} + \Sigma PP_i \rightleftarrows \Sigma\text{glucose-6-phosphate} + \Sigma P_i$$

3. The intracellular concentration of free cytoplasmic PP_i was calcu-

lated from the UDPG-pyrophosphorylase reaction using the relationship

$$PP_i = \frac{1}{K_{UDPG-PP_i\,ase}} \frac{[UTP][glucose\text{-}1\text{-}phosphate]}{[UDPG]}$$

where $K_{UDPG-PP_i\,ase} = 4.5$.

4. Adenine nucleotides and the metabolites of glycolysis were measured in freeze clamped rat liver in three different physiological stages, which were taken to represent different rates of flux through the glycolytic and gluconeogenic pathways.

5. The [glucose]/[glucose-6-phosphate] and [fructose-6-phosphate]/[fructose diphosphate] ratios measured in freeze-clamped tissue are far from equilibrium when compared with the measured ratios of [ATP]/[ADP]. In contrast, the measured ratios of [glucose]/[glucose-6-phosphate] and [fructose-6-phosphate]/[fructose diphosphate] appear to be in near-equilibrium with the measured $[PP_i]/[P_i]$ as judged by comparison of the measured ratios with the equilibrium values obtained *in vitro*. The indicated near-equilibrium between the [glucose]/[glucose-6-phosphate] and [fructose-6-phosphate]/[fructose diphosphate] couples and the $[PP_i]/[P_i]$ couple might be a fortuitous result; that seems unlikely, however, in view of the large changes in metabolite concentrations and the variety of physiological states under which these measurements were made. If these transphosphorylase reactions are in a state of near equilibrium, several consequences follow.

a. The existence of near-equilibrium at glucose-PP_i transphosphorylation and fructose-6-phosphate-PP_i transphosphorylation reactions suggests that the substrate cycling observed by Lardy and co-workers can occur without energy loss.

b. Since glucose freely permeates the liver cell membrane, these data further suggest that the uptake of glucose from the blood and release of glucose from the liver are ultimately determined by only three factors: glucose-6-phosphate, PP_i, and P_i concentrations in the liver.

c. Likewise, changes in the steady-state ratio of [fructose-6-phosphate]/[fructose diphosphate] would be viewed as resulting simply from changes in the $[PP_i]/[P_i]$ ratio rather than as a consequence of complex, coordinate inhibitions of phosphofructokinase and fructose diphosphatase.

d. Their common interaction with the $[PP_i]/[P_i]$ couple would coordinate the glucose/glucose-6-phosphate and fructose-6-phosphate/fructose diphosphate couples for glycolysis and gluconeogenesis.

e. The apparent maintenance of near-equilibrium in these transphosphorylase reactions suggests the evidence of efficient enzymes or sequences of enzymes capable of using PP_i to phosphorylate glucose and fructose-6-phosphate.

f. By analogy to the three previously suggested pyridine nucleotide couples, these results indicate the existence of three phosphorylation states that have different ΔG values and account for important regulatory functions within the cell:

		Values (kcal/mole)	
		ΔG^0	ΔG in cell
$\frac{[NAD]}{[NADH]}$	Cytoplasm	−14.7	−8.7
$\frac{[NAD]}{[NADH]}$	Mitochondria	−14.7	−12.9
$\frac{[NADP]}{[NADPH]}$	Cytoplasm	−14.7	−18.9
$\frac{[ATP]}{[ADP] \times [P_i]}$	Cytoplasm	−7.6	−11.7
$\frac{[ATP]}{[AMP] \times [PP_i]}$	Cytoplasm	−10.0	−19.6
$\frac{[PP_i]}{[P_i]^2}$	Cytoplasm	−5.3	−4.4

REFERENCES

Ashmore, J. and Weber, G. (1968). In *Carbohydrate Metabolism and its Disorders* Vol. I (F. Dickens, P. J. Randle, and W. J. Whelan, eds.), Academic Press, New York, pp. 335–374.

Atkinson, D. E. (1968). *Biochemistry* **7,** 4030–4034.

Atkinson, M. R., Johnson, E., and Morton, R. K. (1961). *Biochem. J.* **79,** 12–15.

Blair, J. McD. (1970). *Eur. J. Biochem.* **13,** 384–390.

Bloxham, D. P. and Lardy, H. A. (1973). In *The Enzymes* Vol. 8, (P. D. Boyer, ed.), Academic Press, New York, pp. 239–278.

Bowen, W. J. and Kerwin, T. D. (1956). *Arch. Biochem. Biophys.* **64,** 278–284.

Clark, M. G., Bloxham, D. P., Holland, P. C., and Lardy, H. A. (1974). *J. Biol. Chem.* **249,** 279–290.

Exton, J. H. (1972). *Metabolism* **21,** 945–990.

Flodgaard, H. and Fleron, P. (1974). *J. Biol. Chem.* **249,** 3465–3474.

Ginodman, L. M. (1954). *Biokhimiia* **19,** 666–676.

Guuma, K. A., McLean, P., and Greenbaum, A. L. (1971). *Essays Biochem.* **7,** 39–86.

Guynn, R. W., Veloso, D., and Veech, R. L. (1972). *Anal. Biochem.* **45,** 277–285.

Guynn, R. W. and Veech, R. L. (1973). *J. Biol. Chem.* **248,** 6966–6972.

Guynn, R. W., Veloso, D., Lawson, J. W. R., and Veech, R. L. (1974). *Biochem. J.* **140,** 369–375.

Guynn, R. W., Webster, L. T., Jr., and Veech, R. L. (1974a). *J. Biol. Chem.* **249,** 3248–3254.

Hanson, R. L., Rudolph, F. B., and Lardy, H. A. (1973). *J. Biol. Chem.* **248,** 7852–7859.

Holland, P. C., Clark, M. G., Bloxham, D. P., and Lardy, H. A. (1973). *J. Biol. Chem.* **248,** 6050–6056.

Junod, A., Lambert, A. E., Stauffacher, W., and Renold, A. E. (1969). *J. Clin. Invest.* **48,** 2129–2139.

Krebs, H. A., Dierks, C., and Gascoyne, T. (1964). *Biochem. J.* **93,** 112–121.

Krebs, H. A. and Veech, R. L. (1969). In *The Energy Level and Metabolic Control in Mitochondria,* (S. Papa, J. M. Tager, E. Quagliariello, and E. C. Slater, eds.), Adriatica Editrice, Bari, Italy, pp. 329–382.

Krebs, H. A. and Veech, R. L. (1969a). *Adv. Enzyme Regul.* **7,** 397–413.

Krebs, H. A. (1972). *Essays Biochem.* **8,** 1–34.

Mansour, T. E. (1972). *Curr. Top. Cell. Regul.* **5,** 1–46.

Meyerhof, O. and Green, H. J. (1949). *J. Biol. Chem.* **178,** 655–667.

Morse, D. E. and Horecker, B. L. (1968). In *Advances in Enzymology,* Vol. 31, Interscience, New York, pp. 125–181.

Newsholme, E. A. and Gevers, W. (1967). *Vitam. Horm.* **25,** 1–87.

Nordlie, R. C. and Arion, W. J. (1964). *J. Biol. Chem.* **239,** 1680–1685.

Nordlie, R. C. and Lygre, D. G. (1966). *J. Biol. Chem.* **241,** 3136–3141.

Nordlie, R. C. (1974). *Curr. Top. Cell. Regul.* **8,** 33–117.

Passonneau, J. V. and Lowry, O. H. (1964). *Adv. Enzyme Regul.* **2,** 265–274.

Paetkau, V. and Lardy, H. A. (1967). *J. Biol. Chem.* **242,** 2035–2042.

Pogell, B. M. and McGilvery, R. W. (1954). *J. Biol. Chem.* **208,** 149–154.

Ramaiah, A. (1974). *Curr. Top. Cell. Regul.* **8,** 297–345.

Reeves, R. E., South, D. J., Blytt, H. J., and Warren, L. G. (1974). *J. Biol. Chem.* **249,** 7737–7741.

Robbins, E. A. and Boyer, P. D. (1957). *J. Biol. Chem.* **224,** 121–135.

Rognstad, R., Clark, D. G., and Katz, J. (1973). *Biochem. Biophys. Res. Commun.* **54,** 1149–1156.

Rutter, W. J., Woodfin, B. M., and Blostein, R. E. (1963). *Acta Chem. Scand.* **17,** Suppl. 1, 226–232.

Scrutton, M. C. and Utter, M. F. (1968). *Ann. Rev. Biochem.* **37,** 249–302.

Sols, A. (1968). In *Carbohydrate Metabolism and its Disorders,* Vol. I (F. Dickens, P. J. Randle, and W. J. Whelan, eds.), Academic Press, New York, pp. 53–87.

Taketa, K. and Pogell, B. M. (1965). *J. Biol. Chem.* **240,** 651–662.

Treveleyan, W. E., Mann, P. F. E., and Harrison, J. S. (1952). *Arch. Biochem. Biophys.* **39,** 440–449.

Van Slyke, D. D. and Neill, J. M. (1924). *J. Biol. Chem.* **61,** 523–573.

Veech, R. L., Eggleston, L. V., and Krebs, H. A. (1969). *Biochem. J.* **115,** 609–619.

Veech, R. L., Raijman, L., Dalziel, K., and Krebs, H. A. (1969a). *Biochem. J.* **115,** 837–842.

Veech, R. L., Guynn, R. W., and Veloso, D. (1972). *Biochem. J.* **127,** 387–397.

Veloso, D., Passonneau, J. V., and Veech, R. L. (1972). *J. Neurochem.* **19,** 2679–2686.

Veloso, D., Guynn, R. W., Oskarsson, M., and Veech, R. L. (1973). *J. Biol. Chem.* **248,** 4811–4819.

Wallace, J. C. and Newsholme, E. A. (1967). *Biochem. J.* **104,** 378–384.

Weber, G., Lea, M. A., Hird-Convery, H. J., and Stamm, N. B. (1967). *Adv. Enzyme Regul.* **5,** 257–298.

Weil-Malherbe, H. and Bone, A. D. (1951). *Biochem. J.* **49,** 339–347.

Williamson, D. H., Lund, P., and Krebs, H. A. (1967). *Biochem. J.* **103,** 514–527.

Williamson, J. R., Scholz, R., Browning, E. T., Thurman, R. G., and Fukami, M. H. (1969). *J. Biol. Chem.* **244,** 5044–5054.

Wood, H. G., Davis, J. J., and Lochmüller, H. (1966). *J. Biol. Chem.* **241,** 5692–5704.

PART IV

Gluconeogenesis in the Human

CHAPTER FOURTEEN

Gluconeogenesis and Its Disorders in Man

NEIL B. RUDERMAN, T. T. AOKI,
AND GEORGE F. CAHILL, JR.

I. INTRODUCTION

The pathophysiology of glucose homeostasis has been best studied in man for a number of reasons: availability, size, standardized techniques for blood sampling, and of course, interest in human disease. This brief chapter outlines certain aspects of the regulation of glucose homeostasis, and in particular gluconeogenesis in man. A description of some disease states in which gluconeogenesis is altered follows.

Normal man, whether eating or fasting, maintains his blood glucose concentration between 2.5 and 7.5 mm. This relatively narrow range can be maintained even with a glucose intake as high as 4 moles/day, a remarkable feat when one considers that total body free glucose is approximately 0.1 mole. Other metabolic fuels such as fatty acids, which can vary in concentration by more than 1 order of magnitude, and particularly ketoacids, which can vary by 2 to 3 orders of magnitude, are obviously less finely regulated. One might ask why the concentration of glucose is

Supported in part by NIH grants AM-15191, AM-15272, and the Clinical Research Center at the Peter Bent Brigham Hospital, FR-31. Neil B. Ruderman is the recipient of Research Career Development Award, AM-15729 from NIH. T. T. Aoki is a fellow, The Medical Foundation, Inc., Boston.

confined to such a narrow range, and only the classical teleologic answers can be given: at very low concentrations, transport across the blood-brain barrier and into brain cells becomes rate limiting for glucose uptake and, as a result, cerebral function is impaired for lack of fuel. Likewise, at very high concentrations, glucose causes an osmotic diuresis accompanied by loss of water and electrolyte, as well as glucose, in the urine. In some individuals, in whom renal glucosuria is not adequate to limit the increase in blood glucose, central nervous system depression and even unconsciousness may result, the syndrome of hyperglycemic hyperosmolar coma (Arieff and Carroll, 1972).

II. THE FIVE PHASES OF GLUCOSE HOMEOSTASIS

A. Absorptive Phase

To relate gluconeogenesis to the broader problem of glucose homeostasis, we have divided the latter into five phases based on the origin of the glucose in blood. This is depicted graphically in Fig. 1, in which the magnitude of glucose utilization is plotted against time in a theoretical man who ingested 100 glucose and then fasted for 40 days. For 3 to 4 hours after glucose ingestion (phase 1: the absorptive period), blood glucose is derived principally from exogenous carbohydrate. The concentrations of insulin and glucose are elevated and glucagon is depressed (Table I). Glucose in excess of the fuel needs of the liver and peripheral tissues is stored as glycogen, primarily in liver or muscle, or is converted to lipid. This is the only phase in which the liver is a net user of glucose and gluconeogenesis is of little consequence for glucose homeostasis.

B. Postabsorptive Phase and Early Starvation

Toward the end of the absorptive phase, insulin and glucagon return to basal levels and the liver is called on to produce glucose, which it derives principally from stored glycogen (phase II). The major user of glucose during this postabsorptive period is the brain, which oxidizes exclusively glucose, as in the absorptive phase. Obligate glycolizers such as the red cell and the renal medulla also use glucose during this period, as do muscle and adipose tissue, which use it at a rate that is lower than the phase I rate (Butterfield and Whichelow, 1965; Owen and Reichard, 1971; Jackson *et al.*, 1973). Liver glycogen provides most of the glucose present in the blood during phase II; however, it is used quite rapidly during the first 24 hours of a fast (Son Nilsson, 1973; Son Nilsson and Hultman, 1973). The 90 g of

Table I. Hormone and substrate levels in man[a]

	N Excretion (g/day)		Insulin (μU/ml)	Glucagon (pg/ml)	FFA (μM)	Alanine (μM)	Pyruvate (μM)	Glucose (mg/100 ml)
	Total	Urea						
Glucose[a] 700 g/day i.v.	2.9	1.2	40	80	120	800	250	100
Glucose[a] 150 g/day i.v.	4.1	3.1	20	80	600	400	80	90
Postabsorptive, 12 hours	15	13	15	100	600	300	60	80
Starvation, 3 days	12	10	8	150	1000	250	45	70
Starvation, 5–6 weeks	4.5	2.5	6	120	1200	125	35	65
Diabetic ketoacidosis	20	17	< 5	500	2500	150	90	500

[a] Adapted from O'Connell *et al.* (1974). Glucose was infused into three normal subjects for 10 days. Other data are average norms from the Joslin Research Laboratory collected from normal weight subjects except for data for 5 to 6 weeks starvation, which was derived from obese subjects. The true insulin value in these is approximately 10 μU/ml, but it was arbitrarily reduced to 6 μU/ml to keep this value in proportion to insulin levels in the more briefly starved normal weight individuals.

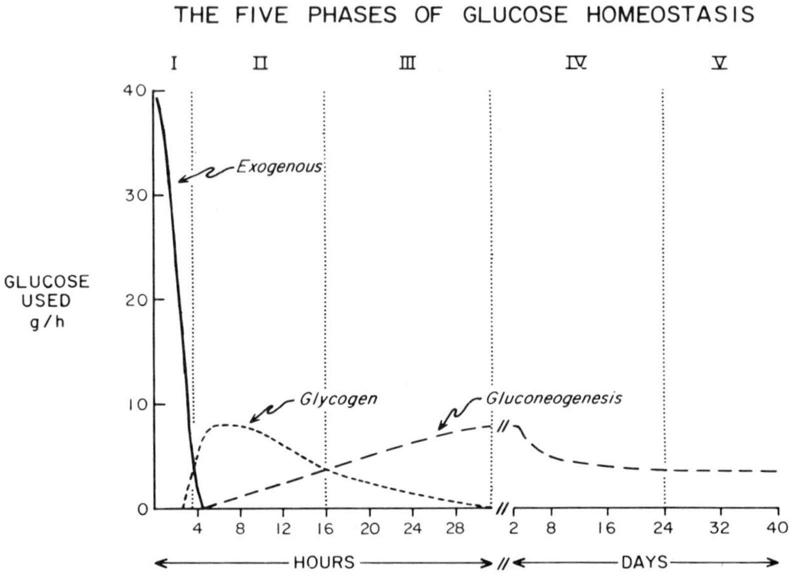

		(I)	(II)	(III)	(IV)	(V)
ORIGIN OF BLOOD GLUCOSE	:	Exogenous	Glycogen Hepatic gluconeogenesis	Hepatic gluconeogenesis Glycogen	Gluconeogenesis, hepatic and renal	Gluconeogenesis, hepatic and renal
TISSUES USING GLUCOSE	:	All	All except liver. Muscle and adipose tissue at diminished rates	All except liver. Muscle and adipose tissue at rates intermediate between II and IV	Brain, rbcs, renal medulla. Small amount by muscle	Brain at a diminished rate, rbcs, renal medulla
MAJOR FUEL OF BRAIN	:	Glucose	Glucose	Glucose	Glucose, ketone bodies	Ketone bodies, glucose

Figure 1. Magnitude of glucose utilization versus time in the five phases of glucose homeostasis.

glycogen present in the liver after an overnight fast is sufficient to meet the requirements of peripheral tissues (about 180 g/day, of which 120 g is oxidized by brain) for only half a day; therefore, gluconeogenesis soon replaces glycogen as the major source of blood glucose. It is during this period—phase III, early starvation—and the early part of phase IV, when glycogen stores are depleted and the brain has not yet begun to utilize ketone bodies in significant quantities—that the demand for gluconeogenesis is at its peak and susceptibility to hypoglycemia because of impaired gluconeogenesis is greatest. As we discuss below, children are more susceptible to hypoglycemia than adults, since they use their glycogen stores and become dependent on gluconeogenesis more rapidly (i.e., phase II is shorter).

C. Intermediate Starvation

The fourth phase of glucose homeostasis, which can be referred to as the intermediate phase of starvation, has been studied predominantly in obese subjects who are fasting for therapeutic purposes (Owen *et al.*, 1967; Cahill and Owen, 1968). It is a transition state during which gluconeogenesis is the sole source of blood glucose. Blood and cerebrospinal fluid levels of acetoacetate and 3-hydroxybutyrate gradually increase (see Tables I and II), and the brain begins to use these substrates as a fuel in place of glucose. Simultaneously the release of gluconeogenic and other amino acids from muscle decreases, leading to a decrease in their blood levels, and *pari passu,* because of the low concentrations of substrate, to a decrease in hepatic gluconeogenesis (Felig *et al.*, 1970). These changes in amino acid metabolism take place at a time when circulating hormone levels are relatively constant (Marliss *et al.*, 1970, see Table I). It is not clear why they occur, although the reason appears to be related to the accelerated rate of fatty acid metabolism in muscle at this time (Cahill *et al.*, 1972). Hypoglycemia does not occur during this phase because the diminished rate of hepatic gluconeogenesis is balanced by a decreased rate of glucose utilization by brain and perhaps other tissues, and by a moderate increase in renal glucose production. Thus the net result of this adaptation is conservation of muscle protein at the expense of adipose tissue triglyceride, which supplies the fatty acid for ketogenesis. Since more than 80 g/day of protein would be required to maintain gluconeogenesis at the same rate as in the first 3 to 5 days of a fast, the survival value of this adaptation is self-evident.

Despite the critical importance of the intermediate phase in the adaptation to starvation, there is relatively little information concerning the magnitude of gluconeogenesis during the initial 4 to 5 days of a fast. Presumably the brain uses almost exclusively glucose during the first 2 to 3 days, since the concentrations of acetoacetate and 3-hydroxybutyrate are not substantially increased (Cahill *et al.*, 1966). Therefore, gluconeogenesis must occur at a minimum rate of 160 g/day—120 g to supply the needs of the brain plus another 40 g for the Cori cycle.

By the fourth or fifth day, however, the concentration of ketone-bodies has increased to 3 to 3.5 mm, about half the level found after 40 days of fasting when brain glucose utilization is diminished to 40 g/day (see Table II). Thus by this time ketone bodies probably account for a significant proportion of the brain's fuel needs, and some decrease in cerebral glucose

Table II. Substrate levels in fasting man[a]

			Levels (mg/100 ml)				
Substrate	Source	Units	Postabsorptive	3 days	7 days	10 days	28–42 days
Glucose	Blood	mM	4.8 ±0.3 (17)	3.8 ±0.1 (6)	3.6 ±0.6 (6)	3.8 ±0.2 (11)	3.6 ±0.1 (32)
Free fatty acids	Blood	mM	0.61±0.03 (17)	1.18±0.06 (17)	1.14±0.05 (6)	1.36±0.11 (11)	1.44±0.08 (32)
Acetoacetate	Blood	mM	0.02±0.01 (17)	0.45±0.04 (17)	0.95±0.08 (6)	1.00±0.11 (11)	1.32±0.09 (32)
β-Hydroxybutyrate	Blood	mM	0.05±0.01 (17)	1.36±0.13 (17)	3.58±0.26 (6)	4.30±0.40 (11)	6.04±0.21 (32)
Lactate	Blood	mM	0.71±0.08 (12)	0.74±0.07 (10)	0.60±0.04 (12)	0.64±0.05 (10)	0.61±0.02 (10)
Pyruvate	Blood	μM	61 ±6 (13)	42 ±5 (10)	36 ±3 (12)	36 ±3 (10)	35 ±2 (10)
α-Ketoglutarate	Blood	μM	12 ±1 (12)	10 ±1 (10)	9 ±1 (12)	10 ±1 (10)	11 ±1 (9)

Glycerol	Blood	μM	134 ±15 (7)	189 ±31 (6)	160 ±20 (8)	171 ±17 (6)	182 ±19 (8)
Blood urea nitrogen	Blood	mg/100 ml	10.9 ±0.5 (6)	16.2 ±1.4 (6)	13.0 ±1.3 (6)	—	5.2 ±0.9 (7)
Glutamate	Blood	μM	127 ±7 (10)	114 ±4 (10)	102 ±5 (12)	105 ±7 (9)	92 ±7 (10)
	Plasma	μM	42 ±4 (12)	28 ±4 (10)	23 ±2 (12)	18 ±2 (10)	18 ±2 (10)
Glutamine	Blood	μM	590 ±33 (10)	515 ±29 (10)	464 ±36 (11)	456 ±28 (10)	476 ±33 (9)
	Plasma	μM	486 ±41 (10)	443 ±54 (8)	385 ±47 (10)	380 ±36 (8)	385 ±32 (9)
Alanine	Plasma	μM	344 ±29 (10)	321 ±38 (3)	—	—	140 ±11 (7)
Leucine	Plasma	μM	112 ±4 (10)	152 ±29 (3)	—	—	71 ±7 (7)
Glycine	Plasma	μM	215 ±8 (10)	189 ±17 (3)	—	—	372 ±14 (7)

[a] Venous levels of substrates from obese subjects undergoing therapeutic starvation. Number of subjects indicated in parentheses.

Table III. Gluconeogenic precursor during first 5 days of a fast[a]

Origin	Quantity used (g/5 days)	Reference
Muscle and liver protein	240	Owen *et al.* (1969)
Muscle glycogen	150	Hultman and Bergstrom (1967)
Adipose tissue glyceride glycerol	85	Cahill and Owen (1968)

[a] In addition to the above, approximately 40 g/day of glucose would be synthesized by way of the Cori cycle. Liver glycogen, approximately 90 g, would be utilized, almost entirely during the first day.

utilization and gluconeogenesis would be expected. Isotope studies suggest that glucose turnover is diminished by 33% in obese subjects after a 7 day fast (Kreisberg *et al.*, 1970).

Table III lists the major gluconeogenic precursors utilized during the first 5 days of a fast. The importance of muscle and liver protein and glycerol derived from the hydrolysis of adipose tissue triglyceride has been noted previously (Cahill and Owen, 1968). Less emphasisis has been placed on muscle glycogen. Hultman and Bergstrom (1967) and others have demonstrated that there are approximately 15 g of glycogen per kilogram skeletal muscle, or a total of about 420 g in a 70 kg man, after an overnight fast. Studies on a limited number of patients indicate that starvation or ingestion of a low-carbohydrate, fat-free diet leads to a 33% decrease in these stores by 5 days. Since lactate and alanine release from muscle are at their peak during this period, it seems likely that a significant proportion of the carbon carried by these gluconeogenic precursors is derived from muscle glycogen.[1]

D. Prolonged Starvation

After approximately 24 days of fasting, blood levels of acetoacetate and β-hydroxybutyrate do not increase further and urine nitrogen excretion (urea-

[1] Alanine derived from glycogen or exogenous glucose would not be adequately accounted for if one calculated gluconeogenesis from amino acid by assuming 3.9 g of amino acid carbon are converted to glucose per g urea-*n* excreted. This is because the factor 3.9 assumes that the composition of the amino acid mix metabolized by liver resembles that of meat protein (see Krebs, 1964).

$N + NH_3$) plateaus at about 4 g/day (phase V: prolonged starvation). At this time ketone bodies account for 60% of the fuel needs of the brain (Owen *et al.*, 1967) and hepatic and renal gluconeogenesis together approximate 80 g/day half the maximum rate observed in phases III and IV (Owen *et al.*, 1969). Of the 80 g of glucose generated by gluconeogenesis, half is derived from recycling of lactate and pyruvate, mainly from red cells, but also from other tissues, including brain. The other half is from both glycerol released from adipose tissue and from amino acids. Amino acids are a far less important source than in the earlier phases of starvation, as evidenced by the decrease in urine nitrogen excretion (Table I). The release of alanine and other gluconeogenic amino acids from muscle may be diminished by as much as 80% (Felig *et al.*, 1970).

Whether all glucose utilization by brain can be displaced is presently not known. Both our group and that of Drenick (1972) have given insulin to patients starved for 4 weeks or more and have lowered glucose concentrations from the usual fasting level of 3.5 to 4 mM to as low as 1 mM without subjective or objective symptoms or signs. This does not mean, however, that the remaining glucose may not be essential. Studies on brain slices have shown that for the optimal oxidation of acetoacetate and β-hydroxybutyrate, a low concentration of glucose may be necessary to provide a continuous input into the tricarboxylic acid by way of pyruvate carboxylation (Itoh and Quastel, 1970; Ide *et al.*, 1969). In this regard it is noteworthy that the 4 to 5 g of nitrogen lost in the urine in phase V is not a minimum nitrogen excretion, since administration of carbohydrate is able to reduce this to 1 to 2 per day (O'Connell *et al.*, 1974), suggesting from a teleologic viewpoint that the slightly increased rate of gluconeogenesis over this baseline that occurs in total starvation may be providing needed glucose substrate.

III. REGULATION OF GLUCOSE HOMEOSTASIS AND GLUCONEOGENESIS

A. Hormonal Control

As we have noted previously (Cahill, 1970; Cahill and Owen, 1968), the dominant controls of fuel metabolism during both the fed and fasted states are hormonal. The increased rate of gluconeogenesis and other alterations in glucose metabolism that occur during a fast are primarily a consequence of the decrease in circulating insulin and, at least in the liver, the increase in glucagon. Glucocorticoids are required, but their role is permissive; they do not appear to have an active modulatory function.

B. Ambient Glucose Concentrations

In general, the direct modulation (i.e., not hormonally mediated) of glucose production and utilization by the ambient concentration of glucose appears to be a secondary form of control, occurring within a framework determined by the hormonal milieu. Although it has been shown that both hepatic glucose balance (Soskin *et al.*, 1934) and gluconeogenesis (Ruderman and Herrera, 1969) and glucose uptake by muscle (Morgan *et al.*, 1959) are affected by glucose concentration when hormonal factors are held constant, in the presence of a hormonal imbalance, this type of control may be lost. For instance, the ability of glucose to depress hepatic glucose output is lost in severe diabetes, and hepatic glucose output is not increased when hypoglycemia is caused by insulin (see Steele, 1966). An excellent example of how subtle may be the hormonal control is the syndrome of hyperglycemic coma (Arieff and Carroll, 1972), in which a patient with a partial deficiency of insulin may develop a blood glucose concentration as high as 100 mM and yet not be ketotic. This disorder is effectively treated with relatively small amounts of insulin as compared to the amounts needed in the treatment of ketoacidosis.

C. Substrate Supply

The major gluconeogenic substrates in man are amino acids, lactate, and glycerol. Of the amino acids, particular emphasis has been placed on alanine and glutamine, which account for approximately 50% of the amino acid released from muscle and an equivalent percentage of the amino acid extracted by the splanchnic bed (Marliss *et al.*, 1971; Felig *et al.*, 1973a; Felig, 1973). Recent studies in both man (Felig *et al.*, 1973b; Felig *et al.*, 1973a) and experimental animals (Elwyn *et al.*, 1967; Ishikawa *et al.*, 1972) indicate that much of the glutamine released by muscle is taken up by the gut, which in turn releases alanine; indeed, they suggest that the gut may release 50 to 100% as much alanine as muscle, and as a result alanine probably accounts for one-half to two-thirds of the glucose derived from amino acid in the liver. Lactate release by the gut was not reported in these studies; however, in an anesthetized rat we have noted an arterial-portal-venous difference for lactate more than twice that for alanine (Table IV). Thus measurements across the splanchnic bed have underestimated the importance of both alanine and lactate as gluconeogenic substrates. The role of the gut in the adaptation to starvation and in the pathogenesis of gluconeogenic disorders attributed to lack of substrate remains to be determined.

The rate of hepatic gluconeogenesis appears to be governed by substrate

Table IV. Arteriofemoral venous and arterioportal-venous differences for alanine and lactate in a rat fasted for 24 hours[a]

	Lactate (μM)	Alanine (μM)
A–FV	−190	−34
A–PV	−210	−90

[a] Results are means of at least seven animals in each group. Rats were anesthetized with pentobarbital (50 mg/kg body wt); minus sign indicates release.

availability during intermediate and prolonged starvation (phases IV and V) when the release of alanine and other amino acids from muscle decreases, their levels in plasma fall and gluconeogenesis is depressed (Felig *et al.*, 1970). An enhanced release of gluconeogenic precursor from muscle occurs during the early phases of a fast but does not initiate the increase in gluconeogenesis; indeed, the concentrations of alanine and lactate, the principal gluconeogenic precursors, are invariably lower in fasted than in fed individuals (Tables I and II), suggesting that increased extraction by liver is the initiating event.

IV. DISORDERS OF GLUCONEOGENESIS

In this section we classify some of the hypoglycemic disorders in man that have been attributed to impaired gluconeogenesis. The classification should be considered tentative, since other factors (e.g., diminished glycogen breakdown) may have contributed to the hypoglycemia in some cases, and in others the mechanism by which gluconeogenesis is inhibited has not been elucidated. We confine our remarks to a description of the general characteristics of the subgroups and to an analysis of some of the difficulties encountered in classification. For a more comprehensive discussion of the pathophysiology of the individual disorders, the reader is referred to the references listed in Table V and to the excellent review by Exton (1972).

A. Disorders due to Insufficient Gluconeogenic Substrate

The hypoglycemia that occurs in pregnant women during starvation (Felig and Lynch, 1970; Felig *et al.*, 1972) and in children with ketotic hypoglycemia, the most common hypoglycemic disorder of childhood (Pagliara *et al.*, 1972), has been attributed to an insufficient supply of gluconeogenic substrate. In large part, this is due to the finding of low concentrations of alanine and other gluconeogenic amino acids in plasma; indeed, these individuals have been likened to subjects during prolonged starvation (phase V) in whom the concentrations of most gluconeogenic amino acids in blood are decreased because of diminished release from peripheral sites. In contrast to these starved individuals, however, the low levels of gluconeogenic substrate in pregnant women do not appear to be secondary to an aboslute decrease in amino acid release from muscle. Urine nitrogen (urea-*N* + NH_3) excretion is normal or increased during a fast (Felig and Lynch, 1970; Herrera *et al.*, Freinkel, 1969). Furthermore, when one considers that the mother is also supplying amino acids to the fetus for protein synthesis, it seems likely that amino acid mobilization from the periphery must be even greater than normal.

There have been no measurements reported of either urine excretion of urea-*N* and NH_3 or amino acid release from muscle in children with ketotic hypoglycemia. On the other hand, many of these children have high plasma levels of branched chain amino acids (Hambreus, *et al.*, 1972). This is usually a sign of increased protein degradation, suggesting that the mobilization of gluconeogenic amino acids from peripheral sites is not diminished. In this regard, a plasma amino acid pattern very similar to that of children with ketotic hypoglycemia, but associated with high rates of gluconeogenesis and amino acid release from muscle and increased urine excretion of urea-*N*, has been observed in rats with ketotic hypoglycemia induced by phloridzin administration (Ruderman *et al.*, 1974).

B. Disorders due to an Impaired Hepatic Gluconeogenic Apparatus

In general, the pathophysiology of gluconeogenesis has been best worked out for disorders in which the ability of the liver to synthesize glucose is impaired. The etiology of those listed in the "Exogenous inhibitors" subgroup of Table V is usually evident from the patient's history, and these individuals generally present little diagnostic problem to the physician. Disorders of gluconeogenesis due to inborn errors (Table V) are of two types—those in which a regulatory enzyme of the gluconeogenic pathway is deficient, and those in which an unrelated metabolic error secondarily leads to a block in gluconeogenesis. High concentrations of alanine and/or lactate may be found in the blood of patients in both the "Exogenous inhibitors" and "Inborn errors" subgroups (a point that could help in differentiating them from children with ketotic hypoglycemia), since release from peripheral sites is not impaired. In at least one disorder, pyruvate carboxylase deficiency, hyperlacticacidemia may be present even when the patient is normoglycemic (Hommes *et al.*, 1968). An exception to the rule is maple syrup urine disease,

Table V. Hypoglycemic disorders associated with impaired gluconeogenesis in man

Disorder	References
Diminished substrate delivery	
Pregnancy	Felig and Lynch (1970), Felig *et al.* (1972)
Ketotic hypoglycemia	Pagliara *et al.* (1973), Garber *et al.* (1974)
Chronic renal insufficiency	
Muscle wasting?	
Malnutrition?	Holt *et al.* (1963), Arroyave *et al.* (1962)
Impaired hepatic gluconeogenic apparatus—adequate substrate	
Inborn errors	
Glucose-6-phosphatase (von Gierke)	Howell (1972)
Fructose-1,6-diphosphatase	Baker and Winegrad (1970), Pagliara *et al.* (1973)
Phosphoenolpyruvate carboxykinase	Lardy (1973)
Pyruvate carboxylase (Leigh's encephalopathy)	Hommes *et al.* (1968)
Maple syrup urine disease	Haymond *et al.* (1973)
Methylmalonic aciduria	Halperin *et al.* (1970)
Hereditary fructose intolerance[a]	Froesch (1972), Kaufman and Froesch (1973), Haymond *et al.* (1975)
Immaturity Small for gestational age infant	
Exogenous inhibitors	
Ethanol	Freinkel *et al.* (1965), Madison (1968)
Biguanides	Haeckel and Haeckel (1972), Davidoff (1974)
Endotoxin	McCallum and Berry (1973)
Hypoglycin—Jamaican vomiting sickness	Corredor *et al.* (1968), Ruderman *et al.* (1971)
Aspirin?[b]	Limbeck *et al.* (1965), David *et al.* (1971)
Tryptophan	Furst *et al.* (1971)
Fructose, sorbitol	Felber *et al.* (1959)
Hormonal alterations (mixed mechanism)	
Glucocorticoid deficiency	Exton (1972), Pagliara *et al.* (1973)
Isolated ACTH deficiency	Pagliara *et al.* (1973)
Excess insulin	Exton (1972), Pagliara *et al.* (1973)
Glucagon lack?	Bleicher (1970), Marliss *et al.* (1972)

Table V. (Continued)

Disorder	References
Mechanism undetermined	
Unique disorders of children	
Reye syndrome	Glasgow *et al.* (1972)
Acute diarrhea	Hirschorn *et al.* (1966)
Chronic diarrhea	Lifschitz *et al.* (1970)
Malnutrition—marasmus	Arroyave *et al.* (1962), Holt *et al.* (1962)
Modest ethanol intake?	Cumming (1961), Madison (1968)
Tumor hypoglycemia	
Tryptophan and derivatives	Silverstein *et al.* (1966)
Insulinlike activity	Silverstein *et al.* (1964), Marks *et al.* (1974)
Glucagon secretion impaired	Silbert *et al.* (1974)
Tumor factor induces gluconeogenic enzymes in liver?	Suda (1964)
Substrate availability?	
Miscellany Congestive heart failure	Block *et al.* (1972)

[a] Hypoglycemia may be due in large part to inhibition of phosphorylase by Fructose-1-P and other intermediates.
[b] Attributed to inhibition of oxidative phosphorylation. The relative levels of importance of increased glucose utilization and impaired gluconeogenesis are not known.

in which the ability of the liver to synthesize glucose is impaired, but low concentrations of alanine may be found in the plasma (Haymond *et al.*, 1973). Transiently impaired gluconeogenesis due to "immaturity" of the gluconeogenic apparatus has been described by Hayward *et al.* (1974) in infants that are small for gestational age. As with other disorders in this category, plasma lactate and alanine were elevated.

C. Impaired Gluconeogenesis due to Hormonal Alterations

Hypoglycemic disorders due to lack or excess of a hormone have been recently reviewed by others (Exton, 1972; Pagliara *et al.*, 1973) and are discussed only in passing. For the most part, the hypoglycemia cannot be attributed to impaired gluconeogenesis alone, since the hormones in question also affect hepatic glycogen metabolism and, in the case of insulin and glucocorticoids, peripheral glucose utilization. In addition, the hormones may affect gluconeogenesis by more than one mechanism. Insulin and glucocorticoids both influence substrate mobilization from the periphery and have direct effects on the liver. Evidence from animal studies suggests that the peripheral effect of glucocorticoids is predominant (Wells and Kendall, 1942). In contrast, studies in man indicate that insulin may affect hepatic alanine extraction and conversion to glucose even when it has no effect on alanine mobilization from the periphery (Felig and Wahren, 1971).

D. Mechanism Undetermined

Hypoglycemia attributed to impaired gluconeogenesis, but for which the precise mechanism has not been clarified, has been noted in children in association with a variety of disorders. Senior and Loridan (1969) have suggested that the unusual predilection of children to hypoglycemia may exist because the child's brain is larger relative to liver and muscle than in an adult. Whereas a 10 kg child at age 1 year has a brain weight of 1100 g, the brain of a 70 kg adult weighs only 1400 g. Thus during a fast, the child uses up his smaller store of liver glycogen and becomes dependent on gluconeogenesis sooner than an adult (i.e., he has a shorter postabsorptive phase). Furthermore, because of his smaller muscle mass and an apparent inability to increase proteolytic rate per unit of muscle mass, the child has a relative shortage of gluconeogenic precursor (muscle amino acids and lactate and other precursors derived from muscle glycogen) and is therefore less able to maintain hepatic glucose production during a fast. The basis for impaired gluconeogenesis in the disorders unique to children, listed in Table V, is not known, although in marasmus, the low concentrations of alanine and other amino acids in blood and the presence of a diminished muscle mass (Arroyave *et al.*, 1962) suggest a deficient supply of substrate.

A second class of disorders in which impaired gluconeogenesis appears to play a role, but for which the mechanism is not known, are the hypoglycemias associated with extrapancreatic tumors. Fasting hypoglycemia has been noted in patients with a wide variety of neoplasms, but in particular large tumors of mesenchymal origin and hepatomas (Silverstein *et al.*, 1964; Marks *et al.*, 1974). The pathophysiology appears to be diverse. In some patients, hypoglycemia occurs despite an adequate supply of liver glycogen; in others glycogen is lacking. As far as is known, in only one instance has a tumor been reported to release insulin (Laurent and Jeanpierre, 1974); however, an increase in nonsuppressible insulinlike activity (NSILA-S) in plasma has occasionally been noted (Silverstein *et al.*, 1964; Marks *et al.*, 1974; Megyesi *et al.*, 1974). Recently, impaired glucagon secretion has been demonstrated in a patient with a mesothelioma (Silbert *et al.*, 1974); however, glucagon has been measured in very few other subjects. In a few cases, high levels of tryptophan metabolites, capable of inhibiting phosphoenolpyruvate carboxykinase, have been found in the urine (Silverstein *et al.*, 1966). Since muscle wasting is very common in patients with tumor hypoglycemia, it is surprising that the role of substrate availability has received little attention.

REFERENCES

Arieff, A. I. and Carroll, H. J. (1972). *Medicine* **51,** 73–94.

Arroyave, G., Wilson, D., deFures, C., and Behar, M. (1962). *Am. J. Clin. Nutr.* **11,** 517–524.

Baker, L. and Winegrad, A. I. (1970). *Lancet* **2,** 13–16.

Bleicher, S. J., Levy, L. J., Zarowitz, H., and Spergel, G. (1974). *Clin. Res.* **18,** 355.

Block, M. B., Gambetta, M., Resnekov, and Rubenstein, A. H. (1972). *Lancet* **2,** 736–738.

Butterfield, W. J. H., and Whichelow, M. J. (1965). *Diabetologia* **1,** 43–45.

Cahill, G. F., Jr. (1970). *New Engl. J. Med.* **282,** 668–675.

Cahill, G. F., Jr., Aoki, T. T., and Ruderman, N. B. (1972). *Trans. Am. Clin. Climatol. Ass.* **84,** 184–202.

Cahill, G. F., Jr. and Owen, O. E. (1968). *Carbohydrate Metabolism and its Disorders,* Vol. 2 (F. Dickens, P. J. Randle, and W. J. Whelan, eds.), Academic Press, New York, pp. 497–522.

Cahill, G. F., Jr., Herrera, M. G., Morgan, A. P., Soeldner, J. S., Steinke, J., Levy, P. L., Reichard, G. A., and Kipnis, D. M. (1966). *J. Clin. Invest.* **45,** 1751–1769.

Corredor, C., Brendel, K., and Bressler, R. (1968). *Proc. Nat. Acad. Sci.* (*D.S.*) **58,** 2299–2306.

Cumming, L. (1961). *J. Pediatr.* **58,** 23–26.

David, D. S., Steere, A. C., P-Sunyer, F. X., Sakai, S., and Clark, S. B. (1971). *Lancet* **2,** 1092–1093.

Davidoff, F. (1974). *New Engl. J. Med.* **289,** 141–146.

Drenick, E. J., Alvarez, L. C., Gabor, C. T. and Brickman, A. S. (1972). *J. Clin. Invest.* **51,** 2757–2762.

Elwyn, D. H., Parikh, C., and Shoemaker, W. C. (1968). *Am. J. Physiol.* **215,** 1260–1275.

Exton, J. H. (1972). *Metabolism* **21,** 945–990.

Felber, J.-P., Renold, A. E., and Zahnd, G. R. (1959). *Modern Problems in Pediatrics,* Vol. 70, S. Karger, Basel, Switzerland, pp. 467–489.

Felig, P. (1973). *Metabolism* **22,** 179–207.

Felig, P., Kim, Y. J., Lynch, V., and Hendler, R. (1972). *J. Clin. Invest.* **51,** 1195–1202.

Felig, P. and Lynch, V. (1970). *Science* **170,** 990–992.

Felig, P., Marliss, E., Pozefsky, T., and Cahill, G. F., Jr. (1970). *Am. J. Clin. Nutr.* **23,** 986–992.

Felig, P. and Wahren, J. (1971). *J. Clin. Invest.* **50,** 1702–1711.

Felig, P., Wahren, J., Karl, I., Cerasi, E., Luft, R., and Kipnis, D. M. (1973a). *Diabetes* **22,** 573–576.

Felig, P., Wahren, J. and Rof, L. (1973b). *Proc. Nat. Acad. Sci. U.S.* **70,** 1775–1779.

Freinkel, N., Arky, R. A., Singer, D. L., Cohen, A. K., Bleicher, S. J., Anderson, J. B., Silbert, C. K., and Foster, A. E. (1965). *Diabetes* **14,** 350–361.

Froesch, E. R. (1972). In *Metabolic Basis of Inherited Disease* (J. B. Stanbury, J. R. Wyngaarden, and D. S. Fredrickson, eds.), McGraw-Hill, New York, pp. 131–148.

Furst, P., Guarnieri, G., and Hultman, E. (1971). *Scand. J. Clin. Lab. Invest.* **27,** 183–191.

Garber, A. J., Bier, D. M., Cryer, P. E., and Pagliara, A. S. (1974). *Diabetes* **23,** 982–986.

Glasgow, A. M., Cotton, R. B., and Dhiensiri, K. (1973). *Am. J. Dis. Child.* **125,** 809–811.

Haeckel, R. and Haeckel, H. (1972). *Diabetologia* **8,** 117–124.

Halperin, M. L., Schiller, C. M. and Fritz, I. B. (1971). *J. Clin. Invest.* **50,** 2276–2282.

Hambreus, L., Westphal, O., and Hagberg, G. (1972) *Acta Paediat. Scand.* **61,** 81–88.

Haymond, M. W., Karl, I. E., Feigin, R. D., DeVivo, D. D., and Pagliara, A. S. (1973). *Pediatr. Res.* **7,** 500–503.

Haymond, M. W., Karl, I. E., and Pagliara, A. S. (1974). *New Engl. J. Med.* **291,** 322–328.

Herrera, E., Knopp, R. H., and Freinkel, N. (1969). *J. Clin. Invest.* **48,** 2260–2272.

Hirschorn, N., Lindenbaum, J., Greenough, W. B., III, and Alam, S. M. (1966). *Lancet* **2,** 128–132.

Holt, L. E., Snyderman, S. E., Norton, P. M., Roitman, E., and Finch, J. (1963). *Lancet* **2,** 1343–1348.

Hommes, F. A., Polman, H. A., and Reerink, J. D. (1968). *Arch. Dis. Child.* **43,** 423–426.

Howell, R. R. (1972). In *Metabolic Basis of Inherited Disease* (J. B. Stanbury, J. R. Wyngaarden, and D. S. Fredrickson, eds.), McGraw-Hill, New York, pp. 149–173.

Hultman, E. and Bergstrom, J. (1967). *Acta Med. Scand.* **182,** 109–117.

Ide, T., Steinke, J., and Cahill, G. F., Jr. (1969). *Am. J. Physiol.* **217,** 784–792.

Ishikawa, E., Aikawa, T., and Matsutaka, H. (1972). *J. Biochem.* **71,** 1097–1099.

Itoh, T. and Quastel, J. H. (1970). *Biochem. J.* **116,** 641–655.

Jackson, R. A., Advani, U., Perry, G., Rogers, J., Peters, N., Dan, S., and Pilkington, T. R. E. (1973). *Diabetes* **22,** 145–159.

Kaufman, U. and Froesch, E. R. (1973). *Eur. J. Clin. Invest.* **3,** 407–413.

Krebs, H. A. (1964). In *Mammalian Protein Metabolism,* Vol. 1 (H. H. Munro and J. B. Allison, eds.), Academic Press, New York, pp. 125–176.

Kreisberg, R. A., Pennington, L. F., and Boshell, B. R. (1970). *Diabetes* **19,** 64–69.

Lardy, H. A. Personal communication.

Laurent, J. X. and Jeanpierre, R. E. (1974). *Ann. N.Y. Acad. Sci.* **230,** 391–402.

Lifshitz, F., Coello-Ramirez, P., Gutierrez-Topete, G. and Gutierrez, M. L. C. (1970). *J. Pediatr.* **77,** 604–612.

Limbeck, G. A., Ruvalcaba, R. H. A., Samols, E., and Kelley, V. C. (1965). *Am. J. Dis. Child.* **109,** 165–167.

Madison, L. (1968). *Adv. Metab. Disord.* **3,** 85–109.

Marks, L. J., Steinke, J., Podolsky, S., and Egdahl, R. H. (1974). *Ann. N.Y. Acad. Sci.* **230,** 147–160.

Marliss, E. B., Aoki, T. T., and Cahill, G. F., Jr. (1972). In *Glucagon: Molecular Physiology: Clinical and Therapeutic Implications* (P. J. Lefebvre and R. H. Unger, eds.), Pergamon Press, Oxford and New York, pp. 123–150.

Marliss, E. B., Aoki, T. T., Pozefsky, T., Most, A. S., and Cahill, G. F., Jr. (1971). *J. Clin. Invest.* **50,** 814–817.

Marliss, E. B., Aoki, T. T., Unger, R. H., Soeldner, J. S., and Cahill, G. F., Jr. (1970). *J. Clin. Invest.* **49,** 2256–2270.

McCallum, R. E. and Berry, L. J. (1973). *Infect. Immun.* **7,** 642–654.

Megyesi, K., Kahn, C. R., Roth, J., Gorden, P., and Neville, D. M. (1974). Program, 66th Meeting American Society for Clinical Investigation, p. 529.

Morgan, H. E., Henderson, M. J., Regen, D. M., and Park, C. R. (1959). *Ann. N.Y. Acad. Sci.* **82,** 387–402.

O'Connell, R. C., Morgan, A. P., Aoki, T. T., Ball, M. R., and Moore, F. D. (1974). *J. Clin. End. Metab.* **39,** 555–563.

Owen, O. E., Morgan, A. P., Kemp, H. G., Sullivan, J. M., Herrera, M. G., and Cahill, G. F., Jr. (1967). *J. Clin. Invest.* **46,** 1589–1595.

Owen, O. E., Felig, P., Morgan, A. P., Wahren, J., and Cahill, G. F., Jr. (1969). *J. Clin. Invest.* **48,** 574–583.

Owen, O. E. and Reichard, G. A. (1971). *J. Clin. Invest.* **50,** 1536–1545.

Pagliara, A. S., Karl, I. E., DeVivo, D. C., Feigen, R. D., and Kipnis, D. M. (1972). *J. Clin. Invest.* **51,** 1440–1449.

Pagliara, A. S., Karl, I. E., Haymond, M., and Kipnis, D. M. (1973). *J. Pediatr.* **82,** 365–379, 558–577.

Ruderman, N. B., MacDonald, M., Neufeldt, N., Park, B., and Berger, M. (1974). *Benzon Foundation Symposium, VI, Liver Metabolism* (F. Lundquist and N. Tygstrup, eds.), Munksgaard, Copenhagen, pp. 291–300.

Ruderman, N. B. and Herrera, M. G. (1968). *Am. J. Physiol.* **214,** 1346–1351.

Ruderman, N. B., Toews, C. J., Lowy, C., Vreeland, I., and Shafrir, E. (1970). *Am. J. Physiol.* **219,** 51–57.

Senior, B. and Loridan, L. (1969). *J. Pediatr.* **74,** 529–539.

Silbert, C. K., Rossini, A., Ghazvinian, S., Wildrich, W. C., Marks, L. J., and Sawin, C. T. (1974). *Clin. Res.* **22,** 479A.

Silverstein, M. N., Wakim, K. G., Bahn, R. C., and Decker, R. H. (1966). *Cancer* **19,** 127–188.

Silverstein, M. N., Wakim, K. G., Bahn, R. C. (1964). *Am. J. Med.* **36,** 415–423.

Slone, D., Taitz, L. S., and Gilchrist, G. D. (1961). *Brit. Med. J.* **1,** 32–34.

Nilsson, L. H. (1973). *Scand. J. Clin. Lab, Invest.* **32,** 317–323.

Nilsson, L. H. and Hultman, E. (1973). *Scand. J. Clin. Lab. Invest.* **32,** 325–330.

Soskin, S., Allweiss, M. D., and Cohn, D. J. (1934). *Am. J. Physiol.* **109,** 155–165.

Steele, R. (1966). *Ergeb. Physiol.* **57,** 91–189.

Suda, M. (1967). *Adv. Enzyme Regul.* **5,** 181–209.

Wahren, J., Felig, P., Cerasi, E., and Luft, R. (1972). *J. Clin. Invest.* **51,** 1870–1878.

Wells, B. B. and Kendall, E. C. (1940). *Proc. Staff Meet. Mayo Clin.* **15,** 565–573.

CHAPTER FIFTEEN

Gluconeogenesis in Normal, Cirrhotic, and Diabetic Humans

OLIVER E. OWEN, MULCHAND S. PATEL, BARRY S. B. BLOCK, THOMAS H. KREULEN, FREDERICK A. REICHLE, AND MARIA A. MOZZOLI

I. INTRODUCTION

A balance between glucose delivery and glucose removal maintains blood glucose homeostasis. Both the liver and kidney have the enzymatic machinery to synthesize glucose, and both the liver and muscle have the ability to store significant quantities of glucose as glycogen. However, in addition to contributing glucose to the bloodstream or storing glucose as glycogen, only the liver has the unique responsibility of maintaining glucose homeostasis by making net contributions of acetoacetate and β-hydroxybutyrate to the bloodstream, thereby sparing glucose utilization by peripheral tissues.

During brief fasting periods the liver of a healthy human assumes the primary role of providing glucose to tissues that utilize glucose as a metabolic fuel. Hepatic glucose production is the result of glycogenolysis and gluconeogenesis. Hepatic glycogen, synthesized and stored postprandially, supplies the bulk of hepatic glucose released during brief fasts (see Chapter 14 for details on the time course of glucose homeostasis). Hepatic gluconeogenesis supplements glycogenolysis during the postabsorptive period. As fasting is extended, gluconeogenesis is augmented and glycogenolysis is diminished because of depletion of glycogen stores. After prolonged starvation, sustained gluconeogenesis is the sole source of glucose production. In addition to hepatic gluconeogenesis, renal gluconeogenesis aids the liver in maintaining blood glucose homeostasis.

Although gluconeogenesis increases during fasting, the quantity of glucose supplied by gluconeogenic process is always far less than the transient quantity of glucose supplied by hepatic glycogenolysis during the brief postabsorptive period. It is important to recognize that blood glucose homeostasis is possible because other substrates, predominantly free fatty acids, acetoacetate, and β-hydroxybutyrate, are preferentially oxidized by the bulk of the body and thereby spare glucose oxidation.

There are two basic methods for determining glucose production and/or oxidation rates. One method depends on the kinetic analyses of data obtained following the injection of large quantities of nonradioactive glucose (Franckson *et al.*, 1962) or trace quantities of radioactive glucose (Franckson *et al.*, 1962; Baker *et al.*, 1954; Shreeve *et al.*, 1956; Reichard *et al.*, 1961; Searle *et al.*, 1959; Manougian *et al.*, 1964; Steele *et al.*, 1956; Searle *et al.*, 1954) or metabolites (Jacobs *et al.*, 1958; DeMeutter and Shreeve, 1963; Forbath and Hetenyi, 1966). The other method is the catheterization technique, wherein circulating glucose concentration differences across organs are combined with regional blood flow rates (Warren and Brannon, 1944; Bondy *et al.*, 1949a). It is surprising that the values obtained with kinetic analyses techniques and those obtained by the catheterization techniques agree as closely with each other as they do (Cahill and Owen, 1968; Owen and Reichard, 1971a), since they fundamentally measure different parameters. Kinetic analyses techniques measure the total flux of a compound throughout the body pools, whereas catheterization techniques measure the net balance of a compound across a specific tissue bed. An advantage of the radioactive tracer techniques is that they allow the calculation of CO_2 derived from glucose oxidation.

Mechanisms by which gluconeogenesis may be controlled by substrate supply, substrate extraction by liver and kidney, mitochondrial metabolite transport, enzymatic activity, and effector actions have been discussed elsewhere in this volume and are not included in this chapter. Here we quantify and compare glucose metabolism in various nutritional states in normal, cirrhotic, and diabetic humans, using the data obtained from catheterization and isotope techniques.

II. GLUCOSE METABOLISM IN THE FED STATE

Reciprocal fluxes in glucose intake, oxidation, and storage influence blood glucose homeostasis. When glucose is supplied in abundance from exogenous sources to normal humans, it is preferentially oxidized as a body fuel, stored as glycogen and converted to lipid for deposition in adipose tissue (Owen and Reichard, 1971a). In the basal state, after an overnight (12–14 hour) fast, splanchnic (hepatic) production is about 0.83 mmole/min (150 mg/min) (Owen and Reichard, 1971a). During this stage, hepatic production rate is matched by peripheral removal rate. Thus endogenous glucose availability and utilization are balanced. This steady-state flux is altered by exogenous glucose administration.

To evaluate the role of splanchnic (hepatic) and peripheral tissues in the disposal of oral glucose, Felig *et al.* (1975) used direct catheterization techniques to examine splanchnic exchange of glucose in the basal state in normal, overnight fasted men before and after oral ingestion of 100 g of glucose. Following glucose intake, splanchnic glucose output rose rapidly, reaching peak values of 2 to 3 times the basal rate after 15 minutes and returned to a basal output rate by 90 minutes. A second increase over basal output peaked at 150 minutes (Fig. 1). Changes in arterial glucose concentrations were largely determined by the amount of splanchnic glucose escape. Potential gluconeogenesis from extracted alanine, lactate, pyruvate, and glycerol was markedly curtailed following oral glucose intake. The arterial concentrations of alanine, lactate, and pyruvate rose by 15, 65, and 80%, respectively. The increases in the arterial concentration were due to altered splanchnic extraction of these precursors. Glucose ingestion resulted in complete inhibition of splanchnic lactate uptake, while pyruvate exchange was reversed from a net uptake to a net output. Splanchnic uptake of alanine fell by 75% within 1 hour of glucose ingestion. The antilipolytic action of insulin caused the arterial glycerol concentration to fall by 50% following oral glucose. A proportionate decline in the splanchnic uptake of glycerol was observed.

Felig and co-workers (1975) suggested that insulin regulated the hepatic extraction of alanine, lactate, and pyruvate but had a peripheral effect by diminishing the availability of glycerol. Other data confirm that insulin has a direct effect on hepatic extraction of a variety of gluconeogenic precursors.

The technique used to study the direct effect of insulin on hepatic me-

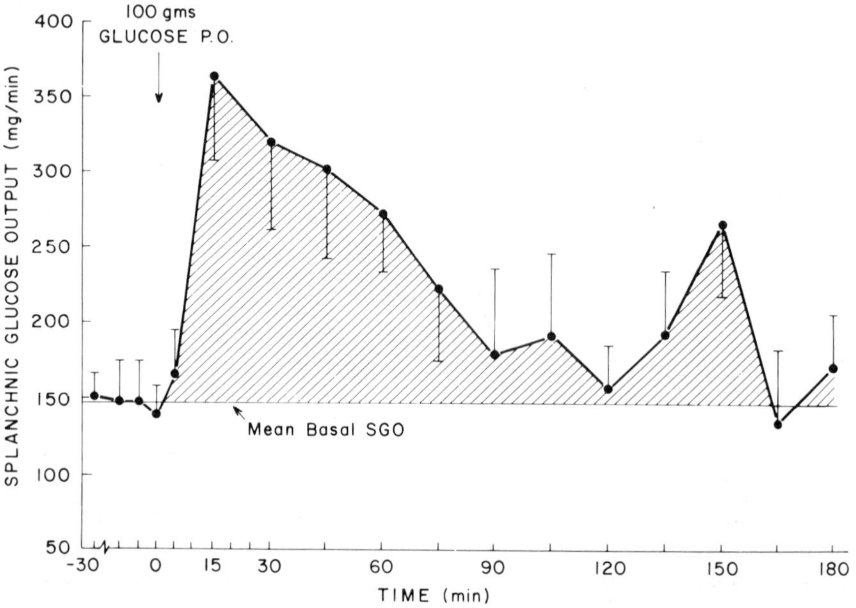

Figure 1. Splanchnic glucose output (SGO) in the basal state and following the ingestion of glucose (mean ± SEM). The horizontal line represents the mean basal splanchnic glucose output. *Source:* Felig *et al.* (1975), by permission of the American Diabetes Association, Inc.

tabolism is illustrated in Fig. 2. An umbilicoportal cannula was inserted during elective abdominal surgery in the human volunteer. After 3 days of caloric deprivation, during the postoperative recovery period, a small catheter was inserted through the umbilicoportal cannula and rested in a hepatofugal position. This created a double lumen device, permitting insulin to be infused through the small catheter and portal blood to be withdrawn from the hepatopedal large cannula orifice. Another catheter was inserted in the antecubital vein and advanced to the right hepatic vein. A Cournand needle was inserted percutaneously into the brachial artery. Insulin was infused into the small inner catheter at a rate of 120 mU/min and arterial, portal, and hepatic venous blood samples were simultaneously obtained, and alanine, lactate, pyruvate, glycerol and glucose concentrations, and hepatic blood flow were determined before, during, and after insulin infusion. This increased the portal venous serum concentration of insulin to about 400 μU/ml during the 30 minute infusion period. The effects of insulin on hepatic glucose and precursor exchange are plotted in Figs. 3 and 4. Insulin converted net hepatic glucose release to net hepatic glucose uptake. This occurred while the arterial glucose concentration fell from a control concentration of 4.2 mmole/liter (76 mg/100 ml) to a 60 minute postinfusion concentration of 2.2 mmole/liter (40 mg/100 ml). This uptake, designated in the shaded area, was not corrected for extrahepatic splanchnic glucose extraction.

During the control period the liver extracted about one-half of the quantities of lactate, alanine, and glycerol presented in the arterial blood (Fig. 4). Insulin infusion inhibited the fractional extraction of lactate by 92% while the arterial concentration increased, probably related to a constant rate of peripheral production (Jackson *et al.,* 1973) but a diminished rate of hepatic removal. The decreased hepatic extraction of lactate may be the primary cause of increased lactate concentration following insulin administration for treatment of diabetic ketoacidosis. Further-

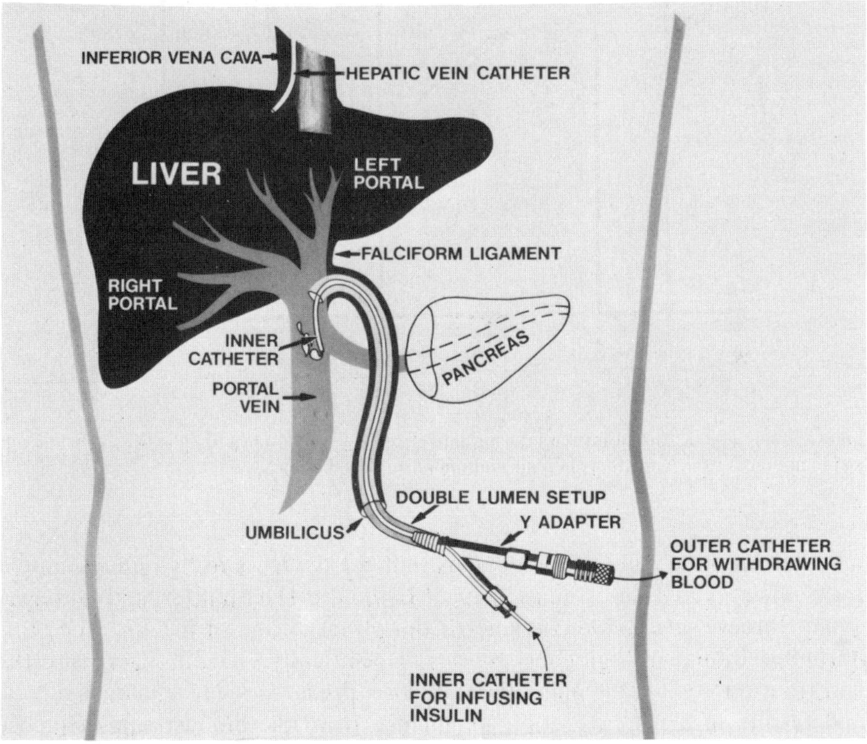

Figure 2. Catheters placed in the obliterated recannulized umbilical vein to obtain portal venous blood from the outer lumen of the double-lumen device and to infuse insulin into the portal vein through the inner catheter. Hepatic vein blood is obtained by passing a catheter from the antecubital vein, down the inferior vena cava, and into the right hepatic vein. A peripheral artery (not shown) is catheterized to obtain arterial blood.

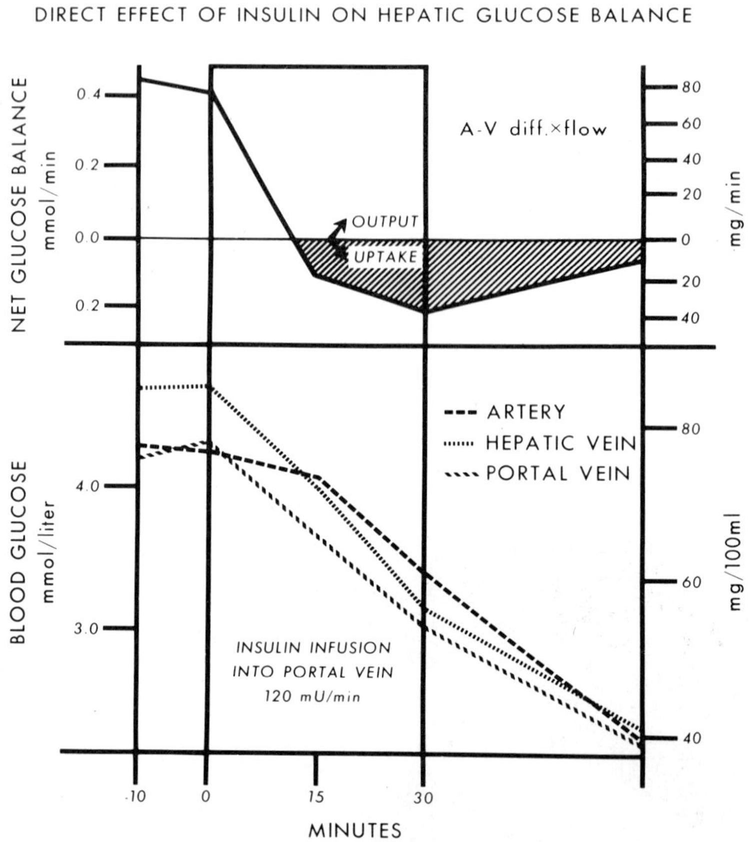

Figure 3. Direct effects of insulin infusion into the portal vein on hepatic glucose exchange and on blood glucose concentrations in a normal man after a 3 day fast.

more, from these studies it is obvious that the decrease in ^{14}C-glucose noted from kinetic analyses studies using ^{14}C-lactate (DeMeutter and Shreeve, 1963; Shreeve *et al.,* 1964) is due to the direct action of insulin, since this hormone diminishes gluconeogenesis. Specifically, insulin decreases the contribution of lactate and other glucose precursors for gluconeogenesis and results in a net extraction of glucose from the bloodstream. Glucose uptake is independent of ambient glucose concentration. Hepatic extraction of glycerol decreased simultaneously as the arterial glycerol concentration decreased. In addition, insulin infusion converted net extraction of alanine to net release of this amino acid while the arterial concentration remained the same.

III. GLUCOSE METABOLISM IN THE FASTED STATE

The liver, kidney, adipose tissue, and muscle have an excellent working relationship during fasting: these organs supply or converse fuels as required by the rest of the body. During the brief postabsorptive period, the liver is transiently the major contributor of fuel to the bloodstream, supplying glucose from glycogenolysis. As fasting progresses the adipose tissue becomes the major contributor by supplying predominantly free fatty acids for oxidation. Although the source of substrates for the liver and types of fuels added to the bloodstream by the liver, and after prolonged starvation by the

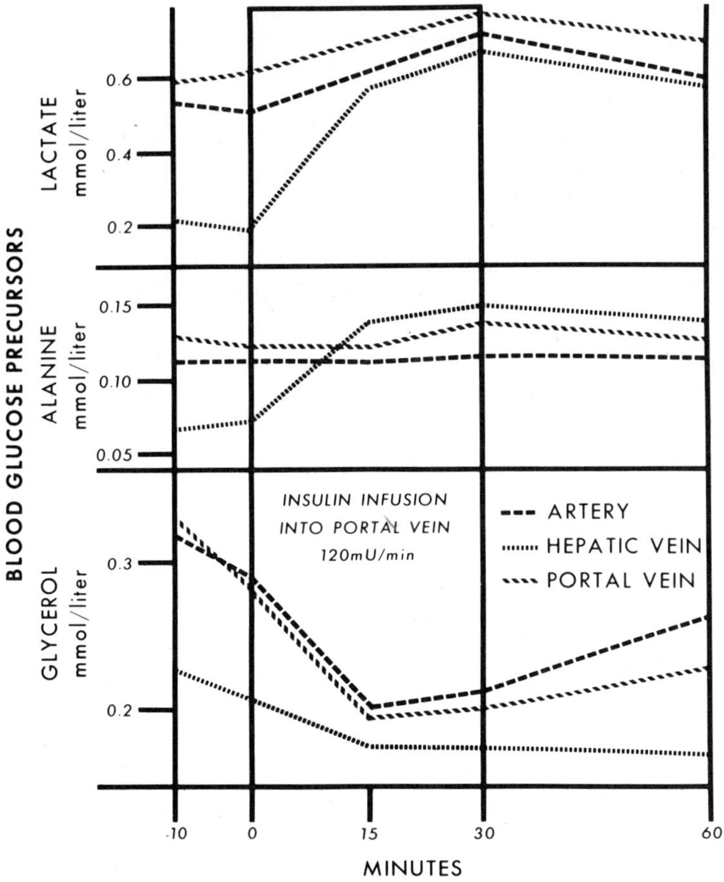

Figure 4. Direct effects of insulin infusion into the portal vein on blood glucose precursors in a normal man after a 3 day fast.

kidneys, change as starvation progresses, the caloric equivalents added to the bloodstream remain rather constant. Thus as the hepatic contribution of glucose diminishes, hepatic contribution of ketone-body (caloric equivalent) increases. Therefore, the total contribution of fuels to the bloodstream by the liver (and kidney) remains reasonably constant as starvation progresses from a brief fast to a prolonged fast.

Direct estimates of net splanchnic glucose production during the brief postabsorptive period (12–20 hours) has been determined by measuring arteriohepatic venous glucose differences and liver blood flow rates. Values from 0.47 to 1.37 mmole/min with an average production rate of about 0.90 mmole/min have been reported for normal humans. These results agree with the glucose turnover rates using kinetic analyses techniques (Owen and Reichard, 1971a). In addition to total glucose turnover rates, the early kinetic analyses techniques allowed for calculated estimates of recycled glucose (Cori cycle). The initial values of 0.10 to 0.22 mmole/min (Owen and Reichard, 1971a) have been confirmed and extended by the more recently acquired data obtained by the catheterization techniques (Wahren *et al.*, 1972). In essence, the catheterization techniques have allowed quantification of the relationship between net splanchnic glucose output and all gluconeogenic substrate (lactate, pyruvate, glycerol, and gluconeogenic amino acids) uptake, thereby providing estimates of the contributions of glycogenolysis and gluconeogenesis to net splanchnic glucose release.

The data obtained by catheterization techniques from humans after overnight, 3 day, and 5 to 6 week starvation periods appear in Fig. 5. The bar on the left gives the balance data obtained from 24 healthy, nonobese, adult male volunteers after an overnight (12–14 hours) fast (Wahren *et al.*, 1972). Mean net splanchnic glucose release and mean net splanchnic glucose precursor extractions are shown. Net splanchnic glucose release was 0.95 mmole/min. From the height of the left bar in Fig. 5 it can be seen that 0.20 mmole/min or approximately 20% of net splanchnic glucose release can be accounted for by gluconeogenesis is all the extracted gluconeogenic precursors are quantitatively converted and released as glucose. After an overnight fast in normal humans, approximately 80% of net splanchnic glucose release can be attributed to glucose derived from glycogenolysis. This brief postabsorptive starvation period is characterized by a high rate of glycogenolysis and a low rate of gluconeogenesis. Furthermore, the quantity of glucose supplied by the liver after this brief fast is enough to account for about one-half the total basal caloric requirements (Owen and Reichard, 1971a).

The balance data obtained from nine nondiabetic adult males and females after a 3 day (80–86 hours) fast appear in Fig. 5 (Garber *et al.*, 1974; Owen *et al.*, unpublished observations). Mean net splanchnic glucose release was

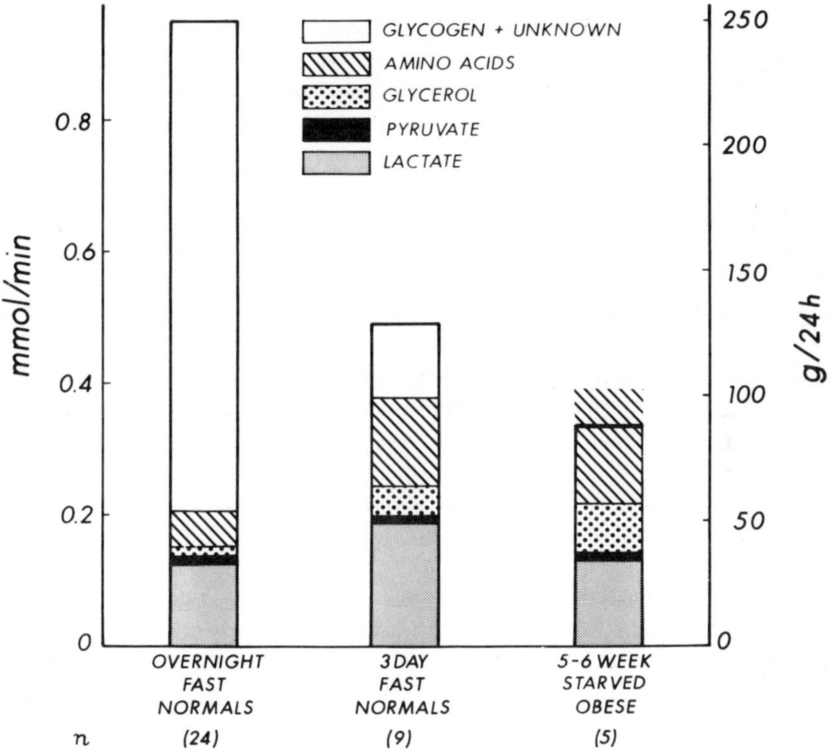

Figure 5. Net splanchnic glucose release and net splanchnic gluconeogenesis from precursor extraction in humans after overnight, 3 day, and 5 to 6 week fasts.

0.49 mmole/min, which is less than that noted after an overnight fast. Previous data have suggested that optimal conversion of gluconeogenic substrates into glucose during prolonged starvation occurs (Owen *et al.*, 1969). If it can be assumed that all extracted lactate, pyruvate, glycerol, and gluconeogenic amino acids are quantitatively converted into glucose, the middle bar indicates that these substrates can contribute approximately 80% of the net splanchnic glucose release. This estimated rate of hepatic gluconeogenesis amounts to 0.37 mmole/min and is approximately twice the rate observed after an overnight fast. Using kinetic analyses techniques, Paul and Bortz (1969) pointed out that the rate of glucose oxidation in lean and obese humans starved between 7 and 33 days was reduced, and that the minimal oxidation rate of 50 g/day was obtained within the first 4 to 5 days of starvation. Thus this period of starvation is characterized by increased hepatic gluconeogenesis, diminished hepatic glycogenolysis, and decreased peripheral glucose oxidation.

The balance data obtained by catheterization techniques from five obese humans starved for 5 to 6 weeks are presented graphically in Fig. 5 (Owen *et al.*, 1969). In addition to net splanchnic glucose production, the data also represent the net renal contribution to glucose production. Total glucose production was approximately 0.33 mmole/min. The splanchnic bed contributed approximately 55% and the renal tissue approximately 45% of the total. Gluconeogenesis, calculated from the measured hepatic-renal uptake of lactate, pyruvate, glycerol, and amino acids, could account for 0.39 mmole/min, or greater than 100% of net hepatic-renal glucose production rate. Using kinetic analyses techniques a similar mean glucose production rate was obtained from the obese humans starved for 5 to 33 days (Paul and Bortz, 1969). Glucose homeostasis during this period of starvation is characterized by maximum conversion of precursors to glucose by the liver and kidney cortex and no contribution of glucose from hepatic glycogenolysis.

Hepatic and renal contributions to caloric homeostasis after overnight, 3 day, and prolonged fasts are shown in Fig. 6. Glucose is a major metabolic fuel contributed to the bloodstream to maintain caloric homeostasis during brief starvation periods. However, the quantity of glucose terminally oxidized to CO_2 and water in healthy humans depends on the duration of starvation. Furthermore, the quantity of glucose burned must be considered in relationship to the quantity of alternate fuels oxidized to CO_2 and water. The monumental studies of Benedict (1915) established that oxygen consumption, thus caloric consumption, per square meter of body surface area, diminishes slightly as starvation progresses. The liver furnishes fuel supplies to peripheral tissues devoid of sufficient endogenous stores (central nervous system) for self-maintenance during prolonged caloric deprivation. Therefore, it can be anticipated that decreased caloric equivalents of glucose added to the bloodstream by the liver will be matched by increased caloric equivalents of alternate fuels added to the bloodstream. It has been established that ketone bodies can serve as alternate fuels for the largest glucose-consuming organ in the body, the central nervous system (Owen *et al.*, 1967), and that the liver is the only organ in humans that makes a net contribution of acetoacetate and β-hydroxybutyrate to the bloodstream (Owen *et al.*, 1969; Owen *et al.*, 1967; Owen and Reichard, 1971b). Thus if the caloric values of ketone bodies contributed to the bloodstream by the liver offset any decrease in the caloric value of glucose from the liver and kidneys, caloric homeostasis is maintained (Owen *et al.*, 1973; Reichard *et al.*, 1974). The caloric values of glucose and ketone bodies added to the bloodstream can be calculated by assessing caloric values of 3.90 kcal/g glucose, 4.69 kcal/g β-hydroxybutyrate, and 4.15 kcal/g acetoacetate (Kleiber, 1961).

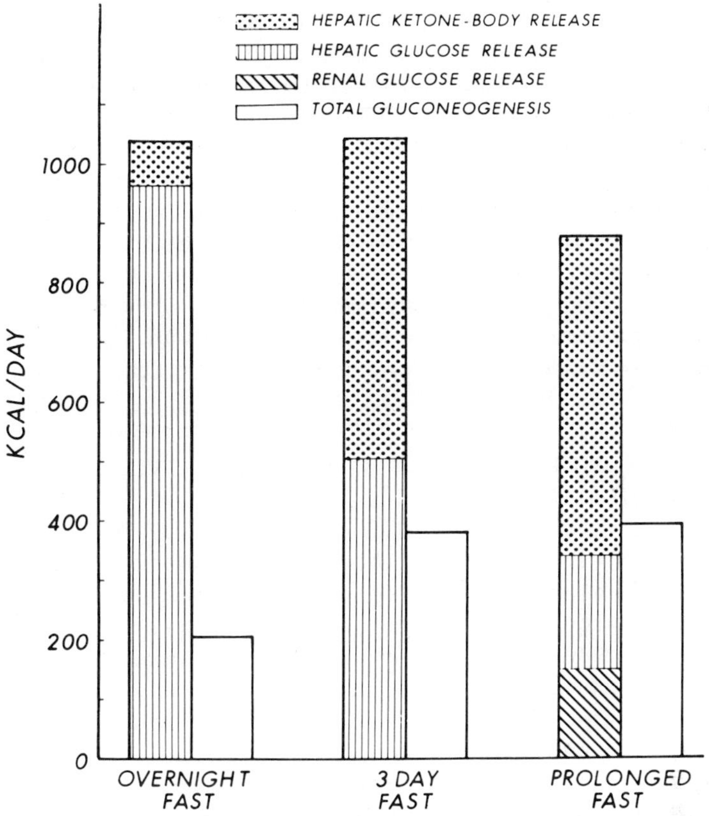

Figure 6. Hepatic and renal contributions to caloric homeostasis in humans after overnight, 3 day, and prolonged fasts.

The relationship among glycogenolysis, gluconeogenesis, and ketogenesis extrapolated to a 24 hour period in maintaining caloric homeostasis as starvation progresses from an overnight fast to a 5 to 6 week fast is presented in Fig. 6. Initially, hepatic glycogenolysis contributes the largest quantity of fuel and, therefore, calories, to the bloodstream. Gluconeogenesis is heightened after 2 to 3 days of starvation but progressively decreases as fasting is prolonged, because of limited precursor supply (Felig, 1973). Gluconeogenesis by the kidneys augments hepatic glucose synthesis after 4 days of starvation. This is based on indirect evidence assuming that renal gluconeogenesis and ammoniagenesis can be related (Owen *et al.*, 1969). Urinary ammonium excretion becomes maximum between 4 and 8 days of starvation and plateaus thereafter (Owen and Cahill, 1973). After 5 to 6 weeks of starvation the kidneys make a significant contribution of glucose to the bloodstream (Owen *et al.*, 1969). After an overnight fast the liver contributes 10 to 20 g/24 hours of acetoacetate plus β-hydroxybutyrate to the bloodstream (Owen and Reichard, 1971a; Owen *et al.*, 1973; Reichard *et al.*, 1974). Using direct catheterization techniques with nondiabetic humans, we have observed a net splanchnic ketone-body contribution of 18 g/24 hours (unpublished observation) after an overnight fast. After 3 days of starvation kinetic analyses and catheterization techniques reveal that hepatic ketogenesis is heightened to a maximum rate of about 122 g/24 hours (Garber *et al.*, 1974; Owen *et al.*, 1973; Reichard *et al.*, 1974) and is maintained as long as the free fatty acid precursor supply is maintained. After 3 days of fasting hepatic ketogenesis is a major process for adding fuels to the bloodstream, thus maintaining caloric homeostasis.

IV. GLUCOSE METABOLISM IN THE FASTED STATE IN CIRRHOTIC HUMANS

Hepatic cirrhosis is the fourth leading cause of death in American males aged 45 to 54 years. Furthermore, death from cirrhosis is increasing, particularly in large American cities. Obviously more knowledge pertaining to hepatic metabolism in patients with cirrhosis is needed. The blood glucose concentration in patients with hepatic cirrhosis can be low, normal, or elevated (Zimmerman *et al.*, 1953). Furthermore, glucose intolerance has frequently been observed in cirrhotic patients. Several studies with large numbers of cirrhotic patients have shown that the incidence of diabetes is 3 to 4 times higher than that observed in the general population (Zimmerman *et al.*, 1953; Creutzfeldt *et al.*, 1962; Conn *et al.*, 1971). However, the changes in concentrations of serum insulin in these patients are not uniformly low, and insulin resistance for glucose disposal by tissues is considered to be one of the possible pathogenic factors (Altszuler and Finegold, 1974). Hypokalemia (Podolsky *et al.*, 1973), elevated growth hormone (Conn and Daughaday, 1970), hemosiderosis (Conn, 1972), and impaired glucose storage as hepatic glucogen (Peters and Van Slyke, 1946; DeMoura and Cruz, 1968) due to excessive liver damage may also be contributing factors. In addition to low, normal, or elevated serum insulin concentrations (Altszuler and Finegold, 1974), it was recently shown that the concentrations of pancreatic glucagon in the plasma are markedly elevated in cirrhosis, especially in patients with naturally acquired portal systemic shunts and surgically induced portocaval anastomosis (Marco *et al.*, 1973; Sherwin *et al.*, 1974; Walker *et al.*, 1974). Furthermore, glucagon response to infusion of alanine into cirrhotic patients is severalfold greater than that observed in control subjects (Sherwin *et al.*, 1974), suggesting that the intact parenchymal cells of the cirrhotic liver may be "set" for augmented gluconeogenesis.

Since alterations in concentrations of plasma insulin and glucagon, hence the insulin:glucagon ratio in cirrhotic patients (Sherwin *et al.*, 1974; Walker *et al.*, 1974) is generally in the direction of that observed in normal subjects fasting for a relatively longer period of time (Marliss *et al.*, 1970), appropriate changes in circulating fuels in the bloodstream are expected. Indeed, increases in blood β-hydroxybutyrate and glycerol and plasma triglyceride and free fatty acids are observed in cirrhotic patients after an overnight fast when compared to normal individuals after a comparable fast (Alberti *et al.*, 1972). Our findings on the mean arterial substrate concentrations in cirrhotic patients and normal subjects after an overnight and 3 day fast are shown in Table I. Although fasting levels of blood lactate and pyruvate in cirrhotic patients are within normal limits, the concentrations of these metabolites in blood are also found to be elevated in patients with more severe cirrhosis (Alberti *et al.*, 1972). Sherwin *et al.* (1974) observed some alterations in plasma amino acid concentrations in cirrhotic patients after an overnight fast. Noteworthy changes are (1) increased levels of methionine, tyrosine, and phenylalanine, (2) decreased concentrations of the

Table I. Mean arterial substrate concentrations (mmole/liter)

	Normal		Cirrhotic		
Substrate	Overnight fast[b] (24)[a]	3 day fast[c] (9)	Overnight fast[c] (6)	3 day fast[c] (3)	Diabetic ketoacidosis[c] (7)
Lactate	0.70	0.56	0.66	0.59	1.98
Pyruvate	0.07	0.06	0.07	0.06	0.13
Glycerol	0.15[c]	0.29	0.10	0.08	0.29
Alanine	0.22	0.18	0.20	0.14	0.17
Glutamine	0.47[c]	0.40[d]	0.50	0.44	0.38
Glutamate	0.16[c]	0.13[d]	0.14	0.13	0.12
Glucose	4.08	3.72	4.13	3.96	22.23
Total ketones	0.22[c]	2.73	0.70	1.49	11.00
Free fatty acids	0.56[c]	1.45	1.26	1.05	1.51

[a] Number of subjects in parentheses.
[b] Wahren *et al.* (1972).
[c] Owen *et al.* (unpublished observations).
[d] Owen *et al.* (unpublished observations), mean of 3 patients.

branched chain amino acids, and (3) no significant changes in alanine concentrations.

Preliminary results (Owen *et al.*, unpublished observations), obtained using the catheterization techniques described earlier, show that net splanchnic glucose release in six cirrhotic patients after an overnight fast (8–14 hours) was diminished compared to normal humans (Figs. 5 and 7). This reduction in glucose release appears to be due to markedly diminished hepatic glycogenolysis in cirrhotic patients. Since we lack information about the exact content of liver glycogen in cirrhotic patients in the fed and

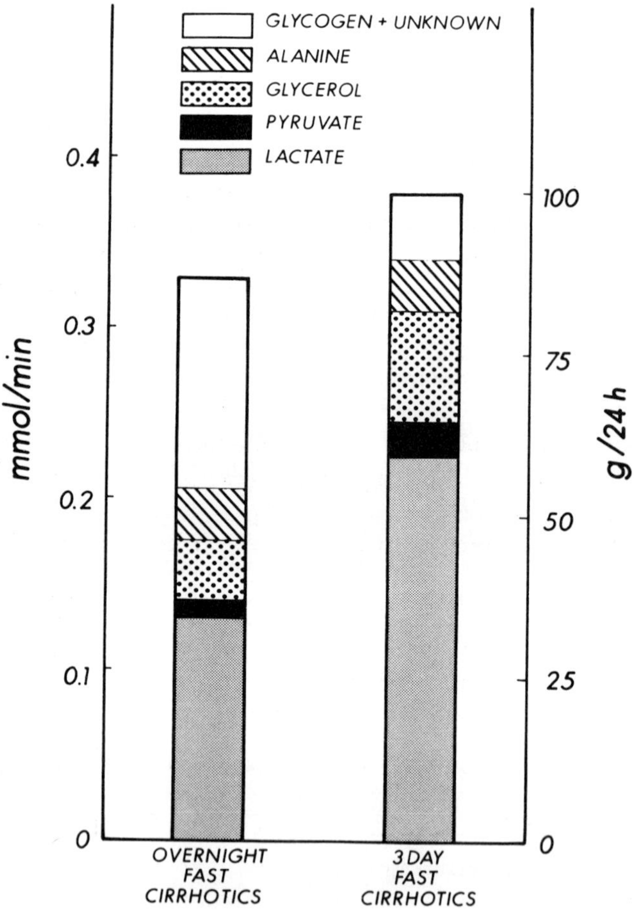

Figure 7. Net splanchnic glucose release and splanchnic gluconeogenesis from precursor extraction in cirrhotic patients after overnight and 3 day fasts.

fasted states, we can speculate that the amounts of glycogen stored in fibrotic livers are limited. In addition, depletion of hepatic glycogen may be the consequence of markedly elevated plasma glucagon concentrations (Marco *et al.*, 1973; Sherwin *et al.*, 1974; Walker *et al*, 1974). In spite of a reduced splanchnic glucose release, there is no decrease in hepatic gluconeogenesis in these patients (Fig. 7). The increased concentrations of plasma free fatty acids and blood ketone bodies in cirrhotic patients (Table I; also see Abterti *et al.*, 1972) indicate accelerated peripheral lipolysis. Not shown are data obtained from portal vein catheterization studies, which indicate that substrate contributions for glucose production from splanchnic extrahepatic sources are negligible, except for a small contribution of alanine and lactate.

In a study of healthy, normal subjects, Son Nilsson *et al.* (1973) reported rates of net splanchnic glucose production of about 0.87 mmole/min with gluconeogenesis contributing about 0.31 mmole/min. The remainder of hepatic glucose production was accounted for by glycogenolysis as verified by serial determinations of hepatic glycogen content in biopsy specimens. In normal subjects consuming carbohydrate-poor diets (6 g/day) for 4 days, the net splanchnic glucose production decreased from 0.87 to 0.30 mmole/min. At this time the liver was depleted of glycogen. The data in Fig. 7 for cirrhotic patients after an overnight fast are similar to the results of Son Nilsson *et al.* (1973) for normal subjects receiving a carbohydrate-poor diet. As with normal humans on a carbohydrate-poor diet, the contribution of glucose by glycogenolysis to the total splanchnic glucose production is minimal.

In normal subjects fasted for 3 days, the contribution of glycogenolysis to the total splanchnic glucose production diminishes markedly and that of gluconeogenesis increases significantly (Fig. 5). Quantitatively similar directional changes were noted in three cirrhotic patients catheterized after a 3 day fast. In these patients approximately 90% (Fig. 7) of net splanchnic glucose release can be attributed to gluconeogenesis and 10% to glycogenolysis. It should be noted that there was no change in the rate of hepatic gluconeogenesis in cirrhotic patients as compared to the control subjects after an overnight or 3 day fast. However, in an experimental animal model, an impairment in hepatic gluconeogenesis has been reported (Henley *et al.*, 1974). The results in Figs. 5 and 7 demonstrate the similarity between the contributions of hepatic gluconeogenesis in normal and in cirrhotic patients after an overnight fast and after a 3 day fast. An important observation, however, is the marked reduction in the contribution of hepatic glycogenolysis to the total splanchnic glucose production in cirrhotic patients after an overnight fast.

V. GLUCOSE METABOLISM IN THE DIABETIC STATE

Despite the fact that diabetes mellitus has been a recognized disease of humans for more than 4000 years and is now one of the leading health problems in affluent societies, there are relatively few detailed studies on the roles of net splanchnic glucose production from gluconeogenesis and glycogenolysis as related to the altered glucose homeostasis in this disease. Furthermore, the limited results available from the investigations are contradictory. It is not fully appreciated, for example, that one of the major reasons for the conflicting findings related to glucose production and oxidation rate in diabetes is that diabetes mellitus and diabetic ketoacidosis are very heterogeneous states. Furthermore, some of the results comparing normal controls to diabetic patients yield similar findings. In some situations this is because the diabetes was very mild or because compensation was secondary to insulin therapy. In addition, most of the available information on glucose metabolism in diabetes has been obtained from experimental animals and extrapolated to humans. Since this method of studying the pathophysiology of human diabetes has many shortcomings, we limit this review to data accumulated from humans.

Abnormal glucose homeostasis in diabetes was recognized early. However, the importance of enhanced hepatic glucose production an impaired peripheral glucose utilization in the genesis of diabetic hyperglycemia is still being debated. This is not surprising because the diabetic patient population varies in age, state of nutrition, degree of insulin deficiency and metabolic decompensation, and treatment regimens. Therefore, it is not difficult to conceive that data related to glucose production, gluconeogenesis, and glucose oxidation form spectra ranging from low rates to elevated rates.

In trying to assess the roles of hepatic glucose production and peripheral utilization in the pathogenesis of hyperglycemia, Bondy *et al.* (1949b) studied five diabetic patients in moderately severe (HCO_3^- = 10–16 mEq/liter) diabetic ketoacidosis. Their patients were well controlled on insulin and a diet containing more than 150 g of carbohydrate per day before being studied. They received no insulin 36 to 72 hours before they underwent at least an 18 hour fast prior to catheterization studies. Arterial glucose concentrations ranged from 13.4 mmole/liter (241 mg/100 ml) to 36.2 mmole/liter (652 mg/100 ml). The mean ± SEM rate of net splanchnic glucose release, not corrected for body surface area, was 2.0 ± 0.5 mmole/min. This rate is approximately twice the value reported by Bondy *et al.* (1949a) for normal individuals. Myers (1950) investigated nine diabetic patients, all of whom had received protamine zinc insulin in adequate or inadequate amounts 24 hours before undergoing catheterization during the postabsorptive period, usually after an overnight fast. None of these diabetic patients was metabolically decompensated beyond having mild glycosuria at the time of study. The subjects' mean ± SEM arterial glucose concentration was 9.88 ± 1.55 mmole/liter (178 ± 28 mg/100 ml). The mean ± SEM rate of net splanchnic glucose release was 0.68 ± 0.02 mmole/min, corrected to 1.73 m^2 body surface area. In these mildly decompensated patients, splanchnic glucose production rate was about one-third that observed in severely decompensated patients of Bondy *et al.* (1949b). Bearn *et al.* (1951) studied 43 unselected diabetic patients, all of whom consumed a diet containing about 250 g of carbohydrate. Twenty-six of the patients had never received insulin, and the remaining individuals received no insulin for at least 48 hours before the study. All fasted for 12 hours prior to catheterization. The mean ±SEM capillary glucose concentration was 13.0 ±1.14 mmole/liter (234 ±20 mg/100 ml). There was no evidence that any of these patients had more than mild ketonemia (Bearn *et al.*, 1951). The mean ±SEM net splanchnic glucose release was 1.24 ±0.14 mmole/min, corrected to 1.73 m^2 body surface area. In their large series of diabetic patients there was no statistical difference in net splanchnic glucose output between normal controls and diabetic individuals. However, from their data (Bearn *et al.*, 1952) it is obvious that diabetic patients have a wide range in net splanchnic glucose production rates. The heterogeneity of glucose production rates became more obvious from the data observed from kinetic analysis techniques.

Using ^{14}C-glucose single-injection tracer (kinetic analyses) techniques, Shreeve *et al.* (1956) compared glucose turnover and oxidation rates in normal controls and eight diabetic patients. The degree of metabolic compensation varied in the diabetic patients. The last regular insulin dose was given from 15 to 120 hours before the ^{14}C studies were initiated. Venous blood glucose ranged from 13.1 mmole/liter (235 mg/100 ml) to 25.4 mmole/liter (457 mg/100 ml). Glucose turnover and oxidation rates in normal controls and diabetic patients were approximately the same when the diabetic patients had hyperglycemia but no ketoacidosis. These results support the hypothesis that impairment of glucose oxidation in diabetic patients was compensated for by hyperglycemia. However, both impaired glucose oxidation and increased glucose turnover rates were noted in diabetic patients with ketoacidosis. Employing similar techniques, Reichard *et al.* (1961) reported the results obtained from 19 elderly patients. Those with mild diabetes, with venous blood glucose concentrations ranging from 5.3 mmole/liter (96 mg/100 ml) to 11.6 mmole/liter (208 mg/100 ml), had glucose turnover rates slightly lower or comparable to normal controls. In contrast, patients with severe diabetes, having venous blood concentrations ranging from 11.6 mmole/liter (208 mg/100 ml) to 30.7 mmole/liter (553 mg/100 ml) also had markedly elevated glucose turnover rates when compared to normal controls. These observations were extended by Forbath and Hetenyi (1966), who pointed out a significant positive correlation between the endogenous rate of glucose production and the fasting plasma glucose concentration between 3.4 mmole/liter (61 mg/100 ml) and 22.7 mmole/liter (409 mg/100 ml). Glucose utilization, corrected for glycosuria, was not diminished in diabetic patients, except for one patient who had severe ketoacidosis. Bowen and Moorhouse (1973), interpreting their data from kinetic analyses techniques, calculated the relative contribution of increased glucose production rate and decreased glucose removal rate to hyperglycemia in maturity onset diabetes. They concluded that overproduction appears to be the greater factor but that underutilization increased as the severity of the hyperglycemia increased.

Peripheral clearance of glucose in decompensated diabetes is primarily channeled into oxidation to CO_2 and H_2O, glycosuria, or recycled glucose by way of the Cori cycle. Using kinetic analyses techniques, Reichard *et al.* (1961) initially suggested that the activity of the Cori cycle might be increased in diabetes. DeMeutter and Shreeve (1963) studied diabetic patients with varying degrees of decompensation. They reported that when lactate-2-^{14}C or pyruvate-2-^{14}C was given intravenously, the appearance of blood glucose-^{14}C within the subsequent 2 hours was directly proportional to the severity of the diabetic state, with about a twofold range between nondiabetic and severely diabetic individuals. Nondiabetic subjects showed 10 to 15% of the injected ^{14}C in calculated extracellular glucose, whereas severe diabetic patients exhibited up to 30%. In the two diabetic patients of Shreeve *et al.* (1964) who were so deficient in exogenous or endogenous insulin that ketoacidosis developed, there was an impairment in oxidation of lactate-2-^{14}C to $^{14}CO_2$. Using similar kinetic analyses techniques, McManus *et al.* (1961) showed that pyruvate-2-^{14}C, which equilibrates with lactate, was preferentially converted to glucose-^{14}C rather than oxidized.

Felig, Wahren, and co-workers, in an outstanding series of studies, have employed arterial and hepatic venous catheterization techniques to identify the quantitative relationship between net splanchnic glucose release and net splanchnic precursor substrate uptake (Felig *et al.*, 1975; Wahren *et al.*, 1972; Felig, 1973; Felig *et al.*, 1969). They recently reported the relative contributions of gluconeogenesis and glycogenolysis to total net splanchnic glucose release in diabetes (Wahren *et al.*, 1972). Seven male, nonobese, insulin-dependent diabetic patients consumed a weight-maintaining diet containing approximately 40% carbohydrate and 25 to 35% protein and fat (caloric values). The last insulin injections were given 24 hours before an overnight (12–14 hours) fast, after which catheterization studies were performed. The mean $\pm$SEM arterial glucose concentration was 13.8 $\pm$0.64

mmole/liter (248 ±12 mg/100 ml). The acid-base status was normal in these patients. The mean ±SEM basal rate of splanchnic output, corrected to 70 kg body weight or 1.73 m² body surface area, was 1.17 ±0.08 mmole/min, which was not significantly different from their 24 normal subjects (Figs. 5 and 8). Marked differences were noted between the diabetic (Fig. 8) and control groups (Fig. 5) with respect to splanchnic extraction of gluconeogenic precursors. Net uptake of lactate and pyruvate was 65 to 115% higher in the diabetic group. The equivalent quantity of glucose

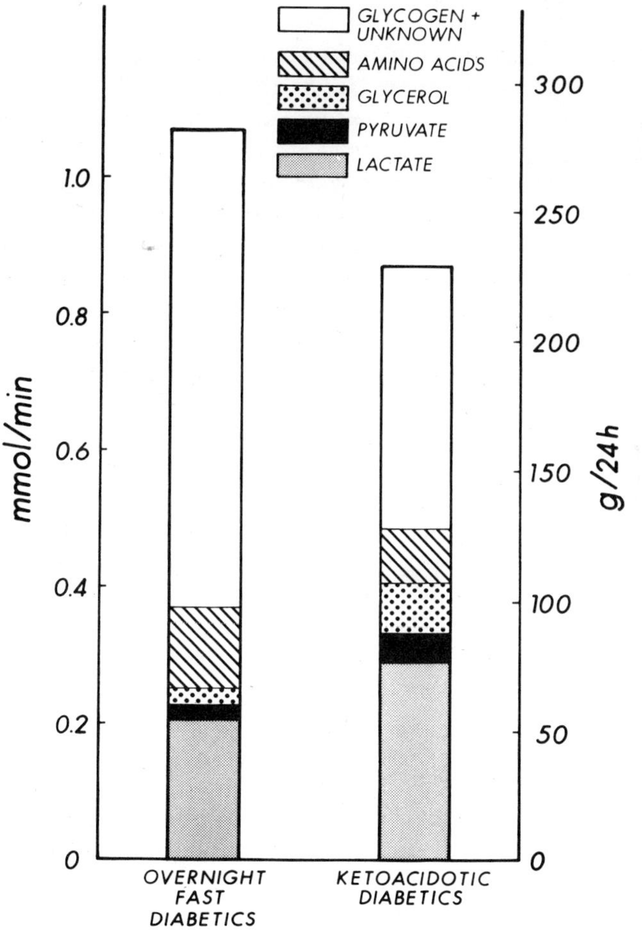

Figure 8. Net splanchnic glucose release and gluconeogenesis from precursor extraction in human diabetics after overnight fast and during ketoacidosis.

formed from the extracted lactate and pyruvate was 0.23 mmole/min. Of the 16 amino acids studied, there was one and one-half to threefold incremental increase in the quantities of gluconeogenic amino acids extracted: alanine, serine, glycine, proline, threonine, tyrosine, cystine, methionine, and phenylalanine. Although Wahren *et al.* (1972) did not measure glycerol, we estimated that the contribution of glycerol to gluconeogenesis is approximately halfway between that observed in normal humans (Wahren *et al.*, 1972) and that observed in diabetic patients during ketoacidosis (Fig. 8). The quantitative data presented in Figs. 5 and 8 indicate that approximately 20% of hepatic glucose output in normal humans after an overnight fast can be accounted for by gluconeogenesis, whereas approximately 33% of hepatic glucose output in diabetic humans in mild metabolic decompensation after an overnight fast can be accounted for by gluconeogenesis.

Using the arterial and hepatic venous catheterization techniques, we have recently investigated net splanchnic glucose release and the uptake of gluconeogenic precursors in seven diabetic patients in severe ketoacidosis. These patients were heterogeneous in that some were newly discovered diabetics, some were thin and others were obese, some had received insulin while others omitted insulin, and some had been consuming carbohydrate-rich beverages while others were fasting. The mean ±SEM peripheral venous blood glucose concentration at the time of admission to the hospital was 25.12 ±1.97 mmole/liter (452 ±36 mg/100 ml). Mean ±SEM of whole blood arterial acetoacetate plus β-hydroxybutyrate concentration was 11.00 ±1.06 mmole/liter, and the mean ±SEM plasma bicarbonate was 7.7 ±1.0 mEq/liter. Mean arterial concentrations of lactate, pyruvate, and glycerol were increased, and mean arterial blood alanine concentration was decreased compared to normal controls (Table I). Of the amino acids present in blood only arterial and venous alanine concentrations were measured. The contribution of amino acids to hepatic gluconeogenesis was estimated by assuming that alanine, the principal gluconeogenic amino acid (Felig, 1973), contributed 45% of total amino acids (Felig *et al.*, 1969). Net splanchnic glucose release was 0.87 mmole/min with a range of 0 to 2.44 mmole/min (Fig. 8). There was a positive correlation between net glucose release and the arterial glucose concentration. Thus the greater the hyperglycemia, the greater is net splanchnic glucose release. If it can be assumed that optimal hepatic conversion of gluconeogenic precursors to glucose occurs during diabetic ketoacidosis, gluconeogenesis contributed about 55% of the total splanchnic glucose release. Heightened Cori cycle activity is indicated by the fact that 0.33 mmole/min glucose equivalent can be contributed to the bloodstream by lactate and pyruvate. Thus during diabetic ketoacidosis, gluconeogenesis is augmented primarily because of recycled lactate and pyruvate, and total gluconeogenesis probably accounts for the majority of hepatic glucose released into the bloodstream. Recent studies have shown that ketonemia suppresses proteolysis, thus diminishes the release of amino acids into the bloodstream from peripheral tissues, primarily muscle (Smith *et al.*, 1975; Sherwin *et al.*, 1975). The lesser contribution of amino acids to gluconeogenesis in diabetic patients in ketoacidosis compared to the overnight fasted nonketotic diabetic patients may be due to the nitrogen sparing effect of hyperketonemia.

From a variety of studies, including our own, it appears that (1) glucose production rates in diabetic patient groups can be low, normal, or high, (2) hyperglycemia facilitates glucose oxidation in the absence of hyperketonemia, (3) hyperketonemia impairs glucose oxidation, and (4) in diabetes, gluconeogenesis is augmented from all precursors including lactate and pyruvate (Cori cycle).

VI. RENAL GLUCONEOGENESIS

The occurrence of gluconeogenesis in kidney cortex slices was first established several decades ago by Benoy and Elliott (1937) and later confirmed by Krebs and co-workers (1963). More recent reports have suggested that acid-base balance is probably a major factor influencing renal gluconeogenesis (Goodman *et al.*, 1966; Kamm *et al.*, 1967; Alleyne, 1970). Data obtained from rat kidney slices and perfusion studies have shown that gluconeogenesis is increased as pH is decreased (Alleyne, 1970; Kamm and Cahill, 1969; Bowman, 1970). Despite the impressive observations made from *in vitro* preparations, the significance of renal gluconeogenesis in maintaining blood glucose homeostasis in humans is not impressive except in specific conditions. On the other hand, it is well established that renal urinary loss of glucose from the body serves to regulate hyperglycemia, thus to aid in maintaining glucose homeostasis. Renal gluconeogenesis from lactate, pyruvate, glycerol, and gluconeogenic amino acids can occur simultaneously with renal glucose utilization and excretion. The balance of these processes determines net renal glucose release or uptake. Net renal glucose exchange in a variety of human conditions is displayed in Fig. 9. A net contribution of glucose to the bloodstream has been observed only in patients with obstructive pulmonary disease (0.32 mmole/min) characterized by respiratory acidosis (bronchitis, overnight fast) (Aber *et al.*, 1966) and in subjects undergoing prolonged starvation (0.15 mmole/min) characterized by metabolic acidosis (obese, 5–6 week fast) (Owen *et al.*, 1969). The kidneys of normal, overnight fasted subjects in the basal state or during physical excercise show insignificant uptake or release of glucose (Wahren *et al.*, 1971). Similarly, insignificant renal exchange of glu-

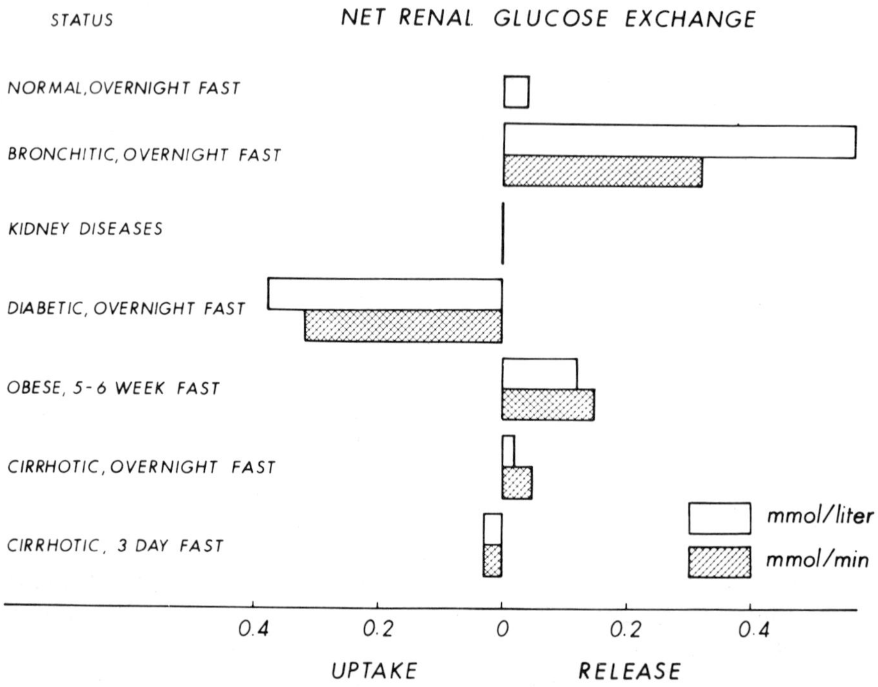

Figure 9. Net renal glucose release in normal and disease states.

cose occurs in patients with kidney disease (Neith and Schollmeyer, 1966) or in patients with hepatic cirrhosis fasted overnight or fasted 3 days (Owen and Patel, unpublished observations). Wahren and Felig (1975) have investigated renal glucose exchange in hyperglycemic (16.8 ±1.5 mmole/liter or 302 ±27 mg/100 ml) insuline-dependent diabetic patients after an overnight fast. The sum of urinary glucose excretion and renal glucose utilization exceeded any concurrent renal gluconeogenesis, resulting in a net renal glucose uptake (0.32 mmole/min). Because urinary glucose excretion can occur at a rate in excess of 1.2 mmole/min during diabetic ketoacidosis with severe hyperglycemia, it is doubtful that the kidneys of diabetic patients with hyperglycemia ever make a net contribution of glucose the bloodstream.

VII. GLUCONEOGENIC SUBSTRATE EXTRACTION

It is well established that alanine is the primary gluconeogenic amino acid, accounting for about 45% of total splanchnic extraction of amino acids

Table II. Fractional extraction (%) of gluconeogenic precursors

	Nondiabetic		Cirrhotic		Diabetic	
Substrate	Overnight fast[a]	3 Day fast[b]	Overnight fast[b]	3 Day fast[b]	Overnight fast[b]	Keto-acidosis[b]
Lactate	30	43	36	37	42	23
Pyruvate	14	33	29	33	34	38
Glycerol	60	28	70	63	—	52
Alanine	38	39	30	14	67	35
Mean	36	36	41	37	—	37

[a] Wahren *et al.* (1972).
[b] Garber *et al.* (1974) combined with additional patient data from Owen *et al.* (unpublished observations).

(Wahren *et al.*, 1972; Felig, 1973; Felig *et al.*, 1969). However, it should be emphasized that the major contributor to hepatic gluconeogenesis is lactate and not amino acids. Recycled lactate and pyruvate account for 61 ±2% (mean ±SEM of six different conditions) of the synthesized glucose. Glycerol accounts for 12 ±2% and amino acids account for 26 ±3% of synthesized glucose (Figs. 5, 7, and 8).

Increased fractional extraction of gluconeogenic precursors has been advocated as a mechanism for heightened gluconeogenesis (Wahren *et al.*, 1972). However, it must be recognized that the bloodstream has limited quantities of lactate, pyruvate, glycerol, and gluconeogenic amino acids. Therefore, increased hepatic substrate extraction must be matched by increased peripheral substrate release. Otherwise, the bloodstream would be rapidly depleted of these metabolites. Table II compares the fractional extraction of gluconeogenic precursors in a variety of conditions. It should be noted that mean fractional extraction of gluconeogenic precursors is approximately equal in all conditions. Therefore, the availability of precursors is of cardinal importance in regulating hepatic gluconeogenesis.

We are out of fuel: this is the end.

ACKNOWLEDGMENTS

The authors wish to express their appreciation to Mrs. Rachelle Matza for her secretarial assistance.

This work was supported in part by U.S. Public Health Service NIH grants NS-11088, AM-16102, HL 05712, HL 18194; 5 MOI RR 349, Geneal Clinical Research Centers Branch, National Institutes of Health; and AHA 73876, American Heart Association.

REFERENCES

Aber, G. M., Morris, L. O., and Housley, E. (1966). *Nature* **212,** 1589.

Alberti, K. G. M. M., Record, C. O., Williamson, D. H., and Wright, R. (1972). *Clin. Sci.* **42,** 591.

Alleyne, G. A. O. (1970). *J. Clin. Invest.* **49,** 943.

Altszuler, N. and Finegold, M. J. (1974). In *The Liver, Normal and Abnormal Functions,* Part A (F. F. Becker, ed.) Dekker, New York, p. 463.

Baker, N., Shreeve, W. W., Shipley, R. A., Incefy, G. E., and Miller, M. (1954). *J. Biol. Chem.* **211,** 575.

Bearn, A. G., Billing, B. H., and Sherlock, S. (1951). *Lancet* **2,** 698.

Bearn, A. G., Billing, B. H., and Sherlock, S. (1952). *Clin. Sci* **11,** 151.

Benedict, F. G. (1915). *Carnegie Institute of Washington, D. C.,* publication 203.

Benoy, M. P. and Elliott, K. A. C. (1937). *Biochem. J.* **31,** 1268.

Bondy, P. K., James, D. F., and Farrar, B. W. (1949a). *J. Clin. Invest.* **28,** 238.

Bondy, P. K., Bloom, W. L., Whitner, V. S., and Farrar, B. W. (1949b). *J. Clin. Invest.* **28,** 1126.

Bowen, H. F. and Moorhouse, J. A. (1973). *J. Clin. Invest.* **52,** 3033.

Bowman, R. H. (1970). *J. Biol. Chem.* **245,** 1604.

Cahill, G. F., Jr. and Owen, O. E. (1968). In *Carbohydrate Metabolism and its Disorders,* Vol. I (F. Dickens, P. J. Randle, and W. J. Whelan, eds.), Academic Press, New York, p. 497.

Conn, H. O. and Daughaday, W. H. (1970). *J. Lab. Clin. Med.* **76,** 678.

Conn, H. O., Schreiber, W., and Elkington, S. G. (1971). *Am. J. Dig. Dis.* **16,** 227.

Conn, H. O. (1972). *Gastroenterology* **62,** 61.

Creutzfeldt, W., Wille, K., and Kaup, H. (1962). *Dtsch. Med. Wochenschr.* **87,** 2189.

DeMeutter, R. C. and Shreeve, W. W. (1963). *J. Clin. Invest.* **42,** 525.

DeMoura, M. C. and Cruz, A. G. (1968). *Am. J. Dig. Dis.* **13,** 891.

Felig, P., Owen, O. E., Wahren, J., and Cahill, G. F., Jr. (1969). *J. Clin. Invest.* **48,** 584.

Felig, P. (1973). *Metabolism* **22,** 179.

Felig, P., Wahren, J., and Hendler, R. (1975). *Diabetes* **24,** 468.

Forbath, N. and Hetenyi, G., Jr. (1966). *Diabetes* **15,** 778.

Franckson, J. R. M., Ooms, H.-A., Bellens, R., Conard, V., and Bastenie, P. A. (1962). *Metabolism* **11,** 482.

Garber, A. J., Menzel, P. H., Boden, G., and Owen, O. E. (1974). *J. Clin. Invest.* **54,** 981.

Goodman, A. D., Fuisz, R. E., and Cahill, G. F., Jr. (1966). *J. Clin. Invest.* **45,** 612.

Henley, K. S., Laughrey, E. G., and Clancy, P. E. (1974). *J. Lab. Clin. Med.* **83,** 175.

Jackson, R. A., Peters, N., Advani, U., Perry, G., Rogers, J., Brough, W. H., and Pilkington, T. R. E. (1973). *Diabetes* **22,** 442.

Jacobs, G., Reichard, G., Goodman, E. H., Jr., Friedmann, B., and Weinhouse, S. (1958). *Diabetes* **7,** 358.

Kamm, D. E., Fuisz, R. E., Goodman, A. D., and Cahill, G. F., Jr. (1967). *J. Clin. Invest.* **46,** 1172.

Kamm, D. E. and Cahill, G. F., Jr. (1969). *Am. J. Physiol.* **216,** 1207.

Kleiber, M. (1961). In *The Fire of Life,* Wiley, New York.

Krebs, H. A., Bennett, D. A. H., De Gasquet, P., Gascoyne, T., and Yoshida, T. (1963). *Biochem. J.* **86,** 22.

Manougian, E., Pollycove, E. M., Linfoot, J. A., and Lawrence, J. H. (1964). *J. Nucl. Med.* **5,** 763.

Marco, J., Diego, J., and Villañeuva, M. L. (1973). *N. Engl. J. Med.* **289,** 1107.

Marliss, E. B., Aoki, T. T., Cahill, G. F., Jr., Unger, R. H., and Soeldner, J. S. (1970). *J. Clin. Invest.* **49,** 2256.

McManus, J. R., Sweeney, P., and Olson, R. E. (1961). *Fed. Proc.* **20,** 191.

Myers, J. D. (1950). *J. Clin. Invest.* **29,** 1421.

Neith, H. and Schollmeyer, P. (1966). *Nature* **209,** 1244.

Owen, O. E., Morgan, A. P., Kemp, H. G., Sullivan, J. M., Herrera, M. G., and Cahill, G. F., Jr. (1967). *J. Clin. Invest.* **46,** 1589.

Owen, O. E., Felig, P., Morgan, A. P., Wahren, J., and Cahill, G. F., Jr. (1969). *J. Clin. Invest.* **48,** 574.

Owen, O. E. and Reichard, G. A., Jr. (1971a). *Progr. Biochem. Pharmacol.* **6,** 177.

Owen, O. E. and Reichard, G. A., Jr. (1971b). *J. Clin. Invest.* **50,** 1536.

Owen, O. E. and Cahill, G. F., Jr. (1973). *J. Clin. Invest.* **52,** 2596.

Owen, O. E., Reichard, G. A., Jr., Markus, H., Boden, G., Mozzoli, M. A., and Shuman, C. R. (1973). *J. Clin. Invest.* **52,** 2606.

Paul, P. and Bortz, W. M. (1969). *Metabolism* **18,** 570.

Peters, J. P. and Van Slyke, D. D. (1946). In *Quantitative Clinical Chemistry: Interpretations,* Vol. 1, Williams & Wilkins, Baltimore, Md., pp. 326, 344.

Podolsky, S., Zimmerman, H. J., Currows, B. A., Cardarelli, J. A., and Pattavina, C. G. (1973). *N. Engl. J. Med.* **288,** 644.

Reichard, G. A., Jr., Jacobs, A. G., Kimbel, P., Hochella, N. J., and Weinhouse, S. (1961). *J. Appl. Physiol.* **16,** 789.

Reichard, G. A., Jr., Owen, O. E., Haff, A. C., Paul, P., and Bortz, W. M. (1974). *J. Clin. Invest.* **53,** 508.

Searle, G. L., Strisower, E. H., and Chalkoff, I. L. (1954). *Am. J. Physiol.* **176,** 190.

Searle, G. L., Mortimore, G. E., Buckley, R. E., and Reilly, W. A. (1959). *Diabetes* **8,** 167.

Sherwin, R., Joshi, P., Hendler, R., Felig, P., and Conn, H. O. (1974). *N. Engl. J. Med.* **290,** 239.

Sherwin, R. S., Hendler, R. G., and Felig, P. (1975). *J. Clin. Invest.* **55,** 1382.

Shreeve, W. W., Baker, N., Miller, M., Shipley, R. A., Incefy, G. E., and Craig, J. W. (1956). *Metabolism* **5,** 22.

Shreeve, W. W., DeMeutter, R. C., and Shigeta, Y. (1964). *Diabetes* **13,** 615.

Smith, R., Fuller, D. J., Wedge, J. H., Williamson, D. H., and Alberti, K. G. G. M. (1975). *Lancet* **1,** 1.

Son Nilsson, L. H., Furst, P., and Hultman, E. (1973). *Scand. J. Clin. Lab. Invest.* **32,** 331.
Steele, R., Wall, J. S., deBodo, R. C., and Altszuler, N. (1956). *Am. J. Physiol.* **187,** 15.
Walker, C., Peterson, W., Jr., and Unger, R. (1974). *N. Eng. J. Med.* **291,** 168.
Warren, J. V. and Brannon, E. S. (1944). *Proc. Soc. Exp. Biol. N.Y.* **55,** 144.
Wahren, J., Felig, P., Ahlborg, G., and Jorfeldt, L. (1971). *J. Clin. Invest.* **50,** 2715.
Wahren, J., Felig, P., Cerasi, E., and Luft, R. (1972). *J. Clin. Invest.* **51,** 1870.
Wahren, J. and Felig, P. (1975). *Diabetes.* In press.
Zimmermann, H. J., Thomas, L. J., and Scherr, E. H. (1953). *A.M.A. Arch. Int. Med.* **91,** 577.

Author Index

Numbers in parentheses refer to the pages on which the complete references are listed.

Subject Index